4th International Conference on Civil Engineering and Materials Science (ICCEMS 2019)

IOP Conference Series: Materials Science and Engineering
Volume 652

Bangkok, Thailand
17 – 19 May 2019

ISBN: 978-1-5108-9848-6
ISSN: 1757-8981

Printed from e-media with permission by:

Curran Associates, Inc.
57 Morehouse Lane
Red Hook, NY 12571

Some format issues inherent in the e-media version may also appear in this print version.

Institute of Physics
Dirac House, Temple Back
Bristol BS1 6BE UK

Phone: 44 1 17 929 7481
Fax: 44 1 17 920 0979

techtracking@iop.org

Additional copies of this publication are available from:

Curran Associates, Inc.
57 Morehouse Lane
Red Hook, NY 12571 USA
Phone: 845-758-0400
Fax: 845-758-2633
Email: curran@proceedings.com
Web: www.proceedings.com

Table of contents

Volume 652

4th International Conference on Civil Engineering and Materials Science (ICCEMS 2019)

17–19 May 2019, Bangkok, Thailand

Accepted papers received: 17 September 2019
Published online: 29 October 2019

Preface

Papers

Civil and Structural Engineering

Materials Science and Engineering

Manufacturing Science

Computer-aided Design and Manufacturing

IOP Conf. Series: Materials Science and Engineering **652** (2019) 011001 doi:10.1088/1757-899X/652/1/011001

Preface

The 4th International Conference on Civil Engineering and Materials Science (ICCEMS 2019) and the 2nd International Conference on Nanomaterials, Materials and Manufacturing Engineering (ICNMMS 2019) were jointly held in Bangkok, Thailand in May 2019. Since their inception, both the ICCEMS and ICNMM have been attracting many delegates from all over the world. This year's delegates, including the world's leading scholars, researchers and expert practitioners represented five continents and 23 countries, i.e., USA, Canada, Mexico, Brazil, China, Japan, Korea, Taiwan, Malaysia, Philippines, Thailand, Singapore, India, Indonesia, Russia, Iran, Iraq, France, Germany, Norway, Poland, South Africa, and Nigeria.

ICCEMS and ICNMM aim to become premier international conferences for in-depth discussions on the most-up-to-date and innovative ideas, research projects and practices in the field of civil engineering, material sciences and manufacturing engineering. Papers published in the ICCEMS 2019 and ICNMM 2019 proceedings cover various topics from civil and structural engineering to nanomaterials science as well as a broad range of related interdisciplinary subjects.

Today, the world develops with unprecedented speed, which is only possible with advances in various fields of sciences and engineering. We are very pleased to contribute to this important global effort. We sincerely hope that the academic community and industrial practitioners will continue to support us in our attempts to provide even more meaningful conferences with numerous critical idea exchanges, diverse opportunities for fruitful networking and future collaborations between the delegates.

Warmest regards,

Conference Chair
Prof. Kyoung Sun Moon
Yale University School of Architecture, USA
ICCEMS 2019 & ICNMM 2019 Committee

IOP Conf. Series: Materials Science and Engineering **652** (2019) 011001 doi:10.1088/1757-899X/652/1/011001

Statement of Peer Review

All papers published in this volume of Materials Science and Engineering have been peer reviewed through processes administered by the proceedings Editors. Reviews were conducted by expert referees to the professional and scientific standards expected of a proceedings journal published by IOP Publishing.

Conference Committees

Honorary Chair

Prof. Dr. -Ing. Thira Jearsiripongkul, Dean, Thammasat University, Thailand

International Advisory Committees

C. W. Lim, City University of Hong Kong, Hong Kong

Richard G. Haverkamp, Massey University, New Zealand

Conference Chairs

Xiaohong Zhu, Sichuan University, China

Kyoung Sun Moon, Yale University School of Architecture, USA

Program Chairs

Chien-Yuan Chen, National Chiayi University, Taiwan

Andrew Boyd, McGill University, Canada

Local Chair

Cattaleeya Pattamaprom, Thammasat University, Thailand

Technical Committees

Suppachai Sinthaworn, Srinakharinwirot University, Thailand

Shiyu Xu, City University of Hong Kong, China

Trong-Phuoc Huynh, Can Tho University, Viet Nam

Elvira P. Chernyshova, Nosov Magnitogorsk State Technical University, Russia

Woosuk Kim, Kumoh National Institute of Technology, South Korea

Ana I. Almerich Chulia, Universitat Politecnica de Valencia, Spain

Abdollah Malekjafarian, University College Dublin, Ireland

Mohammad Hanifehzadeh, University of Southern California, USA

Jun Xia, Xi'an-Jiaotong Liverpoor University, China

Murat Gunduz, Qatar University, Qatar

Alaa M. Ali, Misr High Institute for Engineering and Technology, Egypt

Ali Sarrafi Nik, Islamic Azad University, Iran

IOP Conf. Series: Materials Science and Engineering **652** (2019) 011002 doi:10.1088/1757-899X/652/1/011002

TRONG-PHUOC HUYNH, Can Tho University, Vietnam

Gunalaan Vasudevan, Tunku Abdul Rahman University College, Malaysia

Alaa Morsy, The Arab Academy for Science, Technology & Maritime Transport, Egypt

Karol Sikora, Univeristy of Wollongong in Dubai, UAE

A. N. Ede, Covenant University, Nigeria

Haryati Yaacob, Universiti Teknologi Malaysia, Malaysia

Izobo-Martins Oladunni, Covenant University, Nigeria

Wan Inn Goh, Universiti Tun Hussein Onn Malaysia, Malaysia

Samaila Saleh, Hassan Usman Katsina Polytechnic, Nigeria

B. Kondraivendhan, S.V. National Institute of Technology, India

Nolan Concha, FEU-Institute of Technology, Philippines

Mohamed Mostafa Hassan, Central University of Technology, South Africa

Arash Radman, University of Guilan, Iran

Ubolrat Wangrakdiskul, King Mongkut's University of Tech NB, Thailand

Michał Musiał, Wrocław University of Science and Technology, Poland

Tebogo Mashifana, University of Johannesburg, South Africa

Nastassia Sithole, University of Johannesburg, South Africa

Heru Purnomo, Universitas Indonesia, Indonesia

Dariusz Fydrych, Gdansk University of Technology, Poland

Aleksandra Świerczyńska, Gdansk University of Technology, Poland

Lais Amaral Alves, CEFET/RJ, Brazil

Kong Fah Tee, University of Greenwich, UK

Tetsushi Kurita, Tokyo Electric Power Services Co., Ltd., Japan

Ramesh Singh, University of Malaya, Malaysia

Ahmad Zhafran Ahmad Mazlan, Universiti Sains Malaysia, Malaysia

IOP Conf. Series: Materials Science and Engineering **652** (2019) 011003 doi:10.1088/1757-899X/652/1/011003

Peer review statement

All papers published in this volume of *IOP Conference Series: Materials Science and Engineering* have been peer reviewed through processes administered by the proceedings Editors. Reviews were conducted by expert referees to the professional and scientific standards expected of a proceedings journal published by IOP Publishing.

Innovation diffusion process in the Australian construction industry

W Wipulanusat[1,*], K Panuwatwanich[2], R A Stewart[3], and J Sunkpho[4]

[1] School of Engineering and Technology, Walailak University, Thailand
[2] School of Civil Engineering and Technology, Sirindhorn International Institute of Technology, Thammasat University, Thailand
[3] School of Engineering and Built Environment, Griffith University, Australia
[4] College of Innovation, Thammasat University, Thailand

Email: wwarit@wu.ac.th

Abstract. Innovation diffusion process is an important component for construction companies to remain competitive in today's changing business environment. The aim of this study is to examine how innovations in the Australian construction industry can be understood through innovation diffusion theories and empirical works. To achieve this goal, the author reviewed theoretical frameworks, and then analysed empirical works to shed light on the innovation diffusion in practice. The review found that effective innovation diffusion has played an important role in providing organizations with competitive advantages. From a socio-psychological perspective, leadership and team climate have had a direct effect on organizational culture, which indirectly impacts on innovation diffusion outcomes and business performance. Emphasizing the drivers and obstacles of innovation diffusion is also necessary to effectively develop a conceptual framework of innovation diffusion at project level which begins with knowledge and idea generation, and progresses to implementation and confirmation.

1. Introduction

Innovation is essential for the development of the industrial and economic sectors in Australia because being an important driver for the Australian economy, which is currently one of the fastest growing economies in the world and has witnessed dramatic changes in recent years. Because of these changes, policies on national and regional systems of innovation to encourage this adoption must be coupled with appropriate mechanisms to develop awareness of new innovations available, and to encourage and facilitate their appropriate exploitation [1]. The construction industry is an important sector contributing considerably to the economy of Australia. Therefore, the construction industry needs to adjust in order to innovate in delivering more effective social and economic infrastructure, and to find a more effective way to diffuse innovation.

2. Theoretical backgrounds

Innovation plays an important role in leveraging the competitiveness of companies operating in the construction industry. Stewart and Fenn [2] defined innovation as the profitable exploitation of ideas, practices and technologies novel to the organisation in seeking competitive advantage. Innovation in construction can, therefore, be regarded as the successful development and implementation of new and significantly improved ideas, products, processes, practices and technologies related to their functionality and user characteristics, in order to increase effectiveness and efficiency in organisations. However, the construction industry is confronted with a complexity of project characteristics and technological constraints; therefore, there is a perception that this industry remains slow in adopting innovation, especially when compared to the manufacturing industry that is typically technology intensive, where innovations are rarely large scale or radical, and the adoption and implementation of innovations have been more extensive.

In contrast to manufacturing, construction is perceived as a service-based industry that is inherently labor intensive; thus, innovations in construction are small-scale incremental improvements in services or products. Also, they tend to be more ad hoc than in manufacturing firms as they are usually developed gradually from employee and manager ideas in response to challenges during the service delivery process [3]. Despite this perception, innovation in the construction industry does occur mostly in terms of physical processes and products, and particularly improvements in materials, such as new types of asphalt, concrete, or fiber composites, as well as new processes facilitated by information and communication technology (ICT) [4, 5]. Loosemore [3] developed a theoretical model of innovation in construction using the fifth-generation principles which described innovation in construction as a dynamic process of modification, starting with an organic approach in initial design and tender stages, and continuing to a more systematic approach during the delivery phase, in which tight budgets and programs demand focused discipline. Wipulanusat et al. [6] highlights the importance of organisations providing opportunities for their engineers to use a high degree of creativity and innovation for advancing their engineering expertise.

Innovation diffusion is defined as the process in which a new idea, concept, process or technology is disseminated by communication, relationships and the decision making processes involving acceptance or rejection across a social system over a time period [7]. The term 'innovation diffusion' is eclectically defined in this research as the process by which an innovation is adopted and implemented by firms in the construction industry until the expected users adopt and transfer the knowledge and experience of how to use the innovation. To achieve this process, it is imperative that firms understand how this action occurs within a complex social system of key stakeholders in the construction industry including, but not limited to: architects, engineers, contractors, subcontractors, academic research institutions and product development firms.

3. Study approach

Within the topic of innovation in the built environment, a literature review was performed in regard to the innovation diffusion fields that focus on the Australian construction industry. The examined papers consisted primarily of articles published by the largest world-wide specialists: Emerald, Science Direct (Elsevier), Springer, IEEE-Xplore. Search keywords were "innovation, innovation diffusion, innovation adoption, construction industry, and Australia." The criteria for selecting studies were that the articles contributed to understanding of the factors influencing the innovation diffusion process, while ensuring the selected studies were of acceptable scientific quality. Once selected, the author then sorted the articles into two themes: innovation diffusion at organisational level and the innovation diffusion at project level. The aim of this study is to explore how innovations in the construction industry can be understood through diffusion of innovation theories and empirical works. To accomplish this aim, previously proposed theoretical frameworks and models were critically reviewed; then, empirical works were analyzed to interpret the innovation diffusion literature.

4. Innovation diffusion in the Australian construction industry

Construction scholars have conducted many studies investigating innovation diffusion in the Australian construction industry. For example, Peansupap and Walker [4] identified factors that were found to influence ICT diffusion and adoption in large Australian construction organisations, and grouped these into management, individual, technology and workplace environment categories.

4.1. Innovation diffusion at organisational level

In an attempt to provide a better understanding of the innovation diffusion phenomena from a social perspective, a study was conducted by Panuwatwanich et al. [8] to empirically investigate the impact of the socio-psychological process of innovation diffusion on business performance within architecture and engineering design (AED) firms. Panuwatwanich et al. [9] also investigated innovation diffusion in Australian design firms by studying 181 architectural and engineering design professionals and identified that socio-psychological factors, which might influence innovation diffusion within AED firms, could be divided into three factors: leadership, team climate, and organisational culture.

Each socio-psychological factor has a different role in innovation diffusion within an organisation. A leader plays an important role in inspiring creativity and challenging staff to promote the development of new ideas in the diffusion process which will lead to the innovation initiative in an organisation [9, 10]. When ideas are encouraged and challenged, staff are more likely to be aware of problems and think of new ways of working. A team climate where members actively support each other in developing innovative ideas is considered another requirement for innovation. This climate provides an important motive and determinant to achieve innovative behaviour of staff within the organisational context [8, 11]. The final socio-psychological factor which is an essential element for the successful diffusion of innovation is having an innovation culture within an organization [12]. Within such a culture, creative ideas are more likely to be transformed into innovative solutions, and the adopted innovation is more likely to be diffused because staff perceive encouragement and support from their firms [13].

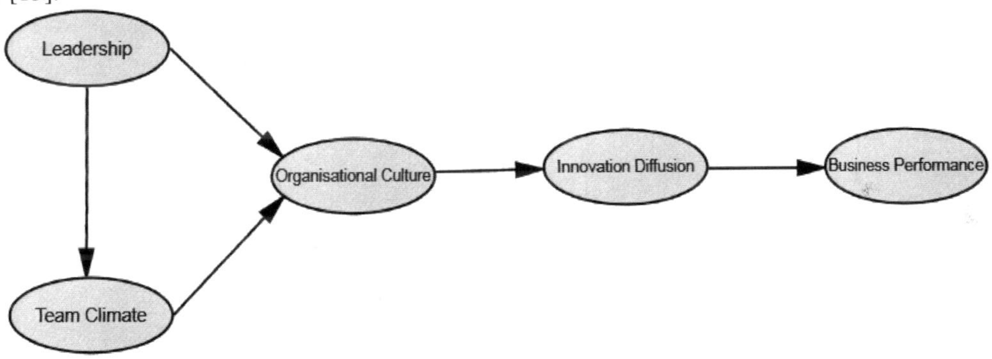

Figure 1. Empirical model for innovation diffusion at organisational level [8]

Panuwatwanich et al. [9] also found that the socio-psychological factor of leadership and team climate influence perceptions of the third factor, an organisation's culture for innovation, which this triad of factors subsequently impact on innovation diffusion as well as business performance as shown in Figure 1. Innovations such as innovative design solutions have a high impact on client satisfaction because the client can appreciate the benefits from the innovative design products when the value of the innovation has been made explicit [9]. As client satisfaction is a critical indicator of business performance, firms should strive for high levels of client satisfaction. This can be achieved by innovation in the firm. Evidence to support this link between socio-psychological factors and innovation diffusion was derived from quantitative and qualitative studies, with data collection methods including survey, documentation analysis, and case studies based on triangulation using multiple sources to ensure the validity of findings [9].

In order to evaluate the innovation diffusion maturity levels within the AED firms, Innovation Diffusion Readiness (IDR) was conceptualized on the basis of three socio-psychological factors; namely, leadership for innovation, team climate for innovation and organisational culture for innovation [14]. The key findings showed that the overall level of IDR among the sampled firms was moderate to high. Using cluster analysis, three main clusters were uncovered: Very High, High, and Moderate IDR. Further analysis indicated that the higher IDR clusters appeared to perform significantly better in diffusing innovative design practices. This finding highlighted that when the firms shifting to a higher IDR cluster, they could achieve a significant improvement in the diffusion of innovation [14].

4.2. Innovation diffusion at project level

Another important factor regarding innovation diffusion highlighted by Rose and Manley [5] is the vital role of governments in promoting the steady flow of product innovations relevant to the Australian construction industry across the supply chain. Despite this strong supply of product innovations, adoption rates in the Australian construction industry remain low, due to six key obstacles that hinder the adoption of innovative products: project goal misalignment; client pressures; weak contractual relations; lack of product trialing; inflexible product specifications; and product liability concerns [5]. Thus, construction firms need to overcome these major barriers to innovation diffusion if they are to be successful.

Rose and Manley [7] also investigated innovation diffusion in the Australian construction industry, drawing on an innovation diffusion model and developing a tailored conceptual framework to guide future empirical work aimed at assessing innovation diffusion in the project-based context. One important aspect of the framework is that the design and operation of innovation diffusion should be dependent upon its intended audience, as what is suitable for innovation diffusion for large firms may be unsuitable for small firms. Hence, input from these studies on how innovation diffuses within an organisation, can assist firms to overcome the complex challenges of innovation diffusion in construction. The conceptual framework consists of five stages: knowledge and idea generation; persuasion and evaluation; decision to adopt; integration and implementation; and confirmation as shown in Figure 2.

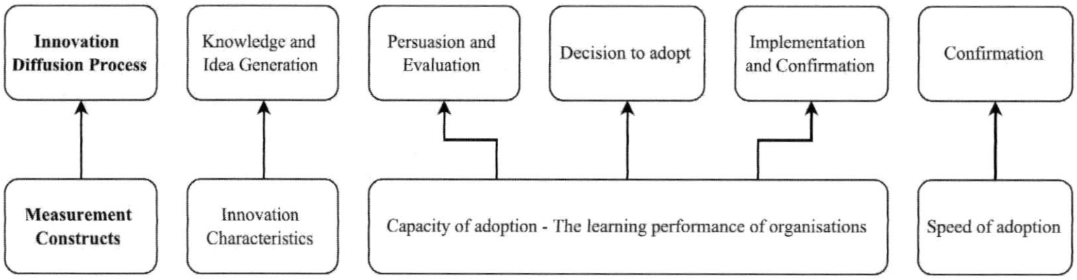

Figure 2. Conceptual framework of innovation diffusion at project level [7]

In the first stage, staff acquire an understanding of an innovation. From a construction innovation perspective, Winch [15] explained that innovation derives from two approach; in the top-down approach, new ideas are adopted by organisational executives; in a bottom-up approach new ideas are generated by team members who have solved the problems during a construction project and gained lessons later in the procurement process. At the second stage, staff determine the advantages or disadvantages of an innovation depending on three measures [7]: the relative advantage in which an innovation might be better than existing practices; compatibility, which considers whether an innovation is perceived to be consistent with the existing values and objectives of the potential adopters; and complexity, the degree to which an innovation is perceived as difficult to understand and apply. The choice for construction firms to adopt innovations can be determined by these three measures.

At the third stage, the decision to adopt, companies decide to adopt or reject an innovation according to the consequences of the selection [7]. Subsequently, an innovation is implemented, and then, the benefits are considered in regard to expanding on a wider scale but its application may be discontinued if poor performance is experienced. It is expected that this framework is expected to improve understanding of innovation diffusion processes on construction projects. Each stage of the conceptual framework is influenced by two elements which are, the internal capacity of team members to consider and implement an innovation according to their past experience and practice, as well as the relationships between organisations which impact the efficiency of knowledge sharing among firms [7]. As a result, these two elements will impact the effectiveness and efficiency of the innovation diffusion process in an organisation.

To assess the effectiveness and efficiency of the innovation diffusion process in the context of construction, Rose and Manley [7] also proposed three measurement constructs to modify their tailored framework for the built environment. The first measurement construct is the innovation characteristics which affect the capability of team members to adopt and understand the innovation. The second measurement construct is capacity of adoption, which refers to the learning performance of organisations. The third measurement construct is speed of adoption, which measures the pace of the implementation of innovation activity in construction firms. Data obtained from these measurement constructs will enable informed policy advice to assist in maximizing the potential for innovation diffusion in construction firms.

In order to interpret their framework, Rose and Manley [16] conducted a quantitative survey to rank the relative importance of the obstacles constraining the adoption of innovative products in Australian road infrastructure projects. The three most important obstacles to address are: overemphasis on up-front project costs during the tender stage; disagreement over who carries the risk of new product failure; and adversarial contract relations. It is important, therefore, for construction firms to focus on procurement systems and industry relationships in order to overcome these obstacles and shape the opportunities for innovation diffusion.

To analyse the support role of clients for improving innovation diffusion in the procurement process, Loosemore and Richard [17] conducted research based on 46 interviews with Australia's leading construction clients, contractors and consultants. The research revealed that many clients were unwilling and unable to employ strategies to foster more innovation because of internal governance constraints, a lack of tools and insight to evaluate innovation in tenders, a poor understanding of how built assets contribute to core business objectives, and a narrow understanding of their central role in driving innovation. The importance of client leadership in driving, encouraging, and facilitating more innovation should be more valued in the construction sector. This can be achieved by clients deploying an integrated project strategy that allows scope and time for innovation to occur, encourages collaboration and early involvement, and distributes risk and opportunity in a balanced and transparent manner [17].

5. Conclusion

Innovation is a dynamic activity that can develop the value of ideas, practices, and technologies through the generation and implementation of new knowledge. Innovation is a significant element for the Australian construction industry to remain competitive in today's changing business environment. Despite a perception that the construction industry is not innovative, there is an evidence of innovation diffusion across construction organisations. Greater attention is required to better understand the role and characteristics of innovation diffusion as elements of firm-level strategies within the complex and interdependent construction industry. Effective diffusion of innovation is essential for construction companies to obtain competitive and operational benefit which can be achieved by completely comprehending the innovation diffusion process. Within this process, there are potential drivers and obstacles to recognise and address. Emphasizing these drivers of and obstacles to innovation diffusion is important in order to effectively develop a conceptual framework of innovation diffusion which starts with knowledge and idea generation and ends in implementation and confirmation. Achieving successful innovation diffusion outcome and business performance is dependent on leadership, team climate and organisational culture. Thus, the adaptation of a conceptual framework of innovation diffusion to the construction context can enable construction companies to integrate information, knowledge, and technology to respond to the emerging demands of the Australian construction industry in the future.

References

[1] Aouad, G., Ozorhon, B., and Abbott, C., "Facilitating innovation in construction: Directions and implications for research and policy," *Construction Innovation,* vol. 10, pp. 374-394, 2010.
[2] Stewart, I. and Fenn, P., "Strategy: the motivation for innovation," *Construction Innovation,* vol. 6, pp. 173-185, 2006.
[3] Loosemore, M., "Construction innovation: Fifth generation perspective," *Journal of Management in Engineering,* vol. 31, pp. 1-9, 2015.
[4] Peansupap, V. and Walker, D., "Exploratory factors influencing information and communication technology diffusion and adoption within Australian construction organizations: a micro analysis," *Construction Innovation,* vol. 5, pp. 135-157, 2005.
[5] Rose, T. and Manley, K., "Adoption of innovative products on Australian road infrastructure projects," *Construction Management and Economics,* vol. 30, pp. 277-298, 2012.
[6] Wipulanusat, W., Panuwatwanich, K., and Stewart, R. A., "Pathways to workplace innovation and career satisfaction in the public service: The role of leadership and culture," *International Journal of Organizational Analysis,* vol. 26, pp. 890-914, 2018.
[7] Rose, T. and Manley, K., "Measurement constructs to explore innovation diffusion in construction," in *Proceedings of the Construction, Building and Real Estate Conference*, 2012, pp. 354-362.
[8] Panuwatwanich, K., Stewart, R. A., and Mohamed, S., "The role of climate for innovation in enhancing business performance: The case of design firms," *Engineering, Construction and Architectural Management,* vol. 15, pp. 407-422, 2008.
[9] Panuwatwanich, K., Stewart, R. A., and Mohamed, S., "Critical pathways to enhanced innovation diffusion and business performance in Australian design firms," *Automation in Construction,* vol. 18, pp. 790-797, 2009.
[10] Wipulanusat, W., Panuwatwanich, K., and Stewart, R. A., "Exploring leadership styles for innovation: an exploratory factor analysis," *Engineering Management in Production and Services,* vol. 9, pp. 7-17, 2017.
[11] Wipulanusat, W., Panuwatwanich, K., and Stewart, R. A., "Workplace innovation: Exploratory and confirmatory factor analysis for construct validation," *Management and Production Engineering Review,* vol. 8, pp. 57-68, 2017.
[12] Wipulanusat, W., Panuwatwanich, K., and Stewart, R. A., "Statistical data analysis of culture for innovation using an open data set from the Australian Public Service," in *Lecture Notes in Computer Science* vol. 10365, ed. Cham: Springer 2017, pp. 78-89.
[13] Panuwatwanich, K., Stewart, R. A., and Mohamed, S., "Validation of an empirical model for innovation diffusion in Australian design firms," *Construction Innovation,* vol. 9, pp. 449-467, 2009.
[14] Panuwatwanich, K. and Stewart, R. A., "Evaluating innovation diffusion readiness among architectural and engineering design firms: Empirical evidence from Australia," *Automation in Construction,* vol. 27, pp. 50-59, 2012.
[15] Winch, G., "Zephyrs of creative destruction: understanding the management of innovation in construction," *Building Research & Information,* vol. 26, pp. 268-279, 1998.
[16] Rose, T. and Manley, K., "Revisiting the adoption of innovative products on Australian road infrastructure projects," *Construction Management and Economics,* vol. 32, pp. 904-917, 2014.
[17] Loosemore, M. and Richard, J., "Valuing innovation in construction and infrastructure: Getting clients past a lowest price mentality," *Engineering, Construction and Architectural Management,* vol. 22, pp. 38-53, 2015.

IOP Conf. Series: Materials Science and Engineering **652** (2019) 012002 doi:10.1088/1757-899X/652/1/012002

Design and construction monitoring and deformation analysis of super-large deep foundation pit

Junshu ZHOU [1] and Guangxiu FANG [2]

[1] Undergraduate in structure engineering of Yabian University, Changchun, Jilin, China

[2] Corresponding author, professor of structure engineering of Yabian University, 977 Gongyuan Road, Yanji, Jilin, China, E-mail: gxfang@ybu.edu.cn

Abstract: Using the large and deep foundation pit of Metro Line 1 of Jinan west railway station as an example, this paper elaborated the design scheme of its retaining structure, formulated a monitoring scheme for foundation pit. In addition, through the real-time on-scene monitoring of horizontal displacement of pile top and pile body, and the monitoring data, we found the law of displacement of deep foundation pit supporting structure. The maximum horizontal displacement of pile body moves down gradually with the advancement of excavation support and the main construction. The maximum displacement occurs around 0.684H. This study has implications for similar engineering geological conditions of foundation pit supporting structure.

Key word: Super-Large deep foundation pit; Supporting structure; Design; Monitoring analysis

1. Project overview[1]

The length of foundation pit of Jinan Metro Line 1 is 207.05m. The standard section width and excavation depth are 21.4m and 18.58m respectively; the width of shield section and excavation depth are 25.3m and 18.58m respectively (±0.00 is equivalent to the absolute elevation of 3036m), which belongs to the super-large deep foundation pit. Its plan diagram is shown in Figure1.

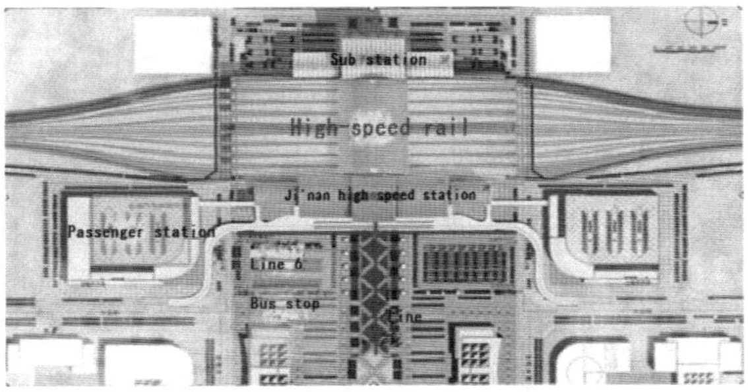

Figure 1. Plane layout of Jinan west railway station.

2 Engineering geological and hydrological conditions

2.1Engineering geological conditions

According to a geological survey report, the strata within the scope of the project site is mainly

Quaternary Holocene Alluvial-diluvial strata(Q4al+pl), which are composed of plain fill, silty clay, pebble, etc. The physical and mechanical properties of the soil in the geological strata are shown inTable1.

Table 1. Mechanical properties of geological soil layer.

Sequence	Name of soils	Depth (m)	Unit weight (kN/m^3)	Cohesive forces c (kPa)	The angle of internal friction Φ (°)	Poisson ratio u	Elastic Modulus E (Mpa)
1	Plain fill	3.20	15.4	15.00	10.00	0.32	6.01
2	Silt	10.00	16.3	23.00	23.00	0.18	24
3	Silty clay	4.10	17.8	28.00	23.50	0.21	26
4	Silty clay	12.20	19.5	34.00	23.40	0.23	35
5	Medium sand	8.80	19.0	0.00	32.00	0.17	55

2.2 Hydrogeological conditions

In the surveyed area, the buried depth of the underground static water level measured from the borehole is 1.3-12.8m. Its absolute elevation is 26.23-27.61m. Underground water is mainly distributed in clayey strata, which is replenished by atmospheric precipitation and underground runoff, with the medium sand bed as the permeable layer. The annual variation of the underground water level is little, the absolute elevation in the high flow period can be considered as 28.5m, and the anti-floating design water level as 29.5m (southwest) -29m (northeast). After executing the corrosive test of soil above underground water level, the results showed that the soil is classified as Grade II environment, which does not cause the corrosion of the reinforced concrete structure.

3 Design and construction of super-large deep foundation pit supporting structure

Based on the field investigation of deep foundation pit project of Jinan Metro Line 1, and the comparative analysis of various supporting schemes, the final determination of the scheme is as follows:

(1) Bored piles (Φ1000@1500, Length 22.88m, C30)+ High pressure rotary injected waterproof curtain+ Three steel tube internal support are adopted for construction of foundation pits, as shown in Figure 2.

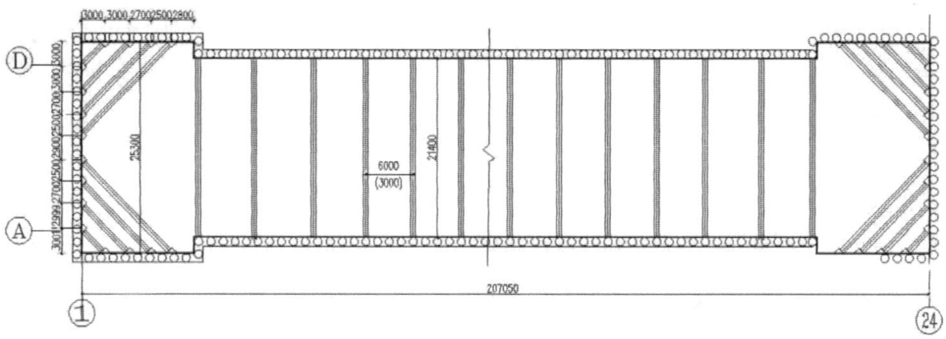

Figure 2. Plane layout of foundation pit.

(2) Three steel tube support system and three steel tube diagonal support system are adopted for the standard and shield section respectively in Internal Support. The steel tube supports are equipped with movable ends to apply prestress, which is 30% of the design value of the steel support axial force, as shown in Figure 3.

The design values of the first, second and third steel support axial force are 1000kN, 2000kN and 1500kN respectively. The first inner-support is Φ609 (t=12) steel tube, which is 2.1m from the top; the second and third inner-support are both Φ609 (t=16) steel tubes, which are 8.35m and 14.88m from the top. The strength grade of concrete top beam is C30 with the section of 1000mm×800mm, and the breast beam are steel beams (Common I-steel of Hot-rolled45b), as shown in Figure 3.

（3）Excavation Supporting Conditions of Foundation pit:

①The first excavation was carried out at a relative elevation of -3.1m; ②Construction of concrete top beam, while setting up the first steel tube (one for every 6m), applying prestress to the steel support; ③The second excavation was dig to a relative elevation of -9.35m; ④Construction of steel breast beam, setting of the second steel tube (one for every 3m), and applying prestress; ⑤The third excavation was dig to a relative elevation of -15.88m; ⑥Construction of steel breast beam, setting of the third steel tube (one for every 3m), and applying prestress; ⑦The forth excavation was dig to a relative elevation of -18.58m, manual excavation of the remaining 300mm earthwork, as shown in Figure 3.

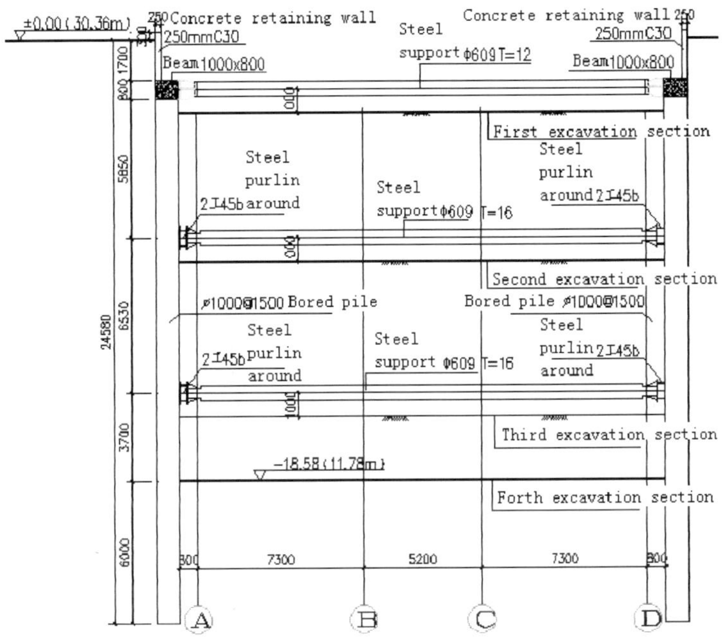

Figure 3. Section layout of supporting structure of Metro Line 1.

（4）In the earthwork excavation of foundation pit, we should follow the idea of "Time-Space Effect", and combine the methods of separating region, separating section,"supporting while digging" in to minimize the exposure time of foundation pit and mechanical excavation as the primary method while manual excavation secondary. In general, the construction is conducted from both north and south ends to the middle. The key procedures and time points are shown in Table 2.

Table 2. The key construction procedures and time points.

No.	Key construction procedures	Completion time	Note
1	Start the excavation	March 20	
2	Set up the first steel support	April 03	In order to reduce the unsupported exposure time of the foundation pit, the steel support should be set up at any time with earthwork excavation.
3	Set up the second steel support	April 27	
4	Set up the third steel support	June 01	
5	Excavation to design elevation	June 17	
6	Remove the third steel support	July 17	Steel support can only be removed when the strength of structural slab (bottom slab, middle slab, top slab) reaches 70% after concrete pouring.[1]
7	Remove the second steel support	September 05	
8	Remove the first steel support	October 21	

（5）Dewatering while scaffolding the pit:

①The foundation pit is surrounded by 918 high pressure rotary injected waterproof curtain (Φ700@500, Length 23.01m). P42.5 ordinary Portland cement is used with water-cement ratio of 1:1. After the completion of piles, hydraulic conductivity should be maintained within the range of 10-7cm/s, and the 28d unconfined compression strength of the piles should be greater than 1.0Mpa.

②In order to construct smoothly, large diameter pipes are adopted to ensure that the underground water level is less than 1m below the design elevation of the foundation pit bottom.

4 Monitoring the construction of super-large deep foundation pit

Project monitoring can not only guide the safe construction of the site, but also be equivalent to a 1:1 experiment to some extent[1]. The monitoring data can truthfully reflect the situation of engineering, which has important implication for the research of foundation pit engineering[2~6].

4.1 Analysis of monitor super-large deep foundation pit project

4.1.1Horizontal displacement of pile top

Pile top displacement monitoring points are laid out as shown in Figure 4. E1-E9 are arranged on the east of the foundation pit, while S is on the south, W1-W9 are on the west and N is on the north.

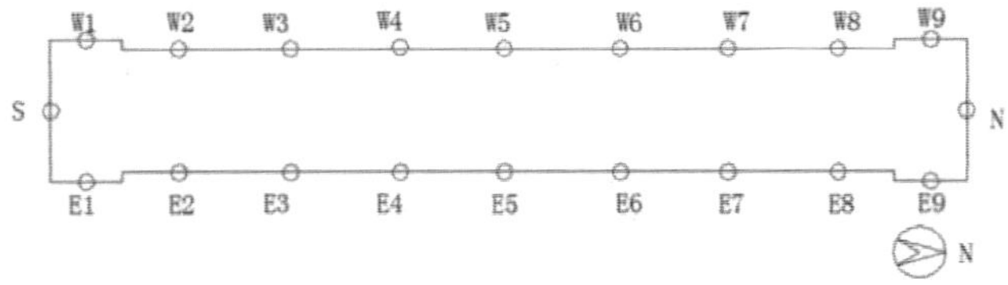

Figure 4. Plane layout of pile top displacement monitoring points.

Due to the limitation of space, this paper only studied and analyzed the west side of representative foundation pit and cross-section E1-W1, E4W-4.

The horizontal displacement-time curve of pile top onthe east and west sides of the foundation pit is shown in Figure 5 and Figure 6.

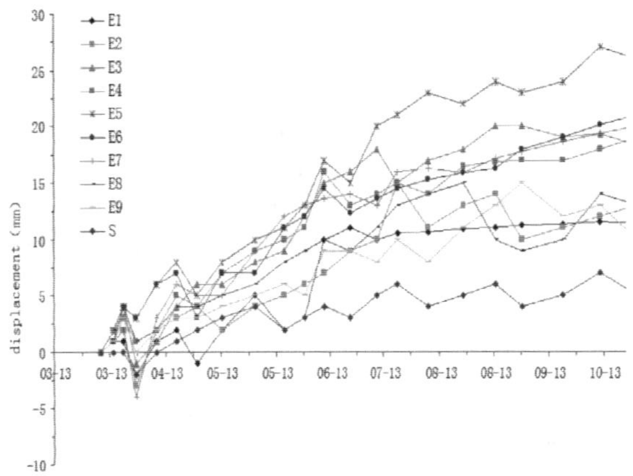

Figure 5. The horizontal displacement-time curve of pile top on the east side of the foundation pit.

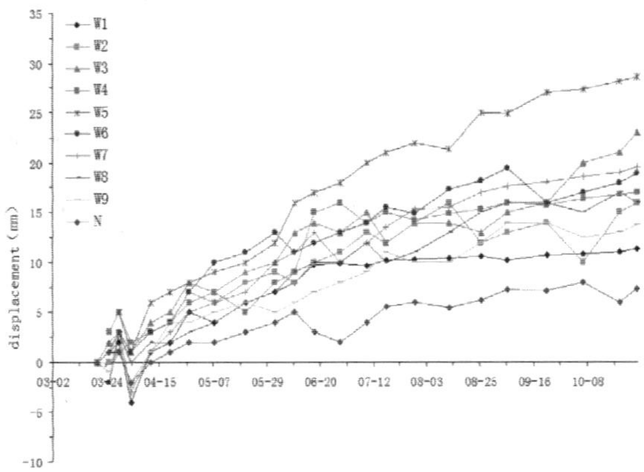

Figure 6. The horizontal displacement-time curve of piletop on the west side of the foundation pit.

As can be shown from Figure 5 and Figure 6, thehorizontal displacement of pile top tends to increase as the excavation depth of foundation pit increases; with the excavation, the maximum deformation of soldier pile top shows "space effect[6,7]. The deformation of pile increases gradually from the north and south ends to the middle part of foundation pit.

It can be seen from the table that the monitoring value of horizontal displacement of pile top at the short-side monitoring points S,N and the long-side corner points (W1, W9, E1, E9) are between 6.5 to 14.1mm, while the middle of the long-side points (W2-W8, E2-E8) are between 16.3 to 28.6mm.Although the monitoring value of horizontal displacement of pile top at each monitoring point is less than the warning limit of 30mm and 0.2H% (37.2mm), the monitoring value of points at the short-side and corner of foundation pit are far less than the warning limit as specified. If the warning limit (30mm or 0.2H%) in specification is adopted for deformation control, it will be too conservative. Therefore, the deformation control index of pile top of the foundation pit supporting structure should be divided into sections to reduce unnecessary waste. As such, the paper suggests that, for strip type deep foundation pit(length:width:depth=10:1:1) of similar geological conditions, the maximum deformation value should be less than 16mm or 0.1H% by the control index of short-side and corner pile top deformation.Horizontal displacement contrast curve of pile top E1-W1, E4-W4under various working conditions is shown in Figure 7 and Figure 8.

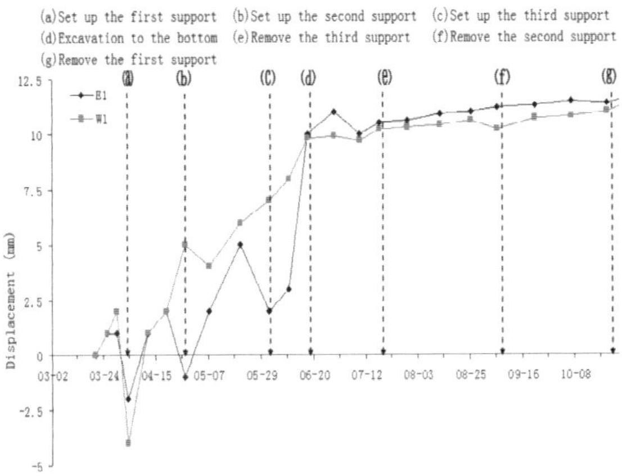

Figure 7. Horizontal displacement contrast curve of pile top E1-W1under various working conditions.

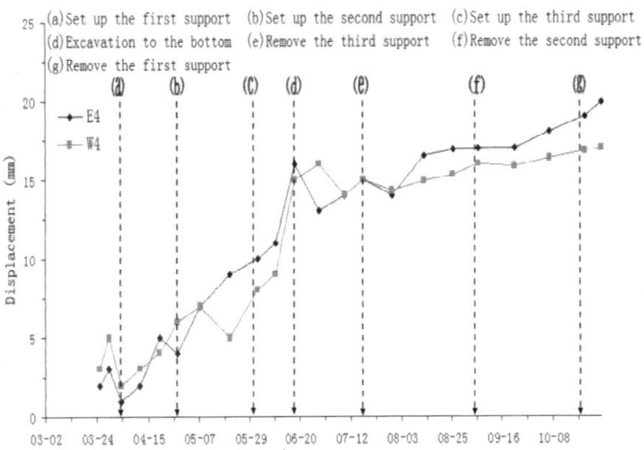

Figure 8. Horizontal displacement contrast curve of pile top E4-W4 under various working conditions.

As shown in Figure 7 and Figure 8, the horizontal displacement of cast-in-place pile top increased quickly at the beginning of foundation pit excavation. After setting up the first support and applying prestress to it, the horizontal displacement of pile top decreased sharply; with the excavation depth of foundation pitincreases, the horizontal displacement rate of pile top tends to increase until the excavation reaches the bottom of the pit. During the pouring period from main bottom slab to main top slab, the displacement change rate of pile top is little, which indicates that the bottom slab, middle slab and top slab of the main structure play a certain supporting role.

4.1.2 Horizontal displacement of pile body

The horizontal displacement of pile body is a physical quantity which varies with the excavation depth of foundation pit. Horizontal displacement monitoring is one of the important monitoring items to ensure the safe construction of large deep foundation pits.

ZXE1-ZXE9 are arranged on the east of the foundation pit, while ZXS is on the south, ZXW1-ZXW9 are on the west and ZXN is on the north. Layout of horizontal displacement monitoring points of pile body is shown in Figure 9. A total of 19 inclinometer boreholes of pile body are arranged in the foundation pit. Due to the limitation of space, this paper only studies and analyses the representative monitoring points ZXS, ZXE4. The horizontal displacement of pile body are shown in Figure 10 and Figure 11.

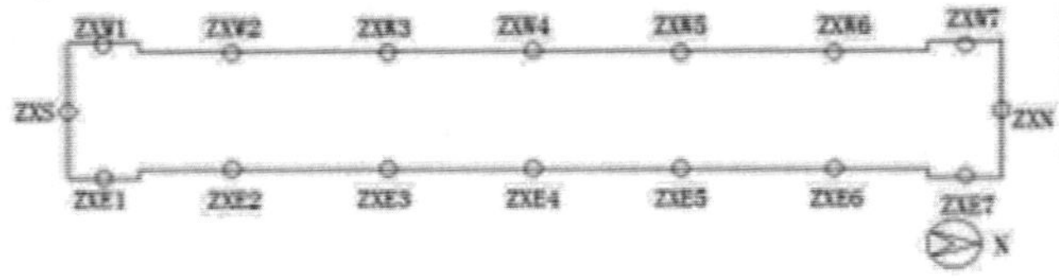

Figure 9. Plane layout of pile body displacement monitoring points.

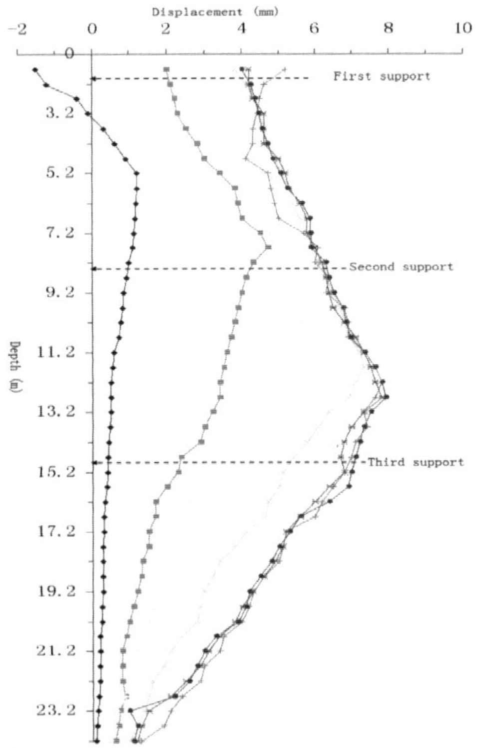

Figure 10. Horizontal displacement of pile body (ZXS).

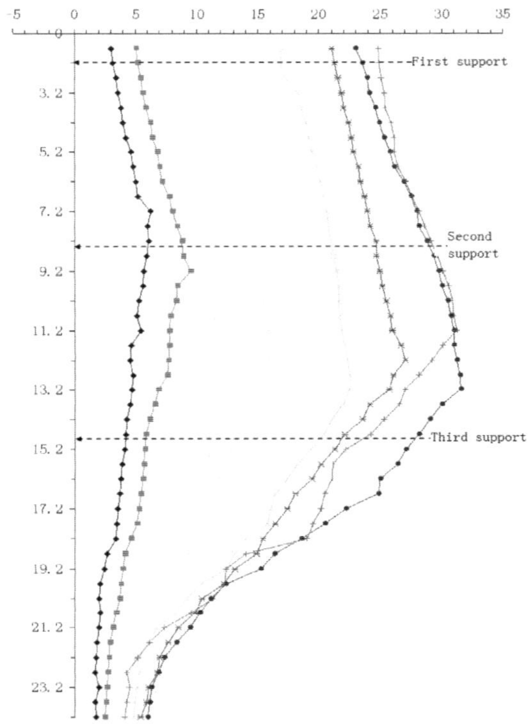

Figure 11. Horizontal displacement of pile body (ZXE).

ZXE1-ZXE9 are arranged on the east of the foundation pit, while ZXS is on the south, ZXW1-ZXW9are on the west and ZXN is on the north. Horizontal displacement monitoring points of pile body are laid out as shown in Figure 9. A total of 19 inclinometerboreholes of pile body are arranged in the foundation pit. Due to the limitation of space, this paper only studies and analyses the representative monitoring points ZXS, ZXE4. The horizontal displacement of pile body are shown in Figure 10 and Figure 11.

As can be shown in Figure 10 and Figure 11, with the earthwork excavation depth of foundation pit increases, greater lateral soil pressure is produced by adjacent soilof foundation pit, which further increases the horizontal displacement of pile body. The horizontal displacement of support pile body mainly occurs in the stage of foundation pit excavation and steel support (excavation to -18.58m), during which the maximum deformation is about 75% of the total maximum deformation of pile body (the specific proportion are shown in Table 3). The maximum deformation of pile body during the main construction stage is about 25% of the total maximum deformation of pile body.

The maximum horizontal displacement of pile body moves down gradually with the advancement of excavation support and the main construction. The maximum displacement occurs around the foundation pit depth of -12.7m (as shown in Table 3), that is, the maximum deformation occurs around 0.684H (H foundation pit excavation depth: 18.58m).

5 Conclusion

（1）The deformation control index of pile top of the foundation pit supporting structure should be divided into sections to reduce unnecessary waste. It suggests that for the maximum deformation value which is the control index of short-side and corner pile top deformation of the strip type deep foundation pit similar to the geological conditions of this project should be less than 16mm or 0.1H%.

（2）The maximum horizontal displacement ofsupporting pile in super-large deep foundation pits shows"time effect". The horizontal displacement mainly occurs in the stage of foundation pit excavation and steel support, which is about 75% of the total maximum deformation of pile body and 25% of the total maximum deformation of pile body. The maximum horizontal displacement of pile body moves down gradually with the advancement of excavation support and the main construction. The maximum displacement occurs around 0.684H.

（3）In the process of large deep foundation pit excavation, unsupported excavation should be avoided as far as possible, and the principle of "supporting while digging" must be followed. Through the implementation of project, the monitoring data is within the controllable range and has no impact on the surrounding environment, which provides a reference for the design, construction and deformation law of large deep foundation pit support structure of similar engineering geological conditions.

Table 3. Statistical table of maximum deformation analysis of pile body.

Measuring points	Maximum deformation A（mm）	Maximum deformation position（m）	Maximum deformation at excavation support stage A_1（mm）	A_1/A（%）	Maximum deformation at main construction stage A_2（mm）	A_2/A（%）
ZXS	7.9	-12.7	7.3	92.4	0.6	7.6
ZXE4	31.5	-13.2	22.5	71.4	9	28.6

Reference

[1] Wang X J. Inspection test and Analysis of axial force in steel brace during construction of Foundation Pit in running tunnel section adjacent to Liming Cultural Club station of Shenyang Metro[J], *Railway Engineering*, **4**, (2007)86-88

[2] Qian Y L, Ou Q, Gao H Y, Wang J, Lv W K, Wang Y X. Analysis of Internal Force and Deformation for asymmetrically loading and irregular Deep Foundation Pite[J], *Journal of Hebei University of Engineering (Natural Science Edition)*, **33(3)**, (2016)14-16

[3] Jin X, Zhao X B. Construction of Deep Foundation Pit Close to the Edge of an Important Building in Downtown Center[J], *Building Construction*, **30(7)**, (2008)532-535

[4] Wang X W, Tong H Y, Li Z Q. Monitoring Analysis on Foundation Excavation Internally Supported with Diaphragm Wall in Complex Environment[J], *Construction Technology*, **38(5)**, (2009)88-91

[5] Qian J G, Huang M S. Effect of non-coaxial plasticity on onset strain localization in soils under 3D stress condition[J], *Chinese Journal Of Geotechnical Engineering*, **28 (4)**,(2006)510-515

[6] Chi P. Research on Field Monitoring of Deformation Rule about Support Structure and It's Optimization in Deep Foundation Fit Engineering [D], (2008)

[7] Li J P, Li Y S. Monitoring Analysis on Construction of Enclosure for Pile Foundation[J], *Underground Space*, **24(4)**, (2004)479-48

IOP Conf. Series: Materials Science and Engineering **652** (2019) 012003 doi:10.1088/1757-899X/652/1/012003

The impact of the anchorage on the shear capacity of reinforced concrete beams

D Marcinczak[1] and T Trapko[2]
[1]PhD student, Wroclaw University of Science and Technology, Faculty of Civil Engineering, Pl. Grunwaldzki 11, 50-377 Wroclaw, Poland.
[2]Associate Professor, Wroclaw University of Science and Technology, Faculty of Civil Engineering, Pl. Grunwaldzki 11, 50-377 Wroclaw, Poland.
Email: dorota.marcinczak@pwr.edu.pl

Abstract. This paper presents an experimental study on shear strengthening of reinforced concrete (RC) beams with PBO-FRCM composites. Key parameter of this study is the presence of composite anchorage. PBO-FRCM (Fabric-Reinforced Cementitious Matrix) composites consist of mineral mortar and PBO (p-Phenylene Benzobis oxazol) composite fibres. The most common failure mechanism of FRCM composites is debonding of the fibers from mineral mortar with no rupture of fibres, due to inappropriate coverage of all fibers by the mortar. To prevent premature debonding of the composite and increase the effectiveness of FRCM strengthening suitable anchorage should be used. In this case, experimental research was conducted on T-shaped RC beams strengthened in shear using PBO-FRCM with anchorages The study consisted of 5 beams, one of which was a control beam, one was strengthened without anchorage and the other three were strengthened with GFRP rebar anchorage. All beams were loaded until failure. Deformation, deflection and the width of cracks were recorded for each level of loading. The failure mechanisms of beams were analysed and described. The results obtained from research showed that the use of FRCM composites increased the shear capacity and had an impact on better use of the properties of the composite material.

1. Introduction

PBO fibers (p-Phenylene BenzobisOxazole) are new generation fibers, which in comparison to other fibers used for strengthening of reinforced concrete structures are characterized by the highest tensile. They are used as a component of the FRCM composite system, where the mineral mortar is used as a matrix, in case to improve its fire resistance. This material is increasingly the subject of research as evidenced by the constantly growing number of publications. The tests concern beams for bending, shearing, plates and columns [1-6]. However, the efficiency of the FRCM system is slightly smaller than that of FRP systems due to the low tensile strength of the mineral mortar and insufficient coverage of all fibers by the matrix. After exceeding this strength, the PBO mesh debonding from the surface of the element occurs, on the matrix-fiber layer [7-10]. All this leads to incomplete use of mechanical properties of PBO fibers in FRCM system.

To increase the efficiency of FRCM reinforcements, anchoring of PBO mesh should be used, which can prevent premature debonding of fibers and increase the utilization of the load capacity of the PBO mesh. Tests conducted on rectangular beams strengthened in bending [11] or shearing [12,13] with PBO-FRCM showed that the anchorage improves the bearing capacity of the element, and PBO fibers are more effectively used (have larger deformations) compared to beams without anchorage. Studies of T-beams with FRCM shear strengthening [14-17] showed that the anchorage has a large contribution in the bearing capacity of the whole element. In fact, there is still a lack of tests on T-shaped RC beams strengthened with PBO-FRCM shear strengthening with anchoring. The subject of the presented research were T-shaped RC beams, strengthened in shear with PBO-FRCM composite with anchorage. The aim of the research was to study the effect of anchoring element on the load-bearing capacity and utilization of composite material.

2. Experimental research

Five reinforced concrete T-shaped beams were designed and tested. One of the beams was the reference beam, one was strengthened without anchorage and other three beams were strengthened with anchorage using PBO-FRCM materials with the same configuration and the same type of anchors.

2.1. Test specimens

The tests were carried out on T-shaped beams with cross-sectional dimensions of 350x400 mm, web width 150 mm and total length of 2300 mm. The main tensile reinforcement was designed to prevent destruction due to bending before exhausting shear capacity. The beams were reinforced at the bottom with 5 bars with a diameter of 20 mm and 2 bars with a diameter of 20 mm at the top. Stirrups with a diameter of 8 mm were made of the same type of steel. The spacing of stirrups was 250 mm (Fig. 1).
The beams and specimens for determining the features of concrete were manufactured in a prefabrication plant during a single concreting and vibrating process. In order to determine the strength qualities of concrete the cylinder specimens were made with 150 mm in diameter and height of 300 mm. The average compressive strength of concrete obtained on day of testing was 42.25 MPa. Mean modulus of elasticity of the concrete was 33.69 GPa. The longitudinal reinforcements consisted B500Sp reinforcing steel bars with a yield strength of 526.6 MPa and modulus of elasticity 206.7 GPa were also obtained from test.

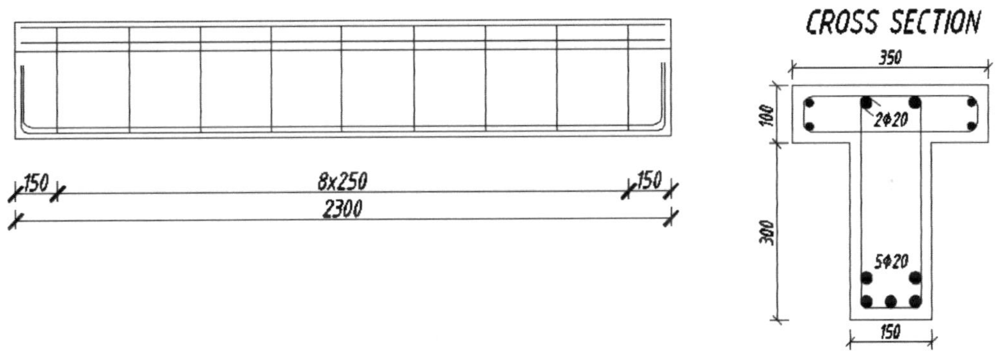

Figure 1. The scheme of steel reinforcement and cross-section of beams.

2.2. Beam preparation and test setup

The beams were shear reinforced with a PBO (p-Phenylene Benzobis Oxazole) Ruredil X Mesh Gold and mineral mortar Ruredil X Mesh M750 (Table 1). The PBO mesh is a bidirectionally woven sheet with four times more fibres in the main direction as in the lateral direction.
Before the application of PBO-FRCM composites, the concrete surface was properly prepared by grinding, dusting and washing. To avoid stress concentration, the two bottom edges of each specimens were rounded (radius equal to 15 mm).

Table 1. Properties of PBO fibers and the PBO-FRCM system [18]

	Tensile strength f_{fz} [MPa]	Young modulus E_f [GPa]	Ultimate tensile strain ε [%]	Thickness of composite [mm]
PBO fibre mesh	5800	270	2.15	0.0455
PBO-FRCM system	1664	137	1.76	-

The same strengthening configuration with the same anchorage was used for each of the strengthened beams. The width of the PBO mesh strips for each beam was 100 mm and their spacing was 100 mm (Fig. 2). All strengthened beams have 20×20 mm cuts made under the slab. After shear strengthening, the ends of the PBO strips were wound on a GFRP bar and glued in a cut under the shelf. The bar had a length equal to the length of the beam and together with the PBO strips were covered with an outer layer of mortar.

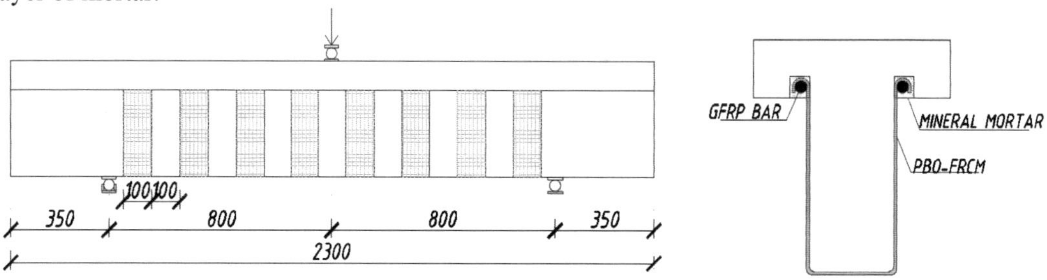

Figure 2. The scheme of shear strengthening with end-anchorage.

The beams were tested under three-point bending monotonic load. The load was applied using material testing machine with a range of 0-6000 kN. An external LVDT was used to measure the vertical displacement at the load supplication position and in the places of support. Deformations of concrete, steel and composite were measured with strain gauges. Strain gauges on concrete were applied in the middle of the beams in the central cross section, at the top (compressed) surface. Strain gauges on the steel were applied before concreting, on stirrups at half height and on the longitudinal bar in the middle of the beam span. Strain gauges were adhered at the external stirrups made of PBO mesh in the middle of the beams in accordance with the direction of main fibres. Strain gauges were also applied at the places of composite anchorage.

3. Results

All tested beams failed in a shear with formation of main diagonal crack. Control beam B_0 reached an ultimate load of 453.67 kN (Fig. 3). Table 2 shows summary of the test results. The shear capacity increase is the ratio of capacity of strengthened beams to the capacity of the control beam. P_{max} stands for the maximum force at the failure, and ε stands for the maximum deformation of the composite.
Shear capacity increase for strengthened beams was 1.05 for beam without anchorage and 1.10-1.12 for beams with anchorages. It means that effectiveness of the anchored composites in increasing the shear capacity of the T-beam was 1.05-1.06 times the effectiveness of the non-anchored composites. It follows that the use of the anchorage improves the shear capacity of RC beams.
The same failure mechanism was observed for all the strengthened beams with anchorages. It consisted in debonding of the PBO mesh from the mortar layer, without the participation of concrete cover (Fig. 4). The first inclined cracks were formed under a load of 80 kN for the control beam and 125 kN for anchored beams. It can therefore be seen that the use of the composite strengthening delayed the appearance of diagonal cracks. Prior to failure an inclined crack was formed in the slab. This crack was develop from the point of the force to the anchorage under the slab. Then the crack would run under the slab along the GFRP bar. This resulted in loosening the anchorage and PBO-FRCM stirrups, and thereby inclined cracks developed on the web of the beam (Fig. 4). Once the tensile strength of the cement mortar was exceeded, the PBO stirrups would debond and the inclined cracks would develop under them until the beam failure. Scratches caused by the slippage of the bundles of PBO fibres on the mortar layer were visible on its surface (Fig. 5). This phenomenon is often found in strengthening, where cement matrix is used as a matrix. This is due to the inaccurate coverage of all fibers in each bundle by the mortar.

Figure 3. Failure of control beam.

Table 2. Test results.

Beam	P_{max} [kN]	ε [‰]	shear capacity increase [-]
B_0	453.67	-	-
B_P100_BZ	477.34	6.39	1.05
B_P100_1	508.10	9.90	1.12
B_P100_2	499.35	9.04	1.10
B_P100_3	499.17	8.27	1.10

Despite the presence of the anchorage no fibres were ruptured in any of the beams. Deformations in the anchorage would abruptly increase in the final stage of loading. This is evidence of the good performance of the anchorage which allowed the load to further.

Figure 4. Failure of strengthened beam with anchorage, debonding of PBO mesh and cracking along GFRP bar.

Figure 5. Failure of strengthened beam with anchorage.

Figure 6. Failure of strengthened beam without anchorage.

For beam B_P100_BZ without anchorage failure mechanism was very similar to beams with anchorage. Debonding of PBO mesh in matrix/fibers layer was also observed. The first inclined cracks were formed under a load of, 100 kN. It can therefore be seen that the diagonal crack was formed earlier in the beam without an anchorage than the anchored beam. The destruction occurred as a result of the loosening of the end of the PBO strip under the shelf (Fig. 6), because it was not held by the anchorage. Prior to failure an inclined crack was formed in the slab. This crack was develop from the point of the force to the PBO stirrups under the slab. Debonding of the ends of the PBO strips occurred only in places where the diagonal crack ran near the shelf.

The PBO-FRCM strengthening would begin to participate in carrying the shearing force while the first diagonal cracks appeared, which resulted in increase of deformations of the PBO fibres. This indicates that the PBO stirrups began to take part in carrying tensile stresses in the cracked cross section. The use of the anchorage has improved the efficiency of using the PBO-FRCM composite material. In the beam without anchorage, smaller deformations of the composite (6.39‰) were noted than in anchored beams (8.27-9.90‰). It means, that the maximum deformation of the composite in anchored beams was 30-55% greater than in the beam without anchorage. Comparing the maximum deformation values of the PBO-FRCM composite in the presented studies to the maximum deformation values of the PBO-FRCM material in the axial tensile test (which amounts to 1.76% according Table 1), it can be concluded that the PBO-FRCM material was used in 36% of the beam without anchoring and 47 - 56% in beams with anchorage.

4. Conclusion

The PBO-FRCM system as a shear strengthening of reinforced concrete beams is a promising solution. Thanks to the use of the PBO-FRCM system the shear strength of RC T-shaped beams increase by 5-12% in comparison with the un-strengthened reference beam. The relatively small increase in load capacity due to the strengthening of that research is associated with a low axial stiffness of the composite. In this research the PBO strips were 100 mm wide with a 100 mm spacing. With a larger amount of composite used, the load increase is greater [13]. For example, using PBO strips with a width of 150 mm, the load increase reaches up to 27%.

The use of the anchorage has a positive effect on shear strength of the beam and on the utilization of composite material. Shear capacity increase for strengthened beams was 1.05 for beam without anchorage and 1.10-1.12 for beams with anchorage. It means that shear capacity of the T-beam with anchorage was 5-6% greater than shear capacity of the beam without anchorage.

For strengthened beams with anchorages the maximum strain in the composite reached 9.90‰, which amounts to 55% of the ultimate tensile strain of PBO-FRCM. For beam strengthened with the PBO-FRCM system without anchorage the maximum strain in the composite reached 6.39‰, which proves that the use of anchorage allows to better usage the properties of PBO-FRCM composite. Despite the use of the anchorage, the ultimate tensile strength of the PBO fibers was not reached. No PBO fibres ruptured in any of the beams.

IOP Conf. Series: Materials Science and Engineering **652** (2019) 012003 doi:10.1088/1757-899X/652/1/012003

The same failure mechanism was observed in each of the strengthened beams. The mechanism consisted in the development of an inclined crack in the web and in the slab and in the propagation of a crack along the anchorage, resulting in the debonding of the latter and the outer stirrups from the surface of the member, in the layer between the fibres and the matrix. The destruction in beam without anchorage occurred as a result of the loosening of the end of the PBO strip under the shelf because it was not held by the anchorage. This proves that anchorages are needed.
The next stage of research should include the possibility of modifying the cement matrix so that premature debonding of the fibers does not occur and that the fibers are better covered with the matrix.

References

[1] Z.C. Tetta, L.N. Koutas, D.A. Bournas, Composites Part B, 77, 338-348 (2015)
[2] T. Blanksvärd, B. Täljsten, A. Carolin, ASCE J. Compos. Construct, 13, 25–34 (2009)
[3] TC Triantafillou, CG Papanicolaou, Mater Struct, 39, 93–103 (2006)
[4] A. D'Ambrisi, L. Feo, F. Focacci, ASCE J. Compos. Construct, 15, 707-720 (2009)
[5] L. Ombres, Compos. Struct., 94, 143-155 (2011)
[6] T. Trapko, Constr. Build. Mater., 73, 332-338 (2014)
[7] T. D'Antino, C. Carloni, L.H. Sneed, C. Pellegrino, Eng. Fract. Mech., 117, 94-111 (2014)
[8] L. Ombres, Composites Part B, 69, 418–426 (2015)
[9] A. D'Ambrisi, L. Feo, F. Focacci, Composites Part B, 43, 2938–2949 (2012)
[10] L. Ombres, Compos. Struct., 12, 316-329 (2015)
[11] T. Trapko, M. Musiał, Composites Part B, 118, 67-74 (2017)
[12] T. Trapko, D. Urbańska, M. Kamiński, Composites Part B, 80, 63-72 (2015)
[13] D. Marcinczak, T. Trapko, M. Musiał, Composites Part B, 158, 149-161 (2019)
[14] E. Tzoura, T.C. Triantafillou, Material and Structures, 49, 17-28 (2016)
[15] D. Baggio, K. Soudki, M. Noël, Constr. Build. Mater., 66, 634-644 (2014)
[16] Z.C. Tetta, L.N. Koutas, D.A. Bournas, Composites Part B, 95, 225-239 (2016)
[17] A. Brückner, R. Ortlepp, M. Curbach, Material and Structures, 41, 407-418 (2008)
[18] Ruredil, X Mesh Gold Data Sheet, Ruredil SPA, Italy,(2009)

Model for predicting cost of rural road projects in Thailand

Wai Phyo Lin[1,2] **and Wasaporn Techapeeraparnich**[1,2]

[1]Civil and Environmental Engineering Department, Faculty of Engineering, Mahidol University, Salaya Campus, Thailand
Email: wphyolin09@gmail.com; wasaporn.tac@mahidol.ac.th

Abstract. Nowadays, construction cost plays an essential role in various projects, which are buildings, roads, railways, and bridges projects. Conceptual cost estimation in feasibility study is require high accuracy and less validation error especially in construction projects at early stages. The more improvement of estimation techniques, it would lead to the lesser problems of cost overrun, and extra expenses. This paper developed a cost estimation model for new constructed rural road projects. Considering the estimation methods for predicting the cost model, parametric method based on regression learner and NN method are applied. Previously, many researchers studied the cost applicable model by using various computer applications, so that this paper differed to compare these methods based on MATLAB. Accordingly, 44 road projects were compiled from DRR database, after that identifying the effective cost parameters referred on collected data. Subsequently, the data implementation process was focusing with regression learner based on automated regression training, and lastly, with NN toolbox concentrated on nftool for simulating the cost models. Another 7 road projects are tested with holdout validation process for these models. Due to this validation, we compare the predicted cost models within these methods. Finally, the developed models are reliable for not only the conceptual phase of future rural road projects but also the related construction fields can be recovered about the cost model creation.

1. Introduction

With a growing population in Thailand, it is difficult to create a well-designed urban public transportation system that meets the needs of the entire population for improving transportation corridors. So, Thailand's government investment in infrastructure has increased each year especially with horizontal infrastructure projects such as roads, railways, and bridges which will lay a foundation for growth and well transportation [1]. Literally, road construction is supporting national development strategies such as tourism, logistics, and agriculture. For these developments, there should be considered not only in urban but also in rural regions as well. Rural roads are mostly built to serve for both economic and social activities that are accessible to local schools, hospitals, and others emergency [2].

Rural roads are a fundamental element in the provision of access in rural areas, so that roads are crucial to economic and social development. It is surprising that the data on rural roads in Thailand region are not difficult to find, but it is hard to find a reliable statistic [3]. Due to the problems of rural roads work with increased funding uncertainties and shrinking budgets, it is more important than ever to ensure proper allocation of funds for highway projects in the preconstruction phase [4].

Although extremely important and often overlooked, project cost estimating has many obstacles to overcome throughout the process [5]. For the above reasons, the following attempts will be considered

IOP Conf. Series: Materials Science and Engineering **652** (2019) 012004 doi:10.1088/1757-899X/652/1/012004

different from of previous studies by developing the reasonable cost models with high accuracy. So, the aim of this research represents:

- Description of road cost influential parameters for predicting cost model, and
- Development of road cost prediction model based on regression learners, and neural network methods

2. Identification factors

According to engineering and management, the crucial part of a road construction project represents its financial demands. Bill of quantities (BOQ) in construction project represents the most effective parameter in the construction project cost [6]. However, precise and feasible studies of construction projects are not simple to do since there have been many factors varying from one project to another such as project size, type, location, the season of the year, material type, structure, and method of construction [7].

In addition, between the several independent variables in highways' projects, the size of the project presents an enormous variation [8]. For instance, location of the project represents important factor correlated to a populated area, ground condition and profile, and environmental impacts. Material alternatives such as several types of asphalt pavements, concrete slabs, and other types of materials contain variation in the value of project costs [9]. Referring to these important variables, the newly constructed project in this paper would be focusing on not only the engineering aspects of the bill of quantities as input factors but also manpower and their costs as outputs.

Consequently, the identification of cost parameters distinguished into six input factors for cost model development in terms of investigation of rural road projects which is highly influenced parameters summarized as below table 1.

Table 1. Input parameters for cost modelling used in this research.

No	Input factors	Descriptions
1	Length	Project road distance (km)
2	Width	Road width of each lane (m)
3	Number of lanes	Number of lanes per direction
4	Pavement type	Asphalt concrete structure, prime coat, tack coat, coating, base, and subbase
5	Earthwork	Embankment, and excavation
6	Miscellaneous	Drainage, guardrail, lighting, precast, retaining wall, seeding, signalization, signage, planting and extra expenses

The above-selected input factors are used in expressing the cost model for the rural road constructed project BOQ data used in this research.

3. Data analysis and cost modelling

A cost database of 51 rural road projects was collected from DRR only for new constructed ones. An analysis of collected data from DRR can be considered into two steps as follows.

3.1. NN development

Proofing the good fitting tools for the practical functions, neural network should simply fit with (nftool) through the dataset and describe the problem in NN toolbox. So, this process passed through using with MATLAB R2018b NN toolbox until models result come out are as follows.

Stage 1 Database Analysis

Stage 2 Neural Network fitting and training
Stage 3 Case Study for Validation
Stage 1 Database Analysis
In this study, having a database of 51 projects; it is decided to utilize about 70% of the database for training, 15% for testing and 15% for validation.

Stage 2 Neural Network fitting and training
Once neural networks fitting, the data set firstly imported as a numeric matrix then classified input and output factors. During this process, the neural network architect came out and filled the acceptable hidden layers which is 25 almost four times greater than input numbers. The below figure 1, is NN toolbox in MATLAB with sigmoid function applied and describes with the data numberings in each segment.

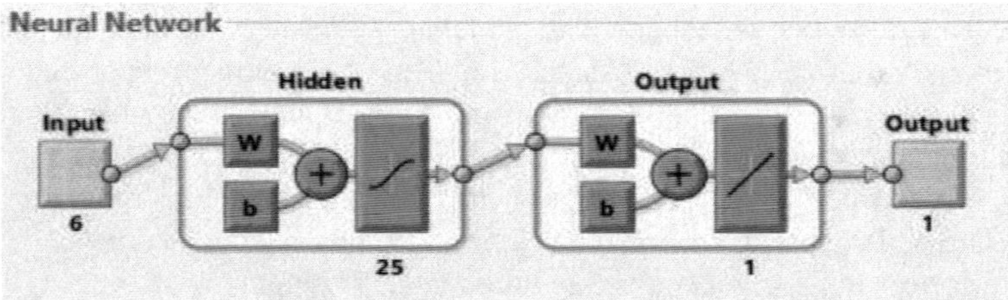

Figure 1. NN Toolbox in MATLAB

In this study, the default Levenberg-Marquardt used as the training test and showed the validation error failed to decrease for twenty-one iterations until its stop.

Stage 3 Case Study for Validation
The MSE display the network outputs with respect to targets for training, validation, and test sets. For a perfect fit, the result data performed to be the best validation at epoch 15.

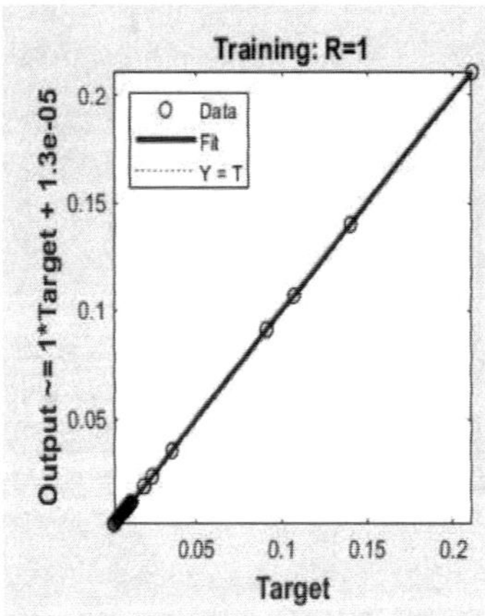

Figure 2. Trained result

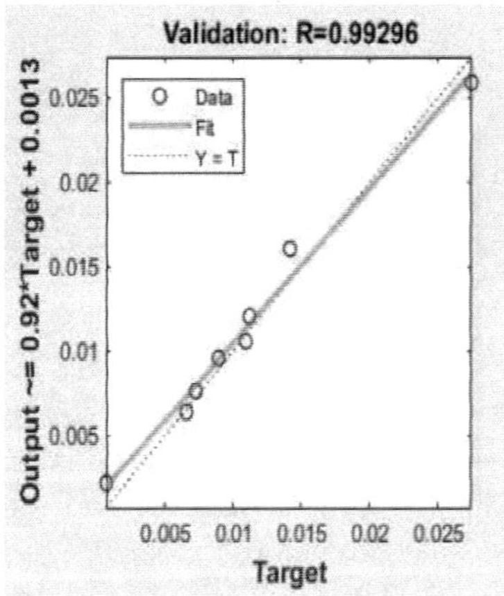

Figure 3. Validated result

Both figure 2 and 3, all input data in training results showed a significance outcome as the form of R=1 and the result for validation also found on the way of line graph around confidential level at 0.05. For the final stage of the model simulation, the MATLAB function applied to analyze in matrix and cell array arguments for the output results. So, the below equation is showing about the cost model processes referred from MATLAB function.
Input weight layer,

$$H = \sigma\,(W.\mathrm{X}^{\mathrm{T}} + b) \tag{1}$$

Non-linear cost model equation,

$$Y = W_2.[\,\sigma\,(\,W_1.X^T + b_1)]^T + b_2 \tag{2}$$

where, σ is sigmoid function, X is input vector, W is controller weight of input as in matrix, and b is the bias.

3.2. Regression analysis

The reason to find the model from using regression analysis based on regression learner, is to test as another way for comparing NN toolbox. The model is developed through the following phases using regression learner in MATLAB summarized as follows:

Stage 1 Database Analysis
Stage 2 Apply in regression learner
Stage 3 Case Study for Validation

Stage 1 Database Analysis

The first stage is used for importing data and selected for all dataset in the column because inputs (modeling factors) are in the first six columns and output (project cost) is the last.

Stage 2 Apply in regression learner

To predict the cost models, the regression learner can also be used to train regression models in MATLAB. Automated regression training was used to search for the best regression model type from the linear, interaction, robust, and stepwise.

Table 2. Best linear regressions

Linear Regression	RMSE
Linear	5.24e-10
Interactions	5.62e-10
Robust	5.24e-10
Stepwise	5.35e-10

As shown in table 2, linear one is suitable as the best regression model of the projects and the above results were implemented only with automated regression model training because the training of this option is fast to fit the multiple models simultaneously.

Stage 3 Case Study for Validation

The training set assesses its performance with holdout validation set. Therefore, this validation is implemented with 15% of the database for all results. From extracting the results, the coefficients of cost model would be printed out from the exported model in the workplace as shown in table 3.

Table 3. Linear model coefficients

Factors	Estimate
(Intercept)	-2.36e-10
X1	-1.02e-08
X2	1.01e-08
X3	4.71e-09
X4	0.7621
X5	0.1456
X6	0.0923

Therefore, the equation can be simulated with these estimations from the coefficients. The general equation is

$$Y = C^T X + b \tag{3}$$

where, Y is output cost, C is the coefficients, X is the factors and b is intercept.

So, the cost model of this projects described as below.

$$Y = -1.02e-08X_1 + 1.01e-08X_2 + 4.71e-09X_3 + 0.7621X_4 + 0.1456X_5 + 0.0923X_6 - 2.36e-10 \tag{4}$$

where, X_1, X_2, X_3, X4, X_5 and X_6 are length, width, lanes, pavement types, earthwork, and miscellaneous.

4. Comparing between linear and nonlinear cost model's performance

All training models from dialog box can be compared the accuracy of the results in regression model process and neural network performance. These test results of two models are shown in figure 4 and 5.

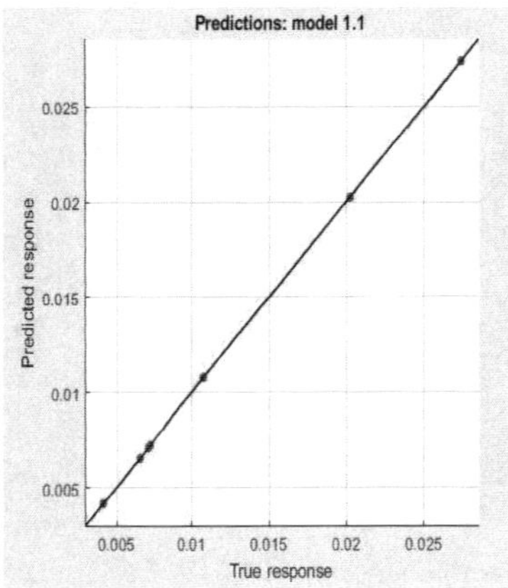

Figure 4. Trained result

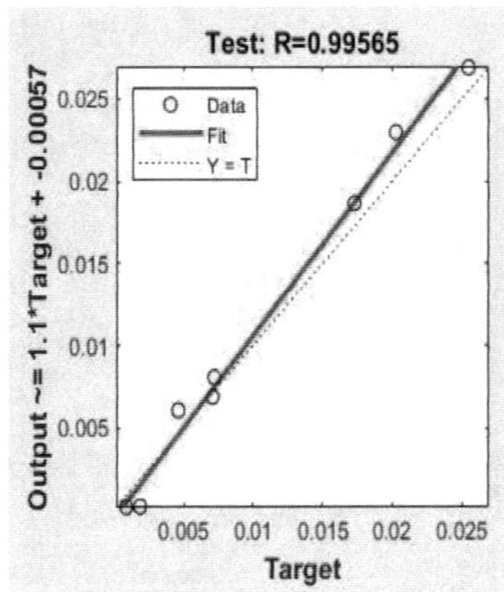

Figure 5. Validated result

Above, linear is rather accurate result than non-linear, so that the below table 4 clearly compared between linear and non-linear of their performance of cost models results.

Table 4. Cost model performance between linear and non-linear

Results	Linear	Non-linear
MSE	2.57e-19	2.21e-6
R	1	9.96e-1
RMSE	5.24e-10	1.49e-3

As cost model results, linear one is the most accurate in this paper. However, the models were developed based on 51 sets of data collected in Thailand and any model's performance is going to be data dependent. For small data sets with less variance, linear model works well, and neural network-based models fail to find meaningful relationships. However, as the complexity and scale of the datasets increase, non-linear models with bigger learning capacity will become more relevant.

5. Conclusion

This research work aimed at formulating a reliable and practical model for conceptual cost estimate that can be used by actual utilizations of the rural road projects. In neural network, the model was trained using Levenberg- Marquardt algorithm. According to [8, 10], the output results were acceptable as mentioned in table 4. As a practical issue, the model was adapted in GUI as MATLAB function to classify the input parameters for the projects and, output results for predicting the cost.

To develop a linear regression models, it is very useful, especially in its simplicity and ability to be handled by simple program as regression learner. It has also a good benefit in estimating project cost at early stages of the project [10].

Accordingly, the relationship between the independent and dependent variables of the developed models are acceptable because of extracting as the best regression results in two methods. Also, the predicted values from a forecast model fit with the actual data, but the root mean squared error of linear model got the less result error than the non-linear one so that the cost model from linear is the best performance in this paper. However, it is also seemed that neural network method based on NN toolbox could be a reasonable tool to solve problems with uncertainties as well [9].

The study is to simulate the cost model with acceptable stage for future rural road projects which are supporting for owners, consultants and contractors under this kind of projects. In the future, if there is a certain amount of distance to estimate the final cost, the cost model will be predicted overall cost. Another recommendation of future work will be developed as cost estimation model by analyzing with other methods such as fuzzy logic, case based reasoning, and other up to date techniques [8, 9].

Acknowledgments

The authors thank professor, colleagues, technical staff and financial support from faculty of graduate studies and department of civil engineering at Mahidol university. Lastly, the authors appreciate for the valuable comments by ICCEMS committee members and staffs.

6. References

[1] Meeampol S and Ogunlan S O 2006 Factors affecting cost and time performance on highway construction projects: evidence from Thailand *J. Financial Management Property Construction* 11 **1** 3-20

[2] Pomlaktong N, Jongwilaiwan R, Theerawattanakul P and Pholpanich R 2014 Road transport in Thailand *In priorities and pathways in services reform: Part II—political economy studies* pp 227-243

[3] Donnges C, Edmonds G and Johannessen B 2007 Rural road maintenance: Sustaining the benefits of improved access *Bangkok International Labour Office*

[4] Hollar D A 2011 *Predicting preliminary engineering costs for highway projects* (Doctoral dissertation, North Carolina State University)

[5] Turochy R E, Hoel L A and Doty R S 2001 Highway project cost estimating methods used in the planning stage of project development *Virginia Transportation Research Council* (No. VTRC 02-TAR3)

[6] Skitmore R M and Ng S T 2003 Forecast models for actual construction time and cost *Building and environment* 38 **8** 1075-1083

[7] Wilmot C G and Mei B 2005 Neural network modeling of highway construction costs *J. Construction Engineering Management* 131 **7** 765-771

[8] Adel K, Elyamany A, Belal A M and Kotb A S 2016 Developing Parametric Model for Conceptual Cost Estimate of Highway Projects *Int. J. Engineering Science 6* **7** 1728-1734

[9] Attal A 2010 *Development of neural network models for prediction of highway construction cost and project duration* (Doctoral dissertation, Ohio University)

[10] Peurifoy R L and Oberlender G D 2002 *Estimating construction costs* (New York: McGraw-Hill)

IOP Conf. Series: Materials Science and Engineering **652** (2019) 012005 doi:10.1088/1757-899X/652/1/012005

Research on the Rural Housing Design to Explore the Rural Labors

Xiuli Sun[1], Chundong Ma[2*] and Cong Pang[1]

1Academy of Fine Arts, Dalian University, Dalian, Liaoning, 116622, China
2 School of Design, Dalian Nationalities University, Dalian, Liaoning, 116600, China
*E-mail: machundong@163.com

Abstract. The rural housing industrialization in China relies on long-term accumulation and is a process of gradual integration with the handicraft Industry. However, the eager to develop deglamorized housing products is hard to adapt to the building capacity of villagers and not favourable for the sustainability of the rural construction industry. Clearly, housing design should not only satisfy the physical functions of housing, but also provide convenient, comfortable and pleasant hardware conditions for the interaction between housing products and rural labors, alleviating the living and consumption burden of villagers and improving the residential satisfaction of villagers. Based on the housing industrialization demands under the strategy of rural revitalization, our work studied the principle of applying rural labor force into housing design and the design requirements for housing products by stressing the importance of technology-labor integration. And then together with the design practice of building components, facilities and ornaments, we probed into the methods of rural housing design. It will be theoretically and practically meaningful for rural housing modernization.

1. Introduction

So far, the developed countries have basically succeeded in the manufacture mode modernization of rural housings depending on the complete housing industrial chain, excellent materials, economic foundation and high quality of rural labor force. For instance, rural areas in Japan have generally manufactured houses online in factories and then transported the houses into in-site fast installation. However, there are no conditions like that to realise fully industrialized architecture in rural China[1,2], due to the relative shortage of architecture resources and the weak economic ability. The large-sized machine for transport and construction, village housing products based on professional and technical personnel and management personnel are often limited to build by the inability or financial shortage of villagers, which restrict the scaled development of rural housing industry.

In view of the dominant position of villagers in current rural housing construction in China, more humanized design needs to make the rural labor resources usable for housing products. In our work, the generation mechanism of rural housing products was studied from the perspective of polyatomic crossover, the principles and methods of housing product design with easy construction and maintenance, learnable and usable to rural labors were explored to offering an innovative path for rural housing industry.

2. Action mechanism of rural labors on housing products

In the majority of rural areas in China, the principal force of residential construction is still villagers, who are following an interdependent and interactional cooperative relationship to the housing

IOP Conf. Series: Materials Science and Engineering **652** (2019) 012005 doi:10.1088/1757-899X/652/1/012005

products[3]. The traditional rural housing design usually, through the manual or semi-manual characteristic of housing products, makes the housing products cooperate with the villagers in all procedures of building, using and maintenance. Then, the rural labour power capitals will convert into a part of values and income of residential construction.

2.1. Coupling efficiency of rural labors

The coordination between rural labors and housing products is essentially a matching and interactive process between technology and human resources, and in the housing technique system, they two possess the characteristic and efficacy of system coupling[4]. Thus, the concept of coupling was used to illustrate the action mechanism of rural labor force in housing products. "Coupling" originated from ancient Greece and means consonance and coordination. System coupling is defined as the tight cooperation and mutual influence between two or more elements, and transmits energy through interaction from one end to the other end, leading to a cooperative, compatible and coexisting process and finally reaching the 1+1> 2 effect[5][6].

Clearly, the rural labor factors considered in housing design can not only reduces the cost expenditure and material depletion by the houses, but also stimulates the principal values and spirits of villagers and enhances their satisfaction with the houses, thereby making the relationship between the villagers and housing products more harmonious and stable.

2.2. Coupling relation between housing products and rural labors

To illustrate the coupling relationship between housing products (e.g. components, facilities, ornaments) and rural labors, the mutual coordinated connotation between that was expressed as follows referencing for the ecological coupling theoretical model[7].

- RHCS ⊆ { S1,S2, Rel, Rst, O, T, L}, Si ⊆ {Ei, Ci, Fi}

Where RHCS is Rural Housing Coupling System coupled from two types of architecture resources, including housing products (S_1) and villager labors (S_2);

- S_i is the i-th architecture system;
- E_i, C_i and F_i are the element, composition and function of S_i, respectively;
- Rel is the coupling set of architecture systems and a coupling relationship set in RHCS, which contains not only the coupling relationship between two types of architecture resources, but also the coupling relationships among resource elements;
- Rel is a coupling relationship set of RHCS, and contains not only the coupling set between E_i, C_i, F_i of S_1 and E_i, C_i, F_i of S_2, but also the coupling set between the architecture function units of S_1 and S_2;
- Rst is the limiting or restraining factor set of house architecture resources (e.g. economic elements, institutional elements, cultural elements in the environment of the housing, and are the parallel elements that constitute the orderly development of this coupling system);
- O is the objective set of the architecture system;
- T, L are the variables of time and space, respectively.

Rural housing products result from the mutual matching among the elements, structures and functions of building materials, technology and rural labors under given space-time and systematic environment. Regardless how new and old technology or materials, the material resources are integrated or replaced by rural labors, this mode will match with the objective as long as a balance within the given coupling relationship for rural housing system can be achieved[8].

IOP Conf. Series: Materials Science and Engineering **652** (2019) 012005 doi:10.1088/1757-899X/652/1/012005

3. To exploit the rural labors on housing products

As relatively low economic capability, low civilized quality and low technical skills for this population, the high-technology housing products obviously do not meet their requirement for villagers' participation, and the traditional housing products with overload labors are not satisfied with their requirements for the efficiency and quality of rural housing.

3.1. Core issue of rural housing design

Although the living environment can be improved within short time though the modern urban housing design, but the comfort of houses is usually achieved at the sacrifice of environment and economic devotion and highly dependent on the external input, and the original diversity and flexibility of living states will be easily lost[9,10].

Moreover, to achieve the recycle and reuse of living materials as much as possible in the traditional rural housing system, the villagers have to cost abundant labors for housing products' running, but housing design are not good at creating more comfortable and efficient hardware environment because that housing products are developed under an instinctive reaction of villager when they handle uncertainty factors in the nature[11].

The degree of using rural labor factors by housing products in the housing system reflects the coupling level between the two, as it largely affects the human-house harmonious relationship and the sustainability of houses. It is unfavourable for the cooperation between housing products and villagers because that an insufficient or overly coupling result in the isolation of technology and labor will be lacking of material exchange, substitution or compensation. So clearly, the rural housing design is at the transition phase to the integration between modern technology and traditional architecture achievements, and thus the core issue of rural housing design is to reasonably utilize rural labors.

3.2. Major tasks of rural housing design

For a certain architecture system, the coupling-based housing design is essentially a process of exchanging, replacing, compensating material and energy between housing products and rural labors, it is aimed to promote the virtuous circle among human and house with materials, energy and information, etc. The final target is to reduce material consumption and economic investment during the construction of rural housing and to achieve incomes at low consumption, low economic costs and high emotional returns[12].

Thereby, the main tasks of housing design are to balance the low and high techniques in a planned way by integrating the traditional and modern architecture achievements to sustainably exchange materials and energy between housing products and labours. The physiological/psychological adaption and experiences of villagers should be paid more attention during rural housing products design, so as to villagers can transit from the passive adaptation to the active conformity on high/new techniques.

3.3. Designing characteristics of rural housing products

The Designing characteristics of rural housing products to be suitable for rural labors are shown as follow:

1) Convenience for manufacture. The building parts/products should be manufactured in a small-batch and quantitative way, which will relieve the labor burden of villagers.

2) Convenience for construction and maintenance. The technical threshold and operational difficulty should be reduced to meet the psychological, physiological and educational levels of villagers. Ergonomics and perceptual technology can be use to make housing products convenient, comfortable and safe to improve the coupling efficiency.

3) Controllability of devoted labor quantity. That is mean more chances for villagers to initiatively control the housing products. For instance, the housing products can be switched among full-auto/mechanized, semi-auto/mechanized or full manual according to their real living demands and economical abilities, so that the building and living materials can be more reasonably controlled by villagers according to their immediate demands.

4) Increment of belongingness. The "experience" housing products design will bring villagers more spiritual happiness during their operating processes and creation activities.

3.4. The guiding idea of rural housing design

The guiding idea in practices of rural housing design is to reasonably utilize rural labors through the housing design, and the effects are reflected in the more-humanistic hardware conditions offered by the housing products for farmer participation, which raises the adaptability of low-level rural labours resources (labor force, life concepts and skills). The rural design should allow villagers to enjoy the modernized material achievements while reserving the traditional rural residence state of low economic costs and low material depletion, moreover, works for enhancing the civilization level of villages are emphasized in our design practices.

4. Design practices for making the most of rural labors on hosing products

The comprehensive benefit of interaction between human and environment is emphasized in our design practices, the constraint conditions and available resources of rural housing construction should be considered to making the most of rural labors, so related architectural elements especially housing products must be re-designed and manufactured.

4.1. The design case of building components

The methods of rural housing products design were studied from the perspective of architecture components considered the rural labors, such as the development of small-sized assembled plate wall structures(figure 1). This wok ensures the wall thermal performance and the lower house construction costs, and significantly improves the cost performance ratio of rural housing construction and maintenance. Moreover, the new type of building blocks and auxiliary components with simple structures and manufacturing process are suitable for industrialized mass-production and for the manual operation by villagers. This system featured by fast construction speed and low construction difficulty well solve the problems of low labor force level, insufficient building resources and low economic levels in most rural areas of China. Moreover, this system can efficiently save building land.

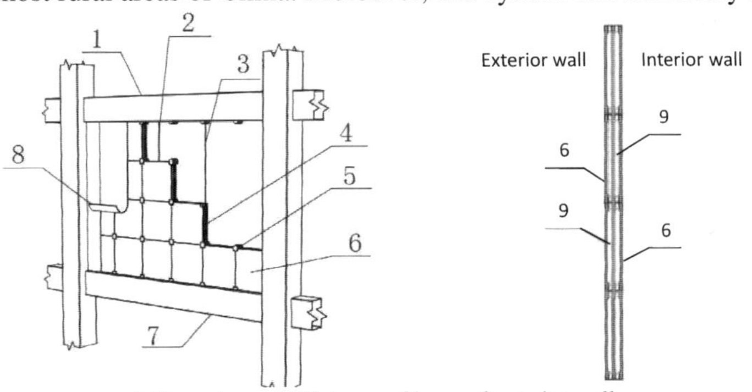

1-Upper beam 2-Waterproof layer of exterior wall
3-Steel rebar 4-Cement mortar
5-Plastic plug-in part 6-Assembled plate（$600x600_{mm}$）
7-Under beam 8-Wallpaper 9- Sir layer

Figure 1. The structure of wall system with small-sized assembled plate

4.2. The design case of building facilities

Our design allows villagers to actively or subconsciously energy-saving behaviors to cooperate with living facilities during daily life, so that they can conveniently and targeted modulate the housing products. For instance, the traditional ground well-cellars (only for food storage) can be changed into cabinet embedded well-cellars, and a liftable storage box was used instead of the refrigerator, so that villagers can take foods conveniently and hygienistically (figure 2). Moreover, the well-cellar liftable facilities were divided into two types of electric (figure 3) and manual (figure 4).

Figure 2. The section of built-in Cabinet

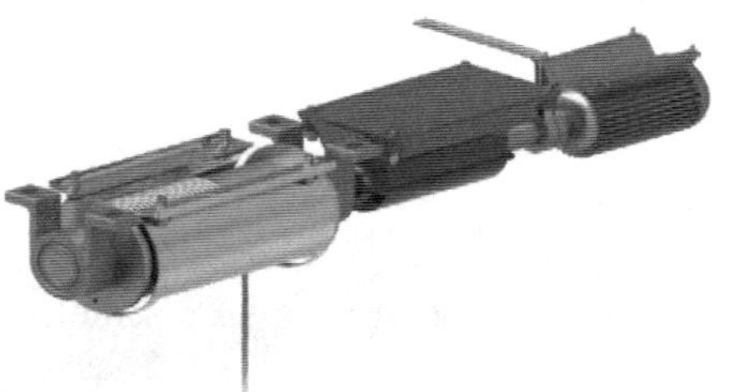

Figure 3. Electric lifting mechanism

Figure 4. Manual lifting mechanism

4.3. The design case of building facilities

The newspapers are usually been pasted on the walls in the traditional rural housing of North China, which is cheap but non-aesthetic. Modern finished wallpapers are aesthetic, but rely on professional construction staff and assisting materials (e.g. special wall protection glues), which is high labor costs and material purchase costs for villagers, especially those living in distant rural.

The new type of wallpaper integrates the advantages of traditional and modern wallpapers(figure 5). In brief, the traditional newspaper pasting and framing skills are used to locally modify the sizes and paper materials of modern finished wallpapers. For instance, A2 krafts can be used instead of finished wallpaper rolls, reducing the difficulty of sticking, so that the labor costs of professional staff can be saved by utilizing the simple works of villagers. Moreover, to allow farmers to lay and stick wallpapers in an aesthetic and skill-less way, the wallpapers can be designed as non-butting patterns, so that farmers can repair at any time, which saves the costs of large-area wallpaper replacement. In this way, the sticking difficulty is reduced by using small patterns and non-continuous design. Computer-aided design and small-batch fast printing ensure the wallpaper quality, and an experiential consumption way is created by using customized patterns, which increase the sense of achievement of villagers.

Figure 5. The new type of wallpaper

5. Conclusions

In this study, the housing design theories and methods targeted at the insufficient labor force resources in rural China were explored. The works of rural housing designer should more frequently investigates to the concrete building conditions of rural families, so as to customize housing products to be suitable for villagers. For such tasks, designer should test the characters of villagers during construction and consider the related factors at early stages in rural housing design, such as the quantity, working time and working skills of available labor resources. Then the architectural elements and with interactive mode of housing products are selected and adjusted, so as to adapt to the participation willingness of villagers.

Above all, it is the requirement and trend of reducing construction consumption and financial investment under the economical and resource background in the rural transition stage, and is the new challenge and responsibility of builders. We hope the complementary advantages between modern techniques and rural labor resources can be utilized to efficiently assimilate the bonus of modern urban development.

Acknowledgments

This research was supported by the special fund of Doctor launched (2019.1-2021.1) in Dalian University, and the fund of Liaoning Province Natural Science (20180550638).

References

[1] Li Y S, Zhang F G and Zhang Y W 2009 *J. Geog. Scie.* **19** 557-67
[2] Wang C G 2008 *J. Soci. Scie. Chi.* **6** 1-2
[3] Assefa G, Glaumann M, Malmqvist T, et al. 2007 *J. Buil.&Envi.* **42** 1458-64
[4] Sun X L, Tang J, Ma C D 2016 *J. Asi. Arch& Buil. Engi. (JAABE),* **15** 357-64
[5] Wan L Q, Hou X Y and Ren J Z 2004 *Chi. J. Eco-Agri.*, (in Chinese) 1 162-4
[6] Huang L K, 2015 *Zhejiang. Univ.* (Doctor Paper in Chinese)
[7] Liu W H, 2009 *Harbin Institute of Technology.* (Doctor Paper in Chinese) 26-31
[8] Sun X L, Ma CD, Tang J 2015 *Proc. 2nd. Int. Conf.* on *Engineering Technology and Application* **3** 1-7
[9] Li T T 2010 *Int. Conf. on Mechanical Automation and Control Engineering (MACE)* p 888-90
[10] Cui M, Qin Z H and Tang C 2006 *J. Urb. Plan. Revi.* **35**, 56-61
[11] Liu R, Zhang Q and Liu J P 2010 *Int. Conf. on Mechanic Automation and Control Engineering (MACE)* 1377-81
[12] Bai X C, Zhang Y Z and He Z Q 2010 *Reso. Scie.* **32**, 1792-8

Implementation of certified intelligent building in Taiwan

Jung Shu Hsiao, Julie Chia-Ping Chen and Lucky Shin-Jyun Tsaih

Department of Architecture, National Taiwan University of Science and Technology, Taiwan

Email: Jasonhsiao1966@gmail.com; D10513001@mail.ntust.edu.tw; akustx@mail.ntust.edu.tw

Abstract. Government of Taiwan since 2002 has been promoting certified and pre-certified intelligent buildings. It was introduced to meet occupant's safety, comfort, convenience, and efficiency concerns with powerful automation and integrating systems in building. Energy saving, ease to maintained and management are the by-product of it. Based on total numbers of intelligent building permit issued and its total floor area, the certified and pre-certified intelligent building seems to be growing fast. However, the proportion between government owned and private sector projects are unbalanced. How to encourage private sector to participate and implement the certification of intelligent building is the aim of this study. To find out the possible factors that contributed the deferring of implementation on intelligent building from private sectors, experts in building construction industry include developers, architects, MEP integrators, real estate agents, and property managers have been invited to share the concerns. Delphi method was first used to develop the possible and consensus factors from the experts. Analytic hierarchy process was then used to rank the important key factors based on experts' responses. The results could be utilized as basis for government agencies to formulate relevant policies and incentives on promoting certified and pre-certified intelligent building in private sectors.

1. Introduction

The rise of intelligent buildings is mainly due to the development of the telecom industry and the rapid advancement of various types of sensing technologies in automation technology. Countries such as United States of America and Japan have introduced these technologies into the construction industry, which has led to the development of intelligent buildings. Smart office buildings, smart shopping centers, smart parking lots, and smart houses are the common types of the intelligent building [1].

Intelligent buildings are the integration and management of automation systems in various fields [2, 3]. It primary includes Communication Automation System (CAS), Building Automation System (BAS), Office Automation System (OAS), Fire Automation System (FAS) and Security Automation System (SAS) [4]. As these systems were developed individually and in different period of time with different technology available then, it is difficult to integrate these systems.

In 1980s, Taiwan began to introduce many related technologies of these automatic systems. Since then, through research and development as well as rigorous tests in market, manufactures gradually explored products that are suitable for the warm and humid environment in Taiwan. At the same time, it has gradually established an industrial chain of intelligent buildings.

In 2003, an executive order was set requiring all government construction projects costing more than 200 Million New Taiwanese Dollar to obtain certified and/or pre-certified Intelligent Building (IB)

status [5]. This regulation was officially launched in July 2003. Since then, the floor area of buildings that applied for the certified and pre-certified IB increased rapidly as shows in Figure 1 in red. The ratio of total floor area for the annual building permit issued versus certified and pre-certified IB projects has increased about 9.5% from 2015 to 2016 [6].

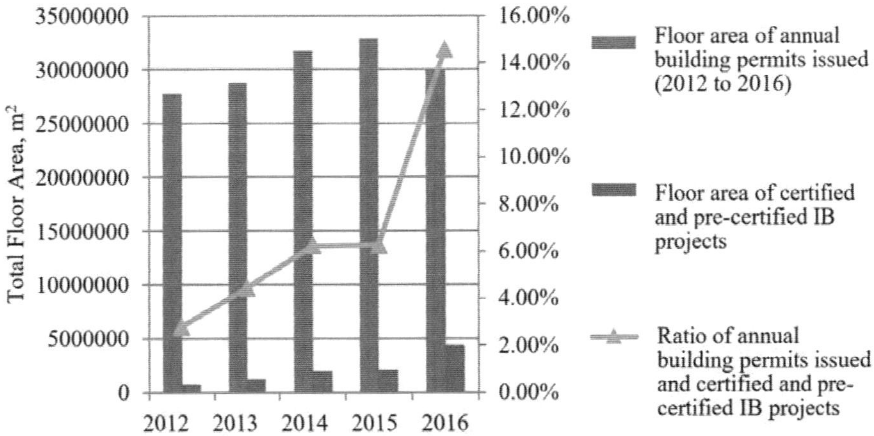

Figure 1. Total floor area comparison between IB certified projects and annual building permit issued (Source: Wen, 2016).[5]

Table 1 indicates that the numbers of annual build permits issued to non-government own commercial offices (Type G) and residential housing (Type H2) projects are between 19257 to 24133. Despite the gradual increasing number of certified and pre-certified IB projects, only about 0.05% of total private sector owned Type G and Type H2 projects in the past 5 years have obtained IB certifications.

Table 1. Certified and Pre-certified IB Projects vs annual Type G and H2 building permits issued

Year	Number of Certified and Pre-certified IB Projects	Building Permits Issued Annually (Type G and Type H2 only)	Ratio of Certified and Pre-certified IB Projects to Annual Type G and Type H2
2012	10	22365	0.03%
2013	20	23471	0.05%
2014	31	24119	0.06%
2015	53	24133	0.04%
2016	58	19257	0.05%

(Source: Wen, 2016)[5]

Figure 2 shows that certified and pre-certified IB projects are primarily government owned. Since most government-owned projects are big projects costing over 2 billion TWD and have larger floor areas, the total number of floor area of IB projects seems to have raised drastically. However with careful review, it is obvious to note that IB certified or pre-certified projects in private sector have actually decreased.

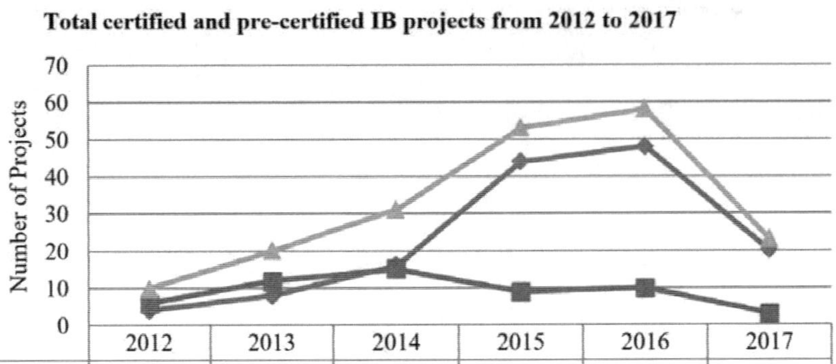

	2012	2013	2014	2015	2016	2017
Government	4	8	16	44	48	20
Private	6	12	15	9	10	3
Overall	10	20	31	53	58	23

Figure 2. Total certified and pre-certified IB projects from 2012 to 2017 (Source: Wen, 2016).[5]

There are five levels of IB certification. They are certified, bronze, silver, gold and diamond. As shows in Figure 3, most of the IB projects achieved basic certified level. In particularly, there are at least 70% IB projects in 2016 acquiring the basic certified level and zero projects obtain gold and diamond certifications.

The above data reflects the willingness of developer to apply for the more stringent Intelligent Building Certificate. Thus, it is the objective of this research to find out possible factors from the building industry experts that can be used to improve or attract stake holders to participate it.

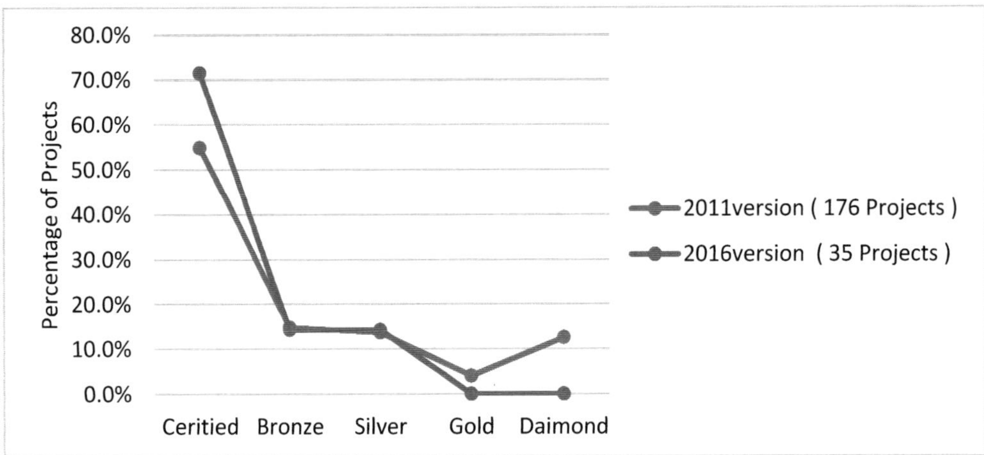

Figure 3. IB certified projects levels in 2011 version and 2016 version (Source: Wen, 2016).[5]

2. Background and marketing of IB

2.1. IB certification

In 2003, Architecture and Building Research Institute (ABRI) of Minister of Interior in Taiwan published the Handbook of Definitions and Evaluation Criteria for Intelligent Building. This handbook sets up the national standardized evaluation criteria and certification procedures for IB projects.

In 2011, the original handbook was revised with updated and advanced technology. The current version was published in 2016 with 8 indicators including unified wiring, information and communication, system integration, facility management, safety and disaster prevention, health and comfort, convenience and friendliness, and energy efficiency and management.

Table 2 shows the five certified levels with the specific points requirement. There is a total of 200 points from the 8 indicators that can be obtained. Project with 50 points and below 90 points will meet the basic certified level. Project with 140 points and above will meet the diamond level [4, 7].

Table 2. IB Certified Level and Scores

Level	Certified	Bronze	Silver	Gold	Diamond
Score, point	Meet 8 indicators basic requirement	50 to 90	90 to 120	120 to 139	140 and above

There are two steps to acquire the IB certificate. Project with building permit and passed the review of IB evaluation criteria before construction completion is classified as pre-certified IB project. After completion, the pre-certified IB project with certificate of occupancy will become certified IB project and received a certificate with specific IB label.

2.2. Life cycle and innovation of product

To promote IB building, it is necessary to understand the development of IB building market. A study into the development of IB building market, the adoption of market life cycle and innovation of product concept is recommended. Consumer products have a rough progress of development, and their competitiveness in the market vary at different phases of the progress. Product developers will also adopt different strategies to meet the needs of consumers according to different phases [8].

Although buildings are long lasting goods and different from the general consumer products, their competitiveness and the need to be innovative are the same. According to Moore (2004), life cycle of market development can be divided into 9 phases [9]. Early market, the chasm, bowling alley and tornado are considered the technology adoption life cycle. Here, the product is introduced. It will be tested and sales will increase. Sense of novelty will be first boosted and disappeared a bit later. Once the product has been re-adjusted and improved at the early product improvement period, the sense of novelty from consumers will increase and sales will increase significantly [10]. It is then followed by the product's mass-production period. Sales are stable. Once the technology is outdated, sales will decrease and eventually reached the end of life for the product.

2.3. Buyer utility map

Kim & Mauborgne (2000) mentioned that companies mostly understand the challenges of engaging in innovation [11]. New products must be able to offer special effects or services to consumers at attractive prices, and bring a considerable profit to enterprises. Many innovators pursue technological breakthroughs and create new products.

However, they often overlook the needs of consumers or the interests of enterprises. In order to avoid such a situation, buyer utility map is proposed for decision making. Essentially, it is a 6-factors matrix with the buyers' experience cycle of purchasing a product versus the utility level. Buyer' experience cycle from purchasing products to abandoning the product is divided into six stages: purchase, delivery (house) goods, use, supplement (expansion) charge, maintenance, and processing.

The utility level of product is also divided into 6 stages: customer productivity, simplicity, convenience, risk, fun and image, and environmental friendliness. Based on market-shaping blue ocean strategy, as a rule of thumb, only provide one utility level factor to one dedicated product cycle to attract customers. For example, during the purchase stage of an IB certified house; convenience can be the product image to attract customer. Likewise, in use stage, simplicity is the image to attract customer. With this buyer utility map in mind, it is easy to see the differences between the new product and exiting product. Marketing strategies can be formulated then.

3. Delphi method and Analytic Hierarchy Process

This study used Delphi method and Analytic Hierarchy Process (AHP). It is so called DHP by Khorranshadhgol and Moustakis in 1988 [12]. For Delphi method, a group of experts will be asked to participate in questionnaires anonymously for couple rounds and until the results are consistent [13, 14, 15].

These experts should be selected with consideration relative to the research topics and have experience in the field. It is also the intent to conduct questionnaire anonymously so that the complicated and divergence factors can be derived to a final list without confrontation between the groups of experts [16]. (Linstone & Murray, 2002)

Feedback control is often used, so that experts have opportunities to re-think the outcome based on trends of other participants. The results will then be used in AHP for the ranking analysis. AHP is introduced by Saaty in 1971. As any complicated issue, there must be an underlying hierarchy order of factors [17, 18]. Pairwise comparison method with 9 scale is used to reveal the importance of one factor to the other and as shows in Table 3. As score 1 means A and B are equal important. Score ± 3 means a moderate importance of a factor than the other. Score 5 means one factor is basically believed and prefer as an important factor than the other. Scores 7 and 9 indicates the strong and absolute preference toward one factor. Intermediated levels are 2, 4, 6 and 8.

Table 3. AHP Fundamental Scale (Source: Saaty, 1990)[19]

Factor A	Equal importance	Moderate importance of one over another	Strong or essential importance	Very strong or demonstrated importance	Extreme importance	Factor B
A	1	3	5	7	9	B

Matrix with the same order of factors in horizontal rows and vertical columns could be used as the template for the questionnaire. Participant is asking to look at one factor from the horizontal row and compare with rest of the factors from the vertical columns. Scale from 1 to 9 can be entered as stated in Table 1 for each pairwise comparison. The diagonal cell eventually will become 1 since it compares the same factor. All data will be normalized, so that the overall comparison scores between one factor to the rest of the factors equal one. Consistency ratio calculation will be performed. A list of factors with mean value and ranking will be derived.

4. Data analyses and results

4.1. Level of importance and agreeing on factors

Nine architecture and construction industry experts participated in three rounds of questionnaires using Delphi method. Table 4 lists the expertise and ratio for the nine experts. The average working experience is over 15 years.

Table 4. Expertise and ratio of the participants

Expertise of Participants	Developer (Including Architect)	MEP Integrator	Real Estate Agent	Property Manager
Number	3	2	2	2
Ratio	33.33%	22.22%	22.22%	22.22%

Eighteen factors such as schedule required, lack of common standard, extra maintenance work, web security, extra costs, high maintenance costs, unknown total costs, low profit rate, products outdated, lack of incentives, perception differences, lack of capable firms, non-perceived extra value, extra installation costs, not attractive enough, budget limitation, and extra transferring costs were derived from first round's responses.

The results were used as the basis for second round questionnaires. Likert scale was used to evaluate the importance of the factors for the second rounds. To further differentiate the level of agreement for responses from second and third rounds, quartile deviation (Q) were used. Table 5 shows the quartile deviation value to the corresponding level of agreement on a factor. According to Faherty (1979) and Holden et al (1993), when Q is equal and or less than 0.6, it indicates the participants agree highly on a factor[14, 20]. Likewise, when Q is greater 1.0, it means there is a low level of agreement on a factor. Greater than 0.6 and less than or equal to 1.0 indicates the participants achieve moderate agreement on a factor.

Table 5. Quartile deviation value to the corresponding level of agreement on a factor

Quartile Deviation (Q)	$Q \leq 0.6$	$0.6 < Q \leq 1.0$	$Q > 1.0$
Level of Agreement	High	Medium	Low

Table 6 shows the eighteen factors from second and third rounds with mean value of importance level and the Q for the agreeing level. It is to note that a "-" is used in third round data as a factor gets $Q \leq 0.6$ in second rounds. It simply indicates that the importance level of this factor has been agreed with by participants in the second round. In addition, to shorten the list of the factors, factors with mean value equal to 4 and above as well as with $Q \leq 0.6$ were selected for the analytic hierarchy process analysis. Ten factors derived. These are lack of common standard, extra maintenance work, extra costs, low profit rate, lack of incentives, perception differences, lack of capable firms, extra installation costs, not attractive enough, and budget limitation.

Table 6. Levels of importance and agreeing of the eighteen factors from round 2 and 3

Factors	Round	Mean	Q
Schedule required	2	3.11	0.5
	3	-	-
Lack of common standard	2	3.67	1
	3	4.11	0.5
Extra maintenance work	2	3.56	0.5
	3	4.00	0
Web security	2	2.78	0.5
	3		
Extra costs	2	3.67	1
	3	4.11	0.5
High maintenance costs	2	3.67	0.5
	3	3.56	0.5
Unknown total costs	2	3.11	1
	3	2.89	0.5
Low profit rate	2	3.78	1
	3	4.11	0.5
Products outdated	2	3.56	0.5
	3	3.56	0.5
Lack of incentives	2	4.44	0.5
	3	-	-
Perception differences	2	3.89	0.5

	3	4.11	0.5
Lack of capable firms	2	4.00	0
	3	-	-
Non-perceived extra value	2	3.33	1
	3	2.78	0.5
Extra installation costs	2	4.11	0.5
	3	-	-
Not attractive enough	2	4.22	0.5
	3	-	-
Budget limitation	2	4.00	0
	3	-	-
Extra transferring costs	2	3.56	1
	3	3.33	0.5

4.2. Results of AHP analysis

Prior to performing AHP analysis on the ten factors, three pre-tests with irrelevant questions to this study were conducted to see if the consistency ratio (CR ≤ 0.1) can be achieved by the participants. First pre-test with three factors were used for evaluation. Second and third pre-tests with five factors were used. According to Saaty & Tran (2007), consistency ratio is the measure of consistency of evaluations for a set of data. It is the ratio of consistency index (CI) to the random index (RI)[14, 20]. Consistency index is calculated from the pairwise comparison matrix for the minimum changes of numbers in a dataset and divided by the actual numbers' changes. Random index is showed in Table 7.

Table 7. Random index (Source: Saaty et al, 2007)

Order	**1**	**2**	**3**	**4**	**5**	**6**	**7**	**8**	**9**	**10**
R.I.	0.00	0.00	0.52	0.89	1.11	1.25	1.35	1.40	1.45	1.49
First order differences		0.00	0.52	0.37	0.22	0.14	0.10	0.05	0.05	0.04

Table 8 summaries the pre-tests' results with the consistency ratio values. It indicated that participant E and G were having inconsistency results on pre-test 1. After communicating with these two participants for the paradox answers, the modified pre-test 1 result showed the value of CR ≤ 0.1.

Table 8. Consistency ratio for the three pre-tests (Source: Saaty et al, 2007)

Participant	Pre-test #1 with 3 factors	Modified Pre-test #1 with 3 factors	Pre-test #2 with 5 factors	Pre-test #3 with5 factors
A	0.00		0.03	0.06
B	0.00		0.04	0.05
C	0.12	0.06	0.07	0.09
D	0.01		0.06	0.05
E	0.06		0.09	0.07
F	0.01		0.06	0.07
G	0.12	0.00	0.07	0.05
H	0.01		0.05	0.06
I	0.03		0.08	0.07

Since all of participants achieved the agreements on the pre-tests, the ten IB factors were then used as the AHP question basis during interview with experts. As shows in Table 9, consistency ratio has been calculated for this AHP interview results and with CR $\leq$ 0.1.

Table 9. Consistency ratio for AHP interview results

Expert	A	B	C	D	E	F	G	H	I
Consistency Ratio	0.078	0.085	0.085	0.091	0.091	0.063	0.070	0.075	0.084

The weighting scores of each expert for the ten factors are showed in Table 10. Higher weighting score indicates the stronger the factor. For example, expert A considered lack of incentives is more important than the other factors for implementing IB certification in Taiwan. It is also agreed by other experts, expect expert I.

Table 10. AHP weighting scores of each expert

Factors	Experts								
	A	B	C	D	E	F	G	H	I
A: Lack of common standard	0.095	0.028	0.032	0.039	0.026	0.062	0.019	0.026	0.016
B: Extra maintenance work	0.035	0.060	0.057	0.036	0.049	0.029	0.081	0.064	0.030
C: Extra costs	0.050	0.133	0.018	0.068	0.018	0.154	0.046	0.123	0.168
D: Low profit rate	0.036	0.229	0.163	0.019	0.131	0.129	0.118	0.063	0.083
E: Lack of incentives	0.314	0.020	0.317	0.289	0.289	0.288	0.255	0.310	0.032
F: Perception differences	0.104	0.256	0.068	0.018	0.107	0.072	0.045	0.142	0.071
G: Lack of capable firms	0.042	0.094	0.085	0.093	0.087	0.023	0.033	0.095	0.114
H: Extra installation costs	0.031	0.037	0.042	0.141	0.047	0.020	0.087	0.062	0.194
I: Not attractive enough	0.105	0.045	0.059	0.165	0.112	0.143	0.162	0.085	0.068
J: Budget limitation	0.190	0.099	0.159	0.132	0.134	0.080	0.154	0.031	0.224

The mean and ranking of the ten IB factors for overall and groups of experts are showed in Table 11. It is obvious to find that lack of incentives, budget limitation and low profit rate are the top three factors that is hindering the implementation of IB certification in Taiwan. However, for developers, architects and property managers, the factor "not attractive enough" weighted a bit more than the factor, "low profit rate". For MEP integrator, "perception differences" ranked as the most important factor. "Extra costs" is weighted as the primary factor for the real estate agent group.

Table 11. Mean and ranking of ten IB factors for the overall and the groups of expert

Factors	Overall		Developer (Including Architect)		MEP Integrator		Real Estate Agent		Property Manager	
	Mean	Ranking	Mean	Ranking	Mean	Ranking	Mean	Ranking	Mean	Ranking
E:Lack of incentives	0.235	1	0.299	1	0.167	2	0.160	2	0.283	1
J:Budget limitation	0.134	2	0.142	2	0.144	3	0.152	3	0.092	4
D:Low profit rate	0.108	3	0.104	4	0.133	4	0.106	5	0.090	5
I:Not attractive enough	0.105	4	0.112	3	0.075	6	0.106	6	0.123	2
F:Perception differences	0.098	5	0.064	7	0.180	1	0.072	7	0.094	3
C:Extra costs	0.086	6	0.034	9	0.091	5	0.161	1	0.084	6
G:Lack of capable firms	0.074	7	0.089	5	0.068	7	0.069	8	0.064	9
H:Extra installation costs	0.073	8	0.077	6	0.034	10	0.107	4	0.074	7
B:Extra maintenance	0.049	9	0.047	8	0.048	9	0.030	10	0.073	8

IOP Conf. Series: Materials Science and Engineering **652** (2019) 012006 doi:10.1088/1757-899X/652/1/012006

work										
A:Lack of common standard	0.038	10	0.033	10	0.061	8	0.039	9	0.023	10

5. Conclusion

Through this study, the primary factors constraining the implementation of IB certification in Taiwan for private sectors have been concluded. Lack of government incentives ranked the most important factor. This could plausibly explain the concern over budget limitation and low profit margin from the developer's perspective.

In addition, the high costs of housing in Taiwan as well as unfamiliarity with IB certification and its advantages could also contribute to this standstill state for purchasing.

Yet, from building industry market analysis, it is true that IB and high-tech consumed product have similar phases in the technology adaptation cycle. During the cycle, the order of four phases such as early market, the chasm, bowling alley and tornado is chaotic. The updated technology along with the new business model and product could exist.

Profitable market during the cycle could not easily be hold on by the developer, as it takes longer to produce a building than a product. Thus, it requires extra effort for building industry companies to consider the necessary support for consumer to own the IB through its entire life cycle. One developer in Taiwan has successfully done the work by providing free operation and management services for the first two years of building with IB certification. It not only solves the possible technical operating issues from the new O&M team, but also ensures that the optimum integrated systems can be executed. Thus, it is believed that with additional effort, IB market can be implemented successfully in both private and public sectors.

References

[1] Ma, Z., Badi, A., & Jorgensen, B.N. 2016 Market opportunities and barriers for smart buildings. *2016 IEEE Green Energy and Systems Conference (IGSEC)*, Long Beach, CA, USA.

[2] Buckman, A.H., Mayfield, M., & Beck, S.B.M. 2014 What is a smart building? *Smart and Sustainable Built Environment*, **3(2)**, 92–109. (2014).

[3] Marche, C., Nitti, M., & Pilloni, V. 2017 Energy efficiency in smart building: a comfort aware approach based on Social Internet of Things. *2017 Global Internet of Things Summit (GIoTS)*, Geneva, Switzerland.

[4] Taiwan Architecture and Building Center, Intelligent Building Handbook, 2016.

[5] Wen, S.L., Chen, Y.H., Wu, Y.F., Lee, Y.S. 2017 Construction and Planning Agency, Taiwan Ministry of the Interior, Annual Report of Evaluation of Certified Intelligent Building. (Translations)

[6] Building Technical Regulations, Construction and Planning Agency, Taiwan Ministry of the Interior, 2016.

[7] Hsu, P.C. 2015 Compare the cost difference of building intelligence levels - Taoyuan case of a residential community, Master Thesis.

[8] Anthony, S., Eyring, M. & Gibson, L. 2006 Mapping Your Innovation Strategy. *HBR*, **84(5)**, PP 104–113.

[9] Moore, A.Geoffrey 2004 "Darwin and the Demon: Innovating Within Established Enterprises", *HBR*, **82**, PP 86-92.

[10] Drucker, F.P. 2002 "The Discipline of Innovation". *HBR*, **80(8)**, PP 95–103.

[11] Kim, W.C. & Mauborgne, R. 2000 "Innovation". *HBR*.

[12] Khorramshahgol, R. & Moustakis, V.S. 1998 Delphi hierarchy process (DHP): A method for priority setting derived from the Delphi method and analytic hierarchy process. European Journal of Operational Research, 37, PP 347–354.

[13] Dalkey, N.C. 1969 The Delphi Method: An experimental study of group opinion, The Rand Corporation, Santa Monica, CA, USA.

[14] Faherty, V. 1979 Continuing Social Work Education: Result of a Delphi Survey. *Journal of Education for Social Work*, **15(1)**, PP 12–19.
[15] Hartman, A. 1981 Reaching consensus using the Delphi technique. *Education Leadership*, **38(6)**, PP 495–497.
[16] Linstone, H.A. & Turoff, M. 2002 The Delphi method: Techniques and applications. Addison Wesley Publishing Company.
[17] Alonso, J.A. & Lamata, M.T. 2006 Consistency in the Analytic Hierarchy Process: A new approach. *International Journal of Uncertainty*, **14(4)**, PP 445–459.
[18] Yang, T. F., Fang, H. & Chen, C. J. 2017 Apply DHP method to study the key indicators of the tenants' interior decoration projects in shopping malls. *JPM*, **8(2)**, PP 1–15.
[19] Saaty, T. L. 1990 How to make a decision: The Analytic Hierarchy Process. *EJOR*, **48(1)**, PP 9-21.
[20] Holden, M.C. & Wedman, J.F. 1993 Future issues of computer-mediated communication: The results of a delphi study. *Educational Technology Research and Development.* **41(4)**, PP 5–24.
[21] Saaty, T. L. & Tran, L. T. 2007 On the invalidity of fuzzifying numerical judgments in the Analytic Hierarchy Process. *Mathematical and Computer Modelling*, **46(7-8)**, PP 962–975.

Preferred criteria in winning budget hotel interior design projects in China

H T Huang[1], C P Chen[2] and S J Tsaih[34]

[1] Department of Architecture, National Taiwan University of Science and Technology, Taiwan
[2] Department of Architecture, National Taiwan University of Science and Technology, Taiwan
[3] Department of Architecture, National Taiwan University of Science and Technology, Taiwan
[4] Corresponding author: akustx@mail.ntust.edu.tw

Abstract. In the competitive Chinese market, to win an interior budget hotel design project, a sound proposal with attractive and preferred criteria is a must. In order to propose and win interior budget hotel design projects, it is necessary to understand client's perspectives. Through literature reviews and interviews with practitioners and architecture professors; twenty criteria from value creation, economic advantage and tender quality aspects were derived. Three groups of a total 30 budget hotel operating and management personnel have participated the survey. Kano's two-dimensional model along with Yang's refined Kano model and Importance and Satisfaction model have been applied to come up with the ranking of the preferred criteria to be included in a design proposal. One-way ANOVA with Schefe's post hoc test were used to see the insight of budget hotel CEOs' views on the twenty criteria. The results show that construction phase scheduling & control, construction cost control, systemized corporate identification, design phase scheduling & control, customer segment identification, easy to maintain design, easy to operate design, cultural heritage connection, room storytelling, and amenity storytelling are the top ten preferred design criteria for an interior designer to win budget hotel projects in China.

1. Introduction

Budget hotel first appeared in the United States in the 1950s. It is the outcome of the rise of American mass consumption and the rapid development of the road network [1]. It refers to small and medium-sized hotel with accommodation and breakfast as the main services. Its central around the idea to provide customer the "cost-effective" and "worthy" services, rather than "cheap" services. Standardized high-grade sanitation, comfort of guest rooms and facilities, and on-site convenience services are the must-haves. The moderate-priced rooms in these budget hotels attracted vast number of businessmen, wage earners, self-funded tourists and students [2-3].

From the early 1960s to the late 1980s, in addition to the substantial increase in room quantity, budget hotels also began to develop diversified products and expanded internationally. The expansion of American hotel chains into Europe resulted the rapid development of local European hotels. To diversify investment strategy, upscale hotel groups created a secondary mid to low scale brand and started to operate budget hotels. Through capital operations, purchases and integrations, these hotel groups quickly established a mature setup of budget hotels by the end of 1980s. However, as competition intensifies, adjustments were made to the branding and management of budget hotels. These hotel groups started to pay more attention to the links of brand image, quality management, market segmentation, and product diversification [4-5].

By the 21st century, these world-famous budget hotel brands even expanded to China, Southeast Asia, and other regions [6]. In China, local Chinese budget hotel brands such as Huazhu Group, Jin Jiang International Company Limited, Home Inns Group, and Atour (ATOUR) were also developing rapidly [7-8]. By now, budget hotels in China included brands such as Accor Group's Ibis, Wyndham Worldwide' Days Inn, Jin Jiang International's Jinjiang Inns, and the Holiday Inn Express of the Intercontinental Group. Thus, to expand in Chinese market, budget hotels pay more attention to brand management and marketing [9]. Hotel operating and management personnel changed their operational goals from not only providing excellent services [10] and quality rooms for stay, but also creating holistic users experience through thematic design and the use of intelligent technology [11-12]. Product innovations and standardization services can be achieved by integrating and cooperating with different trades. And it is also important to incorporate novel business model to win over franchisees and consumers. [13]. Yet, the cost for transformation or brand upgrade must remain the same, thus this becomes a potential winning design strategy to peruse for interior design firms.

The success of a design firm lies on constantly winning projects. In budget hotel design proposal, to win design project and to innovatively achieve value creation with limited resources, a designer is required not only to provide aesthetic and functional spaces, but also to incorporate creative and visionary business model in operating and management (O&M) aspects for a budget hotel. Accordingly, the objective of this study is to rank the proposed elements based on the response of hotel operating and management decision makers such as the chief executive officers (CEOs) and project managers (PMs). The views from the on-site hotel staffs were also included as the comments that might reflect first hand consumer experience. Since it is a preferential orientated study, Kano's model, Yang's refined Kano model and Importance-Satisfaction model were used to analyze the questionnaires that were sent out to the thirty participants.

2. Methods

2.1 Kano's two-dimensional model

Dr. Noriaki Kano's two-dimensional model has been known and used to evaluate customers' expectations or preferences on a service [14-17]. It is often used to study the correlation between service quality and customer's satisfaction. Essentially, there are three primary qualities of conditions that could alter a customer's preference and satisfaction.

First quality is the "must be" condition that customers would expect to have certain quality, or else the customer would be dissatisfying. It is a quality control requirement from a hotel management's perspective. The second quality is the "one-dimensional" requirement as customers will be satisfied and dissatisfied with and without the fulfilled services quality. It is viewed as quality control from a hotel management's perspective. The third quality is to meet the potential needs of customers, called attractive quality creation. Basically, customers receive the unexpected services quality and are satisfied with it. Beside the first three qualities, the fourth and fifth qualities suggested do not need items for the hotel management. Figure 1 shown the completed Kano model. Here, the five qualities with their acronyms are shown. Attractive quality element (A) and one-dimensional quality element (O) are in first quadrant. Indifferent quality element (I) is in second quadrant. Must-be quality element (M) and reverse quality element (R) are in fourth quadrant. [18].

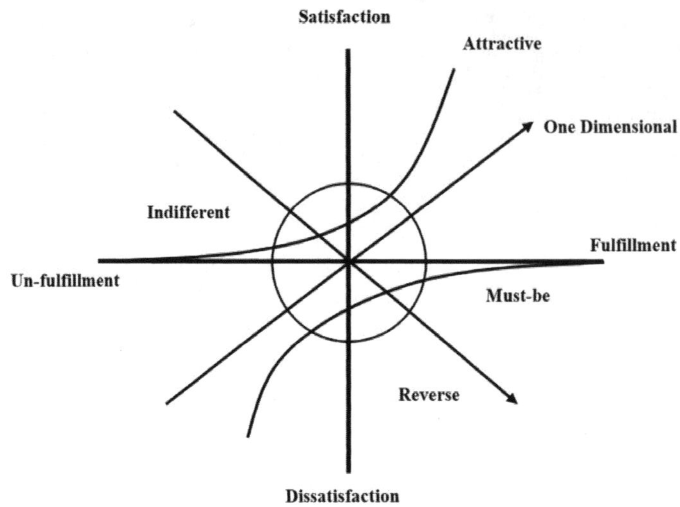

Figure 1. Kano Two-Dimensional Model.

2.2 Refined Kano model and Importance-Satisfaction model

As products and services cannot meet the financial requirements, choices and trade-offs between the importance of quality attributes and customer satisfaction ought to be made [19], a refined Kano's two-dimensional quality model was proposed by Yang in 2005. The refined model as shown in Figure 2 is improved by considering the importance of quality attributes, and distinguishing between attractive quality, one-dimensional quality, must-be quality, and indifferent quality. Attributes of the highly attractive and less attractive qualities can be identified in the second quadrant.

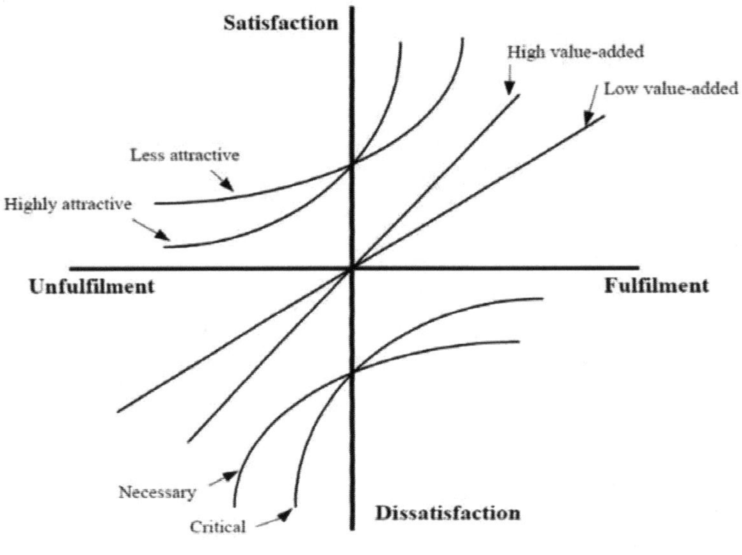

Figure 2. Yang's refined Kano's mode of quality attributes (Source: Yang, 2005)

The highly attractive quality attribute is intended to attract potential customers and with less costs and effort from the hotel management's perspective. Thus, it is a favor item to provide. Likewise, the attributes of the high and low value-added qualities can be identified in the first quadrant. High value-added quality attribute indicates customer satisfaction can be achieved easily with less costs and effort. It is also the best ways to satisfy customers. Meanwhile, despite the less contribution to customer satisfaction, low value-added quality items should be kept, as these are often the expected service items. Must-be quality attributes are identified in third quadrant.

The critical and necessary quality attributes are very important to customers; these items must be provided to avoid dissatisfaction. Additionally, the potential quality attributes will gradually be transformed into attractive qualities. Hotel management should consider providing these attributes as strategic weapons in the future to attract customers. Care-free quality attribute is cost effective with less satisfaction feedback. Hotel management does not need to provide this quality attribute.

3. Data Collection

There are three parts to the questionnaire. First part is the basic information of participants. It was primary used to differentiate the authority level of participants. Chief executive officer (CEO), project manager (PM) and on-site staff were the three groups of the budget hotel participants. Second part is to evaluate the preference of the proposed interior design criteria. Twenty proposed interior design criteria from three aspects were derived based on literature reviews [20-21], interviews and discussion with professional practitioners and architecture professors. The aspects and criteria can be seen in Table 1 below. Third part of the questionnaire is to understand the participant's judgements on how important and satisfaction of the proposed criteria based on their previous project experience.

To discover the attribute of the preferred criterion, Kano's two-dimensional model was used. Each criterion has a positive and a negative question. Figure 3 shown the example of question in part 2. Likert scale of 1 to 5 was used with analogy of one represents dislike the function of criterion and five represents like the function of criterion.

To rank the proposed twenty criteria, Likert scale of 1 to 5 was used. One represents definitely not important or satisfy with the proposed criterion. Five represents very important or satisfy with the proposed criterion. Figure 4 shown the sample of part 3 question for the importance degree and satisfaction level of the criterion. There are a total of thirty questionnaires were collected for this study. The participants included seven budget hotel CEOs, nineteen PMs, and four on-site staffs.

Criterion will be included in the proposal	**Like**	**Must be**	**Neutral**	**Live with**	**Dislike**
1. Systemized Corporate Identification		O			
Criterion will not be included in the proposal	**Like**	**Must be**	**Neutral**	**Live with**	**Dislike**
1. Systemized Corporate Identification					O

Figure 3. Sample questionnaire of part 2.

	Importance Degree				
How important is to include this criterion in the proposal?	**Very important**	**Important**	**Acceptable**	**Not important**	**Definitely not important**
1. Systemized Corporate Identification		O			
Based on your experience, what was the satisfaction level of	**Satisfaction Level**				

How important is to include this criterion in the proposal?	Importance Degree				
	Very important	Important	Acceptable	Not important	Definitely not important
this criterion included in the proposal?	Very satisfy	Satisfy	Acceptable	Not satisfy	Definitely not satisfy
1. Systemized Corporate Identification				O	

Figure 4. Sample questionnaire of part 3.

4. Data analyses and results

4.1 The preferred proposed design criteria and attributes

Twenty criteria from three aspects of a proposed interior design content for the budget hotel in China has been analyzed with Kano's preference attributes and their correlated quality function deployments (QFD). As shown in Table 1, each criterion and the aspect of the proposed design content has its dedicated Kano's preference attribute and QFD. In value creation aspect, since there are four one-dimensional preference criteria, it is considered as the one-dimensional or value-added aspect. The four criteria are cultural heritage connection, amenity storytelling, room storytelling, and business model innovation. These four value-added criteria are recommended to include in a design proposal. For economic advantage aspect, five out of six criteria have must-be preference attribute. These six criteria along with systemized corporate identification and customer segment identification from value creation aspect are considered as the backbone of a design proposal. All six criteria from tender quality aspect have indifferent preference qualities. The attractive criteria such as business model innovation and cross-disciplinary cooperation could use to draw client's attention during the project presentation.

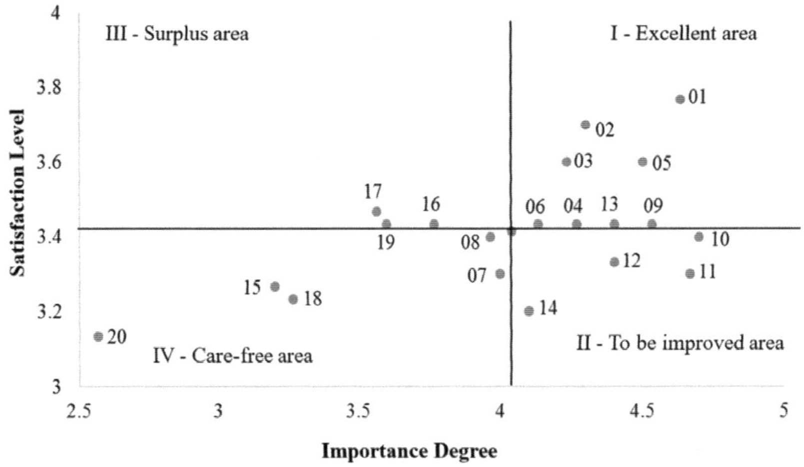

Figure 5. Importance and satisfaction plot for design proposal criteria.

4.2 Importance and satisfaction of the proposed design criteria

As proposed by Yang in 2005, Kano's method cannot differentiate further about the degree of importance and satisfaction level of each criterion, thus the refined Kano model method with the Importance and Satisfaction model were studied. Figure 5 illustrates the twenty design criteria in Yang's proposed importance and satisfaction plot. Mean values of importance degree and satisfaction level are 4.040 and 3.415. It is clearly to see that construction phase scheduling & control, construction cost control, easy to maintain design, and energy and water saving design are the four design proposal criteria that required to be improved, despite the energy and water saving design was in indifferent preference quality from original Kano's model. Six criteria from value creation aspect along with two criteria, design phase scheduling & control and easy to operate design, from economic advantage aspect will be expected to be included in a design proposal and with sufficient supporting materials.

Table 1. Kano's preference attributes and QFD for interior design proposal criteria.

		Kano's Preference Attributes and QFD						
Aspects	Evaluation Criteria	Attractive (A)	One-dimen. (O)	Must-be (M)	Indifferent (I)	Reverse (R)	Deployment of Criteria	Deployment of Aspects
Value Creation	**1. Systemized Corporate Identification**	3.3%	20.0%	50.0%	26.7%	0.0%	M	O
	2. Cultural Heritage Connection	26.7%	33.3%	23.3%	16.7%	0.0%	O	
	3. Amenity Storytelling	20.0%	33.3%	30.0%	16.7%	0.0%	O	
	4. Room Storytelling	20.0%	33.3%	26.7%	20.0%	0.0%	O	
	5. Customer Segment Identification	0.0%	26.7%	50.0%	23.3%	0.0%	M	
	6. Social Media Propagation	10.0%	30.0%	23.3%	36.7%	0.0%	I	
	7. Business Model Innovation	33.3%	33.3%	13.3%	20.0%	0.0%	A/O	
	8. Cross-Disciplinary Cooperation	33.3%	30.0%	13.3%	23.3%	0.0%	A	
Economic Advantage	**9. Design Phase Scheduling & Control**	3.3%	16.7%	70.0%	10.0%	0.0%	M	M
	10. Construction Phase Scheduling & Control	0.0%	20.0%	80.0%	0.0%	0.0%	M	
	11. Construction Cost Control	3.3%	26.7%	63.3%	6.7%	0.0%	M	
	12. Easy to Maintain Design	10.0%	20.0%	50.0%	20.0%	0.0%	M	
	13. Easy to Operate Design	20.0%	26.7%	36.7%	16.7%	0.0%	M	
	14. Energy and	23.3%	20.0%	23.3%	33.3%	0.0%	I	

	Water Saving Design							
Tender Quality	**15. Award Winning Designers**	36.7%	3.3%	0.0%	60.0%	0.0%	I	I
	16. Firm Experiences	13.3%	23.3%	23.3%	40.0%	0.0%	I	
	17. Fancy Presentation	10.0%	13.3%	30.0%	46.7%	0.0%	I	
	18. Presentation with Modern Technology	23.3%	13.3%	6.7%	56.7%	0.0%	I	
	19. Firm Size	10.0%	20.0%	13.3%	56.7%	0.0%	I	
	20. Internationalized Firm	6.7%	3.3%	3.3%	86.7%	0.0%	I	

4.3 Overall ranking and potential of the proposed design criteria

Table 2 summarized and ranked the preferred interior design criteria for budget hotel projects in China. The top ten criteria are the backbone and expect to be included in design proposals. These are categorized in critical and high value-added attributes from refined Kano's model. Construction phase scheduling & control, construction cost control and easy to maintain design are indicated to have area to improve. Social media propagation and energy and water saving are the two potential criteria that should be well addressed in proposals despite these were originally classified as the indifferent criteria.

4.4 Preference of the proposed design criteria with different authority participants

Since the participants have different authority levels that could affect the decision making for a budget hotel interior design project, it is definitely critical to see the preference from the CEOs and PMs' perspectives for the proposed design criteria. Table 3 shows the integrating results of Kano's preference attributes and the ANOVA tests of the importance degree and satisfaction level for each criterion. It is suggested that some criteria have significant differences from three groups of participants. For importance degree, customer segment identification is the only criterion that has a significant difference. For satisfaction level, besides customer segment identification, design phase scheduling & control, construction phase scheduling & control, and the six criteria from tender quality all have significant differences. This implies that participant with different authority level has different prioritized views on budget hotel interior design goal.

Schefe's post hoc tests were conducted to see whether there is significant difference in terms of critical criterion from the participant group for the improvement of importance degree and satisfaction level. Table 4 shows that only eight criteria from satisfaction level have significant difference. This indicated that three groups of participants have reached consensus on the importance degree of all proposed criteria. Likewise, it is also implied that there are improvement areas of the eight criteria to achieve a consensus satisfaction level among the groups. Yet, it is also interesting to find out the paradox from CEOs and PMs on award winning designers and internationalized firm criterion. The average importance degree scores of award-winning designers for CEOs and PMs are 2.86 and 3.21. For internationalized firm, the average importance degree scores for CEOs and PMs are 2.43 and 2.63. Both criteria were classified as indifferent attribute from the original Kano's preference model. However, despite the ECOs group tend to have lower importance degree scores on these two criteria than PMs group, the satisfaction level scores of CEOs are much higher than PMs. This implies that both criteria from tender quality aspect are still the critical proposed factors if they can be improved.

5. Conclusions

The proposed twenty interior design criteria from value creation, economic advantage, and tender quality aspects for a budget hotel in China have been studied for the preference attributes and importance degree and satisfaction level from hotel CEOs, PMs and on-site staffs. Kano's two-dimensional model indicated that four criteria from value creation aspect are the expecting items to be included in a proposal. Additionally, criteria from economic advantage aspects are the "must have" items to be included in a design proposal.

To prioritized the criteria, refined Kano's method along with importance degree and satisfaction level were conducted. The top ten criteria classified as critical and high value-added elements are construction phase scheduling & control, construction cost control, systemized corporate identification, design phase scheduling & control, customer segment identification, easy to maintain design, easy to operate design, cultural heritage connection, room storytelling, and amenity storytelling.
Social media propagation and energy and water saving design are the two potential elements to be included in a design proposal. A winning interior design proposal would need to pay extra attention to these criteria.
On top of all, despite the original Kano's preference attributes for award winning designers and internationalized firm from tender quality aspect were categorized as indifferent elements, Schefe's post hoc test indicated that there are improving areas in terms of satisfaction levels for these criteria from CEOs' perspective. Thus, it is sensible to include these two criteria in a proposal.

Table 2. Ranking and potential of proposed design criteria

Order of Importance	Criteria	Importance Degree	Satisfaction Level	Original Kano Model Attribute	Refined Kano Model Attribute	Area of I-S Model
1	**10. Construction Phase Scheduling & Control**	4.700	3.400	M	Critical	To be improved
2	**11. Construction Cost Control**	4.667	3.300	M	Critical	To be improved
3	**1. Systemized Corporate Identification**	4.633	3.767	M	Critical	Excellent
4	**9. Design Phase Scheduling & Control**	4.533	3.433	M	Critical	Excellent
5	**5. Customer Segment Identification**	4.500	3.600	M	Critical	Excellent
6	**12. Easy to Maintain Design**	4.400	3.333	M	Critical	To be improved
7	**13. Easy to Operate Design**	4.400	3.433	M	Critical	Excellent
8	**2. Cultural Heritage Connection**	4.300	3.700	O	High value-added	Excellent
9	**4. Room Storytelling**	4.267	3.433	O	High value-added	Excellent
10	**3. Amenity Storytelling**	4.233	3.600	O	High value-added	Excellent
11	**6. Social Media Propagation**	4.133	3.433	I	Potential	Excellent
12	**14. Energy and Water Saving Design**	4.100	3.200	I	Potential	To be improved
13	**7. Business Model Innovation**	4.000	3.300	A/O	Less attractive/ Low value-added	Care-free
14	**8. Cross-Disciplinary Cooperation**	3.967	3.400	A	Less attractive	Care-free
15	**16. Firm Experiences**	3.767	3.433	I	Care-free	Surplus
16	**19. Firm Size**	3.600	3.433	I	Care-free	Surplus
17	**17. Fancy Presentation**	3.567	3.467	I	Care-free	Surplus
18	**18. Presentation with Modern Technology**	3.267	3.233	I	Care-free	Care-free
19	**15. Award Winning Designers**	3.200	3.267	I	Care-free	Care-free
20	**20. Internationalized Firm**	2.567	3.133	I	Care-free	Care-free
	Mean	4.040	3.415			

Table 3. Kano's preference attributes and ANOVA tests for importance degree and satisfaction level

Evaluation Criteria	Kano Preference Attributes for Participants			ANOVA of Importance Degree		ANOVA of Satisfaction Level	
	ECOs	PMs	On-site Staff	F	Sig.	F	Sig.
1. Systemized Corporate Identification	O	M	M	.231	.796	1.880	.172
2. Cultural Heritage Connection	O	O/M/I	A/M	.953	.398	.326	.724
3. Amenity Storytelling	O	M	M	1.802	.184	.650	.530
4. Room Storytelling	O	O	M	1.932	.164	.242	.787
5. Customer Segment Identification	O/M	M	M	3.388	.049	8.702	.001
6. Social Media Propagation	O	I	M	.652	.529	.897	.420
7. Business Model Innovation	O	A	A/O/M/I	.349	.708	2.300	.120
8. Cross-Disciplinary Cooperation	A	A/O	A/O/M/I	.229	.797	1.212	.313
9. Design Phase Scheduling & Control	M	M	M	.364	.698	4.867	.016
10. Construction Phase Scheduling & Control	M	M	M	.764	.475	5.826	.008
11. Construction Cost Control	M	M	M	.566	.574	.026	.974
12. Easy to Maintain Design	M	M	M	1.117	.342	1.138	.335
13. Easy to Operate Design	M	O	M	2.435	.107	.280	.758
14. Energy and Water Saving Design	A	I	M	3.172	.058	.120	.887
15. Award Winning Designers	A/I	I	A	2.152	.136	4.559	.020
16. Firm Experiences	A/O/M	I	I	1.028	.371	7.035	.003
17. Fancy Presentation	O	I	M/I	.214	.809	8.019	.002
18. Presentation with Modern Technology	I	I	A	.283	.756	4.066	.029
19. Firm Size	O/I	I	I	1.835	.179	5.310	.011
20. Internationalized Firm	I	I	I	.182	.834	6.369	.005

Table 4. Preference of proposed design criteria with different authority participants.

Evaluation Criteria	Participant (Mean)		Schefe's Post Hoc Test – Sig.
5. Customer Segment Identification	CEOs (4.29)	PMs (3.42)	.002
		On-Site Staffs (3.25)	.011
9. Design Phase Scheduling & Control	CEOs (4.14)	PMs (3.21)	.018
		On-Site Staffs (3.25)	.136
10. Construction Phase Scheduling & Control	CEOs (4.14)	PMs (3.21)	.014
		On-Site Staffs (3.00)	.037
15. Award Winning Designers	CEOs (3.71)	PMs (3.05)	.028
		On-Site Staffs (3.50)	.809
16. Firm Experiences	CEOs (4.00)	PMs (3.21)	.004
		On-Site Staffs (3.50)	.265
17. Fancy Presentation	CEOs (4.00)	PMs (3.26)	.002
		On-Site Staffs (3.50)	.179
19. Firm Size	CEOs (4.00)	PMs (3.21)	.012
		On-Site Staffs (3.50)	.363
20. Internationalized Firm	CEOs (3.57)	PMs (3.00)	.007
		On-Site Staffs (3.00)	.065

References

[1] Woodworth R M and Mandelbaum R 2009 Seventy-Five Years of U.S. Hotel Revenues, Expenses, and Profits *Cornell Hospitality Quarterly* **(51)1** 20-6
[2] Fiorentino A 1995 Budget hotels: not just minor hospitality products *Tourism Management* **16(6)** 455-62
[3] Ren L, Qiu H, Wang P and Lin M C P 2016 Exploring customer experience with budget hotels: Dimensionality and satisfaction *International Journal of Hospitality Management* **52** 13-23
[4] Senior M and Morphew R 1990 Competitive Strategies in the Budget Hotel Sector *International Journal of Contemporary Hospitality Management* **(2)3** 3-9
[5] Peng J P 2015 Improving service management in budget hotels *International Journal of Hospitality Management* **49** 139-48
[6] Guillet B D and Zhang H Q 2011 Interpreting the mind of multinational hotel investors: Future trends and implications in China *International Journal of Hospitality Management* **30(2)** 222-32
[7] Chan W W and Ni S 2011 Growth of Budget Hotels in China: Antecedents and Future *Asia Pacific Journal of Tourism Research* **(16)3** 249-62
[8] Noordzy G, Ricaurte E, James G and Wu M 2016 Environmental implications of hotel growth in China: Integrating sustainability with hotel development *Cornell Hospitality Reports* **16(12)** 3-9
[9] Feng D M 2006 Budget Hotel: Development, Problems, Strategy *Tourism Tribune* **21(7)** 58-62
[10] Huang S, Liu Z and Hsu C 2014 Customer Experiences with Economy Hotels in China: Evidence from Mystery Guests *Journal of Hospitality Marketing & Management* **(23)3** 266-93
[11] Wang Y X 2008 On the Theme Development of Economic-style Hotels *Journal of Hubei Radio & Television University* **28(7)** 88–9
[12] Alfakhri D, Harness D, Nicholson J and Harness T 2018 The role of aesthetics and design in hotelscape: A phenomenological investigation of cosmopolitan consumers *Journal of Business Research* **85** 523-31
[13] Subramanian N, Gunasekaran A and Gao Y 2016 Innovative service satisfaction and customer promotion behaviour in the Chinese budget hotel: an empirical study *International Journal of Production Economics* **171(2)** 201-10
[14] Matzler K and Hinterhuber H H 1998 How to make product development projects more successful by integrating Kano's model of customer satisfaction into quality function deployment *Technovation* **18(1)** 25-38
[15] Yang C C 2005 The refined Kano's model and its application *Total Quality Management* **16(10)** 1127-37
[16] Luor T, Lu H P, Chien K M and Wu T C 2015 Contribution to quality research: a literature review of Kano's model from 1998 to 2012 *Total Quality Management & Business Excellence* **26(3-4)** 234-47
[17] Madzík P and Kormanec P 2018 Developing the integrated approach of Kano model and Failure Mode and Effect Analysis *Total Quality Management & Business Excellence* **29(3-4)** 387-409
[18] Kano N, Seraku N, Takahashi F and Tsuji S 1984 Attractive quality and must-be quality *Journal of the Japanese Society for Quality Control* **14(2)** 147–56
[19] Kristensen K, Kanji G K and DahlGaard J J 1992 On measurement of customer satisfaction *Total Quality Management* **3(2)** 123-8
[20] Jiao M Y 2014 Customer satisfaction measurement of budget-type hotels based on customer value *Tourism Tribune* **29(11)** 80-6
[21] Quan H and Yao L 2018 The Problems and Strategies of Business and Budget Hotel Interior Design in China *Furniture & Interior Design*

IOP Conf. Series: Materials Science and Engineering **652** (2019) 012008 doi:10.1088/1757-899X/652/1/012008

Statistical characterisation of expanded metal mesh strand, to propose a correlation between properties of parent material and finish product, EMM

Jaydeepsingh Solanki[1,2,*] **and Siddhartha Ghosh**[1]

[1]Civil Engineering Department, Indian Institute of Technology Bombay, Powai, Mumbai, 400076, India
[2]Structural Engineering Department, VeermataJijabai Technological Institute, Matunga, Mumbai, 400019, India
*Corresponding author, Email: jgsolanki@st.vjti.ac.in

Abstract. The use of steel plate shear walls (SPSW) for the design of structures under low seismic intensity results in very thin infill SPSW panels. Construction of such structures is infeasible, mainly due to the difficulty in connecting the infill to the boundary members. Thus, a designer is compelled to use a thicker infill section which in turn leads to heavier boundary elements. In order to overcome this constraint of design, a new type of infill panel using Expanded Metal Mesh (EMM) is studied as an alternative in place of a solid infill SPSW panel. Available codes on expanded metal mesh specify requirements of workmanship, quality control, etc. but neither recommend the strength of expanded metal mesh as a whole nor the strength of its components. Thus, it is necessary to perform extensive tests to extract the strength of expanded metal mesh as a whole in terms of the strength of its components and to come out with a correlation between the strength of the parent material (material from which EMM is manufactured) and that of the finished product (EMM). This paper a) presents the statistical characterization of EMM strands, b) recommends nominal design values for basic mechanical properties and c) proposes the correlation between the strength of the parent material and that of EMM strand.

1. Introduction

The use of steel plate shear wall (SPSW) as a primary lateral load resisting system in the construction of new buildings and for upgrading the lateral load resistance of an existing structure, began in the early 1980s in USA and Japan. In its typical form, the SPSW system consists of a steel plate shear panel added as an infill to the building structural frame composed of beams (horizontal boundary elements) and columns (vertical boundary elements). Using the thick or stiffened SPSW is an unattractive option because of its higher cost in comparison to reinforced concrete shear walls. This has led to a gradual and general shift towards the use of thin un-stiffened SPSW. In this arrangement, the steel plate is expected to buckle along compressive diagonal under relatively small shear forces. After buckling, the storey shear force is resisted by the steel plate through the formation of a tension field as explained in pictorial form in figure 1.

Expanded metal mesh (EMM), figure 2b, is a sheet of metal, slit and stretched into a non-raveling unit of uniform-sized, diamond-shaped openings as shown in figure 2a. no metal is lost in the expanding process and the final product is stronger per kilogram and lighter per meter than the original sheet. Expanded metal mesh has no welds, seams or jointed areas. It is more durable and maintenance-free

IOP Conf. Series: Materials Science and Engineering **652** (2019) 012008 doi:10.1088/1757-899X/652/1/012008

than other similar materials. At present, some experts have done research on the expanded metal mesh. The main research areas are as follows:
Investigation on a) natural period of vibration [3] b) its seismic application [4] c) behavior of EMM under shear loading [5, 6] d) use of EMM for retrofitting of RCC moment resisting frames [7] and e) Investigation of energy absorption characteristics of EMM [8]. A study of available codes on expanded metal mesh [9-12] reveals that these codes only specify workmanship, quality control, etc. but do not give information about the strength of expanded metal mesh as a whole or in terms of the strength of its components. Thus, the designer has no idea about the properties of EMM as a whole or in terms of its component. Therefore, it is necessary to perform extensive tests to understand the behavior of EMM as a whole and to propose a correlation between the strength of the parent material (material from which EMM is manufactured) and that of the finished product (EMM). To achieve this objective, the work of statistical characterization of EMM strand has been undertaken. Relevant papers on statistical characterization have been studied for fixing the methodology of statistical characterization of EMM strand [13].

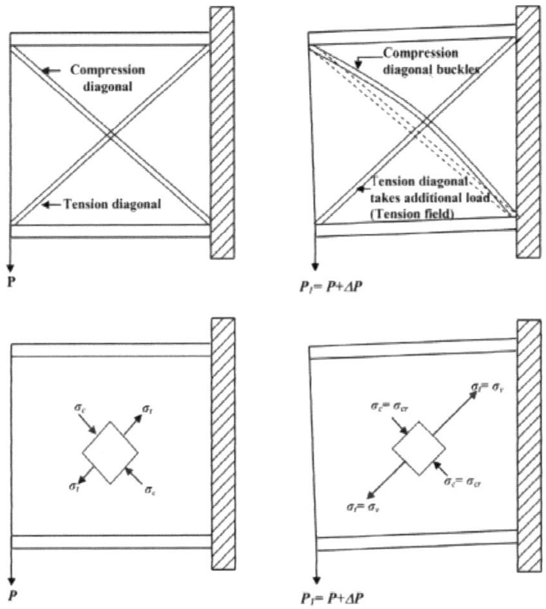

Figure 1. Explanatory sketch of diagonal tension field theory [1].

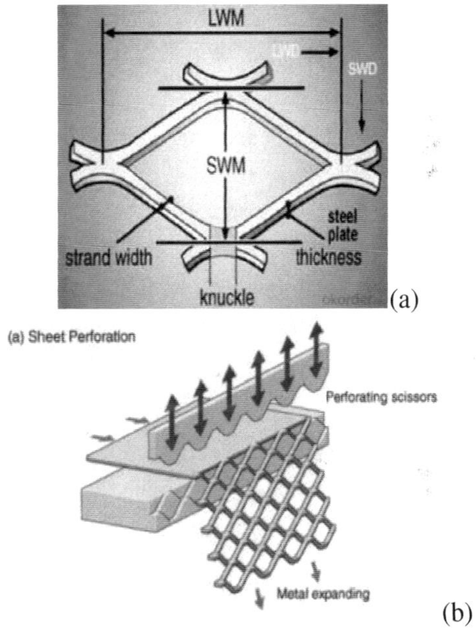

Figure 2. a) Process of making expanded metal mesh [2] b) sketch of expanded metal mesh.

2. Statistical characterisation of expanded metal mesh strand

None of the codes [9-12] specify values of tensile, bending and shear strength of the EMM strand or EMM panel as a whole. Hence, to predict the behavior of EMM from the properties of the parent metal, it is necessary to carry out extensive testing and to establish the statistical characterization of EMM strand. Three sizes of EMM 115x40x4x4mm, 115x40x5x5mm and 115x40x5x6mm were manufactured from a sheet of steel complying with the requirements of IS 2062:2011(specification for hot rolled medium and high tensile structural steel). The present paper consists of two parts viz. laboratory experiments and statistical analyses. The laboratory experiments include a) Coupon tests of the original sheet metal (30 or more number of tests) b) Tension tests on individual strands of EMM (30 or more number of tests) for each of the three sizes (cross-section) of the strand (4x4mm, 5x5mm, and 5x6mm). The statistical analyses focus on recommending nominal/design values for basic mechanical properties of sheet metal and EMM, based on sound statistical methods, applied to the experimental observations. For statistical characterization, the minimum number of tests required to be done is decided

IOP Conf. Series: Materials Science and Engineering **652** (2019) 012008 doi:10.1088/1757-899X/652/1/012008

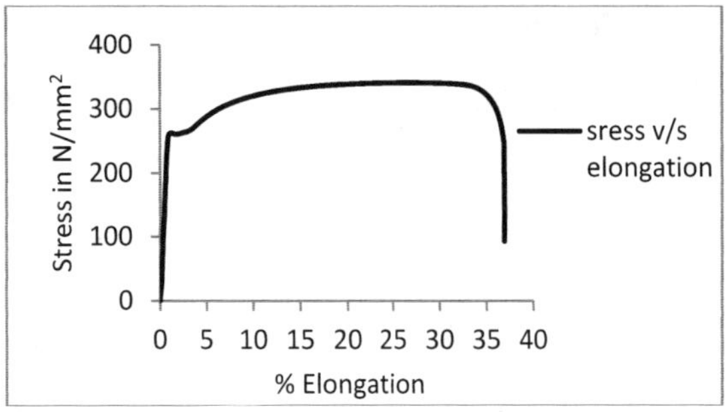

Figure 3. Representative stress vs. % elongation curve for coupon specimen (parent material)

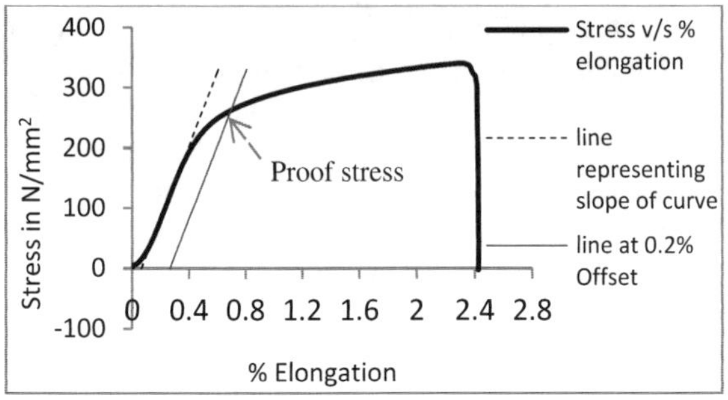

Figure 4. Representative, stress vs. % elongation curve for a strand of EMM (finished product)

as per the recommendation of paper [13] and accordingly, at least 30 numbers of specimens for each size of strand were tested. The stress-strain curve for a coupon of parent material and a strand of the finished product (EMM) is presented in figures 3 and 4 respectively. A photograph of tests being conducted is shown in figure 5a and 5b respectively.

It is a fact that for mild steel parent material, the slitting and stretching process adopted in the manufacturing of EMM leads to strain hardening. Due to this strain hardening, the stress-strain graph of the strand does not indicate a well defined yield point. Therefore, to obtain the yield stress, 0.2% proof stress for the strand is adopted, as per IS 1608:2005. As shown in figure 4, a tangent line with the initial slope of the stress-strain graph is drawn. The point of intersection of this line with the x-axis is obtained. Next, another line is drawn at an offset of 0.2% from this line. The yield point is the y-coordinate of the point of intersection of 0.2% offset line and stress vs. strain (% elongation) curve.

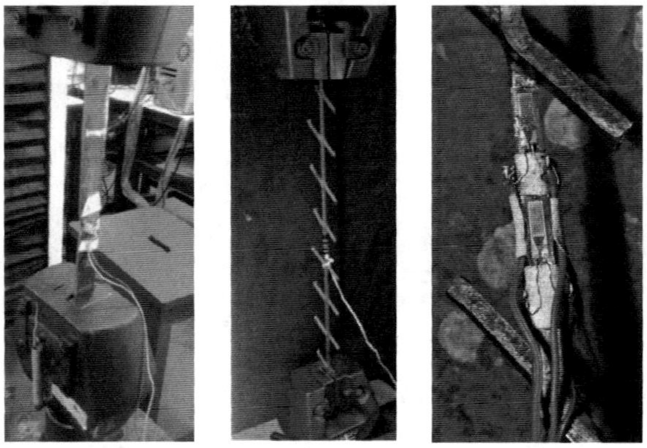

(a) (b) (c)

Figure 5. a) Tension test of the coupon, b) Tension test of 4x4mm strand and c) Two strain gauges fixed on the front side (inner side) and other two on the back side (outer side) of the strand, to study strain variation over the strand.

For experimental data generally 80 to 100% confidence level 'γ' is assumed and 5 percentile 'p' i.e 95% probability of success is assumed similar to the practice adopted by IS code, for finding the characteristic strength of the material. The same has been employed in this study. As per [14] an order of examining probability distributions is to examine first the two-parameter Weibull CDF; if this distribution is not rejected by a statistical hypothesis test, it is chosen and no further distribution is examined. If the Weibull CDF is rejected, then other distributions (lognormal and normal) are examined in a prescribed order. The stepwise procedure adopted is described in the following.

1) The values of thickness, width, yield stress, ultimate stress and maximum % elongation for the set of 30 specimens, for sheet metal and each size of strand were obtained from tension tests on these specimens.
2) The normal, lognormal and Weibull distribution (CDF, PDF) curves were plotted for the above data.

3) Anderson Darling statistic, which is sensitive to discrepancies in the tail region, was found using the equation mentioned in table 1

Table 1. Probability density function

Property	Normal	Lognormal	Weibull
Variable	$-\infty < x < \infty$	$0 \leq x < \infty 0 \leq x < \infty$	
Probability density function $f_X(x)$	$\frac{1}{\sigma\sqrt{2\pi}} exp\left[-\frac{1}{2}\left(\frac{x-\mu}{\sigma}\right)^2\right]$	$\frac{1}{x\xi\sqrt{2\pi}} exp\left[-\frac{1}{2}\left(\frac{lnx-\lambda}{\xi}\right)^2\right]$	$\left(\frac{\beta}{\alpha}\right)\left(\frac{x}{\alpha}\right)^{\beta-1} exp\left[-\left(\frac{x}{\alpha}\right)^{\beta}\right]$
AD Anderson-Darling	$\left[1\frac{4}{n}-\frac{25}{n^2}\right]A_n^2$		$\left[1+\frac{0.2}{\sqrt{n}}\right]A_n^2$
OSL	$\frac{1}{1+\exp[-0.48+0.78lnAD+4.58AD]}$		$\frac{1}{1+\exp[-0.10+1.24lnAD+4.548AD]}$

4) The observed significance level (OSL) is obtained by applying the equation mentioned in table 1

$$A_n^2 = \sum_{i=1}^{n}\left[\frac{1-2i}{n}\{\ln[F_0(x_i)] + \ln[1-F_0(x_{(n+1-i)})]\}\right] - n \quad (1)$$

$$\frac{\sum_{i=1}^{n} x_i^{\hat{\beta}}\ln(x_i)}{\sum_{i=1}^{n} x_i^{\hat{\beta}}} - \frac{1}{\hat{\beta}} - \frac{1}{n}\sum_{i=1}^{n}\ln(x_i) = 0 \quad (2)$$

$$\hat{\alpha} = \left(\frac{\sum_{i=1}^{n} x_i^{\hat{\beta}}}{n}\right)^{\frac{1}{\hat{\beta}}} \quad (3)$$

5) If OSL < 0.05, then that particular distribution was rejected (this is statistical hypothesis test)
6) Maximum likelihood estimator (MLE) of parameter α (scale parameter) and β (Shape parameter) are $\alpha \& \hat{\beta}$ are defined by expression (2) and (3) respectively.
7) Using values of $\hat{\alpha} \& \hat{\beta}$ obtained from Eq. (2) and (3), the mean and cov were evaluated.

$$\text{mean} = \hat{\alpha}\Gamma\left(1+\frac{1}{\hat{\beta}}\right) \quad (4)$$

$$\text{cov} = \left(\frac{\sqrt{\Gamma\left(1+\frac{2}{\hat{\beta}}\right)-\Gamma^2\left(1+\frac{1}{\hat{\beta}}\right)}}{\Gamma\left(1+\frac{1}{\hat{\beta}}\right)}\right) \quad (5)$$

Where 'Γ' is the gamma function.

8) Next, M LE of the 5th percentile value $X_{0.05}$ is obtained from equation (7).

$$x_p = \hat{\alpha}[-\ln(1-p)]^{\frac{1}{\hat{\beta}}} \quad (6)$$

$$x_{0.05} = \hat{\alpha}[0.0513]^{\frac{1}{\hat{\beta}}} \quad (7)$$

9) The data confidence factor Ω (function of cov of the data and sample size n) is then obtained from table no. 2 given in [13].

$$\Omega = e^{\left(-U_\gamma/\sqrt{n\hat{\beta}}\right)} \quad (8)$$

10) The nominal value for which there is 80% confidence that 95% of the population is above this value can be evaluated from equation (10).

$$x_{p,\gamma} = \Omega\, x_p \quad (9)$$

$$x_{0.05,0.80} = \Omega\, x_{0.05} \quad (10)$$

IOP Conf. Series: Materials Science and Engineering **652** (2019) 012008 doi:10.1088/1757-899X/652/1/012008

3. Results

From the repetitive tensile test carried out on coupon specimen and on each size of EMM strand, values of proof stress/yield stress, ultimate stress and maximum % elongation are obtained. This data set of each property is checked for the best distribution it fits and following the statistical characterisation procedure, given in section 2, nominal/design values of each material property are obtained. Sample histogram of proof stress, ultimate stress and % elongation for coupon tests and

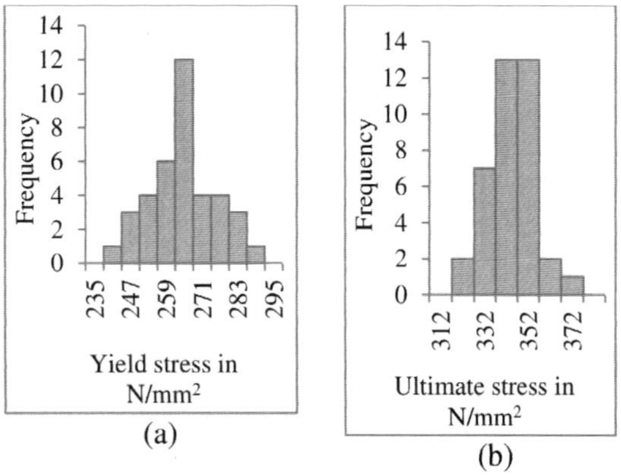

Figure 6. Histogram of a) yield stress of coupon and b) ultimate stress of coupon

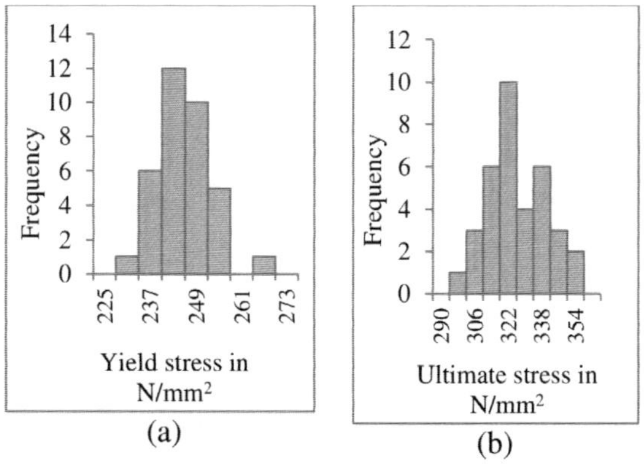

Figure 7. Histogram of a) proof stress of 6 mm strand and b) ultimate stress of 6 mm Strand

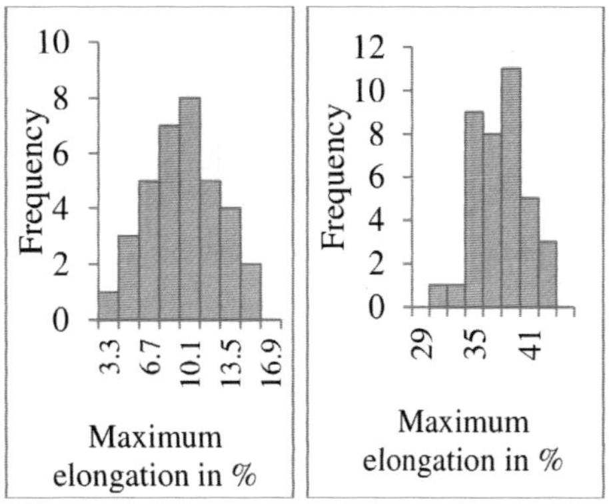

Figure 8. Histogram of a) maximum % elongation of 6 mm stand b) maximum % elongation of coupon

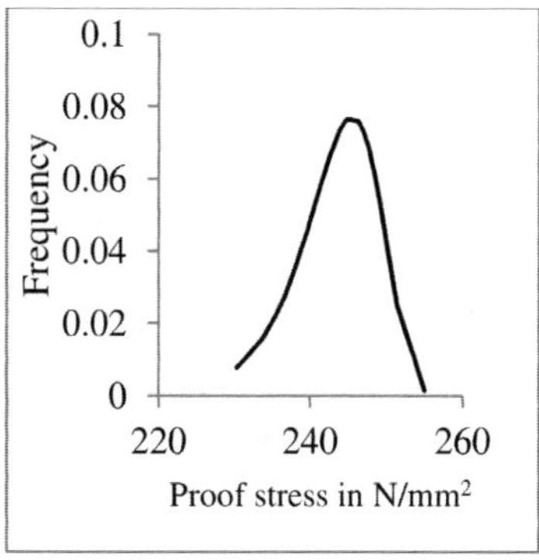

Figure 9. PDF of Weibull distribution for proof stress of 6 mm strand specimen

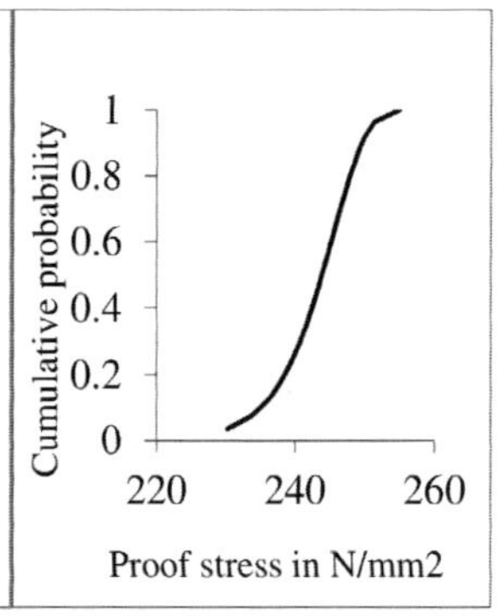

Figure 10. Representative, CDF Weibull distribution for proof stress of 5x6 mm strand specimen

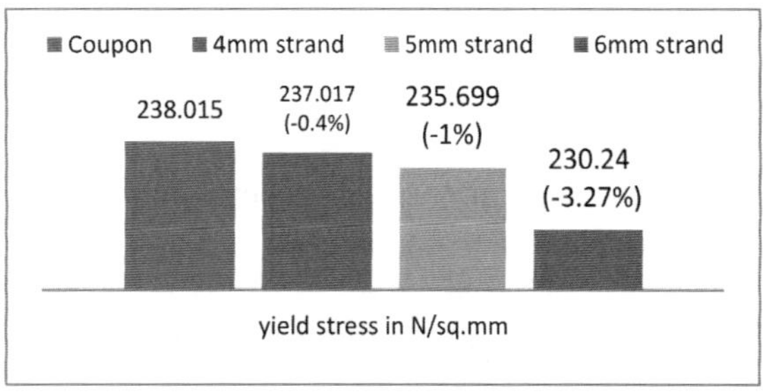

Figure 11. Comparison of the statistical characteristic nominal yield stress of strand with parent material

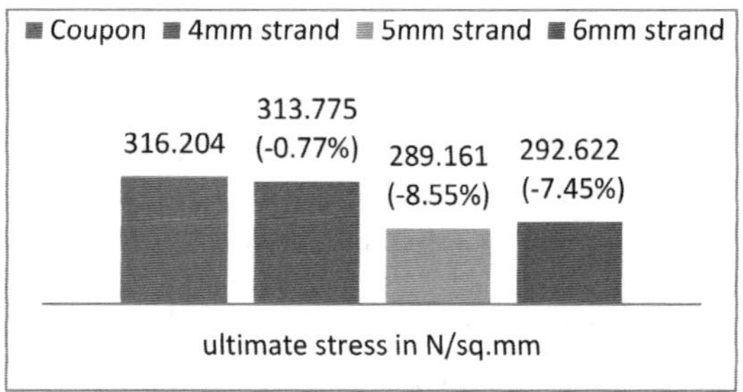

Figure 12. Comparison of statistical characteristic ultimate stress of strand with parent material

strand tests is presented infigure 6, 7 and 8 respectively.All the specimens fit into the Weibull distribution. Details of the statistical parameters of each of the Weibull distributions are given in table 2 and 3 respectively. Comparison of statistical characteristic nominal yield stress, ultimate stress and % elongation of the strand with the coupon is presented in figure 11, 12 and 13 respectively. Representative, PDF, and CDF for Weibull distribution of 5x6mm strand is shown in figure 9 and 10 respectively.

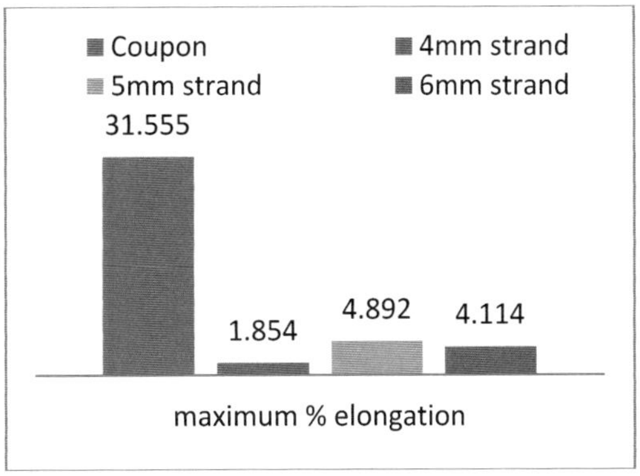

Figure 13. Comparison of statistical characteristic % elongation

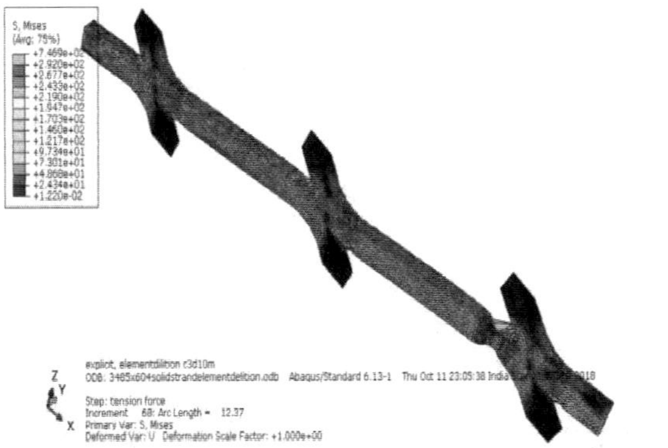

Figure 14. finite element analysis, a tensile test of strand simulated in ABACUS software to understand the effect of the size of the strand on its yield stress and ultimate stress

3.1 Observations and discussion

It is observed that all the strands had broken close to the knuckle. This is understood from the strain values obtained from tension test on 4 numbers of strands with 4 numbers of strain gauges, two each at the center and near the join (knuckle), refer figure 5c. Because the adjacent strands are eccentric (eccentricity e = width of the strand), the joint is subjected to axial force and bending moment (Fe) due to this eccentric force. So there is the highest strain at the inner side of each joint and lowest on the opposite side (almost compressive), this indicates that failure is due to combined axial and bending effect.

The strand was manufactured from IS 2062:2011 grade of steel. For coupons, the statistical characteristic yield stress and statistical characteristic ultimate stress obtained were 238.02 N/mm^2 and 316.2 N/mm^2 respectively. For strands, the statistical characteristic 0.2% proof stress for 4x4, 5x5 and 5x6 mm size of strands obtained were 237.02, 235.7 and 230.24 N/mm^2 respectively. The statistical characteristic ultimate strengths obtained for 4x4, 5x5 and 5x6 mm size of strands were 313.78, 289.16 and 292.62 N/mm^2 respectively. The 0.2% proof stress and the ultimate strength of strands are less than the corresponding coupon stresses. This may be due to the combined effect of axial and bending forces at the joint.

It is also observed that the stresses decrease with an increase in the strand width. This may be due to an increase in the moment at the knuckle because an increase in the width leads to higher eccentricity between the line of action of forces (i.e*e*). To verify this, a tension test of the strand was simulated in ABAQUS 6.13[15] and parametric studies were done for strands of increasing widths. It was found that the stresses decrease with the increase in the strand width.

Though the strand is manufactured from mild steel, its stress-strain graph did not show a well-defined yield point. The coupon tests on parent material have shown average maximum elongation of 31.55% while the tests on 4x4, 5x5 and 5x6 mm strands have shown maximum elongations of 5.4%, 9.72%, and 9.29% respectively. This may be due to cold form process of manufacturing EMM.

Table 2. Summary of Weibull distribution for

	(a) yield/proof stress of 4,5,6mm strand and 4mm coupon specimen (N/mm^2)				(b) ultimate stress of 4,5,6mm strand and 4mm coupon specimen (N/mm^2)			
	4x4mm strand	5x5mm strand	5x6mm strand	Coupon	4x4mm strand	5x5mm strand	5x6mm strand	Coupon
Number of data points	31	35	35	38	31	35	35	38
Sample mean	249.039	259.321	242.954	261.946	352.34	318.26	321.907	340.33
Sample Sigma	5.298	11.667	5.967	10.655	18.572	14.955	13.305	10.41
cov	0.021	0.045	0.025	0.041	0.053	0.047	0.041	0.03
AD test statistic	0.557	0.574	0.647	0.717	0.709	0.251	0.615	2.00
AD* test statistic	0.577	0.594	0.669	0.740	0.734	0.259	0.636	2.07
OSL	0.137	0.129	0.083	0.053	0.054	0.444	0.097	0.36
α	251.406	264.484	245.616	269.240	360.532	324.874	327.804	347.426
β	59.567	27.802	51.510	30.824	23.633	26.591	30.323	41.203
$\hat{\alpha}$	251.518	264.755	245.799	267.057	361.048	325.062	328.239	345.305
$\hat{\beta}$	50.016	25.550	45.420	25.798	21.164	25.378	25.858	33.735
Nominal value in N/mm^2	237.017	235.699	230.240	238.015	13.775	289.161	292.622	316.204

Table3.Summary of Weibull distribution for % elongation of 4,5,6mm strand and coupon specimen (N/mm^2)

	4x4mm strand	5x5mm strand	5x6mm strand	coupon
Number of data points	31	35	35	35
Sample Mean	5.460	9.720	9.285	37.063
Sample Sigma	2.417	2.992	3.291	2.791
COV	0.443	0.308	0.354	0.075
AD test statistic	0.658	0.632	0.720	0.590
AD* test statistic	0.682	0.654	0.745	0.609
OSL	0.078	0.091	0.054	0.118
α	6.159	10.786	10.385	38.280
β	2.406	3.609	3.085	16.348
$\hat{\alpha}$	6.164	10.783	10.344	38.313
$\hat{\beta}$	2.472	3.758	3.221	15.305
Nominal value in %	1.854	4.892	4.114	31.555

Table 4. Correlation, statistical characteristic nominal design value of strand in % as compare to coupon values

	Coupon test N/mm^2	4x4mm strand %	5x5mm strand %	5x6mm strand %
Yield stress	238.015	-0.42	-0.97	-3.27
Ultimate stress	316.204	-0.77	-8.55	-7.46
Maximum% elongation	31.555	-94.12	-84.50	-86.96

4. Conclusions

The yield stress, ultimate stress and % elongation of finished product i.e strand of EMM, is less than that of the parent material. This is mainly due to the combined axial and bending effect at the knuckle, cold form process of manufacturing EMM and irregular geometry of EMM. From table 4 it is evident that the stress of strand also decreases with an increase in its width. This is due to higher eccentricity with an increase in width, which results in inflation of bending effect near the knuckle. Thus, more the eccentricity lesser will be stress.

On the basis of statistical characterisation done in this study and comparative values recorded in table 4, it can be concluded that after applying an appropriate reduction factor to properties of parent material, sufficiently accurate nominal design stress values of the finished product can be estimated. This estimated value can be used by the designer to either predict the behaviour of EMM as a whole or to design the structure with EMM as lateral load resisting system, for low seismicity situations. In continuation of this study, further experimentation on the whole EMM is carried out. As per this study, a reduction factor of 0.95 for yield stress and 0.90 for ultimate stress is proposed to be used for predicting the nominal design values of EMM strand (up to strand width of 6 mm) from respective stress of parent material.

References

[1] Ghosh S and Kharmale S 2009 *Nova Science Publishers, Inc.* ISBN:978-1-61728-104-4
[2] Graciano C. et al., 2015 *Magazine Faculty of Engineering*, 1 (74): 132-142
[3] Ghosh S, Kharmale S and Gupta K 2011 *8th International Conference on Structural Dynamics EURODYN*, Leuven, Belgium.
[4] Vinay K, Kharmale S and Ghosh S 2010 *14th Symposium on Earthquake Engineering*, Roorkee, India.
[5] Dung Pand Plumier A 2010 *SDSS*
[6] Dung Pand Plumier A 2009 *9th National Congress on theoretical and Applied Mechanics*, Brussels.
[7] Dung P and Plumier A 2010*14th ECEE*, Orchid.
[8] Smith D, et al, *LATAM J Solid Stru*, **13**, 3145-3160
[9] ASTM F1267-07 *Standard Specification for Metal, Expanded, Steel*
[10] BS:4592 part1 *Expanded Metal Grating Panels for use in Flooring, Walkway and Stair treads*
[11] IS 412:1975 *Specifications for Expanded metal steel sheets for General purposes*
[12] EMMA 557-99, *Standard for expanded metal, National Association of Architectural Metal Manufacturers*, NAAMM
[13] Zureick A, Bennett R M andEllingwood B.R 2006 *Journal of Structural Engineering*, 132, 1320–1327.
[14] ASTM Int 2002 Composite Materials Handbook-MIL17
[15] ABAQUS Simulia 2013 *Abaqus FEA User's Manual, Version 6.13*. V´elizy-Villacoublay, France

IOP Conf. Series: Materials Science and Engineering **652** (2019) 012009 doi:10.1088/1757-899X/652/1/012009

Variation Prognostic Model on Port Infrastructures: Employment of Principal Component Analysis for Factor Extraction Instigating Disparity on Planning and Implementation Phase

Maria Virginia Cabalag[1] and Dr. Dante L. Silva[2]

[1] School of Graduate Studies, Mapua University, 658 Muralla St., Intramuros, Manila, 1002, Metro Manila, Philippines
[2] Mathematics Department, Mapua University, 658 Muralla St., Intramuros, Manila, 1002, Metro Manila, Philippines

Email: mavie.cabalag.15@gmail.com; dantesilva2000@yahoo.com

Abstract. Port is a transportation and communication marine infrastructure that is established for ship berthing and dockage, loading and unloading of passenger, storage of goods and other operational agendas that would impact a region or country in terms of social and economic aspects. Variation order is the modification on the design whilst construction is on-going. This is inevitable in a project. Through the utilization of Principal Component Analysis with a Cronbach's Alpha of 0.930 and Kaiser-Meyer-Olkin Value of 0.763, nine components weighted, these are Actual Concreting and Precasting on-site for Structure formation, Meteorological Conditioning affecting Weather and Tides, Test Pile Apportionment on Both design and construction, Off-shore soil in-situ Properties, Behaviour and Suitability on fills, Inaccuracy of Geotechnical, Topographical and Hydrographical surveys, Project Release and Initiation Timing on engineering departments, Sudden unforeseen Turbulences during Pile Driving, Coordination of Planning and Implementation Departments, and Unavailability of Material Specification to Actual site from design. After extracting the regressor, assessment of completed port projects with variation and multiple linear regression was then employed to generate the model. The model had an R^2 of 0.777 and an R of 0.882. Finally, the model was evaluated and has shown a significant value. Hence, a satisfactory breakthrough in an attempt to develop a model regarding variation order on port projects was established.

Index terms: Port, Design and construction variation, Project management, Statistical model, Principal Component Analysis.

1. Introduction

Port design and construction engineering is not an ordinary methodology. Design considerations and foundation procedurals are dependent on the morphological, geotechnical, and sea level assessment of the probable area. Planners and designers have to conduct a site visit, topographical and hydrographical survey, geotechnical assessment of both the offshore and inland area that are to be affected and lastly the social, economic and environmental need for structuring or improvement of a port. A port foundation is commonly created through pile driving or reclamation. Variation order occurs when there is a change of scope or change of work in the planned and designed port improvement or port expansion. Change is inevitable. Hence, this research was intended to provide a linear model and paradigm that could evaluate and approximately predict the percent variation of a port project. The percent variation that the researcher is pertaining is simply the quotient of the variation order amount over the project's final cost, wherein the variation order amount is the difference of the actual cost of implementation and the projected cost. Principal Component Analysis is a statistical method that was utilized for highlighting and showing off strong patterns of set of data. This statistical procedure is usually employed in empirical data analysis and generation of predictive models. The eigenvalues, extraction, communality and factor loadings was analysed to weigh in the significant factors that will be beneficial for the model generation. The model generation will be done thru multiple linear regression and shall be evaluated.

2. Port projects' variation order on financial cost and delays

Construction management in port and harbour is not a common field in engineering. Nearly all construction projects not only the port type one, would be expected to differentiate from the plan, scope and definition. From analysis, it revealed that the lower the project cost is, the higher in percentage was the variation order and it seemingly presented an approximately indifferent disperse in both negative and positive ways. Additionally, the R2 of the points below is 0.0075 which is close to zero, stating that there is no clear and direct relation between the two variables. The main implication it can tell is that at any amount of project, variation order is still existent and unavoidable. Since percent variation formula is derivative from the cost itself, then a higher variation order would mean that a larger final project cost will come out and that has no argument already. Evidently, the hypothesis of W. Smith [1] that the amount of variations indeed affects the magnitude of the cost under/overruns was also proven in his study. Majority of construction contracts gives space to the extension of construction period or duration. This is simply called as an extension of time. Variations may but not necessary affects directly to durations. According to the data gathered and its scatter plot, the variation order has no direct effect on the project delays, even project with minimal to almost no variation order experiences project delays, and they are of varying timeframe. Having a coefficient of correlation of -0.264 and a coefficient of determination of 0.07 or close to zero, delays and variation order has no affect with each other. Thus it would be suggested to further investigate the two variables' influence and effect to each other.

3. Validation, demographics and principal component analysis

Out of 53 factors, 40 factors had passed the content validation. The Cronbach's Alpha was achieved with a value of 0.907 for pilot test and 0.930 for the final instrument used which clearly signifies that the instrument is strong and is suitable for PCA. The Kaiser-Meyer-Olkin and Bartlett's Test of Sphericity are intended to check the sampling adequacy if it is fit for principal component analysis utilization. An assessment close to 1.00 shows that configurations of correlation are comparatively intact. On the other hand, studies suggests that the Kaiser-Meyer-Olkin value should not be less than 0.50 to state the adequacy of sample size [2]. KMO of the Variation Factor Assessment on Port Projects had sustained a good value of 0.763 and the Bartlett's test was significant too based on level of confidence of 0.05 implying that it was an identity matrix. The respondents that participated in this survey had significant working experience in both design and construction phase of a port project, which are significantly port engineering professionals.

Principal Component Analysis starts by recognizing a group of variables in which its variance in accordance to the researcher's objective judgment could be embodied further significantly through a lesser group of factors or components. Hence, the outcome of PCA displays what variables may be characterized by which components, and which variables shall stay as individual variables due to inadequate representation of their information by the factor solution. The average communality of factors upon excerption was 0.751 by which the conventional applicable standards about communality values is an extraction value of not less than 0.50 in first iteration. Rotation proposes the characteristic of the variables which are in extreme behaviour and brings out the maximum potential of the bearing of every factor on one of the extracted variables whilst reducing the bearing on the entire remaining factors. Varimax is the most accepted rotation that is being utilized for factor analysis, at most. In varimax, a basic solution implies that every variable is composed of a fewer amount of heavy bearings and a huge number of zero or minor bearings. Furthermore, the factors may frequently be construed from the disagreement of less variables with positive loadings to less variables with negative loadings [3]. To create significant components and avoid the complexity, the final factor loading is set to 0.7（table 1）.

Table 1. Summarized Rotated Component Matrix - Varimax Rotation under Kaiser Normalization

Comp. 1	Comp. 2	Comp. 3	Comp. 4	Comp. 5	Comp. 6	Comp. 7	Comp. 8	Comp. 9
(37) .83	*(16)* .88	*(21)* .84	*(31)* .84	*(9)* .79	*(7)* .75	*(26)* .79	*(8)* .79	*(4)* .74
(39) .79	*(17)* .87	*(22)* .78	*(32)* .77		*(5)* .75	*(27)* .74		
(35) .76	*(15)* .76							
(36) .70	*(18)* .70							

4. Extracted Factors

Actual Concreting and Precasting On-site for Structure Formation (ACPS) is the first component that was generated, consisted of four variables carrying a factor loadings of above set standards which are incorrect selection of type of formworks to be used (82.8%), corrosion of reinforcement (79.1%), type of concrete pouring (75.8%), and cold-joints on precast (70.4%). Concreting procedurals is an integrated part in all type of civil engineering structuring, port areas are mostly constructed using concrete and precasting, for beams, decks, slab, RoRo ramps and other integral parts of a port. Meteorological Conditioning Affecting Weather and Tides (MCWT) is a formed component consisted also of four variables which are calamities such as typhoons (88.3%), sudden climate changes in area (86.5%), changes in sea or tidal waves (76.4%), and changes in sea level or elevation itself (70.1%). Since port is an off-shore structure, civil engineering methodologies are much more taken care of in this specialized area of a project as compared to vertical structures. Tidal settings and elevations are easily affected on changes of weather and so it is evidently certain that the said group of factor will cause a variation order. . A study regarding causes of variation order conducted also signified that weather conditions are top factors that causes variation order in construction projects, which was agreed by the consultant and contractor itself [4]. Test Pile Apportionment on Both Design and Construction (TPAB) is a component composed of two variables only that relates solely to test pile only, which are no assigned test pile on design plans (83.5%) and no test prior driven prior to construction (78.4%). Test pile is very significant prior to final driving of pile in actual implementation. In this way, an adjustment can be proposed prior to final purchase up to drive of the set of piles. The pile testing is to examine such flaws in both pile and mistakes in bedrock projection from surveys before turbulence must exist on the final pile driving. Off-shore Soil In-situ Properties, Behaviour and Suitability on Fills (OPBS) is a component having also two variables that has a loading of greater than 0.7, these are instability of soil strata underneath water (83.9%) and not suitable rock and sand and gravel fill for the soil (76.9%). These properties are being collected from a subsoil investigation or geotechnical investigation. This investigation shall be executed to find description for characteristics, natural properties, load bearing capacity and settlement capacity of the soil before designing of pile or reclamation rock, gravel and selected fills. Inaccuracy of Geotechnical, Topographical and Hydrographical Surveys (IGTH) has a factor loading of 79%, geotechnical surveys are much needed for pile type projects which shall determine the maximum elevation by which the bedrock will be reached, this is the basis of the designer to determine the applicable length and bearing capacity of a pile shall be specified. Hydrographical survey is like a topographical surveys wherein it shows the contour and elevation of the oceanographic surface and as well as the in-land part. From here the area of the proposed project will be assigned, elevations with less disparities at every horizontal proximity is better. Since these surveys are highly needed and highly studied for basis of design, an error or inaccuracy with the actual would mean it will largely affect the designated pile lengths or fill volumes. Project Release and Initiation Timing on Engineering Departments (PRIT) is a component pertaining to construction management technicalities and not to port structuring itself, it contained two variables which are overlapping of projects (75%) wherein new projects were to be implemented while some are still on-going on the said port and rush projects in both department occurs (74.8%). Although if to look at it objectively, it is not merely dependent or pin-pointed to the departments per se, it's just that the frequency, distribution and timing of port project for initiation are at times not in balance based on the merging of the two factor. A previous study had shown also that part of variation order consequences is project time pressure which can synonymously pertain to this component [5]. Sudden Unforeseen Turbulences during Pile Driving (STPD) is a component regarding unexpected instances or trouble that happens on the site itself during driving and pitching of pile which are crane/barge movement when driving affecting misplacement of pile (79%) and pile breakage when driving (74%). Crane/barge movement of the pile driver could affect the bearing capacity to be transferred to the pile and possibly misplace or misalign the positioning of pile if encountered, the movement is majorly caused by the waves of sea but can sometimes be affected by the gravity of the barge, pile and machine itself. There are also times in construction site where excessive strength in driving of pile happens thus resulting to a significant breakage or damage to pile. Hence, these damages will count for variation order. Coordination of Planning and Implementation Departments (CPID) has a 0.7 factor loading and it was entitled by the factor alone itself already which is coordination of the two engineering departments responsible for the port projects (78.9%). The coordination of the two department will be seen only during a project commencement if there is a variation order occurrences that will to alter the design to fit for the actual conditions. This was somehow showed in a study previously, that reasons for variation order are disputes between management parties, conflicts on projects and lack of value engineering or merely the coordination and unambiguity between the two significant department [6]. Unavailability of Material Specification to Actual Site from Design (UMSA) pertains to material specification unavailability in the actual construction (74.3%), both in the pileworks and reclamation activity in constructing a port, this issue may arise. This could be tedious in the part of the designer since a pile's change in cross-sectional area would also mean a modification on a bearing capacity and thus a new plan for pile layout as well, evidently will account for a variation order. In terms of reclamation, armour rocks and core rocks sizes and quality differ from the designated versus the actual available.

5. Variation Prognostic Model

Through assessment of completed port projects with variation order. A model generation has been initiated thry multiple linear regression. All factors was tested and has significantly above zero R^2 and adjusted R^2 which signifies its positive correlation to the independent variable and thus, all is accepted for final modelling. Multiple regression permits to establish the overall fitness. It is probable to determine true parametric values in factor analysis with Likert scale data, if presumptions on skewness, and amount of classifications were satisfied [7]. The generated variation prognostic model has a coefficient of determination of 0.777 and adjusted value of it at 0.677 and coefficient of correlation of 0.882. There is also an indication that the generated prognostic model from the extracted components predicts the dependent variable significantly well with $F(9,20) = 4.830$, $p < .000074$ which shows overall that variation prognostic model scientifically forecasts the outcome percent disparity and displays significant fitness for the gathered data. Below is the model generated based on multiple linear regression utilizing the nine emerging factors from the Principal Component Analysis Extraction using beta coefficients.

$$\% \, Var \, on \, Port \, Proj = 0.112ACPS + 0.172MCWT - 0.115TPAB + 0.464OPBS + 0.252IGTH - 0.004PRIT + 0.047STPD + 0.157CPID + 0.189UMSA$$

IOP Conf. Series: Materials Science and Engineering **652** (2019) 012009 doi:10.1088/1757-899X/652/1/012009

The model was evaluated using a different set of projects. Hence, the p-value is less than 0.05 wherein $p<0.037$ and is within the confidence interval level. It also signifies that the entire methodology used is a good fit for the type of study and can be re-done to other infras with regards to variation order. This is a breakthrough in employing factor analysis processes as regressor extraction for modelling percent variation order, which is a turmoil that cannot be avoided in all types of construction projects. Based on the paradigm on Figure 1., the variables that affect variation order can be categorized into five areas, this just simply illustrates the wide scope of a variation order since all aspects of a project will indeed affect the modification from the initial plan to the finality of the implementation. It can be stated that variation order is a complex thing to solve in construction industry but in anyways can be mitigated if turbulences could be deeply evaluated. The study is a different approach in dealing for variation order for the reason that it has attempted to quantify such. Previous literature had attempted to assess only qualitatively variation order in the sense that causes and effects were evaluated. In a deep matter, variation order is also a quantitative aspect, since each changes are clearly being felt by the cost of the project. The study opened a way for more profound evaluation of variation order in the path for it to be projected scientifically thru the use of Principal Component Analysis and Multiple Linear Regression.

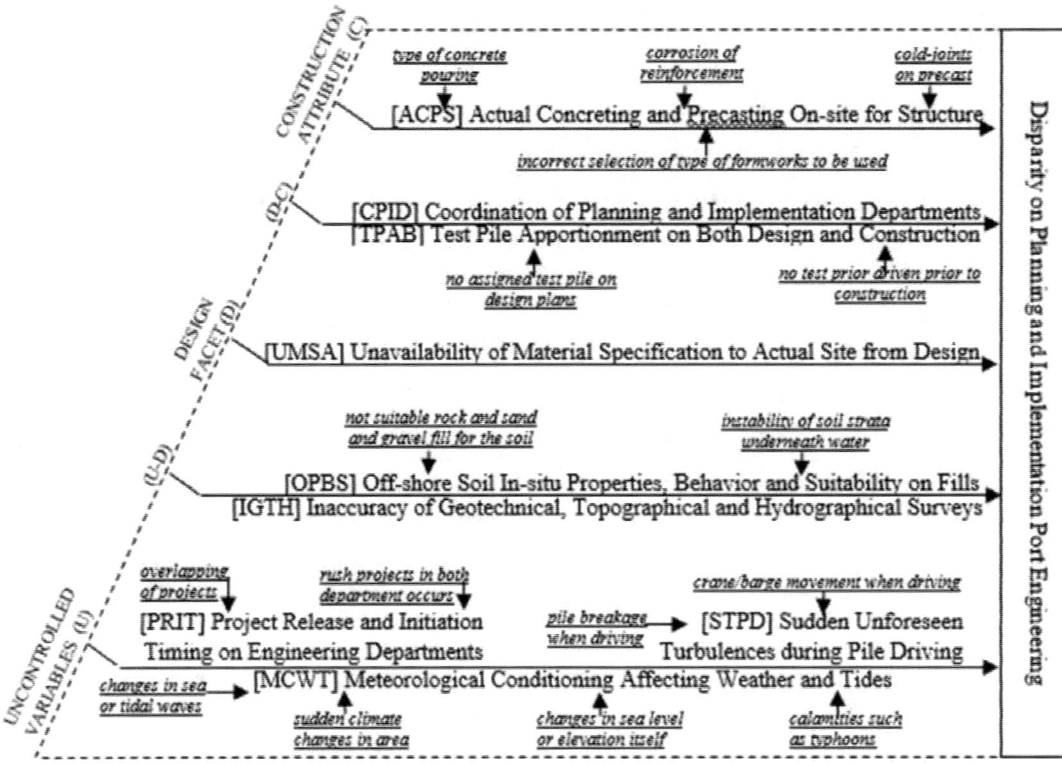

Figure 1. Variation Prognostic Paradigm on Port Infrastructures

6. Conclusion

Considering all the input of the study, it is considered as an accomplishment to achieve the main objective which is to generate a model through utilization of Principal Component Analysis that will determine the percent variation of a port project between the design and construction phase. The other most extensive provision of the study is the revelation of the components and factors of such that influences variation order on port projects and the attempt to quantify it scientifically to provide approximation on projects. Granting that the statistical outcomes established on this research pertaining to variation orders, this paper is however capable as a reliable reference or guide in observing and investigating variation from the design phase and construction phase of any type of a

IOP Conf. Series: Materials Science and Engineering **652** (2019) 012009 doi:10.1088/1757-899X/652/1/012009

project, purposely moreover for a port structure. Nonetheless, the discovery and apprehension of such influential causative factors and scheming for contingencies and preclusion of the said obstructions prior to its occurrence can be beneficial and contributory to a project's accomplishment and accuracy of its planning, design, scheduling and projected costing to its actual implementation, construction and financing of a port project.

References

[1] Smith W. 2016 The Effect of Variation Orders on Project Cost and Schedule Overruns
[2] Field, A. 2005 Factor Analysis Using SPSS: Theory and Application
[3] Abdi. H., and Williams, L.J. 2010 Principal component Analysis
[4] Ismail A., Pourrostam T, Soleymanzadeh A. and Ghouyounchizad M. 2012 Factors Causing Variation Orders and their Effects in Roadway Construction Projects.
[5] Ndihokubwayo R. 2009 Variation Orders on Construction Projects: Value Adding or Waste?
[6] Shehu Z., Endut I., Akintoye A. and Holt G. 2014 Cost Overrun in Malaysian Construction Industry Projects: A Deeper Insight.
[7] Lubke G. H., Muthen B. O. 2004 Structural Equation Modelling

Predicting the service life of buildings and facilities to minimize the risk of losses in the conditions of natural and technogenic emergency situations

Elena Smirnova[1*] and **Sergey Savin**[1]

[1]Saint Petersburg State University of Architecture and Civil Engineering, Vtoraja Krasnoarmejskaja ul. 4, St. Petersburg, 190005, Russia

Abstract. This article discusses the solution to the task of predicting the consequences of catastrophic earthquakes for operational buildings and facilities. We suggest using such basic parameters for verifying the calculation models as natural frequencies, oscillation forms and decrements of objects under earthquake loads. At the same time we substantiate the possibility to use micro-oscillations for that purpose, even though buildings and facilities are cushioned systems. We present the methods and means to conduct tests to estimate the dynamic parameters of buildings, as well as an example of such a test and the process of verification of the calculation model in the example of a panel multi-storey building. We set an example of evaluation of the consequences of an accident with a building in St. Petersburg, which was partially destroyed as a result of non-uniform setting of soils in the base. We drew the conclusion that using dynamic parameters makes it possible to solve various tasks for current predicting the remaining service life of damaged objects.

1. Introduction

In the modern world, the risk of natural disasters and catastrophes has greatly increased. In this situation, we have to be able to predict the possible consequences of the earthquake in advance. In this context, the "possible consequences" mean the reaction of the building to seismic loads. On the one hand, we can note a significant progress in the calculation methods to predict the possible seismic load, for example, using synthetic accelerograms. On the other hand, we can model the reaction of the building itself with high accuracy. However, without detailed inspection of the building, it is practically impossible to take all its particular features into account, due to the possible presence of concealed defects and damage, non-observance of project characteristics etc. that do not manifest themselves under operating loads but can cause destruction under seismic impact. That is why the task of quick assessment of the technical state of the building is actual to answer the question about the consequences of emergencies [1–7].

2. Materials and Methods

At the current moment, the most widely used method to estimate the dynamic parameters of buildings is called the "Method of free oscillation" (MFO) [8–12].

In the MFO, the following operations are used:

- exciting and recording free or forced micro-oscillations;
- calculating their Fourier specters;

* Corresponding author: esmirnovae@yandex.ru

IOP Conf. Series: Materials Science and Engineering **652** (2019) 012010 doi:10.1088/1757-899X/652/1/012010

- analyzing the Fourier specters to find the resonance peaks corresponding with different waveforms of free oscillations;
- using the Fourier reverse transformation, obtain the impulse realizations of the detected resonance peaks for each waveform of free oscillations;
- identification and graphical representation of different oscillations waveforms.

Using the MFO made it possible to develop a universal mechanism to elaborate the dynamic parameters of calculation models for buildings using different software packages (Scad, Abacus, Ansys, Lira).

3. Results

As an example, let us discuss a high-rise panel building in Moscow district: how can we compile a reliable calculation model for predicting the results of any external influence, including the seismic one (in this case, the application of the method of free vibrations is limited to specific building geometry, i.e. buildings with load-bearing brick walls, not with flexible geometry). The measuring system consisted of sensors installed in the vertical cross-section at different levels in the building (from the basement to the roof) and connected with the recording equipment. Oscillations were excited in the upper and the middle part of the building by applying multiple strikes, with pauses during the attenuation period of natural oscillations of the building (**Figure 1**).

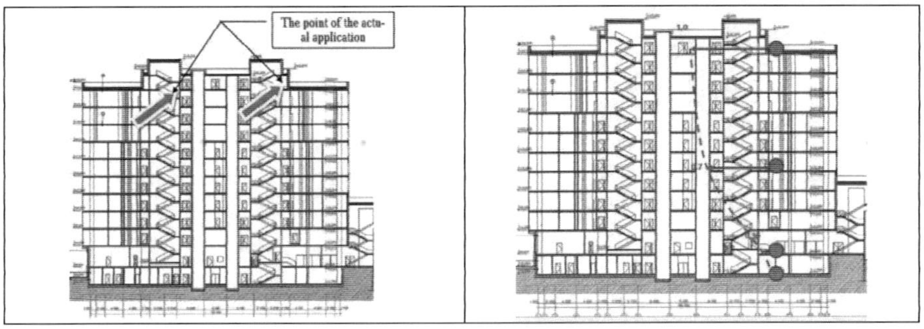

Figure 1. Scheme of of load application, sensors arrangement and form of oscillations

The test results were graphically represented as the oscillations of the building in the longitudinal and the transverse directions. The obtained dynamic parameters were used to adjust already constructed FE model of the building in the FEA program Ansys with an ultimate goal to use this model for predicting the responses of the building in future strong earthquakes (see **Figures 2, 3** and **4**). The mass, damping, and stiffness matrices of a building are obtained by spectral approach – decomposition motion into natural oscillation forms, based on the Duhamel integral. The vertical axis denotes the height of the building in meters, and the horizontal one shows the relative movement recorded with the sensors (points in the diagram) installed across the height of the building, for different forms of oscillations:

- Bending by the height; transverse direction (first tone) – 1.25 Hz;
- Bending by the height; transverse direction (second tone) – 5.96 Hz;
- Bending by the height; transverse direction (third tone) – 14.89 Hz;
- Bending by the height; longitudinal direction (first tone) – 1.92 Hz;
- Bending by the height; longitudinal direction (second tone) – 6,98 Hz;
- Bending by the height; longitudinal direction (third tone) – 18.99 Hz;
- Bending by the front (first tone) – 1.94 Hz;
- Bending by the front (second tone) – 7.812 Hz.

For convenience, the maximum deflection of the sensor during oscillations for each form was assigned a value of one. The readings of other sensors are expressed as a fraction of the maximum deflection.

IOP Conf. Series: Materials Science and Engineering **652** (2019) 012010 doi:10.1088/1757-899X/652/1/012010

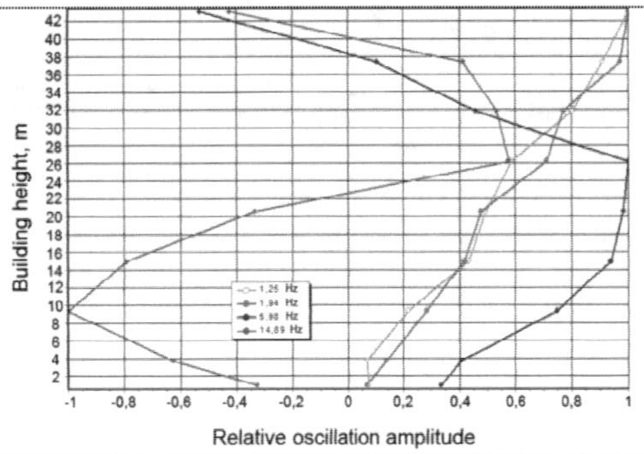

Figure 2. Oscillation forms. Vertical cross-section, transverse direction

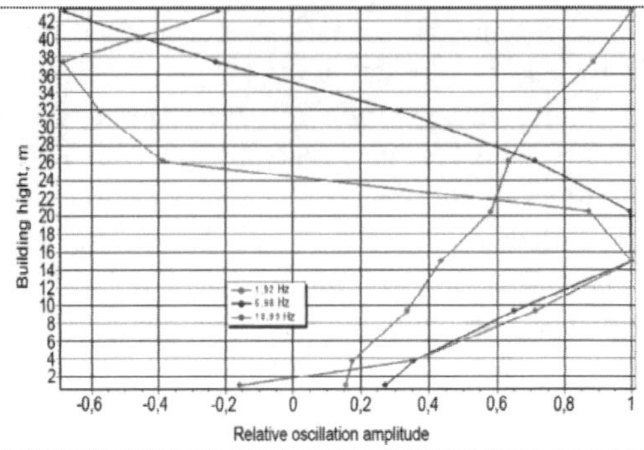

Figure 3. Oscillation forms. Vertical cross-section, longitudinal direction

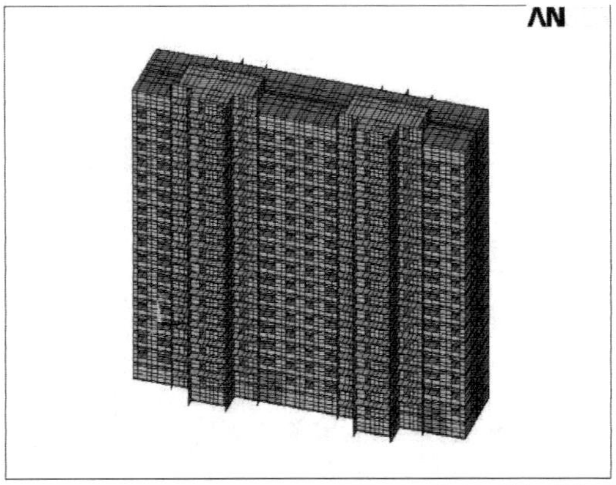

Figure 4. A fragment of the end-element model of the building compiled in the software complex Ansys

How can dynamic parameters be obtained with such a weak impact? The responses are summarized using extremely sensitive sensors (100 mV/m/sec^2). This significantly raises the dynamic range of measurements (the excess of the useful signal over interference). Obviously, the second and the third oscillation forms are compiled with the use of the possibilities of the principle of superposition of oscillations and are quite a rare result for a residential building (higher forms of oscillations are very difficult to record due to their small amplitude). The obtained forms help elaborate the character of building deformation and investigate it without consideration of conditions of its "sticking" in the ground, which largely determine oscillations frequency of the first form.

Further on, the parameters of the calculation model were elaborated in the following manner. During the first stage, a modal analysis was carried out and the lowest frequencies were estimated as well as their corresponding forms of natural oscillations. These values were analyzed and compared with similar data obtained from tests. In our opinion, the refusal to take into account changes in the geometry (coordinates of nodes) can cause a significant reduction in the quality of the calculation. But in the current practice of computational justification, it is considered quite acceptable not to allow for the change in the geometry of the computational model at different stages of the existence (installation) of the structure. On condition that the geometric dimensions, weight and structural features of the building were modelled correctly, the changes were applied to the parameters describing the generalized stiffness of the structure materials. As a result of a series of corrections to the values of generalized material stiffness, we were able to achieve satisfactory correspondence between the calculated and the experimental values of natural oscillations frequencies. The actual measured values of natural oscillations frequencies recorded during dynamic tests well correspond between themselves (**Table 1**).

Table 1. Comparative data on oscillations frequencies of the examined building

Oscillation forms acc. to the calculation model	Oscillation frequencies recorded during tests, Hz	Oscillation frequencies acc. to the calculation model adapted by dynamic characteristics, Hz	Difference of the calculated oscillation frequencies
1	1.25	1.291	3.1 %
2	1.92	1.944	1.2 %
3	1.94	2.045	5.1 %
4	5.96	5.604	5.9 %
5	6.98	6.832	2.1 %
6	7.812	7.456	4.5 %
7	-	11.716	-
8	-	12.587	-
9	14.89	-	-
10	18.99	-	-

Thus, we can state that we were able to create an adequate calculation model that can be used for reliable predicting the results of emergency influences, including seismic impacts.

On the other hand, specialists of emergency services often have to assess the remaining service life of partially damaged or destroyed buildings in order to make decisions on restoring or reinforcing the damaged structures. An example of such investigations is the accident in Dvinskaya street (St. Petersburg, 4 July, 2002), when the Southern 17-m long part of the 9-level building suddenly collapsed. Four lives were lost. Immediately after the accident, the natural oscillations of the building in the following frequencies were detected:

1.8 Hz – bending oscillation in the transverse direction (vertical cross-section);
2.18 Hz – swaying on the foundation in the transverse direction;
2.25 Hz – bending oscillations in the longitudinal direction (vertical cross-section);
5.4 Hz – bending oscillations, second form, transverse direction (horizontal cross-section).

Analysis of the obtained dynamic parameters showed that the building had a typical frequency of natural oscillations for such objects by the first tone [5], which is the evidence of sufficient spatial stiffness of the entire structure. Relative movement of the foundation during oscillations was more than 10%, i.e., the foundation has high mobility. As the main features, it was noted that there is low stiffness of connections between sections at frequencies from 6 Hz (this frequency is usually 15 Hz and more for buildings with deformed links) and a large input (up to 55 %) into the bending form of oscillations of the building swaying on the foundation (this value is usually within 15%).

4. Discussion

The most important finding the authors made in this study is that the dynamic responses of a building when subjected to the impact of a mass of 30 to 50 kg over aduration equal to half of the period of the first vibration mode is sufficient to evaluate the frequencies and mode shapes of the free vibrations of the building. At the present time, there are two methodical approaches to carry out measurements in the active and passive modes of exciting oscillations in the object. During the "passive" recording, the response is recorded of the inspected structures to the background micro-seismic impact. During the "active" recording, the response is recorded to the deliberate impact onto the structure. If such impact is applied to the strength nod in the upper part of the structure (with a weight of up to 50 kg), and the duration of the application is about 0.5 period of oscillation of the building by its first form, then the amplitude of the recorded response is good enough to evaluate the frequencies and waveforms of free oscillations [4-5]. At the same time, there are more methods to increase the dynamic range of the recording equipment as well as achieve the maximum amplitude of oscillations with the required form. They are: exciting and recording oscillations under a pointed impulse load applied at different locations of the structure; adding up (with consideration of the load vector) the oscillations recorded during the impact of the load at different locations of the structure (imitating simultaneous application of loads at several locations). The damping/stiffness matrices of a building may change significantly if the building is damaged. It is true that the NDT techniques can only estimate the damping/stiffness of

a building at its present state (which may have slight to no damage at the moment when the NDT test is performed). However, if destructive changes occur, we will see this by changing the natural frequency of oscillation and take them into account in the design model as, for example, reducing the stiffness of the damaged element. We agree without pushover curves of the individual columns and/or the entire building (i.e., the nonlinear responses of the building at various ductility levels), the current mass/damping/stiffness matrices of a building could not be used to predict the responses of the building after it is severely damaged in strong earthquakes. It is important to accent the values of frequencies and the corresponding vibration mode shapes being those parameters that "feel" any "features" of both the model and the real object. They are compared to establish the reliability (correctness) of the assembled model [13–14].
With a strict approach to the theoretical substantiation of correctness of division between the actual oscillations of buildings into harmonics, it is known that, in the general case, the attenuating systems cannot be completely divided, because it is impossible to completely negate the multiplication of coefficients of dampening different forms [15]. Accepting the fact that the nonlinear factor will be too small for standard constructions it will not be erroneous to assume that the damping factor of each mode is linearly dependent on the system frequency. From the physical viewpoint, there are no isolated own forms of oscillations, and they are not in-phase with oscillations with any form of movement of different points. However, if the attenuation is small enough, this feature can be ignored and it can be assumed that a cushioned system has classical own forms of oscillations when all the masses reach their maximum deviations simultaneously. Therefore, if the building oscillation decrement (oscillations attenuation) is small, the proposed method can be reasonably used for diagnostics. Practically, it means that, if there are isolated peaks in the specter of recorded realizations, the attenuation can be ignored. An indirect evidence of the “small” attenuation is the possibility to clearly divide between different forms of oscillations [16–17].

5. Conclusions
1. Using dynamic parameters in calculations not only helps the best model of the facility in each case, but also elaborates its parameters to achieve its authenticity to the strength of load-bearing structures.
2. The authors have made other contributions in simplifying the procedure for constructing the FE model of a building using the data collected from NDT tests. The main theoretical provision allowing reliable assessing of dynamic parameters using the MFO method is the assumption about the linearity of the examined systems. In this case, the method of solving the differential equation systems we should use is the method of decomposition by own forms; with this method, instead of a coherent system of differential equations, we obtain a number n of independent equations, with which, using constant coefficients, we can establish the input of each own form number j into the overall response of the system to the external influence. If we know the frequency and form of oscillations, we can reliably estimate the properties of all the elements in the system.
3. Linearity of the discussed systems makes it possible to use the principle of frequency superposition during measurements; this allows artificial “amplification” of the oscillations of the examined forms for massive and high objects.

References
[1] M. Döhler, F. Hille, L. Mevel. Vibration-based monitoring of civil structures with subspace-based damage detection. In *Mechatronics for Cultural Heritage and Civil Engineering* (Ser: Intelligent Systems, Control and Automation: Science and Engineering, ISCA, vol 92). E. Ottaviano, A. Pelliccio, V. Gattulli (eds.). New York: Springer, Cham, 2018, pp. 307–326. DOI https://doi.org/10.1007/978-3-319-68646-2_14
[2] M. Döhler & F. Hille. Subspace-based damage detection on steel frame structure under changing excitation. *Structural Health Monitoring*,(2014), **5**: 167–174. DOI https://doi.org/10.1007/978-3-319-04570-2_19
[3] S. Das, P. Saha, S. K. Patro. Vibration-based damage detection techniques used for health monitoring of structures: a review. *J Civil Struct Health Monit*, (2016), **6**(3): 477–507. DOI https://doi.org/10.1007/s13349-016-0168-5

[4] I. L. Danilov & S. N. Savin. Service safety of buildings and structures at high mechanical loads. *Natural and man-made risks (physical, mathematical and applied aspects)*,(2015), **3**(15): 24–29.

[5] Y. Y. Shatilov & A. A. Lyapin, Vibration-based damage detection techniques for health monitoring of construction of a multi-storey building. *Materials Science Forum*,(2018), **931**: 178–183. DOI https://doi.org/10.4028/www.scientific.net/MSF.931.178

[6] L. Ramos, L. Marques, P. Lourenço, G. De Roeck, A. Campos-Costa, J. Roque. Monitoring historical masonry structures with operational modal analysis: two case studies. *Mech Syst Signal Process*,(2010), **24**(5):1291–1305. DOI https://doi.org/10.1016/j.ymssp.2010.01.011

[7] A. Mosallam, T. Zirakian, A. Abdelaal, A. Bayraktar. Health monitoring of a steel moment-resisting frame subjected to seismic loads. *J Constructional Steel Research*,(2018), **140**: 34–46. DOI https://doi.org/10.1016/j.jcsr.2017.10.023

[8] S. N. Savin & I. L. Danilov. *Seismic safety of buildings and territories*. Saint Petersburg: Lan, 2015. ISBN: 978-5-8114-1880-0

[9] S. L. Sushyov, V. A. Larionov et al. On the practical application of the method of dynamic tests for assessing the category of the technical state and seismic stability of buildings and facilities under weak and strong impulse influences. *Seismic-Resistant Construction. Safety of facilities*,(2014), **3**: 52.

[10] S. N. Savin. The use of elastic oscillations of different wavelengths to evaluate the dynamic parameters of buildings and structures and assess the strength of stone walls. *Seismic-Resistant Structures. Construction Safety*,(2017), **4**: 43–55.

[11] S. Savin & V. Tsakalidis. The use of elastic oscillations of different wavelengths to evaluate the dynamic parameters of buildings and structures and assess the strength of materials of the building construction. In *COMPDYN 2015. 5th International Conference on Computational Methods in Structural Dynamics and Earthquake Engineering*, 25–27 May 2015 Crete Island, Greece. M. Papadrakakis, V. Papadopoulos, V. Plevris (eds.). Hersonissos, Crete: National Technical University of Athens, 2015, pp. 706–720. ISBN: 978-960-99994-7-2

[12] S. Savin & E. Smirnova. Evaluation of mechanical safety of building structures using elastic vibrations varying in wave length. *World Appl Sciences J*,(2013), **23**(11): 1448–1454. DOI https://doi.org/10.5829/idosi.wasj.2013.23.11.13161

[13] M. Radzien´ski, M. Krawczuk, M. Palacz. Improvement of damage detection methods based on experimental modal parameters. *Mech Syst Signal Process*,(2011), **25**(6): 2169–2190. DOI https://doi.org/10.1016/j.ymssp.2011.01.007

[14] S. Park, Y. B. Kim, N. Stubbs. Nondestructive damage detection in large structures via vibration monitoring. *Electron J Struct Eng*, (2002), **2**: 359–374.

[15] T. Veijola, G. De Pasquale, A. Somà. Comparison between damping coefficients of measured perforated micromechanical test structures and compact models. In *Proceedings of DTIP Symposium on Design, Test, Integration and Packaging of MEMS and MOEMS*. 9–11 April 2008 Nice, France. V. Bright, T. Bourouina, B. Courtois et al. (eds.). Grenoble: EDA Publishing, 2008, pp. 236–241. DOI https://doi.org/10.1109/DTIP.2008.4752991

[16] Xian-yang Zhao, Nian-song Zhang, Ai-min Wang. Modeling and simulation technology of high frequency vibratory stress relief treatment for complex thin -walled workpiece. In *ICCEMS 2018. The 3rd International Conference on Civil Engineering and Materials Science*. 13–15 April 2018 Chengdu, China (Ser: MATEC Web Conf., vol. 206. 04001). C. W. Lim and Xiaohong Zhu (eds.). Les Ulis: EDP Sciences, 2018. pp. 1–6. DOI https://doi.org/10.1051/matecconf/201820604001

[17] D. J. Inman. *Engineering Vibration*. 4th edn. Upper Saddle River, NJ: Pearson Prentice Hall, 2008, pp. 43–48. ISBN 0-13-228173-2

Marks rating system – an approach in condition assessment and prioritization of sewer rehabilitation – a case study

Esam Hewayde

PhD, P.Eng., Dep. of Civil Engineering, College of Engineering, Al Muthana University. Iraq. e.hewayde@gmail.com

Abstract: Marks rating system is a simple system that can be used to evaluate the condition of sewer pipes and to determine the prioritization for rehabilitation purposes of sewer pipes and sewer infrastructures. The system provides a consistent approach for assessing the impact of pipe failure, coding defects and assigning priorities for rehabilitation of sewer pipes. Each sewer segment is separately evaluated based on several considerations such as inspection results, impact rating factors, frequency of defects, and engineering and administrative factors. The marks of defects were assigned in such a way that the worst condition has the highest mark. The factors considered by applying marks system were identified from questionnaires that were prepared and distributed by the author to technicians, engineers, heads of departments and managers of municipalities of several cities in Iraq. The case study includes seventeen concrete sewer pipes in the municipality of Al Diwaniyah city – Iraq. CCTV inspection was carried out to determine the marks of the structural and performance defects. Results of marking rating system applied on the seventeen sewer pipes showed that MP#12 has the worst condition with highest marks of 38. Sewer pipes MP#1 and MP#7 have the best condition with lowest marks of 18. The prioritization of rehabilitation should have descending ordered from the highest to lowest mark.

Keywords: Sewer Pipes, Rehabilitation, Priorities, Infrastructures.

1. Introduction

Countries spend hundreds billions of dollars annually in building infrastructures [1]. Projects of sewer infrastructures occupy the biggest part of the budget of the infrastructures. Although, the useful life of the sewer systems can be sometimes extended to more than fifty years, the sewer systems need to be rehabilitated and maintained periodically. The failure of the suitable and timing rehabilitation of sewer systems may result in disasters that threat the humans' life. In USA, the annual failures of sewer pipes owing to the lack of suitable and timing rehabilitation cost about 30 billion dollars and death of some people [2]. More information about the impact of the sewer systems on the economy and life can be found elsewhere [3, 1].

The priority of rehabilitation of sewer pipes represents a substantial challenge to decision makers in municipalities especially when the budget of the rehabilitation is limited. It should be stated that there is no engineering principles can be used to determine the priority of rehabilitation [4, 5]. The use of "fix when damage" rule can be very risky when the failure occurs in critical sewer infrastructure that may threat the routine life of people. Many of municipalities rely on the experience of engineers and personnel of the municipality in determining the priority of the rehabilitation [6]. The absence of the priority system of rehabilitation can be attributed to the many factors that can affect the service life

and the rate of degradation of the sewer pipes. These may include but are not limited to the material of the sewer pipes, the type of sewage, the location and depth of sewer pipes and others [5].
It is worth mentioning that researches about the priority of rehabilitation of sewer pipes is very limited, if any, in the open access literature. Some attempts were made to connect the failure of sewer pipes with the functionality of the pipe, burial depth and pipe size [6]. Many researches discussed the relationship between the sewage type and the rate of degradation in concrete sewer pipes [7-10]. Nonetheless, a comprehensive study about the factors that affect the useful life of the sewer pipes was carried out by others [8].

2. Factors affecting priority of rehabilitation

Based on literature review related to failures of sewer pipes, factors affect the situation of sewer pipes and those may determine the priority of rehabilitation were classified into three categories [9]:

2.1 Structural defects

Structural defects in sewer pipes can be the main factors that identify the priority of rehabilitation. These defects may include cracking in the pipes, deflection of pipes, corrosion and others. To determine the extent of these defects in pipes, many techniques of inspection can be used. One of the most common techniques used to investigate the structural defects of sewer pipes is Closed Camera Television (CCTV). Although the CCTV technique is widely used by municipalities to determine the defects, this technique is not very efficient in measuring some defects such as the corrosion extent. It may provide a relative indication of how rapidly corrosion is progressing by comparing conditions in the same segment over time [11-13].
The structural defects that may occur in the sewer pipes can include the following: a) cracking and leakage, b) deflection, c) missing bricks for brick sewer pipes, d) joint opening, and e) corrosion. It should be noted that joint opening and corrosion defects are only applied to concrete sewer pipes while brick missing is applied to brick sewer pipes.

2.2 Performance defects

Performance defects may include infiltration of pipes, root intrusion and the existing of debris. These performance defects can affect the service life of the sewer pipes and accelerate the failure. The CCTV technique is efficient in determining the debris and root intrusion while it is not very efficient in determining the infiltration of pipes.

2.3 Impact factors

Impact factors refer to those factors that are related to the installation of pipe and its function. The impact factors may include a) location of pipe, b) burial depth, c) pipe size, and d) functionality. The location factor can be subdivided into land use, traffic intensity, access for repair, and environmental effect.

3. Marks rating system

The marks rating system applied for each type of defects were mainly obtained from the art-of-state especially research performed by Zhao and McDonald [10-15]. The author also prepared a questionnaire that was distributed to technicians, engineers, heads of sewer departments and mangers of municipalities. The questionnaire is simply including questions about the probable defects and about marking range of each possible factor that may affect the prioritization of rehabilitation of sewer pipes. The details and the statistical analysis of the questionnaire can be found elsewhere [16]. However, results of the questionnaire were gathered with those obtained from the literature review to identify the marking rating of each factor:

3.1 Marking rating system of structural defects

Table 1 shows the marking rating system of structural defects that may affect the prioritization of sewer pipe rehabilitation. As previously discussed, the inspection technique of CCTV is not very efficient to determine the extent of some defects. Thus, it was proposed to multiply the marks of each defect by an Inspection Method Factor (IMF) which ranges from 1.0 to 1.2. The IMF of 1.0 should be used when the results of inspection technique is perfect while IMF of 1.2 should be used when the inspection results are not clear enough to accurately evaluate the defects.

Table 1. Marking rating system of structural defects

Defect	Description
Corrosion	• 10 marks for minor corrosion (less than 10% of concrete thickness) • 10 – 50 marks for moderate corrosion (steel reinforcement of pipe is not visible) • 50 – 100 marks for severe corrosion (steel reinforcement of pipe is visible)
Joint Opening (JO)	• 10 marks for JO up to 10 mm • 10 – 50 marks for JO between 10 – 50 mm • 70 marks for JO more than 50 mm • 10 marks for gasket off • 10 marks for leakage related to JO
Cracks and leakage	• 5 marks for each crack (longitudinal, diagonal or circumferential) • 15 marks for each leakage related to cracks
Deflection	• 10 marks for change in diameter up to 5% • 10 – 20 marks for change in diameter between 5% and 10% • 20 – 50 marks for change in diameter between 10% and 25%

3.2. Marking rating system of performance defects

The performance defects included in this study are debris and root intrusion and infiltration. As it can be seen in Table 2, both defects were ranked from 0 to 100. The IMF can also be applied for marks of performance defects because it is difficult to measure the reduction in diameter of pipes owing to stagnation of debris and root intrusion.

Table 2. Marking rating system of performance defects

Defect	Description
Debris and root intrusion	• 0-20 marks for reduction in dia. less than 10% • 20-50 marks for reduction in dia. between 10% and 25% • 50-100 marks for reduction in dia. more than 25%
Infiltration	• 0-20 marks for seeping and dripping • 20-50 marks for running and trickling • 50-100 marks for gushing and spurting

3.3. Marking rating system of impact factors

Table 3 depicts the marking assigned to each of the four impact factors. The marks of each factor were ranged from 0 to 100. The IMF is not applied to impact factors as information related to these factors can be accurately obtained from the municipality office. Because the location factor includes many sub-factors including traffic intensity, land use and others, it was proposed that the location factor to have its own marking rating system which also ranged from 0 to 100 as shown in Table 4.

Table 3: 100 Marks rating system of impact factor

Factor	Description
Burial Depth	• 10-20 marks for depths less than 3 m • 20-50 marks for depths between 3 and 10 m • 50-100 marks for depths more than 10 m
Pipe Diameter	• 10-20 marks for dia. less than 900 mm • 20-50 marks for dia. between 900 and 1800 mm • 50-100 marks for dia. more than 1800 mm
Functionality	• 10-30 marks for collectors sewer and storm sewer • 30-60 marks for major trunk, sanitary and combined • 60-100 marks for major interceptor and from to wastewater plant

Table 4: 100 Marks rating system of location factor

Factor	Description
Land Use	• 30 marks for industrial use • 60 marks for residential use • 100 marks for commercial use
Traffic Intensity	• 0-30 marks for 1 to 2 traffic lanes (15 marks for each lane) • 30-75 marks for 3 to 5 traffic lanes • 75-100 marks for 6 traffic lanes or more
Under or adjust to	• 0-30 marks area clear of low risk utilities • 30-60 marks area of high risk utilities • 60-100 marks area of railway, canals, building, rivers and etc.
Access for repair	• 30 marks for unrestricted access • 60 marks for limited access • 100 marks for restricted access
Environment	• 0-30 marks for non-sensitive area • 30-60 marks for environmental conservation zones • 60-100 marks for environmental protected zones

It should be stated that the weight of impact factors are different depending on the case of rehabilitation. Based on results of the questionnaire, the weight of each factor was assigned as follows: Weight of burial depth is 28%; Weight of location is 30%; Weight of pipe size 25%; and Weight of functionality is 17%. These weights were identified based on the experience of the participants in the questionnaire.

3.4. Engineering and administrative factors

It can be noticed that some engineering factors were not included in the assessment of sewer. These may include; historical data of sewer, frequency of defects, progress of the problem and previous inspection. In addition, the experience of the municipal engineer and staff can be an important factor in identifying the priority of rehabilitation. Hence, an Engineering Factor (EF) was proposed to be used to correlate the final marks of sewer. The EF ranges and from 0.8 to 1.2 and can be applied on all or some of the sewer depending on the available data.

The future plans of expanding and the point of view of the decision-maker should also be considered in the assessment and prioritization of the sewer rehabilitation. Hence, an Administrative Factor (AF) can be applied to correct the final marks of sewer. The AF ranges from 0.8 to 1.2 and can be applied on some or all sewer.

4. Case study

Al Diwaniyah city in governorate of Al Qadisiyah – Iraq was considered as a case study for the purpose of condition assessment and prioritization of rehabilitation of sewer pipes. Al Diwaniyah city is located about 180 km south of Baghdad and it is the capital city of Al Qadisiyah governorate. The population of the city is about 250,000. The city is not entirely served by a sewer system. The investigated area is located south-east of the city and include five sectors with a population of approximately 100,000. The investigated area is classified as a residential area with some commercial areas. Figure 1 shows the location of Al Diwaniyah city and the location of the investigated area.

The sewer system of the investigated area is consisted of force main and gravity concrete pipes of different diameters, uplift stations and two sewage treatment plants. The sewer system was installed in early eighties of the last century. The entire sewer system is currently suffering from repeated blockage and major structural and performance defects. The city spent a lot of money in rehabilitation of the sewer system. These include replacing of some old and broken pipes, in addition to the replacement of mechanical instruments of the lift stations and treatment plants. Because the limited earmarked budget of the maintenance and rehabilitation, the municipality is following the rule of "fix when damage".

In 2014, the municipality decided to assess the sewer system using CCTV technique. The assessment included seventeen concrete pipes of various sizes. The main purpose of the assessment was to identify the breakage of pipes and investigate the main reasons of the repeated blockage in the sewer system. The movies of the CCTV were analyzed by the author with the assistance of the municipal engineers to determine the marks of the structural and performance defects. Information related to impact factors was provided by the municipality office while the engineering and administrative factors were determined by the former engineer and the manager of the municipality, respectively.

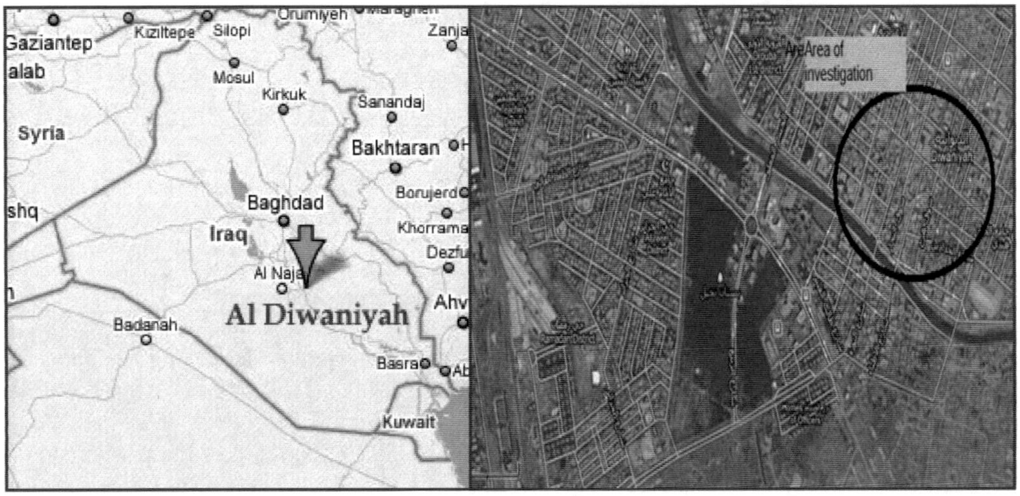

Figure 1. Locations of Al Diwaniyah city and investigated area

4.1 Structural defects

Based on results of the CCTV techniques, marks of the structural defects were determined as shown in Table 5. It should be stated that corrosion of concrete pipes was occurred in the invert of pipes.

Table 5. Marks of structural defects of the seventeen sewer pipes

Pipe ID	Corrosion		Joint Opening		Cracks and Leakage		Deflection		Structural Rating Marks
	Marks	IMF	Marks	IMF	Marks	IMF	Marks	IMF	
MP #1	10	1.1	0.0	1.0	10	1.0	10	1.1	8
MP #2	15	1.1	0.0	1.0	20	1.0	10	1.1	12
MP #3	40	1.1	15	1.0	35	1.0	10	1.1	26
MP #4	30	1.1	10	1.0	15	1.0	10	1.1	17
MP #5	30	1.1	10	1.0	30	1.0	10	1.1	21
MP #6	20	1.1	0.0	1.0	15	1.0	10	1.1	12
MP #7	15	1.1	0.0	1.0	15	1.0	10	1.1	11
MP #8	45	1.1	20	1.0	25	1.0	15	1.1	28
MP #9	25	1.1	15	1.0	25	1.0	10	1.1	20
MP #10	45	1.1	20	1.0	25	1.0	15	1.1	28
MP #11	45	1.1	20	1.0	30	1.0	15	1.1	29
MP #12	60	1.1	25	1.0	60	1.0	15	1.1	42
MP #13	60	1.1	25	1.0	35	1.0	15	1.1	36
MP #14	60	1.1	20	1.0	35	1.0	15	1.1	34
MP #15	40	1.1	20	1.0	30	1.0	15	1.1	28
MP #16	45	1.1	15	1.0	25	1.0	15	1.1	27
MP #17	60	1.1	20	1.0	30	1.0	15	1.1	33

Structural Rating Mark = Σ(defect mark x IMF)/Number of defects

4.2 Performance defects

Marks assigned to structural defects of the seventeen concrete pipes were relied on results of CCTV inspection. Table 6 shows the final marks of performance defects of the inspected pipes.

Table 6. Marks of structural defects of the seventeen concrete sewer pipes

Pipe ID	Debris and Root Intrusion		Infiltration		Performance Rating Marks
	Mark	IMF	Mark	IMF	
MP #1	15	1.1	25	1.0	21
MP #2	15	1.1	30	1.0	23
MP #3	20	1.1	30	1.0	26
MP #4	20	1.1	40	1.0	31
MP #5	20	1.1	45	1.0	34
MP #6	30	1.1	20	1.0	27
MP #7	20	1.1	10	1.0	16
MP #8	30	1.1	30	1.0	32
MP #9	25	1.1	30	1.0	29
MP #10	25	1.1	20	1.0	24
MP #11	30	1.1	45	1.0	39
MP #12	30	1.1	20	1.0	27
MP #13	25	1.1	20	1.0	24
MP #14	30	1.1	20	1.0	27
MP #15	30	1.1	20	1.0	27
MP #16	25	1.1	15	1.0	21
MP #17	25	1.1	20	1.0	24

Performance rating marks =Σ(defect mark x IMF)/number of defects

4.3 Impact factors

Prior to assigning the marks of impact factors, marks of location factors including land use, traffic intensity, location (under or adjust to), access for repair, and environment were determined. No correction factor was applied on the location factor because this data was accurately obtained from the municipality. Table 7 depicts marks of location factors assigned to the seventeen sewer pipes.

Table 7. Marks of location factors of the seventeen concrete sewer pipes

Pipe ID	Land Use	Traffic Intensity	Location (adjust to or under)	Access for repair	Environment	Location Factor Mark
MP #1	60	15	30	30	10	29
MP #2	60	15	30	60	10	35
MP #3	60	30	60	60	10	44
MP #4	60	30	60	60	10	44
MP #5	60	15	30	30	10	29
MP #6	60	15	30	30	20	31
MP #7	100	30	60	60	30	56
MP #8	100	30	30	60	30	50
MP #9	60	15	30	60	30	39
MP #10	60	15	30	30	30	33
MP #11	60	15	30	30	10	29
MP #12	60	30	60	60	10	44
MP #13	100	30	100	60	10	60
MP #14	60	30	60	30	10	38
MP #15	60	15	30	30	10	29
MP #16	60	15	30	30	10	29
MP #17	100	30	30	60	30	50

IOP Conf. Series: Materials Science and Engineering **652** (2019) 012011 doi:10.1088/1757-899X/652/1/012011

Subsequently, marks of impact factors were assigned based on information provided by the municipality. Table 8 shows the marks of the impact factors accompanied by the weighting factor (WF) of each impact factor.

Table 8. Marks of impact factors of the seventeen concrete sewer pipes

Pipe ID	Location		Burial Depth		Pipe Size		Functionality		Impact Rating Mark
	Mark	WF	Mark	WF	Mark	WF	Mark	WF	
MP #1	29	0.30	10	0.28	15	0.25	30	0.17	20
MP #2	35	0.30	10	0.28	15	0.25	30	0.17	22
MP #3	44	0.30	15	0.28	20	0.25	30	0.17	28
MP #4	44	0.30	15	0.28	20	0.25	30	0.17	28
MP #5	29	0.30	20	0.28	15	0.25	30	0.17	23
MP #6	31	0.30	20	0.28	20	0.25	30	0.17	25
MP #7	56	0.30	15	0.28	25	0.25	40	0.17	34
MP #8	50	0.30	15	0.28	20	0.25	30	0.17	29
MP #9	39	0.30	20	0.28	15	0.25	30	0.17	26
MP #10	33	0.30	10	0.28	20	0.25	30	0.17	23
MP #11	29	0.30	15	0.28	15	0.25	30	0.17	22
MP #12	44	0.30	15	0.28	15	0.25	30	0.17	26
MP #13	60	0.30	20	0.28	25	0.25	40	0.17	37
MP #14	38	0.30	20	0.28	25	0.25	30	0.17	28
MP #15	29	0.30	20	0.28	15	0.25	30	0.17	23
MP #16	29	0.30	20	0.28	15	0.25	40	0.17	25
MP #17	50	0.30	20	0.28	20	0.25	40	0.17	32

Impact rating mark = Σ(impact factor x WF)

4.4 Final condition assessment

The final condition assessment of prioritization of rehabilitation was determined based on the average marks of structural defects, performance defects and marks of impact factors. He average rating marks of the sewer pipes were corrected using the engineering factor (EF) and administrative factor (AF) which were determined by the former engineer and the manager of the municipality, respectively. Table 9 depicts the final marks of the seventeen sewer pipes.

Table 9. Final rating marks of the seventeen concrete sewer pipes

Pipe ID	Structural Rating Marks	Performance Rating Marks	Impact Factor Marks	EF	AF	Final Rating Marks
MP #1	8	21	20	1.1	1.0	18
MP #2	12	23	22	1.0	1.2	23
MP #3	26	26	28	1.0	1.0	27
MP #4	17	31	28	1.0	1.0	25
MP #5	21	34	23	1.2	1.0	31
MP #6	12	27	25	1.0	0.9	19
MP #7	11	16	34	1.0	0.9	18
MP #8	28	32	29	1.0	1.0	30
MP #9	20	29	26	1.2	1.0	30
MP #10	28	24	23	1.0	1.0	25
MP #11	29	39	22	1.0	1.0	30
MP #12	42	27	26	1.2	1.0	38
MP #13	36	24	37	1.0	1.0	32
MP #14	34	27	28	1.0	1.0	30
MP #15	28	27	23	1.0	1.0	26
MP #16	27	21	25	1.0	1.0	24
MP #17	33	24	32	1.1	1.1	36

Final Rating Marks = (Structural Rating +Performance Rating +Impact Rating) x EF x AF/3

As it can be seen in Table 9, the sewer pipe (MP#12) has the highest final rating marks of 38 while sewer pipes MP#1 and MP#7 have the lowest final rating marks of 18. This means that the prioritization of rehabilitation should be given to sewer pipe MP#12 because it has the worst condition while sewer pipes MP#1 and MP#12 have the best condition and should occupy the end of the prioritization list.

5. Conclusions

In this study, factors that may affect the condition of the sewer system were determined based on information available in the open access literature and questionnaires prepared by the author. Factors were divided into three main categories including structural defects, performance defects and impact factors. Marking system was assigned on the factors in such a way that the sewer with the worst condition has the highest marks. The marking system was applied on seventeen concrete sewer pipes in Al Diwaniyah city – Iraq as a case study. The marks of the structural and performance defects were selected based on results of CCTV inspection while those of the impact factors were determined based on information provided by the municipality of Al Diwaniyah city. The final rating marks of the seventeen concrete sewer pipes ranged from 18 to 38. The highest mark refers to the worst condition while the lowest mark refers to the best condition of sewer pipes. Accordingly, sewer pipe MP#12 comes first while sewer pipes MP#1 and MP#7 come the last in the prioritization of rehabilitation of the seventeen concrete sewer pipes.

References

[1] Hewayde E 2005 "Investigation on Degradation of Concrete Sewer Pipes by Sulfuric Acid Attack", *PhD thesis*, The University of Western Ontario, London, Ontario, Canada.

[2] Hewayde E, Allouche E and Nakhla G 2003 "Experimental Investigations of the Effect of Selected Admixtures on the Resistance of Concrete to Sulfuric Acid Attack", *Proc. of the ASCE Int. Con. on Pipeline Engineering and Construction*: New Pipeline Technologies, Security, and Safety, pp. 504-513, Baltimore, USA.

[3] Lui Z and Kleiner Y 2013 "State of the Art Review of Inspection Technologies for Condition Assessment of Water Pipes", *J. of Measurements*, **45**, 1, pp. 1-15

[4] EPA 1985 "Odor and Corrosion Control in Sanitary Sewage System and Treatment Plants", *EPA/625/1-825/018*, Oct.(1985)

[5] Galeziewski T, Edmondson S and Webb R 1995 "Condition assessment and rehabilitation program for large diameter sewers in Phoenix, Arizona", *Advances in Underground Pipeline Engineering II*, Bellevue, Washington, June 25-28, pp. 560 – 571

[6] ASCE 1994 "Existing Sewer Evaluation and Rehabilitation", *ASCE Manuals and Reports on Engineering Practice No. 62*. American Society of Civil Engineers

[7] EPA 1992 "Detection, Control, and Corrosion of Hydrogen Sulfide Corrosion", *EPA/1992-312-014/40169*, Sep.

[8] Daczko A, Johnson S and Amey S 1997 "Decreasing Concrete Sewer Pipes Degradation Using Admixtures", *Mat. and Perf. J.*, Jan. pp. 51-56

[9] Kampfer W and Berndt M 1999 "Estimation of Service life of Concrete Pipes in Sewer Network", *Conf. of Durability of Building Materials and Components*. National Research Council of Canada, **1**. pp. 36-45

[10] Shock W and Bell L 2003 "Corrosion Control in Concrete Pipe and Manhole", *Technical Presentation, Water Environmental Federation*, Orlando, Florida, USA.

[11] Zhao J 1998 "Trunk Sewers in Canada", *APWA Int'l Public Works Congress Seminar Series*, American Public Works Association. Las Vegas, Sept. 14-17.

[12] Hewayde E, Nakhla G, Allouche E and Mohan P 2007 "Beneficial impact of coatings on biological generation of sulfide in concrete sewer pipes", *J. of Structure and Infrastructure Engineering*, **3** (3) pp. 267-277

[13] McDonald Sand Zhao J 2001 "Condition Assessment and Rehabilitation of Large Sewer", *Institute for Research Canada*

[14] City of Edmonton 1996 "Standard Sewer Condition Rating System Report", *City of Edmonton Transportation Department,* Edmonton, AB., Canada.
[15] Hewayde E 2014 "Results of CCTV Inspection in Diwaniyah City", *Special Report submitted to Al Diwaniyah City*, Al Diwaniyah, Al Qadisiyah, Iraq
[16] Fenner R 2000 "Approaches to sewer maintenance: a review", *J. Urban Water* **2**, pp. 343-356

Critical Criteria for Customer Satisfaction of Interior Design

Pen-Kai Wang[1*] and Yeng-Horng Perng[1]

[1]Department of Architecture, National Taiwan University of Science and Technology, 6F, No. 40-1, Alley 57, Lane 12, Sec. 3, Pa Te Rd, Songshan Dist., Taipei Taiwan

*Email: vincentw.design@gmail.com

Abstract. Because of the wide scope of interior design work, the satisfaction of interior design service customers is easily affected by perception differences. This study developed a set of indicators that can be used to evaluate interior-design-related customer satisfaction. Potential factors were first identified by conducting a literature review and in-depth interviews. The modified Delphi method was then employed to establish the analytic hierarchy process (AHP) framework. Lastly, the AHP method was used to identify the key indicators. The study results revealed 4 dimensions and 15 factors. The results of this study can serve as a reference for the interior design industry.

1. Introduction

Because of the increasing emphasis people place on their living environment, the interior design industry has become an important economic domain. This trend can be observed through the increased rate of housing improvement activities compared with overall economic activities [4]. Despite this change, customer perception problems caused by the diversity of interior design have not received the attention they deserve. In interior design, customer satisfaction is affected by various factors, including economic advantage, reputation, and environmental protection awareness (e.g., energy conservation and carbon emissions reduction). The level of customer satisfaction thus depends on not only whether customers are happy with the quality of design output but also subjective factors such as perception of the overall service process and experience. Therefore, identifying the factors affecting customer satisfaction in the diverse sector of interior design has become a critical issue. This study aimed to use both qualitative and quantitative research methods to develop a set of indicators that can be used to evaluate interior-design-related customer satisfaction. First, potential factors were identified by conducting a literature review and in-depth interviews. The modified Delphi method was then used to establish the analytic hierarchy process (AHP) framework. Lastly, the AHP method was used to identify the key indicators affecting interior-design-related customer satisfaction.

The research objectives of this study were thus as follows:

1. To assemble a set of interior design satisfaction indicators by considering the practical experiences , diversity, and culture and folk beliefs of interior design.
2. To combine the modified Delphi method and AHP and subsequently examine the appropriateness of this combination.
3. To obtain the importance rankings of interior design satisfaction indicators from practical and academic experts, the results of which will provide a reference for operations and management in the interior design industry.

2. Literature review

2.1 Scope of interior design services

The work of an interior designer involves using the budget effectively and schedule control (economic advantages dimension), as well as ensuring regulation compliance and giving impressive presentations (designer characteristics dimension) [1]. With reference to Maslow's hierarchy of needs theory, in addition to fulfilling customers' basic needs such as physiological and safety requirements, interior designers must consider how they can fulfill high-level needs such as social, esteem, and self-actualization needs. Many customers consider these high-level aspects to be crucial and adding high value, because interior design can express their identity and confer a status symbol.

Additionally, awareness of energy conservation and carbon emissions reduction and ensuring construction materials are environmentally have become social responsibilities of concern to modern residents [5]. Being environmentally conscious is thus now a crucial part of the modern interior designers' job and means using green materials in indoor spaces and designing or adopting energy-saving, water-saving, and electricity-saving facilities in interior design.

2.2 Customer satisfaction

Holbrook [2] divided Customer Satisfaction into four components, namely interactivity, relativity, preference, and experience. From the perspective of interior design, the factors related to customers' satisfaction include Customer participating experience, low operation cost, easy maintenance, optimal solution for current housing conditions , service accessibility, good customer–designer communication, and Designer Reputation & word of mouth. On the basis of the scope of work for interior design services and the literature related to customer satisfaction, 4 dimensions and 17 factors related to customer satisfaction with interior design were identified:

1.Value Creation: Social Status Symbol,Customer participating experience,Identity representation, optimal solution for current housing conditions.

2. Economic Advantages: Budget Effectiveness,Schedule Control,easy maintenance, Low Operation Cost.

3. Designer Characteristics: Designer Reputation & Word of Mouth, Service Accessibility, Impressive Presentation, Regulation Compliance, good customer–designer communication.

4. Environmental Conscious: Green Material, Energy Saving, Water Saving, Electricity Saving.

3. Methodology

By using both qualitative and quantitative research methods, this study developed a set of indicators for evaluating interior-design-related customer satisfaction. First, the potential factors were identified by conducting a literature review and in-depth interviews. The modified Delphi method was then employed to establish the AHP framework. Lastly, the AHP method was used to identify the key indicators affecting interior-design-related customer satisfaction.

3.1 Examining the dimensions and factors of customer satisfaction using in-depth interviews

Interviews were conducted with (1) two interior design company owners who had more than 10 years of operating experience (2) two legal professionals who had more than 5 years of experience in providing expert opinion in interior design dispute resolution; and (3) two customers who had previously received interior design services. To ensure that the interview content was realistic, the interviews were guided by the 4 dimensions and 17 factors identified through the literature review. During the interviews, three interviewees mentioned that depending on the culture and folk beliefs of where they live, most people consider certain types of living arrangement taboo. Interior design must thus have some connection with the service values of this domain. In response, culture and folk beliefs was included as a factor. Additionally, the interviewees mentioned that the good customer–designer communication factor and service accessibility factor are excessively homogeneous. Therefore, the good customer–designer communication factor was deleted.

3.2 Establishing the AHP framework through the Modified Delphi Method

The following 5-point Likert scale was employed to score the expert questionnaire: 5very important, 4important, 3moderately important, 2unimportant, and 1very unimportant. The target respondents of the questionnaire were interior designers, building material merchants, construction workers, professors, and customers. A total of 45 questionnaires were distributed, and 35 were successfully retrieved, a return rate of 78%. Among the returned questionnaires, 8 were invalid, whereas 27 were valid. The evaluation criteria in the present study were mean > 3 and coefficient of variation ≤ 0.5 [3]. Additionally, the criterion was used that the standard deviation and absolute values should be ≤1. According to analysis of the returned expert questionnaires, all dimensions fulfilled the evaluation criteria; their standard deviation and absolute values were <1, mean values were >3, and coefficients of variation were <0.5. Such results indicated that the experts were relatively consistent in their identification of the indicators in the 4 dimensions (Table 1).

Table 1. Evaluation of the dimensions using the Modified Delphi Method.

Dimension	Standard deviation	Mean	Coefficient of variation	Absolute value	Evaluation results
Value creation	0.565	4.2	0.136	0.833	○
Economic advantages	0.506	4.1	0.124	0.9	○
Designer characteristics	0.620	3.9	0.159	0.1	○
Environmental consciousness	0.679	3.6	0.189	0.4	○

However, in the sub-dimensional factor layer, the mean values of optimal solution for current housing conditions and electricity saving were < 3. These two factors thus did not meet the evaluation criteria and were eliminated (Table 2).

Table 2. Evaluation of the factors using the Modified Delphi Method.

Factor	Standard deviation	Mean	Coefficient of variation	Absolute value	Evaluation result
Social status symbol	0.643	4.1	0.158	0.933	○
Customer participating experience	0.8	3.7	0.216	0.3	○
Identity representation	0.718	3.7	0.192	0.267	○
culture and folk beliefs	0.877	3.6	0.244	0.4	○
optimal solution for current housing conditions	0.859	2	0.422	0.033	×
Budget effectiveness	0.447	4.3	0.105	0.733	○
Schedule control	0.577	4.1	0.141	0.9	○
Easy maintenance	0.58	4.1	0.143	0.933	○
Low operation cost	0.675	3.5	0.191	0.467	○
Designer Reputation & Word of mouth	0.465	4.2	0.11	0.767	○
Service accessibility	0.518	3.6	0.145	0.433	○
Impressive presentation	0.506	3.4	0.149	0.6	○
Regulation compliance	0.62	3.3	0.188	0.7	○
Green materials	0.641	3.8	0.169	0.2	○
Energy saving	0.48	4.2	0.114	0.8	○
Water saving	0.58	4	0.144	0.967	○
Electricity Saving	0.77	2.6	0.3	0.433	×

3.3 Using the AHP to identify evaluation indicators

Next, the AHP questionnaire was used to survey experts. The target respondents of the questionnaire were interior designers and customers. A total of 80 questionnaires were distributed, and 61 of them were returned, a return rate of 76%. Among the returned questionnaires, 9 were invalid, whereas 52 were valid.

The local weight of each indicator was obtained using the AHP matrix calculation process. As indicated in Table 3, the rankings of the indicators according to their importance (weight) are Economic Advantages 0.397, Designer Characteristics 0.223, Value Creation 0.192, Environmental Consciousness 0.188. This ranking indicated that U in the dimension layer, the most influential item is economic advantages.

Table 3. Weight analysis of the interior-design-related customer satisfaction dimensions.

Dimension	Local weight	Ranking	λmax	C.I.	R.I.	C.R.
Value creation	0.192	3				
Economic advantages	0.397	1				
Designer characteristics	0.223	2	4.0513	0.0171	0.58	0.0295
Environmental consciousness	0.188	4				

Based on the local weight analysis of the value creation dimension, the rankings of the factors are Social Status Symbol 0.391, Customer participating experience 0.229, culture and folk beliefs0.21, Identity representation 0.17 (Table 4). Among the factors in this dimension, social status symbol appears to be the most influential.

Table 4. Weight analysis of the factors in the value creation dimension.

Factor	Local weight	Ranking	λmax	C.I.	R.I.	C.R.
Social status symbol	0.391	1				
Customer participating experience	0.229	2				
Identity representation	0.17	4	4.1022	0.0341	0.58	0.0587
culture and folk beliefs	0.21	3				

Based on the local weight analysis of the economic advantages dimension, the rankings of the factors are Budget Effectiveness 0.391, Schedule Control 0.229, Low Operation Cost 0.21, Easy maintenance 0.17 (Table 5). Budget effectiveness thus appears to be the most influential factor.

Table 5. Weight analysis of the factors in the economic advantages dimension.

Factor	Local weight	Ranking	λmax	C.I.	R.I.	C.R.
Budget effectiveness	0.391	1				
Schedule control	0.229	2				
Easy maintenance	0.17	4	4.1116	0.0372	0.58	0.0641
Low Operation Cost	0.21	3				

The local weight analysis of the designer characteristics dimension reveals that the rankings of the factors are Reputation & Designer Reputation & Word of mouth 0.411, Service Accessibility 0.238, Impressive Presentation 0.184, Regulation Compliance 0.167 (Table 6).

In the designer characteristics dimension, Designer Reputation & word of mouth thus appears to be the most influential factor.

Table 6. Weight analysis of the factors in the designer characteristics dimension.

Factor	Local weight	Ranking	λmax	C.I.	R.I.	C.R.
Designer Reputation & Word of mouth	0.411	1	4.1203	0.0401	0.58	0.0692
Service accessibility	0.238	2				
Impressive presentation	0.184	3				
Regulation compliance	0.167	4				

Based on the local weight analysis of the environmentally conscious dimension, the rankings of the factors are Green Material 0.445, Energy Saving 0.3, Water Saving 0.255 (Table 7). Among the factors in this dimension, green materials appear to be the most influential factor.

Table 7. Weight analysis of the factors in the environmentally conscious dimension.

Factor	Local weight	Ranking	λmax	C.I.	R.I.	C.R.
Green materials	0.445	1	3.0326	0.0163	0.58	0.0281
Energy saving	0.3	2				
Water saving	0.255	3				

Regarding the consistency of the expert questionnaire, the results of reliability analysis revealed that the consistency index and ratio were both <0.1. Thus, consistency exists between same-level questions in the expert questionnaire. This also indicates that the questionnaire has favourable reliability [6].

4. Results and analysis

The results obtained from the weight analyses for each level are summarized in Table 8. Using these results, the overall weight of each factor was calculated for further analysis.

Table 8. Summary of the factors affecting interior-design-related customer satisfaction.

Layered framework	Dimension layer	Local weight	Factor layer	Local weight	Overall weight	Ranking
Critical Criteria for Customer Satisfaction of Interior Design	Value creation	0.192	Social status symbol	0.391	0.075	6
			Customer participating experience	0.229	0.044	11
			Identity representation	0.17	0.033	15
			culture and folk beliefs	0.21	0.04	13
	Economic advantages	0.397	Budget effectiveness	0.426	0.169	1
			Schedule control	0.195	0.078	5
			Easy maintenance	0.246	0.098	2
			Low Operation Cost	0.133	0.0527	9
	Designer characteristics	0.223	Designer Reputation & Word of mouth	0.411	0.092	3

		Service accessibility	0.238	0.0529	8
		Impressive presentation	0.184	0.041	12
		Regulation compliance	0.167	0.037	14
Environmental consciousness	0.188	Green materials	0.445	0.084	4
		Energy saving	0.3	0.056	7
		Water saving	0.255	0.048	10

The data in Table 8 indicate that in the experts' opinion, four factors most crucially affect interior-design-related customer satisfaction: budget effectiveness, easy maintenance, Designer Reputation & word of mouth, and green materials. These results reveal that the element that customers are most concerned with when using an interior design service is budget. Effectively, customers wish to get more than they pay for. In addition, customers also wish that their interiors will be easy to maintain. As such, the easy maintenance factor is the second most crucial factor affecting customer satisfaction. If a designer were to focus solely on budget effectiveness, concerns might arise regarding the quality of the design. Customers thus consider Designer Reputation & Word of mouth (which is related to quality assurance) after considering budget effectiveness. That Designer Reputation & word of mouth is the third most crucial factor affecting customer satisfaction indicates that customers wish to employ designers with favorable business reputation. Lastly, the rise in environmental consciousness has caused green materials in interior design to be ranked the fourth most crucial factor affecting interior-design-related customer satisfaction.

5. Conclusion

To assemble a set of interior-design-related satisfaction indicators that reflect practical experiences , diversity,. and culture and folk beliefs, this study first identified 4 potential dimensions and 17 potential factors by conducting a literature review and in-depth interviews with experts. The Modified Delphi method was then employed to establish the AHP framework. Lastly, AHP was used to identify 4 dimensions and 15 factors that affect interior-design-related customer satisfaction. The study results indicate such combined application of the Modified Delphi method and AHP was practical, and the evaluation of the factors was highly relevant and objective. Therefore, this approach was deemed adequate. Of the dimensions identified, the economic advantages dimension has the highest weight of 0.397. Regarding the indicator factors, budget effectiveness (from the economic advantages dimension) ranks first among all factors, with an overall weight of 0.169. These results indicate that the element customers are most concerned with when using an interior design service is budget, followed by ease of maintaining the design (also related to economic advantages). After considering the economic advantages, customers next consider Designer Reputation & Word of mouth (related to quality assurance), followed by the greenness of materials (related to environmental protection compliance consideration).

This study discovered four dimensions as follows, in order of their importance: Economic Advantages, Designer Characteristics,Value Creation,Environmental Consciousness. Additionally, the 15 relevant factors identified were as follows, in order of importance: Budget Effectiveness,Easy maintenance, Designer Reputation & Word of mouth ,Green Material,Schedule Control,Social Status Symbol, Energy Saving,Service Accessibility, Low Operation Cost,Water Saving,Customer participating experience,Impressive Presentation,culture and folk beliefs,Regulation Compliance,Identity representation. The results of this study can help interior designers understand the needs of customers and demands of the market. Furthermore, the results can serve as a reference when developing strategies for increasing the satisfaction of interior design customers.

References

[1] CIDA 2019 https://accredit-id.org/
[2] Holbrook M and Morris B 1999 *Consumer Value: A Framework for Analysis and Research* (New York: Routledge)
[3] Hsu PF and Chen BY 2007 *Qual. Quant.* 41(2) 275-290
[4] Joint Center for Housing Studies of Harvard University 2019 *Improving America's Housing* (Boston: Harvard Design School).
[5] O'Neill N 2011 *J. Coll. Character* 12(2)1940-163
[6] Saaty TL 1980 *The Analytic Hierarchy Process* (New York: McGraw-Hill)

Scour and potential earthquake damage on bridges

D F Vidriales[1,2], R Gómez[1,*], and J A Escobar[1]

[1]Institute of Engineering, National Autonomous University of Mexico, Structural Engineering Department, México CDMX, México
[2]Section of Postgraduate Studies and Research ESIA-UZ, Zacatenco CDMX, México
*Corresponding author's email: RGomezM@iingen.unam.mx

Abstract. The vast majority of total and partial bridge collapses in México are due to scour damage. Bridges are not repaired promptly after a hurricane so the bridge is exposed to both, scour and earthquake hazard. This paper presents the results of a research project oriented to study the vulnerability-based hazard risk assessment of highway bridges subjected to extreme rainfalls due to hurricane events and its effect on bridge lateral strength. Scour hazard is computed using robust statistical procedures to minimise the intrinsic uncertainty of the extreme rainfalls. Earthquake hazard is computed accordingly to Mexican codes and well-known methodologies. A case study is used to present the methodology used in the research project, which was funded by the National Center for Disaster Prevention CENAPRED through the Institute of Engineering of the National Autonomous University of Mexico.
Keywords: Scour, bridge, earthquake, damage

1. Introduction

Scour damage is the most frequent cause of bridge collapses in México, an average of 2.5 bridges per year [1] ;in the last 20 years a total of 36 bridges suffered total or partial collapse during extreme rainfalls, and 99 more were damaged [2].

In the Mexican Pacific Coast, a survey of the most damaged bridges due to scour has shown that these structures are located in seismic prone regions [3] .In this paper, simplified methods are used to assess the bridge vulnerability under scour and seismic hazards.

When an extreme scour develops in a bridge, several failure stages must be addressed, such as local failure due to inelastic buckling of piles [4], failure due to high overturning moments, and formation of plastic hinges.

Because data analysis played a key role in the estimation of the design flow [5], the first part of this paper focuses on uncertainty analyses and the characterization of the extreme rainfall hazard. The second part is dedicated to present well-known simplified calculations of scour depth, using Yaroslavtziev [6] and HEC 18 [7] methods, for an idealized hydraulic section of a specific bridge.

The potential earthquake damage on a case study bridge foundation is addressed using Song and Chun [8] methodology, and is presented in the third part of this paper. Finally, conclusions are described, and recommendations are given.

2. Uncertainty Analysis

2.1 Data analysis

In order to derive mathematical models for the prediction of return periods for different intensities of extreme rainfall events, meteorological and hydrological information was collected including up to

fifty years of maximum water flow in river crossings. This was performed using data bases of BANDAS system [9].
Frequently, when dealing with data bases the problem of missing data needs to be addressed. A data augmentation toolbox for MATLAB [5], was used to estimate a complete data set and minimise the possible errors. After dealing with the missing data, different probability distributions were fitted to the estimated complete data set and tested for goodness of fit using Kite's test and probability plots, and a best fit distribution was selected.

2.2 Frequency analysis
Once the best fitted distribution was chosen, a frequency analysis was carried out considering the following premises:

1. Extreme events arise from serially independent time series with a known underlying probability distribution, whose parameters can be calculated and are stationary.
2. Data fits to a Bernoulli series, which is a simplification of the analysis, however, more powerful analysis can be performed by using Markov models [10].

If these premises are true, then the T year event that occurs at least once in a N year period is:

$$F(x)^{-1} = 1 - \left(1 - \frac{1}{T}\right)^{N} \quad (1)$$

where F(x) is the cumulative probability function, T is the return period and N is the planning year period. In most of the cases reviewed in this project, the underlying probability function in the data set was that of the General Extreme Value (GEV) type [11]. The design stream flow for a given return period for this distribution is:

$$x_n = \mu + \frac{\sigma}{\xi}\left[-Ln(1 - T^{-1})^{-\xi} - 1\right] \quad (2)$$

where x_n is the flow for a given return period T, μ, σ, ξ are the unknown location, scale and shape parameters of the distribution, respectively. These parameters of the distribution are computed using Maximum Likelihood Estimation (MLE). Figure 1 shows a comparison of the design stream flow obtained using a GEV distribution for the processed and original data. The maximum return period shown is a hundred years since this is the assumed service life of bridges in México.

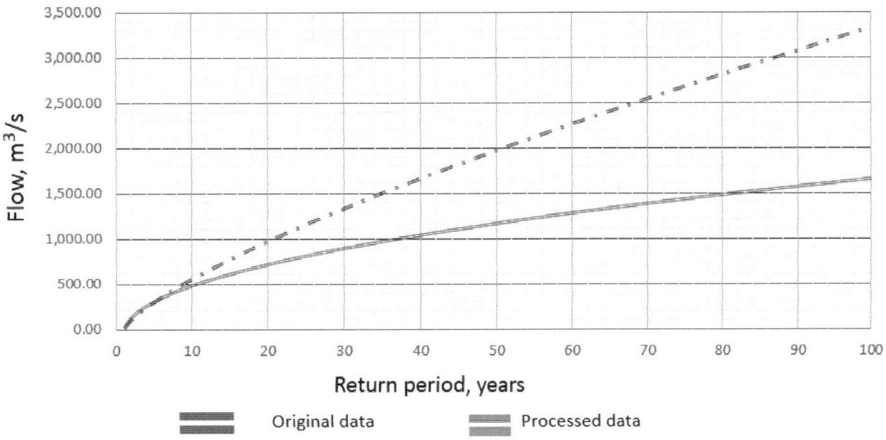

Figure 1. Design flow for different return periods.

3. Scour Analysis

Figure 2 shows the idealized section of the bridge crossing that will be used in the scour model for the case study. For this section, the hydraulic area ($\boldsymbol{A_h}$), wetted perimeter ($\boldsymbol{P_m}$), and hydraulic radius ($\boldsymbol{R_h}$) were obtained and used to calculate the speed and water depth [12] as follows:

$$V_s = \frac{1}{n} R_h^{\frac{2}{3}} S_h^{\frac{1}{2}} \tag{3}$$

where $\boldsymbol{n = 0.05}$, is the Gauckler-Manning coefficient and $\boldsymbol{S_h = .06862}$, is the slope of the hydraulic grade line.

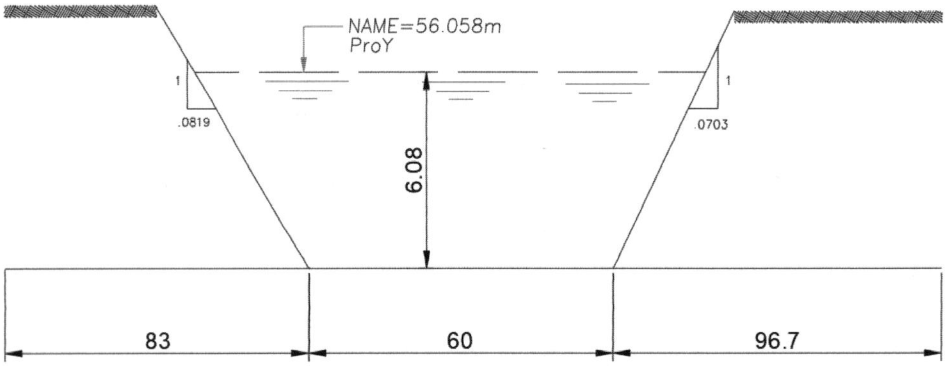

Figure 2. Idealized hydraulic section. Dimensions in m.

The local scour depth evolution for this section is estimated using Yaroslavtziev [6] and HEC 18 methods for normal and complex piers [7], and the stream design flow computed in section 2.2 Furthermore, using the value of V_s obtained with eq.(3), the design stream flow is obtained for different return periods (0-100 years) as shown in figure 3. Figures 4 show the annual probability of exceedance for different scour depths and methods.

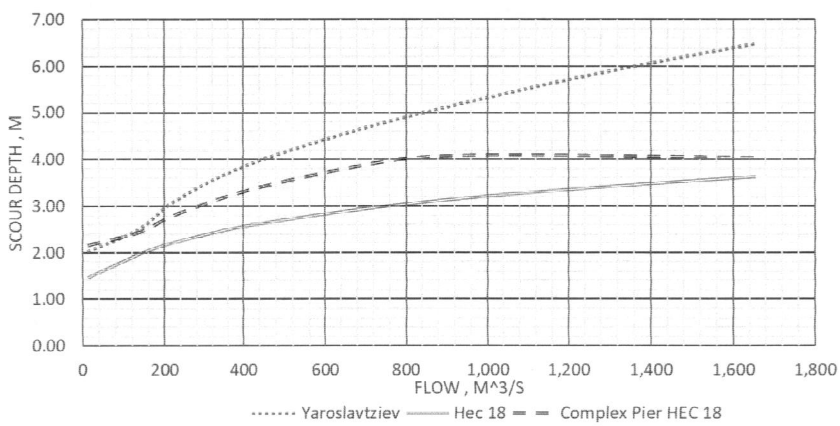

Figure 3. Scour depth/ flow plot for Yaroslavtziev, HEC18 method, and HEC 18 complex pier method.

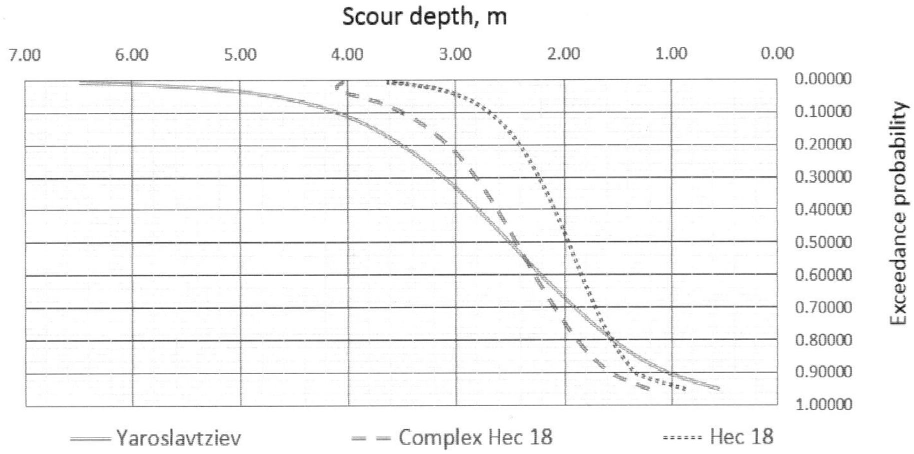

Figure 4. Annual probability of exceedance for different scour depths and methods.

4. Case study

The case study bridge is located in the state of Oaxaca in the highest seismic prone zone in México. This is a four span, simply supported bridge, each span has a length of 56 m. Bridge piers made of reinforced concrete in a multi-column bent type, with circular columns of diameter $\boldsymbol{D_p = 1.2\ m}$; foundation piles have the same diameter. Soil mechanics information as well as mechanical properties of the bridge columns and piles were available and are used in the mathematical model.

4.1. Potential Earthquake Damage Assessment

The approach presented in the following paragraphs is based on the premises defined by Priestley on bridge seismic modelling [13] and the methodology of Shin-Tai [14]. Accordingly, the pier is idealized as two-degree freedom system as shown in the following figure 5:

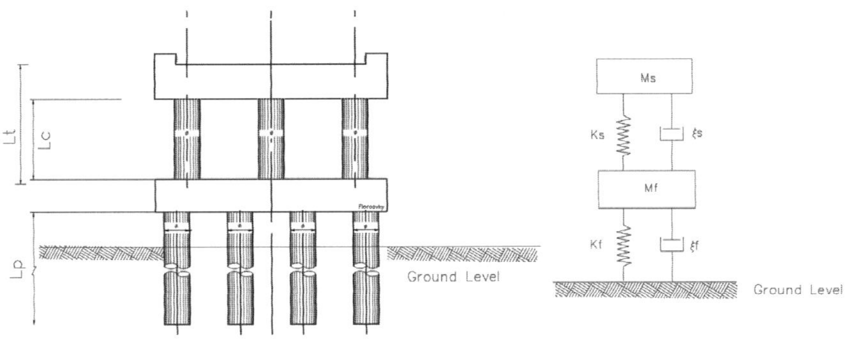

Figure 5. Pier and correspondent two degree of freedom mathematical model.

The lateral stiffness of the foundation piles and the tributary masses of piles, columns and superstructure are used to calculate the first and second natural periods of the model, which are then used with a constant acceleration spectrum and equivalent damping in order to derive the seismic forces acting on the pile group [14] [15].

4.2. Shear demand

Song and Chun [8] demonstrated that the elastic seismic shear on the pile cap connection is given by:

IOP Conf. Series: Materials Science and Engineering **652** (2019) 012013 doi:10.1088/1757-899X/652/1/012013

$$V_{f\ dem} = \sqrt{\left[\frac{(2\beta_m+\lambda_b+\lambda_c)^2}{4\beta_m+(\lambda_b+\lambda_c)^2} S_{a1} m_s^* T_1\right]^2 + \left[\frac{(2\beta_m+\lambda_b-\lambda_c)^2}{4\beta_m+(\lambda_b-\lambda_c)^2} S_{a2} m_s^* T_2\right]^2} \tag{4}$$

where S_{a1} and S_{a2} are the acceleration magnitudes derived from first and second natural periods and the acceleration spectrum; m_s is the seismic tributary mass of the super structure; β_m is the ratio of the generalized mass at the centroid of the pile-cap and at the center of the superstructure; β_k is the ratio of the lateral stiffness of the bridge column and the lateral stiffness of the soil; and:

$$\lambda_b \equiv \beta_k - \beta_m + 1; \tag{5}$$

$$\lambda_c \equiv \sqrt{{\beta_k}^2 - 2\beta_k(\beta_m - 1) + (\beta_m + 1)^2} \tag{6}$$

Figure 6 shows the change of natural period because of the scour depth modification.

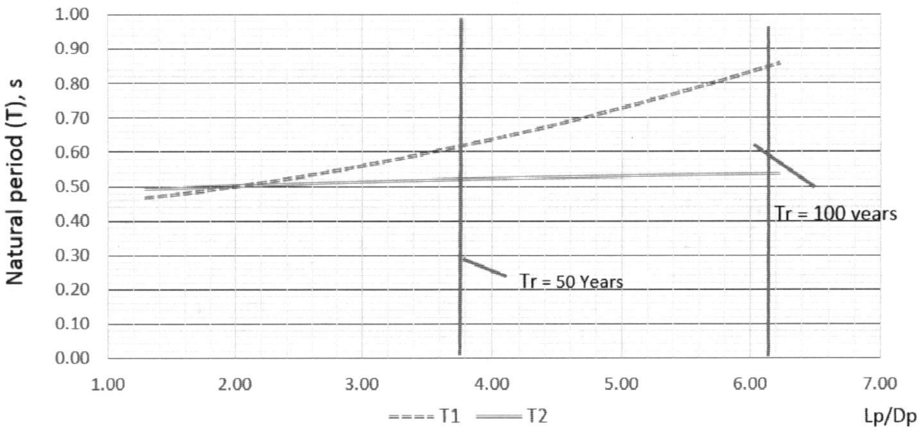

Figure 6. Natural period variation for a series of normalized scour depths.

4.3. Shear strength

On the other hand, the first yield limit state of a pile foundation is identified by its bending moment at the pile/pile-cap connection of the piles, and is associated to the shear capacity [8]:

$$\mathbf{V_f} = \frac{1 + \xi_a(\xi_e)_2 + \frac{7}{4}(\xi_e)_2}{.5{\xi_a}^2(\xi_e)_2 + \frac{7}{4}\xi_a(\xi_e)_2 + \frac{13}{8}(\xi_e)_2} \frac{\mathbf{M_u}}{(\mathbf{R_n})_1} \tag{7}$$

where V_f is the lateral force required to cause plastic hinge formation in the first-row of piles, M_u is the flexural strength of the piles at the pile/pile-cap connection, and R_n is the length of the pile. The shear demand/capacity ratio of the studied bridge is shown in Figure 7 for a normalized scour depth $\frac{Lp}{Dp}$, where Lp is the exposed length of the pile and Dp is the pile diameter.

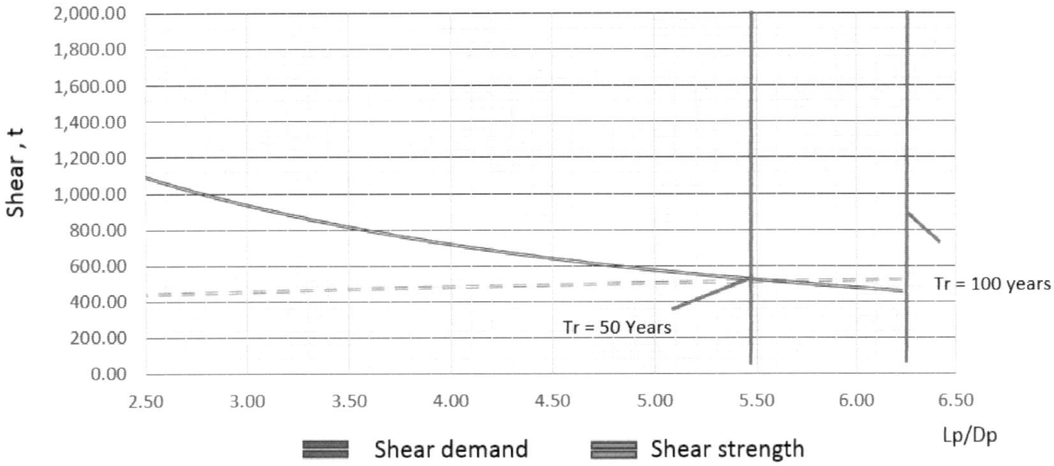

Figure 7. Shear demand/ capacity for a series of normalized scour depths.

5. Conclusions

In this paper the proposed approach takes special consideration in characterizing and evaluating an extreme rainfall hazard as the fundamental variable in the multi-hazard evaluation of bridges. Scour depth is computed using approximated semi-empirical methods that are commonly used in engineering practice.

The proposed methodology considers the degradation of the lateral strength of the bridge piles as the scour depth increases. The approach is robust and easily accessible can be easily used by engineers to estimate the critical scour depth that demands corrective maintenance of the bridge.

It can be seen that nonlinear behaviour is expected for a single extreme rainfall event with a return period near 50 years, which is half the service life of the bridges in México.

In México, minimum budget is devoted to the preventive maintenance of civil infrastructure, a situation which leads inevitably to the damage accumulation, which in turn is associated to the vast majority of bridge collapses.

Since the scour damage accumulation has a stochastic behaviour that depends heavily on the scour – sedimentation process, this paper should be taken as a first approximation to a more complex problem. More research is needed to develop procedures that take in to account the damage accumulation and its effects.

References

[1] Gómez S C, Cruz J G and Frias R 2018 *A revision on bridge collapses due to scour* Proc Mex Soc Stru Eng, México

[2] Frias R 2014 *Experiences of bridge disasters in México* in Spanish, International Seminar on Bridges, Mexico

[3] Gómez R, Flores-Vidriales D and Arenas-García L M 2017 *Bridge Structure Vulnerability in Zones of Great Influence of Tropical Cyclones*, Internal report, Institute of Engineering, UNAM, Mexico

[4] Liang F, Zhang H and Haung M 2016 *Extreme scour effects on the buckling of bridge piles considering the stress history of soft clay*, Natural Hazards, **77**, pp 1143-1159.

[5] Folch-Fortuny A, Arteaga F, Ferrer A 2016 *Missing data imputation toolbox for MATLAB* Chemometr Intell Lab Sys, pp 93-100

IOP Conf. Series: Materials Science and Engineering **652** (2019) 012013 doi:10.1088/1757-899X/652/1/012013

[6] Yánez-Díaz C J 2012 *Comparative analysis of the functional relationships for the scour depth calculation around obstacles*, Thesis, in Spanish Universidad Central de Ecuador, Quito, Ecuador

[7] Federal Highway Administration 2012 *Evaluation Scour at Bridges*, Publication No. FHWA-HIF-12-003 HEC 18 Washington, USA

[8] S.-T. Song, C.-Y. Wang and W.-H. Haung 2015 *Earthquake damage potential and critical scour depth of bridges exposed to flood and seismic hazards under lateral seismic loads* Earthquake Engineering and Engineering Vibration, **14**, no. 4, pp 578-594

[9] National Board of Water 2019 *BANDAS* 15 02 [Online]. Available: http://www.conagua.gob.mx/CONAGUA07/Contenido/Documentos/Portada%20BANDAS.htm. [Accessed 2019].

[10] Castellarin A, Voguel M R and England J 1987 *An assessment of exceedance probabilities of envelope curves* Water resource research, **113**, pp 563-579

[11] Pawitan Y 2001 Statistical Modelling and Inference using Likelihood (New York: Oxford University Press Inc)

[12] Manning R 1981 *On the flow of water in open channels and pipes* Transactions of the Institution of Civil Engineers of Ireland, **20**, pp 161-207

[13] Priestley M and Calvi G 1996 *Seismic Design and Retrofit of Bridges* (New York: Wiley–Interscience)

[14] Song S and Chai Y 2008 *Performance Assessment of Multi-column Bents with Extended Pile-Shafts under Lateral Earthquake Loads* The IES Journal A: Civil & Structural Engineering, **1**, pp 39-54

[15] Manual of Civil Works 2015 *Structures Design Manual* National Bureau of Electricity México

IOP Conf. Series: Materials Science and Engineering **652** (2019) 012014 doi:10.1088/1757-899X/652/1/012014

Investigation of infrastructure maintenance cost of intercity railway in Thailand

Geerati Kruajun[1] and Boonchai Sangpetgnam[1,2,*]

[1]Department of Civil Engineering, Faculty of Engineering, Chulalongkorn University, Bangkok, Thailand
[2]Infrastructure Management Research Unit, Chulalongkorn University, Bangkok, Thailand

Email: boonchai.sa@chula.ac.th

Abstract. Infrastructure in the rail system requires considerable investment in construction and maintenance. In many countries, railway operators have been able to share the capacity of the railroad to achieve efficient infrastructure usage. For the development of Thailand's railway network system in the future, it is possible for separating management between infrastructure and the rolling stock operations. Therefore, it is necessary to know the costs and factors that affects the cost of infrastructure, including government budget subsidies, to be the basis for track utilization charges. In this study, the costs of infrastructure construction and maintenance on 5 route sections representing different traffic characteristics were studied by using records from 5 years. It was found that the maintenance cost of the telecommunication and signalling system ranges from 1.3 to 1.7 million baht per year per station. The average annual maintenance cost of the railway track was 280,000 to 310,000 baht per track per kilometre. The traffic density was found to be the main factor that influenced the maintenance cost.

1. Introduction

Rail transport has many advantages over other modes of transport, such as lower transport unit cost comparing to transport on roads, faster than transport on waterways, less environmental impact than air transport [1]. Currently, Thai government has put tremendous investment on improving the services and expanding the network of rail transport to gain competitiveness over other modes of transport [2]. In the future, utilization of the railway network at the full capacity will be needed to gain the worthiness from the government investment [3]. To allow for multiple operators sharing the railway infrastructure usage and finance, it is necessary to understand the cost components and factors that include in the construction and maintenance of infrastructure [4].

Costs incurred in the maintenance of infrastructure used in regional railways in Thailand have been carried by State Railway of Thailand (SRT). While the road transport mode has received the government budget for infrastructure maintenance [5].

There are differences in the agreement of infrastructure usage charge in each country. It depends on railway organization structure and the factors in consideration. Railway infrastructure charge should represent the actual cost carried by infrastructure managing authorities to make their financials viable. In addition, there should be an agency supervising the charges to ensure fairness amongst the infrastructure providers and users. Some example cases are presented as follows.

IOP Conf. Series: Materials Science and Engineering **652** (2019) 012014 doi:10.1088/1757-899X/652/1/012014

In France, the railway charge is calculated from the administration fee maintenance and railway rehabilitation costs. This fee is charged annually and the charge is split into 2 categories: freight and passenger transportation. However, the fee is only charge to the passengers [6].
In Spain, the railway charge cost is calculated annually like in France. And there is a separation of the fee with freight train and passenger train. But the source of the calculation is not clearly stated [6].
In Slovenia, railway charge depends on the total cost of track maintenance. It does not include profit. There is a designated public agency regulating and determining the infrastructure charge cost. It is calculated from the coefficient of train weight, track deterioration and the facility cost for traffic management service [6].
The main concept of infrastructure charge is to consider all costs involved that are close to reality and must be consistent with the current situation. The cost should be broken into category appropriately to keep in order and manageable [7].
There was a study of cost estimation using modelling prototypes that could be modelled for access cost estimation. The infrastructure access cost was calculated from the maintenance fee and management fee. The goal of the prototype was to find the relative relation of the cost to the resulting cost. Those mentioned factors affecting the cost of infrastructure provider can be used to calculate the access charge cost. The selection of some or all of the factors in order to calculate the cost to collect from the infrastructure users would depend on the measures and agreement of the organization [8,9].
The purposes of this study are to compile and analyse the cost of construction and maintenance of railway infrastructure. The infrastructure consists of railway, bridge, and signalling and telecommunication system. The factors that influence the costs will be included in this study.

2. Research methodology

In this study, the infrastructure cost was separated into two parts: construction cost of railway and signalling system; maintenance cost of railway and signalling system. The construction cost consisted of a new railway track and station. The railway was designed to carry 20-ton axle load. The railway track included bridges, embankment, substructure, and superstructure such as rails, sleepers fastening devices, ballast, switches, and additional components which provided safety such as guard rail and safety rail etc. [10]. The station was designed mainly for passengers. The signalling systems used to control the traffic was based on three-aspect wayside signalling system.
Railway maintenance cost comprised of monitoring and maintenance activities that occurred on the infrastructure so that all track components remained within the standard tolerance. Activities included track inspection, rail maintenance, track geometric maintenance, ballast sleeper and fastening maintenance, and emergency services.
This study determined the costs of railway infrastructure by investigating the actual costs of constructing and maintaining railway and signalling system including influencing factors. The construction cost and maintenance cost of five representative routes of State Railway of Thailand (SRT) were studied. The five routes covered passenger and freight traffic. They were selected from SRT railway network to cover various traffic density from low level to high level. The five routes were ballasted tracks with BS100 rails on mono-block concrete sleepers. They were under the same maintenance standard [11]. Based on data collected from SRT in the fiscal year of 2015, the average traffic density of train operation of SRT was 26 trains per day per track as shown in Figure 1. And the five routes were selected to represent various traffic density as shown in table 1.

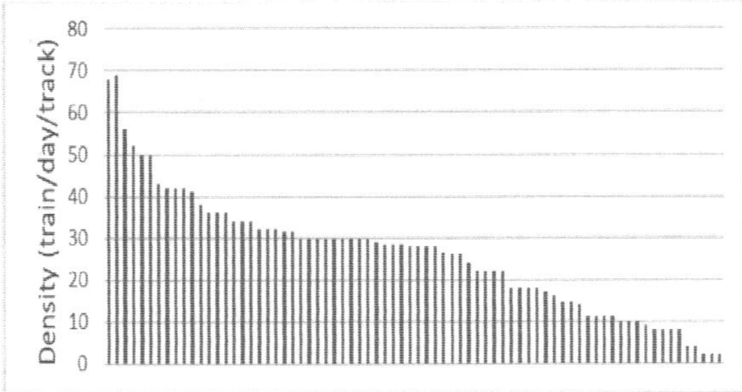

Figure 1. Traffic density of SRT routes.

Table 1. The representative routes of the intercity railway used in this study.

Route	Density (trains/day/track)	Traffic type
Nong Pla Duk junction – Ratchaburi	38	Passenger
Bua Yai junction – Khon Kean	41	Passenger &freight
Thonburi – Taling Chan junction	22	Passenger
Nakhon Ratchasima – Bua Yai junction	18	Passenger
Lat Krabang – Laem Chabang	17	Passenger &freight

3. Analysis and discussion

3.1. Construction cost

The construction costs of the five routes were obtained based on the past construction project document from SRT in 2009-2013. The average costs per kilometre of single track are shown in table 2. The brief details of each route are as follows.

Table 2. Construction cost of 5 intercity railway route.

Route	Construction cost (million baht/km-track)	Percentage by components				
		Track and signalling	Station	Bridge	Switch	Others
Nong Pla Duk junction - Ratchaburi	77.26	71%	21%	5%	2%	1%
Bua Yai junction - Khon Kean	72.39	76%	11%	9%	1%	3%
Thonburi - Taling Chan Junction	116.99	46%	29%	12%	4%	9%
Nakhon Ratchasima - Bua Yai junction	68.00	81%	13%	5%	1%	<1%
Lat Krabang - Laem Chabang	74.31	71%	11%	16%	2%	<1%

The construction costs varied due to different conditions. For examples, some routes had longer distance of bridges than other routes. It made the construction cost higher. Number and size of stations on some route were bigger than the others. By average, the construction cost of railway infrastructure was 72.9 million baht per kilometre per one track.

3.2. Maintenance cost of signaling system

The maintenance cost of signalling system was broken down according to the group of works, such as communication work, level crossing, traffic signalling and overhead. The overhead part included staff cost and overhead office costs, In each fiscal year, the costs were allocated differently depending on the needs and work priorities in that year. The maintenance costs for each route were as follows:

- Nong Pla Duk Junction - Ratchaburi in the range of 10,653,923 to 13,899,912 baht per year.
- Bua Yai Junction - Khon Kaen in the range of 11,985,669 to 15,637,401 baht per year.
- Thonburi - Taling Chan Junction in the range of 2,663,482 to 3,474,978 baht per year.
- Nakhon Ratchasima – Bua Yai Junction in the range of 15,980,892 to 20,849,868 baht per year.
- Lat Krabang - Laem Chabang in the range of 15,980,892 to 20,849,868 baht per year.

This research has presented the maintenance cost of signaling and telecommunication system per station, because most of maintenance and supervisory activities occur at the station [12]. Table 3 presents the average maintenance costs per station from 3 years.

Table 3. Signalling and telecommunication maintenance costs.

Year	Signalling & telecommunication cost (million baht/year/station)	Percentage by components			
		Overhead	Level crossing	Signalling	Communication
2013	1.74	24%	18%	48%	10%
2014	1.74	28%	21%	41%	10%
2015	1.33	39%	17%	31%	13%

3.3. Railway maintenance costs

Railway maintenance activities were conducted to keep the operation continuous safe and effective. The study found that the maintenance costs were varied year by year. It was found that in 2012, the cost of maintenance was lower than those of other years. In 2012, the State Railway of Thailand had lacked of liquidity so it was necessary to cut off some budgets. Therefore, in cost analysis some cost items of this year did not appear. The railway maintenance costs of five routes is shown in figure 2.

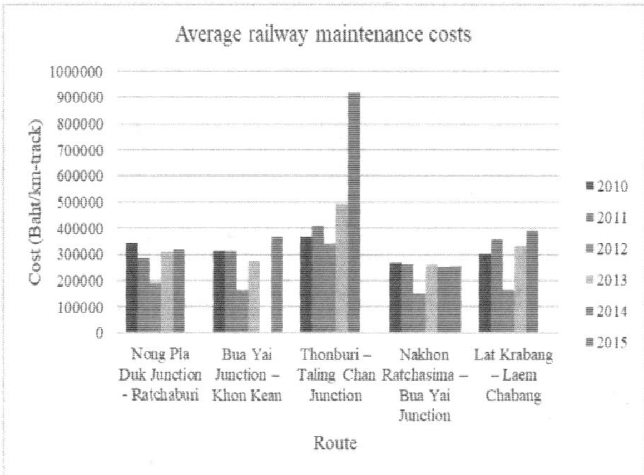

Figure 2. Average railway maintenance costs of the five routes.

The railway maintenance cost of SRT consisted of four main expenses: staff cost, contracting-out cost, material cost and miscellaneous cost. It was found that the staff cost was the largest portion. The details of maintenance costs are as follows. Nong Pla Duk Junction

- Ratchaburi has an average railway maintenance cost of 313,448 baht/ km per one track.
- Bua Yai Junction – Khon Kaen has an average railway maintenance cost of 318,330 baht/ km per one track.
- Thonburi – Taling Chan Junction has an average railway maintenance cost of 629,067 baht/ km per one track.

- Nakhon Ratchasima – Bua Yai Junction has an average railway maintenance cost of 260,537 baht/ km per one track.
- Lat krabang – Laem Chabang has an average railway maintenance cost of 364,004 baht/ km per one track.

3.4. Factors affecting railway maintenance costs

In this study, there were three parameters that could be collected from the past data and were likely to have influences on railway maintenance cost [11]. Those three parameters were load tonnage, traffic density, and track quality index.

3.4.1. Load tonnage. Tonnage is a measure of the cargo-carrying capacity of the train. Tonnages on the five routes are shown in table 4.

3.4.2. Traffic density. According to the sensitivity of the railway system rises as the capacity utilisation increases in traffic demand [13]. It has been found that number of train density is one of all factor affects the cost of maintenance work. If one route is used more than other routes, the structural maintenance cost of that route is higher than other routes. Traffic density on the five routes are shown in table 4.

Table 4. Load (tonnage) and Number of train density (train per day per one track)

Route	Load (tons/day/track)	Density (Train per day per one track)
Nong Pla Duk junction - Ratchaburi	29,440	38
Bua Yai junction – Khon Kaen	43,520	41
Thonburi – Taling Chan Junction	14,080	22
Nakhon Ratchasima – Bua Yai junction	16,640	18
Lat Krabang – Laem Chabang	15,872	17

Table 5. Level of track quality index, the State of railway of Thailand.

Level of track maintenance	QI (%)
Immediate maintenance	> 40
Rapid improvement	31-40
Poor	21-30
Good	11-20
Very Good	< 10

3.4.3. Track quality index (QI). SRT has surveyed geometry conditions of its railway track network by using track survey vehicles. At one sample point on a railway track section, five parameters are measured including track gauge, cross level, twist, longitudinal alignment, and vertical alignment. The measured data of sampling points in a given track segment distance is used to calculate QI. [14, 15] QI is percentage of probability that a number of sampling points whose geometry parameter measurements fall outside ±3 mm to the mean value of all sampling points in a track segment. By definition, QI expresses the amount of track geometrical deviation from the allowance criteria. The larger the QI, the worse the track segment in some quality aspect. Table 5 illustrates a general guide for track condition used by SRT. SRT has planned the track maintenance works based on QI values obtained from track inspection once or twice a year.

Figure 3 illustrates QI graphs of the five routes during 2009-2015. The graph shows that QI of Thonburi – Taling Chan route in 2011 is missing because the major flood in that year had obstructed the track survey. The route had been rehabilitated after the flood crisis. So, the QI value had returned to the acceptable level.

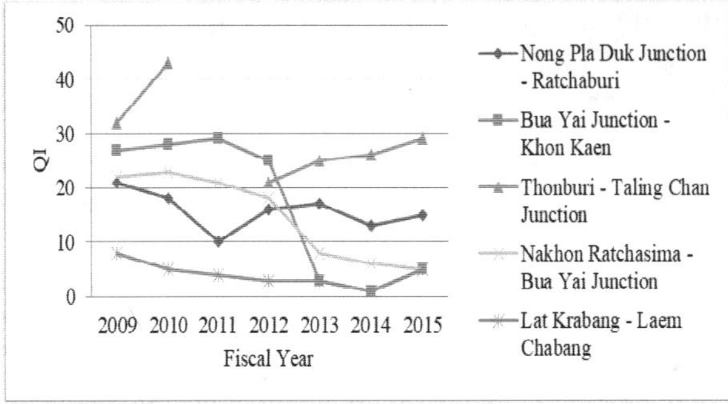

Figure 3. Chart of track quality of 5 routes in 2009 – 2015.

3.4.4. Results of statistical analysis. Since it was likely that railway maintenance cost was influenced by various factors, the multiple regression analysis was conducted using the data from 5 routes in 6 years. The railway maintenance cost on one considered year was the dependent variable. Traffic density, load tonnage, QI, and maintenance cost of previous year were assigned as the independent variables. The route number 2 to 5 and year 2010 to 2015 were added into the analysis as dummy variables to reduce the impact of variables that were not statistically analyzed. The results of the analysis is presented in Table 6 and 7.

Table 6. Results of multiple regression analysis of railway maintenance cost and related variables.

Number of obs	=	12
F (8,3)	=	28.97
Prob>F	=	0.0093
R-squared	=	0.9872
Adj R-squared	=	0.9531
Root MSE	=	12231

Table 7. Results of multiple regression analysis of railway maintenance cost and related variables.

Railway maintenance cost	Coef. std.	Err	t	P>t	[95% conf. Interval]	
QI current year	1880.52	693.16	2.71	0.073	-325.44	4086.50
Traffic Density	4810.33	1302.07	3.09	0.054	666.56	8954.11
Load	-2.779	0.974	-2.85	0.065	-5.878	0.320
Dummy Route 2	Omitted					
Dummy Route 3	Omitted					
Dummy Route 4	Omitted					
Dummy Route 5	Omitted					
Dummy FY 2010	Omitted					
Dummy FY 2011	Omitted					
Dummy FY 2012	-126531	11457.48	-11.04	0.002	-162994	-90068.3
Dummy FY 2013	-41919.2	42493.51	-0.99	0.397	-177152.5	93314.1
Dummy FY 2014	8732.53	13894.36	0.63	0.574	-35485.5	52950.6
Dummy FY 2015	7328.27	18854.9	0.39	0.723	-52676.6	67333.1
_cons	300679.	74347.07	4.04	0.027	64073.5	537284.7
Maintenance cost in previous year	-0.3963	0.3017	-1.31	0.28	-1.3566	0.5639

The result showed that the dummy variables of year 2010, year 2011 and the routes did not affect the analysis. Because those relative variables were omitted from the analysis. The result showed that the traffic density, QI, load tonnage and maintenance cost in previous year had some influences to the railway maintenance cost. Amongst the four independent variables, traffic density had the highest significance to railway maintenance cost. However, these four variables did not achieve at 95% significant level. And it was found that the maintenance cost in year 2012 was a dummy variable which was significantly different from other years. The constant of the regression model showed that the average railway maintenance cost per year per track was 300,679 baht per kilometer.

4. Conclusion

This study had investigated the costs of constructing and maintaining railway infrastructure in Thailand. Five railway routes connecting regional cities in Thailand were selected for this study. The past six-year records of maintenance costs were input into the statistical analysis. It was found that traffic density, track quality index, load tonnage and maintenance cost spent in previous year had some influences to the railway maintenance cost. The results from this study had revealed the average unit costs of railway construction, signalling system maintenance, and railway maintenance that should be useful for calculating railway infrastructure usage charge or infrastructure subsidy from the government in the future.

Lastly, the authors would like to mention on some limitation of this study. The infrastructure components comprised of railway track, bridges, civil works, signalling and telecommunication system that were constructed according to SRT standard and conditions in Thailand. Therefore, the unit costs presented may be largely different for other railway authorities or other countries.

Acknowledgments

The author would like to thank to the Department of Civil Engineering, Chulalongkorn University for Civil Engineering Scholarships, Chulalongkorn University's 100th Anniversary and thank to the State Railway of Thailand for providing data records and supports to this study.

References

[1] Warunee A 2012 *Sugarcane Transportation Management using Network and Multi-Objective Decision Analyses*

[2] Tungtrongjitr P 2015 Transportation system development the national railway transport system according to the Ministry of Transport's strategic plan *EAU Heritage Journal Science and Technology* **3 (1)** 64-75

[3] Baumol W J and Bradford D F 1970 Optimal departures from marginal cost pricing *The American Economic Review* **60(3)** 265-283

[4] Hay W W 1982 *Railroad Engineering* (John Wiley & Sons)

[5] Chanthasorn N 2011 *Railway Trainers General Knowledge of Railway Engineering* (Bangkok: National Science and Technology Development Agency)

[6] Puetter U 2012 Europe's deliberative intergouvernementalisme: the role of the Council and European Council in EU economic governance *Journal of European Public Policy* **19(2)** 161-178

[7] Prok J D 2009 High Speed Rail: Planning and Financing the Next Fifty Years of American Mobility *Transportation Law Journal* **36(1)** 47-74

[8] Independent Regulators' Group – Rail 2012 *Initial review of charging practices in Europe, Technical Report*

[9] Crozet Y and Chassagne F 2013 Rail access charges in France : Beyond the opposition between competition and financing *Research in Transportation Economics* **39(1)** 247-254

[10] Indraratna B, Salim W and Rujikiatkamjorn C 2011 *Advanced rail geotechnology–ballasted track* CRC press

[11] Harris R G 1977 Economies of traffic density in the rail freight industry, *The Bell Journal of Economics* 556-564

[12] Dekker R 1996 Applications of maintenance optimization models: A review and analysis. *Reliability Engineering & System Safety* **51(3)** 229-240

[13] Lingfeldt A 2015 *Railway capacity analysis: Methods for simulation and evaluation of timetable, delays and infrasructure* (Stockholm: KTH Royal Institute of Technology)

[14] Liu R, Xu P, Sun Z, Zou C and Sun Q 2015 Establishment of Track Quality Index Standard Recommendations for Beijing Metro *Discrete Dynamics in Nature and Society*, Article ID 473830

[15] State Railway of Thailand 1974 *Periodical Track Maintenance Manual*

Measurement of the displacement response of a structure using line detection

S W Kim[1*], B G Jeon[1], D U Park[1] and D W Yun[1]

[1]KOCED Seismic Research and Test Center, Pusan National University, 49 Busandaehak-ro, Mulgeum, Yangsan, Kyungnam 50612, Republic of Korea
*corresponding author, Email: swkim09@pusan.ac.kr

Abstract. The shaking table test is required to examine the seismic safety of a structure as well as the structure's unique functions before and after the occurrence of an earthquake. That is, the data on the displacement responses obtained from the shaking table test are required to assess the seismic performance of a structure. In this study, the shaking table test was conducted to verify the validity of a method of measuring displacement responses using images. In this method, the lines of a structure were detected, and the responses were measured using the amount of change in the intersection of the detected lines. In addition, image filter processing was used to facilitate the detection of the lines of the structure in the shaking table test.

1. Introduction

As strong earthquakes and seismic damage have become more frequent of late, the interest in the seismic design of various social infrastructures is also increasing. Moreover, in the cases of seismic damage, damage and destruction of the structural and non-structural elements of the concerned social infrastructures occurred. Therefore, it is necessary to develop technologies for securing the seismic safety of social infrastructures, and for improving their usability in the global situation where the number of earthquakes is increasing [1].

For the assessment of seismic performance, the shaking table test is currently being conducted on major structures. Data on the displacement responses [2], which represent the overall behavior of a structure, are important for safety assessment, and seismic performance is determined based on these. In general, when the shaking table test is conducted, the displacement responses are measured by installing contact-type sensors like linear variable differential transformers (LVDTs) and ring gauges on a structure or by applying costly laser equipment like the laser Doppler vibrometer (LDV). For LVDTs, the accessibility must be secured, and temporary facilities for fixing them need to be installed. Therefore, they are not economical. In addition, the reliability of the measurements varies depending on the fixing conditions of the temporary facilities. For ring gauges, cables are used instead of temporary facility installation, which is the drawback of LVDTs, and separate facilities capable of minimizing external influences must be installed. When the cables are shaken by external factors, the measurement accuracy is decreased. In the case of LDV [3], it is difficult to apply it to the shaking table test because it is expensive precision equipment despite its outstanding performance.

The method of using images [4-6] is simple, economical, and suitable for extracting the displacement responses of a structure that is hard to access. The existing method of measuring the displacement responses in the shaking table test involves the installation of targets on a structure and measures the changes in the targets using an image processing method, to measure the displacement responses. In this study, a method of measuring the displacement responses by detecting the lines [7-8] of a structure without installing targets on the structure was proposed. In addition, the shaking table test was conducted to verify the validity of the proposed method.

2. Algorithm for measuring the displacement responses of a structure using line detection

The purpose of the image filter processing [9] is to process the original image and then to convert it in accordance with a specific application purpose. When the test was conducted, the Gaussian filter was used to reduce the noise generated while acquiring optics, lighting, and images. The filter reduces noise and connects the disconnected edges of the image while reducing its sharpness. Due to these characteristics, noise reduction using the Gaussian filter is used in the pre-processing of object recognition sensitive to noise. Figure 1 shows the 7×7 matrix obtained by applying discrete approximation to the Gaussian function with a standard deviation of 0.4, and a weight was given to the image using the convolution mask.

In the shaking table test, unsharp-mask processing was used to sharpen the geometry of the structure. Unsharp-mask processing is the process of removing a blurred image from the image itself. The discrete approximated 7×7 matrix can be expressed as shown in Figure 2, and a weight was given to the image using the convolution mask. The boundary of the structure with an improved geometry is the boundary between two regions with different information, and can be detected by calculating the difference between the two neighboring regions.

$$\frac{1}{142}\begin{bmatrix} 1 & 1 & 2 & 2 & 2 & 1 & 1 \\ 1 & 2 & 2 & 4 & 2 & 2 & 1 \\ 2 & 2 & 4 & 8 & 4 & 2 & 2 \\ 2 & 4 & 8 & 16 & 8 & 4 & 2 \\ 2 & 2 & 4 & 8 & 4 & 2 & 2 \\ 1 & 2 & 2 & 4 & 2 & 2 & 1 \\ 1 & 1 & 2 & 2 & 2 & 1 & 1 \end{bmatrix}$$

Figure 1. 7x7 Gaussian function

$$\begin{bmatrix} -1 & -1 & -1 & -1 & -1 & -1 & -1 \\ -1 & -1 & -1 & -1 & -1 & -1 & -1 \\ -1 & -1 & -1 & -1 & -1 & -1 & -1 \\ -1 & -1 & -1 & 60 & -1 & -1 & -1 \\ -1 & -1 & -1 & -1 & -1 & -1 & -1 \\ -1 & -1 & -1 & -1 & -1 & -1 & -1 \\ -1 & -1 & -1 & -1 & -1 & -1 & -1 \end{bmatrix}$$

Figure 2. Unsharp-mask processing

$$\begin{bmatrix} -1 & -2 & -1 \\ 1 & 1 & 1 \\ 1 & 2 & 1 \end{bmatrix}$$

(a) Horizontal mask

$$\begin{bmatrix} -1 & 0 & 1 \\ -2 & 0 & 2 \\ -1 & 0 & 1 \end{bmatrix}$$

(b) Vertical mask

Figure 3. Geometry extraction mask

$$\begin{bmatrix} 0 & 1 & 0 \\ 1 & 1 & 1 \\ 0 & 1 & 0 \end{bmatrix}$$

Figure 4. Dilation structure

The boundary can provide information on the image, such as the position, geometry, and size of the object, and it is the pre-processing stage for enhancing the boundary of the structure. Figure 3 shows the 3×3 mask used for extracting the geometry of the structure. The mask consists of a horizontal mask and a vertical mask.

The dilation structure expands the outermost pixels of the object. The size of the object is expanded, and the background is reduced. To facilitate line detection by expanding the pixels of the structure geometry installed on the shaking table, the outer pixels were expanded using the 3×3 dilation mask structure, as shown in Figure 4.

$$y = ax + b \tag{1}$$

The line of the geometry was extracted from the structure geometry improved through the above pre-processing processes using the linear regression equation, as shown in equation (1).

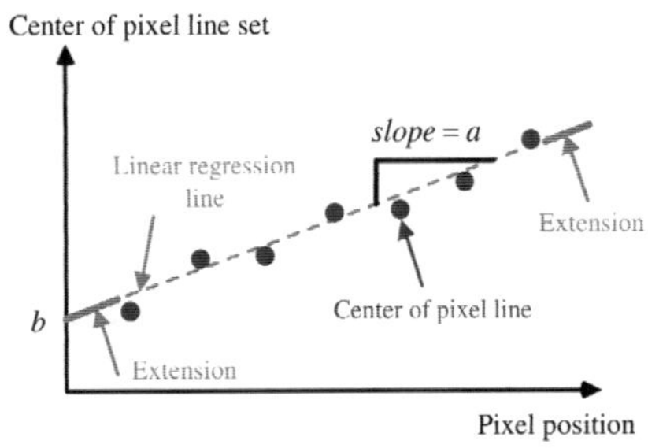

Figure 5. Line detection process

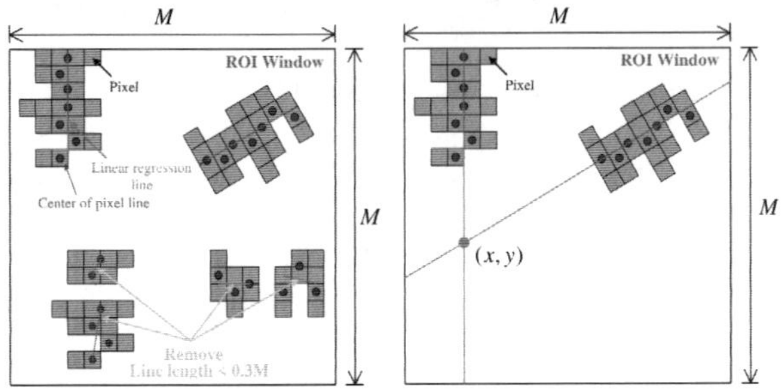

Figure 6. Line detection

Figure 5 shows a linear regression curve for the center of the pixel line at the pixel position of equation (1). It is possible to obtain equation (1), a first-order regression curve, for the center of the pixel line at the position of the pixel judged to be a line, and the regression curve is expanded to the selected region of interest (ROI) window.

The pixel information was set in the ROI window of Figure 6, and the coordinates (displacement responses) were measured using the information on the intersection of the two straight lines. The ROI window represents the square M×M pixel set, and the pixel length detected as a line by the window was set to at least 0.3 M. In addition, to minimize the errors in detecting lines from the image due to noise, the ROI window was set to a minimum of 50×50 pixels and a maximum of 100×100 pixels.

Through equation (2), valid lines corresponding to the geometry of the structure as well as various and many lines caused by noise are detected. For the lines that could be judged to be a shape, the pixel length was designated as 0.3 M or longer in the ROI window, which is the square of the M×M pixel set size, in equation (2), and the lines with less than 0.3 M pixels were removed. When such lines are removed, the lines created by noise and other environments are also removed.

$$\begin{aligned} &Linear\ regression\ line\ >\ 0.3\times ROI\ window\ pixel\ Length \\ &else\ =\ remove \end{aligned} \qquad (2)$$

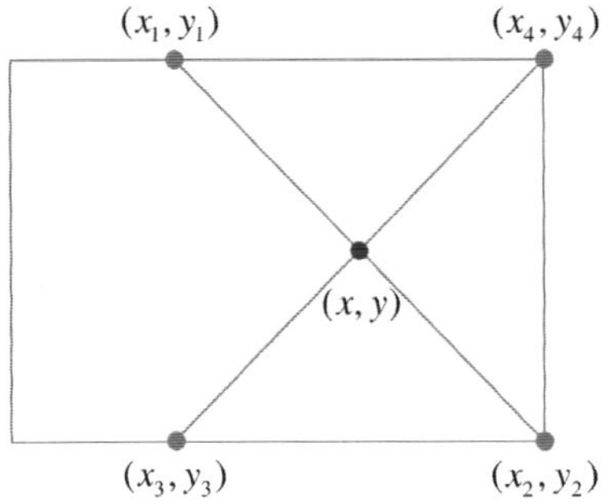

Figure 7. Calculation of the coordinates of the intersection

In Figure 7, the pixel line was extended from its center for each pixel position using the linear regression curve. In the linear regression curve that was used, the line was extended to the end point of the ROI window, and the coordinates (x, y) of the intersection of the two regression curves were calculated using equation (3).

$$x = \left(\frac{(x_1 y_2 - y_1 x_2)(x_3 - x_4) - (x_1 - x_2)(x_3 y_4 - y_3 x_4)}{(x_1 - x_2)(y_3 - y_4) - (y_1 - y_2)(x_3 - x_4)} \right)$$
$$y = \left(\frac{(x_1 y_2 - y_1 x_2)(y_3 - y_4) - (y_1 - y_2)(x_3 y_4 - y_3 x_4)}{(x_1 - x_2)(y_3 - y_4) - (y_1 - y_2)(x_3 - x_4)} \right) \tag{3}$$

Figure 8 shows the displacement response measurement algorithm that uses line detection. The algorithm has a total of eight steps. From the acquired image, a square ROI window is extracted. The luminance plane of the image is extracted from the extracted ROI window. For the information on the extracted luminance plane, the Gaussian filter is used to reduce the noise in the image and to improve the disconnected points. The geometry of the structure is sharpened by applying unsharp-mask processing to the geometry for which the disconnected points were improved and the noise was reduced. Horizontal and vertical masks are applied to the improved geometry for boundary extraction, and the outer pixels are expanded using the dilation structures of the extracted boundaries. In the ROI window of the designated M×M pixel size in the expanded image, the lines with 0.3 M pixels or more are detected and extended while the lines with less pixels are removed. The displacement responses of the structure are measured by calculating the displacement responses in the *x*- and *y*-axis of the two-dimensional plane using the changes in the coordinates of the intersections of the extended lines.

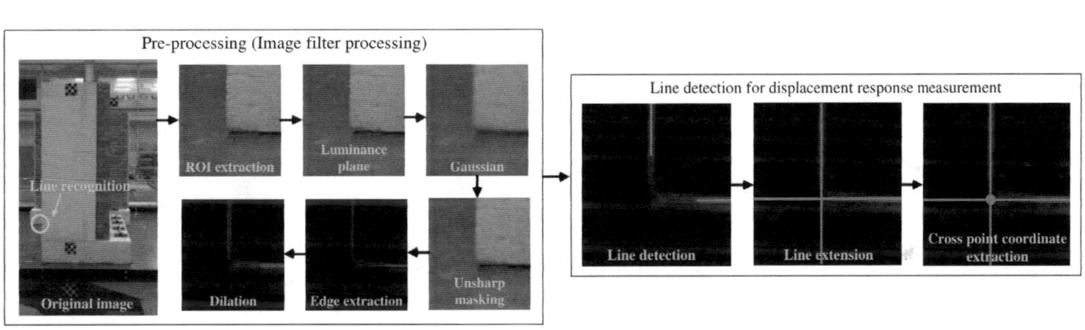

Figure 8. Algorithm summary

3. Shaking table test

To verify the algorithm for measuring the displacement responses of a structure using line detection, the shaking table test was conducted. Figure 8 shows the masonry wall installed on the shaking table. The displacement responses were measured at the point using the line detection algorithm. At the point, the table motion measured by the DAQ of the shaking table was compared with the displacement responses measured using the line detection algorithm. To extract the time history of the displacement responses for the intersection of the detected lines, the algorithm presented in Figure 8 was applied. To examine the accuracy and precision of the algorithm for measuring the displacement responses using line detection, error analysis was conducted using the percent error and the root mean square (RMS) error [7]. Figure 9 compares the table motion measured by the DAQ of the shaking table with the displacement responses extracted by the algorithm for measuring displacement responses using line detection. The error rate for the percent error was less than 1%, and the RMS error was less than 1 mm, indicating that the errors were very small. This confirms that the algorithm for extracting displacement responses using line detection has excellent reliability.

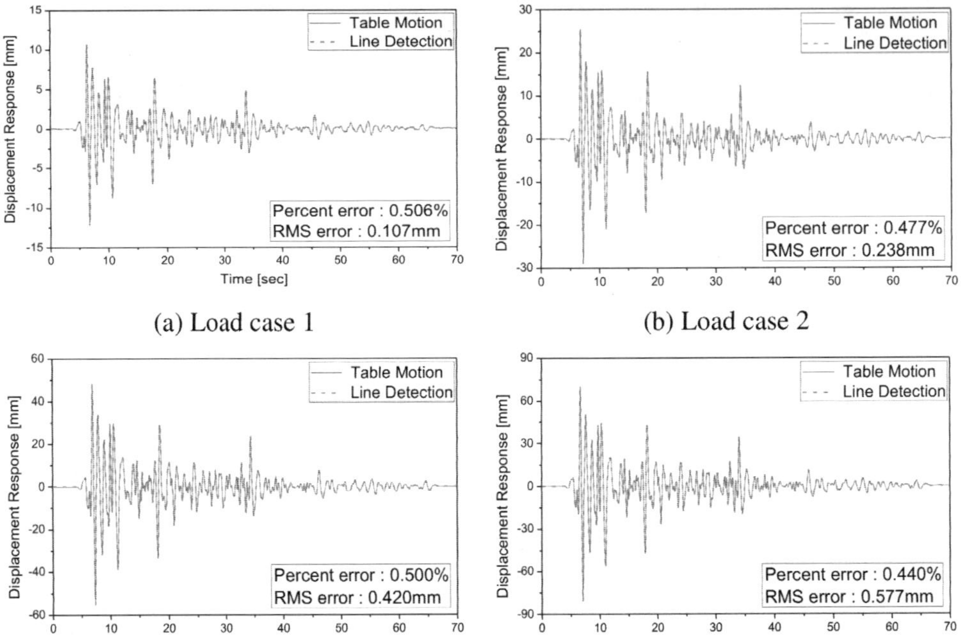

(a) Load case 1 (b) Load case 2

(c) Load case 3 (d) Load case 4

Figure 9. Comparison of the table motions

4. Conclusions

The differences between the table motion measured by DAQ in the shaking table test and the displacement responses measured by the line detection algorithm were small, thereby verifying the validity of the image processing data. In addition, it was confirmed that it is possible to measure the displacement responses of a structure using the structure's geometry, without installing arbitrary targets in the shaking table test. The proposed method of measuring displacement responses in the shaking table test using the line detection algorithm is simple and will make it possible to efficiently measure the displacement responses of a remotely located structure.

Acknowledgments

This work was supported by the Korea Institute of Energy Technology Evaluation and Planning (KETEP) and the Ministry of Trade, Industry & Energy(MOTIE) of the Republic of Korea (No. 20171510101910).

References

[1] Oh SH and Shin SH 2016 Correlation analysis of Gyeongju Earthquake waveform and structural damage scale *Journal of the Architectural Institute of Korea Structure & Construction* **32** 33-44
[2] Kim SW and Kim NS 2011 Multi-point displacement response measurement of civil infrastructures using digital image processing *Procedia Eng* **14** 195-203
[3] Kim JH, Kim KY and Sohn H 2013 In situ measurement of structural mass, stiffness, and damping using a reaction force actuator and a laser Doppler vibrometer *Smart Mater Struct* **22** 85004
[4] Kim SW, Lee SS, Kim NS and Kim DJ 2013 Numerical model validation for a prestressed concrete girder bridge by using image signals *KSCE J Civ Eng* **17** 509-517
[5] Kim SW, Jeon BG, Kim NS and Park JC 2013 Vision-based monitoring system for evaluating cable tensile forces on a cable-stayed bridge *Struct Health Monit* **12** 440-456

[6] Kim SW and Kim NS 2013 Dynamic characteristics of suspension bridge hanger cables using digital image processing *NDT E Int* **59** 25-33
[7] Kim SW, Choi HS, Jeon BG, Hahm DG and Kim MG 2018 Strain and deformation angle for a steel pipe elbow using image measurement system under in-plane cyclic loading *Nucl Eng Technol* **50** 190-202
[8] Kim SW, Choi HS, Jeon BG and Hahm DG 2019 Low-cycle fatigue behaviors of the elbow in the nuclear power plant piping system using the moment and deformation angle *Eng Fail Anal* **96** 348-361
[9] Kim SW, Choi HS, Park DU, Baek ER and Kim JM 2018 Water level response measurement in a steel cylindrical liquid storage tank using image filter processing under seismic excitation *Mech Syst Signal Process* **101** 274-291

IOP Conf. Series: Materials Science and Engineering **652** (2019) 012016 doi:10.1088/1757-899X/652/1/012016

Construction and demolition waste aggregates: analysis of the physical and mechanical properties of mortars

G Calcado[1], L Alves[2,5], E Vazquez[3] and R D Toledo Filho[4]

[1] Civil Engineering Program, Fluminense Federal University (UFF), 156 Rua Passo da Pátria, Rio de Janeiro, Brazil
[2] Department of Mechanical Engineering, Federal Center of Technological Education Celso Suckow da Fonseca (CEFET/RJ), 229 Av. Maracanã, Rio de Janeiro, Brazil
[3] Department of Civil Construction, Federal University of Rio de Janeiro (UFRJ), 149 Av. Athos da Silveira Ramos, Rio de Janeiro, Brazil
[4] Civil Engineering Program, Federal University of Rio de Janeiro (UFRJ), 149 Av. Athos da Silveira Ramos, Rio de Janeiro, Brazil
[5] Corresponding author: lais.alves@cefet-rj.br

Abstract. Civil construction is responsible for the excessive consumption of natural resources and the generation of the largest share of solid urban waste. In this context, recycling of construction and demolition waste (CDW) is presented as an alternative to minimize the impacts generated by construction. This study aimed to evaluate the influence of recycled CDW aggregates on the behavior of Portland cement mortars with three distinct mixtures. In addition to the reference mixtures, mortars with three different substitution levels of the natural aggregate for recycled aggregate, which were submitted to physical and mechanical tests in the Laboratory of Structures and Materials (LABEST) - COPPE/UFRJ. The results showed that for some substitution levels there was no loss on the properties of the mortars. Therefore, their performance is a function of the mixture adopted and the substitution content of the natural aggregate for the recycled.

1. Introduction

In recent years, the sharp industrial growth in emerging countries members of the BRICS (Brazil, Russia, India, China and South Africa) suggests that demand for raw materials will increase further in the coming decades [1]. The most developed countries have already been suffering from the scarcity of natural resources such as gravel and sand. In view of these developments, the search for ways to increase material use efficiency becomes increasingly important.

It has been acknowledged that the construction industry is one of the main contributors for environmental degradation around the world. The industry has produced large amount of waste that have a great impact on the ecosystem and human health [2]. Demolition and construction activities are responsible for the generation of 31% of waste among the monitored countries and accounts for 12% of municipal waste [3]. Concern for the environment has lately heightened awareness about the need for recycling in the construction industry [4]. In addition, some research dated back in 1996 [5] and 1998 [6] showed that about 40 to 50% of all the recycled material becomes fine aggregates, which suggests that it can replace natural sand.

Concrete recycled aggregates are characterized by high water absorption and loss of mechanical properties when compared to natural aggregates [7]. These characteristics can be explained by the high

porosity of the outer layers of the particles [8]. Several studies show the feasibility of using recycled aggregates in both large aggregate [9, 10] and fine aggregate [11, 12] for the production of concrete. The results obtained by Martinez et. al. [13], Corinaldesi et. al. [14] and other researchers show that partial substitution of natural aggregates by recycled mortars can present satisfactory performance.
Some countries like the Netherlands, Germany, Denmark and Switzerland, can reuse and recycle 50% to 90% of generated CDW [15]. However, despite the high rate of recycling of waste generated, less than 20% of natural aggregate is replaced by recycled aggregate in these countries, indicating that the regularization of land and ground are the main destinations for such waste [16].
This article aims to present the evaluation of the influence on testing tensile strength by diametrical compression, wicking and absorption by immersing the mortar mixtures, as well as characterization of aggregates of mortars produced with recycled aggregate and those that have only added natural. From the analysis, we intend to expand knowledge in this area and verify the technical feasibility of using recycled aggregate from RCD as raw material, allowing in this way, insertion of new materials on the market.

2. Materials and Methods

The cement selected for the experimental mortars was type CPI-32. Deionized was used water for moulding and curing of the specimens. The recycled aggregate was obtained by crushing the CDW from the demolition of the South Wing of the University Hospital Clementino Fraga Filho in Rio de Janeiro.
An experimental analysis of different substitution levels of natural aggregate for recycled aggregate was performed resulting on twelve mixtures, blended with three distinct mixture and water/cement ratio (w/c). Mixture M1 had trace 1:2 and w/c of 0.45, mixture M2 had trace 1:4 and w/c ratio of 0.68 and mixture M3 had trace 1:6 and w/c 1.05.
For each of these classifications different percentages of substitution were defined, based on the volume of cement, natural aggregate (AGN) and recycled aggregate (AGR): 0% (no replacement of natural aggregate), 15%, 25% and 50% of replacement of AGN.
The experimental campaign consists in determining physical properties of the aggregate (particle size analysis and density) and the mechanical (tensile strength by diametrical compression) and physical (total absorption and capillarity absorption) properties of mortar samples dated 28 days.

3. Results and discussion

The results of the first experimental program refer to particle size analysis tests and density of natural and recycled aggregates. Table 1 presents a comparison of parameters of the granulometric curves of natural and recycled aggregate. The recycled aggregate presents a specific mass of 2,45 g/cm^3, slightly lower than the natural aggregate (2.67 g/cm^3) and, therefore, it is expected that the density of the mortar containing AGR is lower.

Table 1. Granulometric parameters of natural aggregate x recycled aggregate.

Properties	AGN	AGR	Unit
FN:	2.55	2.89	-
ϕ max.:	4.80	4.80	mm
C_{nu}	3.90	8.00	-
D_{10}	0.15	0.90	mm
D_{50}	0.43	0.57	mm
D_{80}	0.99	1.73	mm
D_{90}	1.54	2.55	mm

Table 2 presents the results of mechanical tests by tensile strength by diametrical compression for 28 days. For data analysis, the average tensile strength and standard deviation were calculated.
For M1, a reduction in tensile strength of 3.4%, 2.1% and 3.4% can be observed as the percentage of substitution increases, which indicates that there was no significant reduction in this property. For mixture M2, at 15% and 25% of substitution, the tensile strength increases 11.6% and 12.0%, respectively, and for 50% substitution, a decrease of 1.4% can be observed.

Table 2. Tensile strength of the mortar mixtures.

Mixture	MPa	+/-	Mixture	MPa	+/-
M1	3,88	0,38	M2-25%	2,43	0,36
M1-15%	3,75	0,45	M2-50%	2,14	0,22
M1-25%	3,80	0,21	M3	1,01	0,12
M1-50%	3,75	0,22	M3-15%	0,83	0,16
M2	2,17	0,27	M3-25%	1,15	0,15
M2-15%	2,42	0,29	M3-50%	1,00	0,11

Considering the standard deviation, it can be said that there was no significant change in the tensile strength of the median mortars with different replacement levels. As for M3, the mixtures at 15% and 50% showed an increase of 17.8% and 1.0%, and for 25% a decrease of 13.9%. Despite the results obtained for mixtures of 25% and 50%, where tensile strength increased or remained constant, the mixture for 15% showed a decrease, making it difficult to get a clear conclusion for the test performed.
To evaluate the durability of mortars made with aggregates from CDW, physical absorption tests were performed. In the wicking test, the measured pores were only connectable. In the total absorption test, pluggable pores and non-pluggable were measured. The results showed that, for first stage absorptivity coefficient, the mixed mortars presented an increase when compared to the reference mixture M1 of 66.8%, 21.3% and 49.4% for 15%, 25% and 50%, respectively. For the absorptivity coefficient at second stage there was a reduction of 46.9% in the mixture 15% and an increase of 49.6% and 64.0% in the mixtures 25% and 50%. The saturation point was increased compared to the reference mixture by 75.5%, 40.8% and 87.8%.
When comparing to the reference mixture M2, the first stage absorption coefficient increased 24.1% and 95.5% in the mixtures 15% and 50% and decreased by 12.7% in the mixture 25%. The absorptivity coefficient of the second stage was reduced by 6.7%, 24.2% and 71.1%. The saturation point increased when compared to the reference mixture M2 of 21.1%, 40.0% and 83.2%, respectively. As expected, there was a gradual increase in the value of the saturation point higher substitution levels of the natural aggregate for the recycled.
For the reference mixture M3, the first stage absorptivity coefficient increased 25.2%, 29.8% and 40.8%. The absorptivity coefficient of the second stage was reduced by 21.6%, 11.8% and 21.6%. The saturation point increased when compared to the reference M3 by 5.5%, 11.8% and 19.3%.
Figures 1, 2 and 3 shows the mass gain curves versus the square root of time for all the mixtures studied, respectively M1, M2 and M3. It can be concluded that the poor results obtained for mortar mixtures are as expected, considering that there is an increasing water absorption by capillarity as the percentage of substitution of natural aggregate for the recycled increases. Thus, the mortar with increasing substitution content also has higher porosity and a poorer quality.

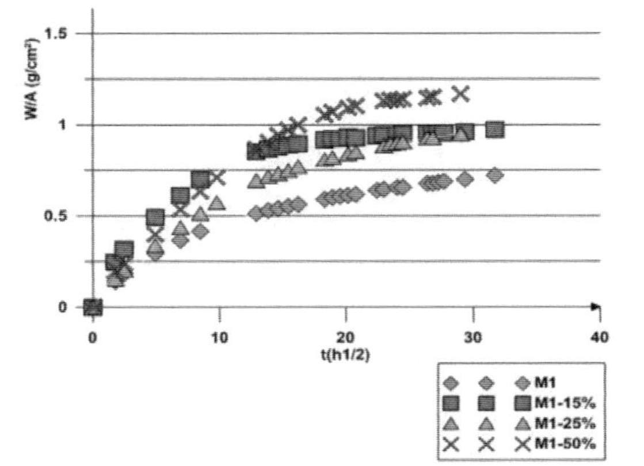

Figure 1. Mass gain curves versus the square root of time for mixture 1.

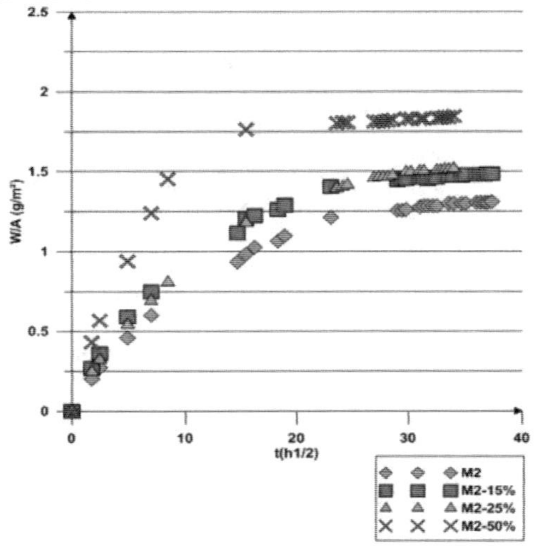

Figure 2. Mass gain curves versus the square root of time for mixture 2.

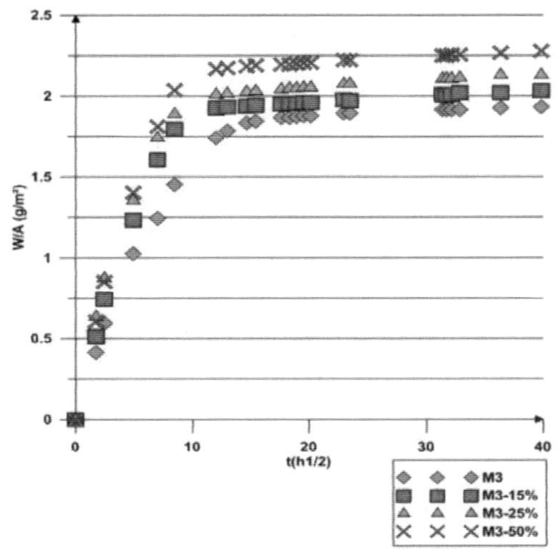

Figure 3. Mass gain curves versus the square root of time for mixture 3.

Table 3 shows the results for the physical experiments. In total absorption test in mixture1 there was a gradual increase of absorption (6.3%, 35.4% and 50.2%) as an increase in the substitution level occurred.

Table 3. Absorption (A) and voids (V) of mortar mixtures.

Mixture	A (%)	+/-	V (%)	+/-	Mixture	A (%)	+/-	V (%)	+/-
M1	4.30	0.24	8.98	0.48	**M2-25%**	8.91	0.87	17.14	1.38
M1-15%	4.57	0.27	9.40	0.54	**M2-50%**	9.82	0.68	18.44	1.02
M1-25%	5.82	0.38	11.75	0.71	**M3**	11.37	0.69	21.34	1.20
M1-50%	6.46	0.59	12.73	1.05	**M3-15%**	13.22	0.58	24.03	0.85
M2	7.10	0.32	13.85	0.54	**M3-25%**	13.11	0.88	24.16	1.27
M2-15%	7.75	0.42	14.94	0.74	**M3-50%**	14.93	0.88	26.72	1.24

IOP Conf. Series: Materials Science and Engineering **652** (2019) 012016 doi:10.1088/1757-899X/652/1/012016

The fact that the mixtures 25% and 50% have presented significant increase in absorption is related to a greater presence of recycled aggregate, more porous than natural aggregate, affecting the porosity and quality of the mortar. For M2, the increase in absorption was of 9.2%, 25.5% and 38.3%, as the substitution levels increase. Regarding the M3 mixture, the increase was 16.3%, 15.3% and 31.3%. The mixtures that showed variation of more than 10% shows a significant change in absorption.
By analysing the percentage differences of voids in mixture M1, the increase observed is 4.7%, 30.9% and 41.8%, as the substitution levels are higher. Although the specific mass has not been changed, the mixtures containing more than 15% replacement of AGN for AGR show a considerable increase in void volume. Thus, become more porous, having therefore a higher absorbency. Through analysis of the percentage differences of the voids in the mixture M2, an increase of 4.7%, 30.9% and 41.8% is observed. For mixtures 25% and 50%, growth in the void ratio is considered significant, which was expected. As for M3, the increase on the void ratio was of 12.6%, 13.2% and 25.2%. Analysing Table 3, it can be observed that all mixtures had a significant change in the void ratio, a considerable increase in porosity and, therefore, in the void volume, as the replacement content of natural aggregate by recycled increased.

4. Final considerations

This article presents the evaluation of the influence on the compression tests to different mixtures for mortars, due to the partial substitution of natural aggregate by recycled aggregate. Mixtures were produced using 15%, 25% and 50% substitution levels, and the reference mixtures. It aimed to better understand the behaviour of mortars produced with the partial replacement of natural aggregates by recycled.
The particle size of the recycled aggregate obtained in the test was satisfactory. It was observed that the recycled aggregate used in this research has a specific mass lower than natural aggregate, which led to believe that the mortar admixture with natural aggregate possess also lower specific mass than the reference mixtures. However, all the mixtures with recycled aggregates showed no change in density when compared to the reference mixture.
For the mechanical properties of tension by diametrical compression, a progressive reduction was observed as the level of substitution increases. Regarding the physical properties of absorption, all mixtures tested showed an increase absorptivity. As for the total absorption test of the mortars it was noted a reduction in density and an increase absorption and void ratio in the mixtures.
Based on the information presented on this study, recycling is a real possibility of allocations for construction and demolition waste and the use of recycled aggregates in the production of mortars can be feasible technically and economically. The test results obtained showed that the recycled aggregate does not affect the properties of mortars, and the performance is a function of the adopted mixture and substitution level of natural aggregate by recycled aggregate.
As possible future further study of this paper other tests, such as the permeability to water and gas and compactness of the mixtures, are suggested. It is also relevant to perform an analysis of rheological behavior and physical characterization through microscopic study, by shape, roughness and compressive strength of the natural and recycled aggregates used in the study. This is recommended to better understand the influence of the recycled aggregates on the performance of the mortars.

References

[1] Moll S, Bringezu and Schutz H 2005 Resource use in European countries – material flows and resource management Environment and Energy (Wuppertal Institute for Climate)
[2] Amal E M, Yusof N and Ishak N I 2015 The compliance of green criteria into the life cycle of building material: a theoretical review Advances in Environmental Biology 4 126
[3] Pepe M, Toledo Filho R D, Martinelli E and Koenders E A B 2013 Designing Concrete with recycled ecological aggregates Proceeding of SB13 Coventry 90-99

[4] Rodríguez C, Parra C, Casado G, Miñano I, Albaladejo F, Benito F and Sánchez I 2016 The incorporation of construction and demolition wastes as recycled mixed aggregates in non-structural concrete precast pieces Journal of Cleaner Production 127 152-16
[5] Pera J 1996 State of the art report – use of waste materials in construction in Western Europe Reciclagem e Reutilização de Resíduos como Materiais de Construção 1-20
[6] Sagoe-Crentsil K, Taylor A and Brown T 1998 Properties of concrete incorporating fly ash and recycled demolition waste Materials and Technologies for sustainable construction – CIB World Building Congress 1 443-449
[7] Yong P C and Teo D C L 2009 Utilization of Recycled Aggregate as Coarse Aggregate in Concrete UNIMAS E-Journal of Civil Engineering. 1:1 1-6
[8] Kwan W H, Ramli M, Kam K J and Sulieeman M Z 2011, Influence of the amount of recycled coarse aggregate in concrete design and durability properties Construction and Building Materials 26 565-573
[9] Li T, Xiao J, Zhu C and Zhong Z 2016 Experimental study on mechanical behaviours of concrete with large-size recycled coarse aggregate Construction and Building Materials 120 321-328
[10] Etxeberria M, Vázquez E, Mari A and Barra M 2007 Influence of amount of recycled coarse aggregates and production process on properties of recycled aggregate concrete Cement and Concrete Research 37 735–742
[11] Evangelista L and Brito J 2007 Mechanical behaviour of concrete made with fine recycled concrete aggregates Cement & Concrete Composites 29 397–401
[12] Evangelista L and Brito J 2010 Durability performance of concrete made with fine recycled concrete aggregates Cement & Concrete Composites 32 9–14
[13] Martínez I, Etxeberria M, Pavón E and Díaz N 2013 A comparative analysis of the properties of recycled and natural aggregate in mansory mortars Construction and Building Materials 49 384-392
[14] Corinaldesi V and Moriconi G 2009 Behaviour of cementitious mortars containing different kinds of recycled aggregate Construction and Building Materials 23 289–294
[15] Banias G, Achillas Ch, Vlachokostas Ch, Moussiopoulos N and Papaioannou I 2011 A web-based Decision Support System for the optimal management of construction and demolition waste Waste Management 31:12 2497-2502
[16] Mueller A Closed Loop of Concrete Rubble Bauhaus (Universitat Weimar) Available on: www.uniweimar.de/Bauing/aufber/Lehre/Gastvorlesung/Barcelona/Lecture07.pdf.

Development of construction blocks from alkaline activated basic oxygen furnace slag and Fly Ash

.N.T Sithole[*] and TP Mashifana

Chemical Engineering Technology, University of Johannesburg, South Africa
[*]Email: nastassias@uj.ac.za

Abstract. The objective of this study was to develop basic oxygen furnace slag (BOFS) blended with Fly Ash (FA) based masonry blocks that can be used in building and construction. The effect of variation of FA, potassium hydroxide (KOH) concentration, curing temperature, solid-liquid ratio on the Unconfined Compressive Strength (UCS) was investigated. The optimum conditions for synthesis of BOFS/ FA based masonry blocks was found to be 9 M KOH concentration, 20% Solid- liquid (S/L) and 1:5, FA:BOFS mixing ratio and curing temperature 80 °C. The masonry block prepared under aforementioned optimum condition achieved unconfined compressive strength (UCS) of 9.67 MPa. Characterization of the masonry blocks was done using X-Ray Diffraction (XRD) analysis, X-Ray Fluorescene (XRF) analysis. The major phase identified in XRD pattern for BOFS was larnite whereas for FA was glassy aluminosilicate. Optimum BOFS/FA masonry block had new phase which is zeolite (Philipsite) responsible for higher UCS of the construction block. The produced construction block met the minimum requirements for ASTM C34-13, C129-14a and South African standard (SANS227: 2007).

1. Introduction

The rapid growth of industrialization has resulted in the generation of large quantities of Basic Oxygen Furnace Slag (BOFS) and Coal fly ash (CFA) of which the available land for land-filling of large quantities of waste is reducing all over the world [1]. In South Africa alone, 35 million tons of coal ash is generated annually, where 29 million is FA and 6 million is bottom ash [2]. The main sources of fly ash (FA) are coal fired power stations and cement producers are the main consumers of FA [2]. Previous studies on utilization of FA has found that the cement making industry uses about 72% of the total FA volume (2.5 million tons) consumed yearly in South Africa [2]. Out of the 29 million tons produces, only 2.5 million tons is consumed and the remaining annually piles up in our landfills. It is estimated that by 2020 42 million tons of FA will be generated yearly on average [2]. On the other hand, approximately 5.7 million tons of slag is produced annually in South Africa of which less than 50% is utilized and the remainder being disposed in landfills [3]. In other countries like China 120 million tons of slag is produced annually and only 22% is utilized which leads to the occupation of farmland and serious pollution to the environment [4]. In the United States 10- 15 million tons of steel slag are produced every year. Approximately 40% of the steel slag output is initially stockpiled in the steel plants and, eventually, sent to slag disposal sites [5]. Alkaline activated BOF slag utilization in road constructions has decline due to high concentration of free lime. It has been reported that less than 50 percent of the BOF slag is utilized, and the remainder is dumped. Previous studies on the properties of BOFS for construction applications, showed that free lime (CaO) causes expansion at early ages and form Ca (OH)2 resulting in dimensional instability of building or material constructed

[6]. BOFS and Fly ash application technology in South Africa has been primarily been used in civil engineering works. However, the use of BOFS is not suitable for road base course due to its high concentrations of free lime. The soluble lime hydrate in BOFS causes expansion and disintegration of roadbeds and embankments. However, recent studies reported that slag expansion could be avoided if the slag particle sizes of less than 13.2 mm for use in building and construction [7]. Sithole et al. [11] investigated development of lightweight construction blocks by alkaline activation of BOF slag. It was found that optimum synthesis conditions were 5M NaOH, 80°C and 1:1 Sodium Silicate: NaOH ratio with a minimum unconfined compressive strength (UCS) of 4.1 MPa. Furthermore, [10] investigated the development of lightweight construction blocks from alkaline activated desilicated fly ash. It was found that FA could be alkali activated using KOH to produce competent lightweight construction block. The developed construction block satisfied the minimum requirements for ASTM C34-13 and South African standards (SANS227: 2007) for non-facing building masonry with a minimum UCS of 14. 25 MPa. The feasibility to synthesize construction blocks from alkaline activated basic oxygen furnace slag and Fly Ash that meet the minimum requirements for ASTM C34-13, C129-14a and South African standard (SANS227: 2007) was explored. In addition, in view of declining investments in construction projects, including public works projects in recent years, promoting effective utilization of BOFS and fly ash while responding to environmental demands is of paramount importance.

2. Materials and Methods

Fly Ash and Basic Oxygen Furnace Slag samples were obtained from Eskom power station and AcerloMittal South African respectively. Rochelle Chemicals supplied the alkaline activator KOH. X-ray Fluorescence (XRF- Rigaku ZSX Primus II) was used to determine the elemental composition of the material. The samples of FA and BOFS were dried in an oven at the temperature of 50 °C before conducting any experimental work. The material was then pulverized to fine particles size of -75µm. Alkaline activation of FA and BOFS were carried out varying the BOFS: FA ratio, alkaline activator concentrations, curing temperature and solid to liquid ratio. The alkaline activated BOFS/FA paste were cast and cured for 7 days. The chemical and geotechnical properties of FA and BOFS are shown in Table 1.

Table 1. Chemical and geotechnical properties of BOFS and Fly Ash

Parameter	BOFS	FA
pH	12.7	10.98
Specific gravity	3.27	2.47
%Na_2O	0.191	0.905
% CaO	51.81	12.66
% Al_2O_3	3.5	16.07
% SiO_2	7.7	40.4
% MnO	4.188	0.25
% Fe_2O_3	27.58	24.6
% Gravel	0	0
% Sand	77.92	45
% Fine	11.12	55
% Silt	10.96	54.7
%Clay	0	0.4
Liquid limit	76	54
Plastic limit	non plastic	3.65
Shrinkage limit	non shrinking	4.2

MDD (kg/m^3)	2265	1205
OMC (%)	10.58	24.3

2.1 Preparation of the paste

280 g of BOFS and FA were mixed using KOH solution. After forming a workable slurry, the slurry was poured into a 50 x 50 x 50 mm3 mould. The cast sample was allowed to set and was removed from the mould when it was stiff enough for demolding. The hardened samples were then cured until they were dry. Three specimens were casted to investigate each variable per parameter.

3. Results and Discussion

3.1 Effect of FA variation on the UCS

Figure 1 below shows the effect of FA variation on the UCS prepared with different FA content.

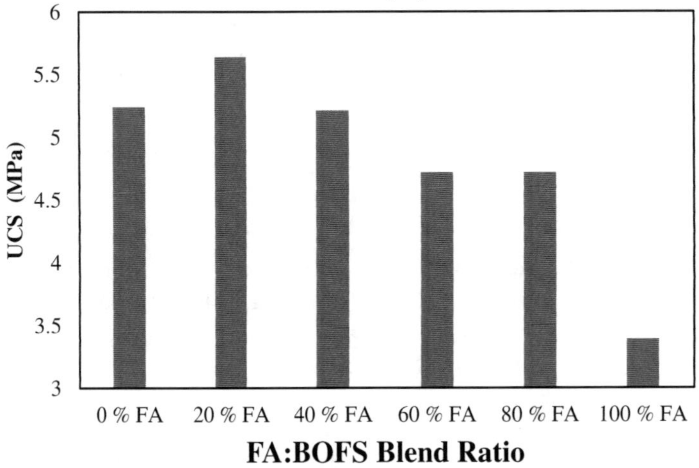

Figure 1. The effect of variation of FA on the UCS

Figure 1 shows that BOFS construction blocks prepared with FA/BOFS of 20% had the highest UCS of 5.63 MPa; followed by a construction block prepared with 100% of BOFS and 0% of FA (5.24 MPa). As Si/Al ratio plays a huge role during alkaline activation, researchers had shown that the addition of FA to BOFS based construction blocks helps to maintain the ratio of Si/Al to be at 1 to 3 [8]. This range gives an optimum UCS for development of construction block that meet the minimum requirements for ASTM C34-13, C129-14a and South African standard (SANS227: 2007) [1]. XRF results (Table 1) supports the aforementioned as the 9 M BOFS based geopolymer prepared with 20 % FA maintain a ratio Si/Al to be 1.9. Construction blocks prepared with higher FA: BOFS ratio resulted in an increase in CaO, thus results in a decrease in UCS. Another reason to account for an increase in UCS of BOFS/FA based construction block is the BOFS: FA (80:20), the reactivity of FA is higher than that of BOFS which results in an increase in UCS of geopolymer [8].

Materials with higher reactivity allows more silica and alumina to be dissolved during alkaline activation [8]. The construction blocks prepared with 40%, 60%, 80% and 100% FA/BOFS had UCS of 5.21 MPa, 4.71 MPa, 4.71 MPa and 3.39 MPa respectively. Figure 1 also shows that there was a decrease in UCS from a construction block prepared with 40% FA to 100 % FA.

3.2. Effect of KOH concentration on the UCS of geopolymer

Figure 2 shows that a construction block prepared with 9 M KOH concentration had highest compressive strength compared to the construction block prepared with 6M and 3M. UCS for construction block prepared with 3 M, 6 M and 9 M KOH solution were 5.90 MPa, 5.96 MPa and 6.63 MPa respectively. There was an increase in UCS with an increase in KOH. This is because of higher

KOH concentration contributes to dissolution of more silica and alumina and formation of larger amount of geopolymeric gels and consequently results in higher UCS [8].

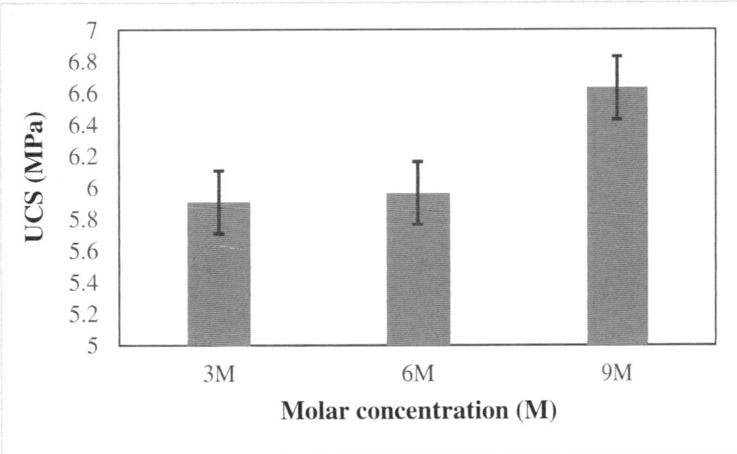

Figure 2. The effect of molar concentration on UCS of the construction block

3.3 Effect of solid-liquid ratio on the UCS

Figure 3 shows the effect of solid-liquid (S/L) ratio on the UCS

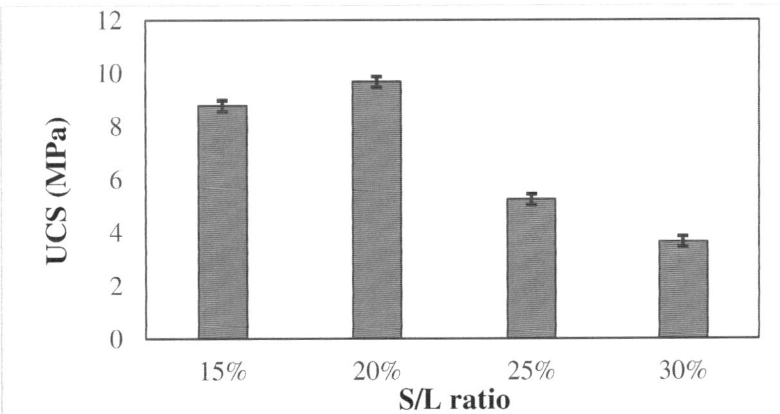

Figure 3 : Effect of solid-liquid (S/L) ratio on the UCS of construction block.

Figure 3 shows that geopolymer prepared with 20% S/L had the highest UCS which was 9.67 MPa. The UCS for S/L ratio of 15%, 25% and 30% were 8.77 MPa, 5.24 MPa and 3.65 MPa respectively. The construction block paste prepared with 25-30% S/L ratio was less workable and exhibits segregation compared to the construction blocks paste prepared with 20% S/L ratio. At 15% S/L ratio, the UCS of geopolymer is slightly close to that of optimum one (20% S/L ratio), the slight decrease can be accounted by the fact that the construction blocks mix with 15% S/L ratio was slightly low in liquid content which makes the geopolymer paste less workable and thus difficult to mix resulting in slight decrease on UCS as compared to construction block with 20% S/L ratio.

3.4. The effect of curing temperature on the UCS of geopolymer

Figure 4 shows the effect of curing temperature on the UCS of construction blocks. Curing temperatures were 25 °C, 40 °C, 80 °C and 100 °C respectively.

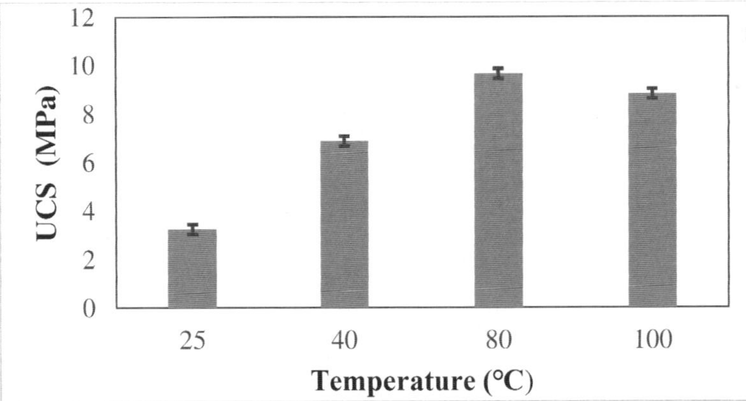

Figure 4: Effect of temperature variation on the UCS construction blocks.

Figure 4 shows that the construction blocks cured at 80 °C had the highest UCS (9.67 MPa). Construction blocks cured at 25 °C, 40 °C and 100 °C had a UCS of 3.24 MPa, 6.89 MPa and 8.83 MPa respectively. Figure 4. shows that UCS increases with an increase in curing temperature from 25 °C up to 80 °C. Beyond 80 °C, UCS begins to decrease. This agrees with the studies conducted by Jaarsveld and Deventer, which studied the effect of temperature between ranges from 50°C to 100 °C [2]. Curing at 100 °C resulted in a decrease in UCS because at high temperature the construction block begins to lose its moisture resulting in the construction block cracking and losing its structural integrity [2, 3]. Studies by Deventer supported that at 80 °C (optimum curing temperature), construction block is still retaining some moisture which prevents cracking thus makes its compressive strength higher [2].

3.5 XRF analysis before and after alkaline activation effect of curing temperature on the UCS of geopolymer

Table 2 shows the XRF results of raw FA, raw BOF slag and 9M BOF/FA geopolymer. The constituent of each elements in oxide is shown in percentage.

Table 2: Chemical composition of FA and BOF slag

Chemical composition (%)	Raw FA	Raw BOFS	9M BOFS/FA
Na_2O	0.47	0.16	8.26
MgO	1.60	3.63	2.45
Al_2O_3	20.34	4.21	6.90
SiO_2	44.98	10.51	18.35
P_2O_5	0.41	0.79	0.47
SO_3	1.68	0.97	1.35
K_2O	1.00	0.1	0.3
CaO	12.55	47.50	32.88
TiO_2	1.72	0.37	0.71
Cr_2O_3	0.11	0.20	0.22
MnO	14.6	3.95	2.85
Fe2O3	0.11	-	25.85
SrO	0.23	-	0.1

Table 2 shows that the main dominant elements in the FA oxides are aluminium (20.34 %) and silicon (44.98 %) hence FA is the rich source of aluminosilicate and was used to supplement BOFS to produce alkaline activated BOFS/FA based construction blocks. The main oxide element present in BOFS is Ca (47.50 %) followed by silicon (10.51 %). XRF results of 9 M BOF/FA based geopolymer shows that CaO % has reduced from 47.5 to 32.9 % which shows that alkaline activation reaction has taken place. During alkaline activation Ca^{2+} reacts with OH- to form $Ca(OH)_2$. In the alkaline activation chain of BOFS/ FA paste, there is negatively charged ions chain of -Si-O-Si-O- , the free cation Ca^{2+} react with negative charged silicate to form $Ca_2(SiO_4)$ which is the phase to be expected in XRD pattern. There was more dissolution of aluminosilicates with increase in concentration of KOH [9]. After dissolution, Si^{4+} and Al^{3+} ions are released, since SiO_2 exist in greater percentage than Al2O3 (from XRF results table 1) the rate of leaching is higher in Si^{4+} than in Al^{3+} [9]. This means during depolymerisation reaction, a greater number of ions Si4+ took part in the formation of tetrahedral oligomeric silicate units than aluminate units. After depolymerisation, poly-condensation reaction took place and geopolymer chain is thus produced [9].

4. Conclusion

BOFS/FA based construction blocks prepared with 20 % FA accounted for high UCS because of high reactivity of FA and stabilisation of Si/Al ratio range during alkaline activation by FA. Solid-liquid ratio of 20 % contributed to an increase in UCS of the construction block because of more workability. The optimum curing temperature was found to be 80 °C because the UCS was enhance and the structural integrity of the construction block was maintained. The optimum KOH concentration for developing a construction block with high UCS was found to be 9 M with UCS of 9.39 MPa. The produced construction block met the minimum requirements for ASTM C34-13, C129-14a and South African standard (SANS227: 2007).

References

[1] M. Reutter, Y. Xiao, and U., Boin. J. South. Afr. Inst. Min. Metall. (2004).
[2] R. A., Kruger. Nashville: World of Coal Ash (2015).
[3] Department of Environmental Affairs (2012). National Waste Information Baseline Report. Department of Environmental Affairs. Pretoria, South Africa.
[4] L. Huang and D., Lin. Int. J. Pavement Res. and Technol. 6(4), 347-355 (2011)
[5] L.Z., Yildirim and M., Prezzi, Advances in Civil engineering (2011).
[6] R.M., Novais, L.H., Buruberri, G., Ascensão, M.P. Seabra, and J.A. Labrincha J CLEAN PROD, 119, 99-107(2016).
[7] L., Yüksel, İ. Development and Sustainability, 19(2), 369-384 (2017).
[8] L., Zhang, S., Ahmari, & J., Zhang, Constr. Build. Mater. 25(9), 3773-3781 (2011)
[9] T. S., Singh, & K. K., Pant, J. Hazard. Mater, 131(1-3), 29-36 (2006)
[10] T., Falayi, F. N., Okonta, and F. Ntuli, International Journal of Environment and Waste Management, 20(3) 233-253(2017)
[11] N.T., Sithole, F. N., Okonta, and F., Ntuli Solid waste management, in press (2019).
[12] ASTM. (2013). Standard Specification for Structural Clay Load-Bearing Wall Tile. C34-13 Ayala, J., Blanco, F., Garcia, P., Rodriguez, P., Sancho, J., (1998). Asturian fly ash as a heavy metals removal material. Fuel, 77, pp.1147-1154.
[13] South African National Standards (SANS 227:2007), Burnt clay masonry units.

Deformation of cantilever foundation pit engineering in sand ground: centrifuge model test

Q H Fu[1,*], L X Li[2,*] and X H Chen[1]

[1]School of Transportation, Southeast University, Nanjing 211189, China
[2]School of Civil Engineering, Shandong University, Jinan 250061, China

*E-mail: shandafqh@163.com; jk_doctor@163.com

Abstract. A centrifuge model test was conducted to investigate the excavation-induced deformation characteristics of the foundation pit engineering in sand ground. Laser displacement sensors (LDSs) and the particle image velocimetry (PIV) technology were jointly used to obtain the horizontal deflection (δ_H) of the retaining wall and ground surface settlement (δ_V), and the relationship between δ_H and δ_V. Test results show that the profile of δ_V is a spandrel type with the maximum values (δ_{Vm}) close to the wall, which is consistent with Peck's result. While, the profile of the retaining wall deflection is a straight line with the maximum values (δ_{Hm}) at the top. The δ_{Vm} is equal to 1.05~1.75 δ_{Hm} for the cantilever foundation pit engineering in sand ground. Although the deformation obtained by PIV method is smaller than that obtained by LDSs because of the inevitable boundary effect of the model box, PIV is an effective method to get the deformation characteristic in the deep zone.

1. Introduction

Controlling ground surface settlement around the excavation zone and the deflection of supporting structures is an essential task in the design of an excavation project. Excessive deformation of the foundation pit frequently damages the adjacent properties in urban areas. The magnitude of the deformation is related to many factors, such as soil properties, structural properties, and excavation geometry [0]. Currently, a number of researches have been reported to give the relationship between those factors and the deformation of foundation pit. Research methods mainly include statistical analysis, theoretical analysis, laboratory model tests and finite element numerical simulation [0-0]. Many deformation prediction curves and empirical formulas have been put forward to estimate the values and influence ranges of the horizontal deflection and ground surface settlement [0, 0, 0]. However, these researches focus on the deformation of foundation pit in soft clay ground, and there is only a limited amount of researches available regarding the projects in sand ground. Besides, research methods also have certain limitations. For example, statistical analysis method has engineering conditions and geographical dependence; theoretical methods have many simplifications; laboratory small-scale model tests cannot create the stress state in prototype due to the self-weight stress dissimilarity; numerical simulation has a large dependence on the selection of constitutive model and soil parameters.

Geotechnical centrifuge model test is the most advanced and effective method in geotechnical engineering research. Its principle is based on creating the same self-weight stress in the model as in a prototype by accelerating a model in a centrifuge *N* times of the Earth's gravitational acceleration (1g). On this basis, the mechanical deformation characteristics and interaction of soil and structures are studied. Therefore, a centrifuge model test was conducted to investigate the excavation-induced deformation characteristics of the foundation pit engineering in sand ground. Laser displacement sensors (LDSs) and the particle image velocimetry (PIV) technology were used to obtain the horizontal deflection of the supporting structure and the ground surface settlement. Besides, in order to make test results more prominent, a cantilever wall was adopted as the supporting structure and the deep excavation was conducted. Research results can be used to predict the deformation characteristics of similar foundation pit projects, which is useful in protecting adjacent buildings.

2. Centrifuge model test setup and procedure

The test was carried out at 40g on geotechnical centrifuge ZJU-400 in Zhejiang University.

2.1. Excavation method

Currently, three methods are always used in centrifuge model tests on foundation pit engineering. (1) Model excavation is carried out at 1g and then the centrifuge model is subjected to an increasing centrifugal acceleration field [6]. (2) Draining heavy liquid in-flight to simulate the soil excavation [7] or (3) excavation in-flight with an advanced robot excavator [8]. However, these methods have some disadvantages, such as the inability to accurately simulate the excavation process or the difficulty of implementation. Therefore, the excavation device in reference [9] was applied in this research, which was more accurately to simulate the unloading process in-flight.

2.2. Foundation pit model preparation

Fujian standard sand (FS) was used to prepare the foundation pit model. The physical and mechanical parameters of FS are shown in Table 1. The sand pluviation method was used to build the model. The falling height was determined to be 1010 mm to get a relative density of 85% according to the calibration result of the pluviation device, which is shown in Figure 1.

Table 1. Physical and mechanical parameters of FS.

Items	*Dr* (relative density)	γ_d / (kN/m^3)	*c* / (kPa)	d_{50} / (mm)	*e*	e_{min}	e_{max}	φ/(°)
Value	85%	15.9	0	0.17	0.663	0.612	0.957	32

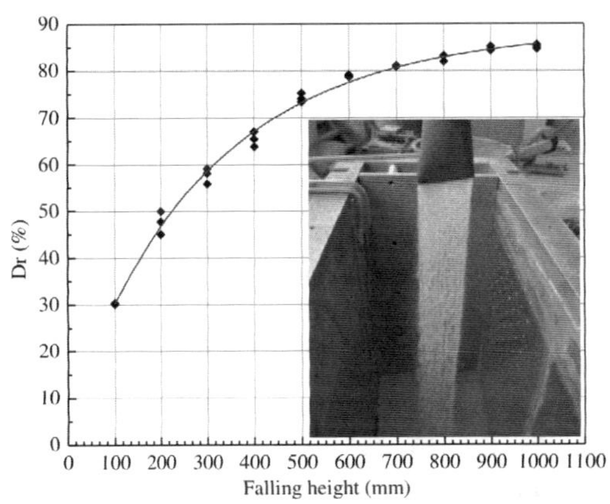

Figure 1. Calibration of the sand pluviation device.

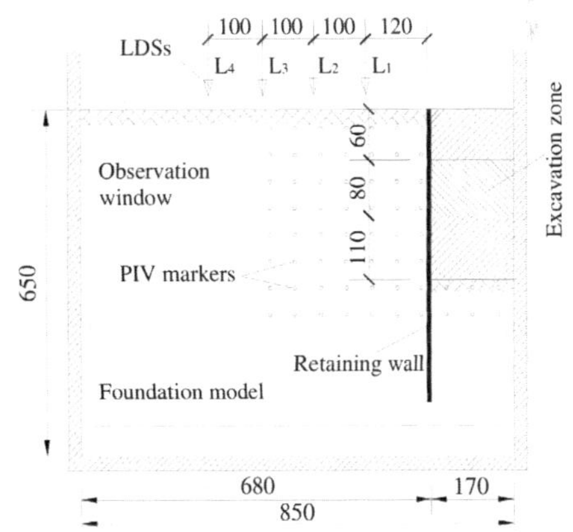

Figure 2. Model setup (unit: mm).

The model test was designed based on a foundation pit engineering with the pile-anchor supporting structure in north China. As shown in Figure 2, the height of the foundation model (the prototype) was 650mm (26m), the length was 850mm (34m), and the width was 400mm (16m). The height of the model retaining wall was 550mm (22m). The depth of the pit was 250mm (10m), which was filled with three earth-filled bags with the thickness of 60mm (2.4m), 80mm (3.2m) and 110mm (4.4m). The model retaining wall was made of an aluminum alloy plate with a thickness of 9.5 mm and a modulus of elasticity of 68.9 GPa according to the principle of equivalent flexural stiffness ($E_p \times I_p = N^4 \times E_m \times I_m$, where E_p and E_m=elastic modulus of the prototype concrete and aluminum alloy, respectively; I_p and I_m=moment of inertia of the concrete pile and aluminum alloy plate, respectively; N=40 here.). Four LDSs (L_1, L_2, L_3, and L_4) were used to get the settlement of the ground behind the retaining wall, as shown in Figure 2. Besides, a high-speed camera arranged in front of the observation window was used to get model photos for PIV analysis.

3. Test results

All the laboratory-measured deformation were converted to counterpart results in the field as well as time and dimensions according to similarity scales.

3.1. Measured surface settlement by sensors

The induced surface settlement along with the excavation process (actually, time) obtained by LDSs are shown in Figure 3. It can be seen that the ground surface settlement (δ_V) gradually increased during the excavation process. L_1, which is closest to the retaining wall, has the maximum value of the measured settlement. By contrast, L_4 has the minimum value and the least change in the process. Furthermore, based on the analysis, induced δ_V at different positions from the retaining wall after each excavation stage can be obtained. As shown in Figure 4, the maximum measured ground surface settlements (δ_{Vm}) at the position of L_1 are -5mm, -22.5mm and -48.8mm in stage 1, stage 2 and stage 3, respectively. It should be noted that stage 0, stage 1, stage 2 and stage 3 refer to the moment before the excavation and the first, second, third excavation stage, respectively.

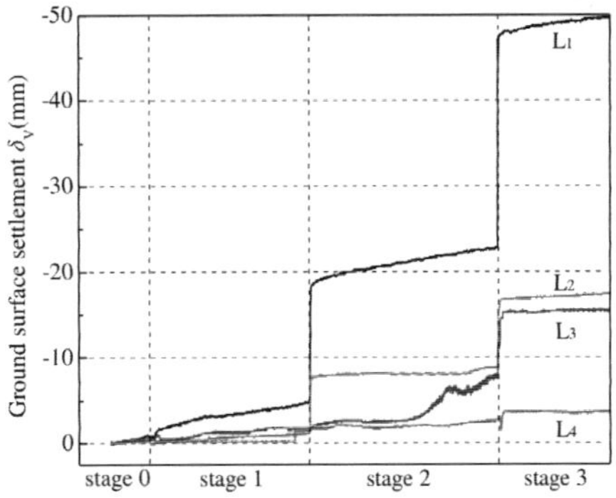

Figure 3. Measured surface settlement during the excavation process.

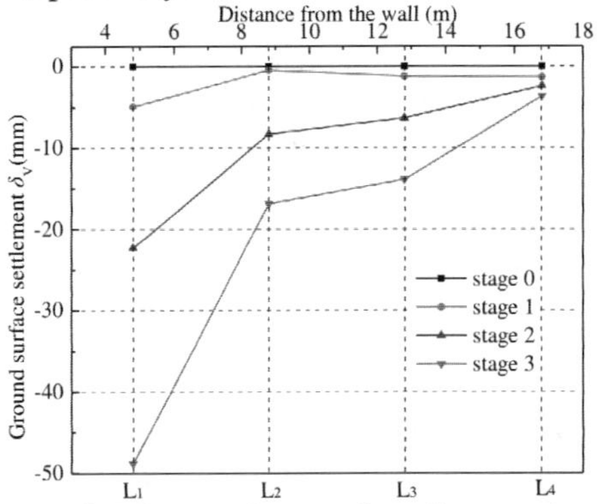

Figure 4. Measured ground surface settlement in different stages.

3.2. Foundation deformation by PIV

In many model tests, PIV technology was often used to obtain the ground deformation effectively. Therefore, this technology was taken to get the ground deformation caused by foundation pit excavation in this research. Considering the accuracy and focus of the test, only a model area of 12 m × 13 m was used for PIV analysis.

Figure 5 comparatively presents the soil displacement vectors behind the retaining wall during the excavation process. Results are magnified 10 times to ensure a better visualization of the displacement field. It can be seen that the induced foundation deformation is not obvious in stage 1, but significant in stage 2 and stage 3.

The induced δ_V, as shown in Figure 6, is obtained according to the deformation parameters of the horizontal grids in the model surface. It can be seen that the profile of δ_V is a spandrel type, which is similar to Peck's result [0]. δ_V at 4.8 m distance (L_1) from the retaining wall are -2 mm, -7.5 mm and -15 mm in stage 1, in stage 2 and in stage 3 respectively. Obviously, these values are much smaller than the measured ones (-5 mm, -22.5 mm and -48.8 mm, respectively). Because the inevitable boundary effect of the model box reduces the soil deformation in the model side. And the measured δ_V by LDSs are actually in the middle of the ground surface.

Similarly, the horizontal deflection (δ_H) of the retaining wall or the soil near the wall can be obtained by extracting the deformation parameters of the vertical grids. As shown in Figure 7, the induced horizontal deflection profile of the retaining wall can be linear fitted. Obviously, δ_H is also affected by the boundary effect.

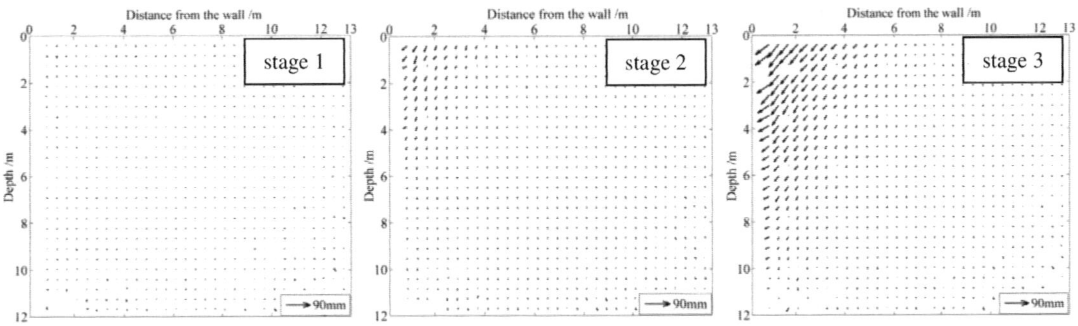

Figure 5. Soil displacement vectors behind the retaining wall by PIV in different excavation stages.

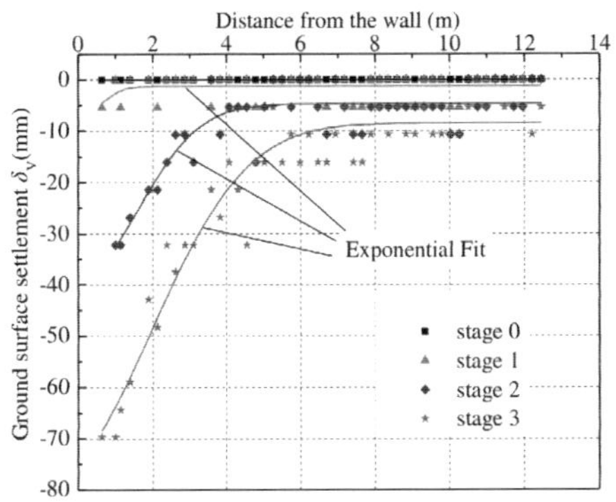

Figure 6. Ground surface settlement by PIV.

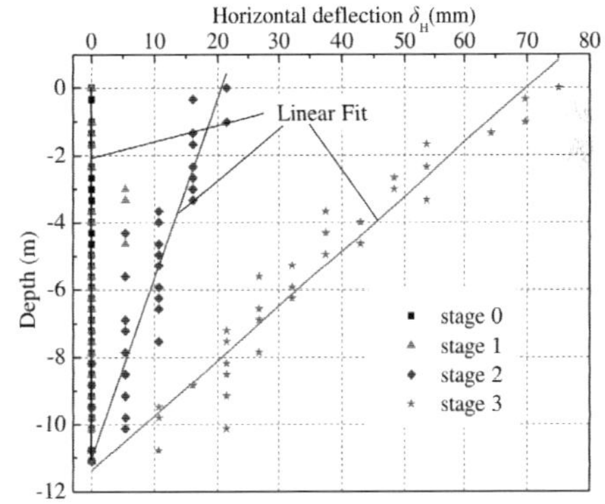

Figure 7. Deformation of the wall by PIV.

3.3. *Comparison of δ_V between test results and theoretical predictions*

As can be seen above, the deformation of the foundation pit is significant in the third excavation stage, so this section focuses on δ_V in stage 3. There are two general types of settlement profile caused by excavations: spandrel type and concave type. Many researchers have proposed empirical methods for predicting both types of settlement profile [2, 4, 5]. Figure 8 shows δ_V in stage 3 by different prediction methods. It can be seen that the values by LDSs are between that by Peck's method and Bowles's method [4, 10]. Therefore, we can assume that the maximum ground surface settlement ($\delta_{Vm\text{-}3}$) is 110mm near the retaining wall, which is about 1.1 % of the final excavation depth.

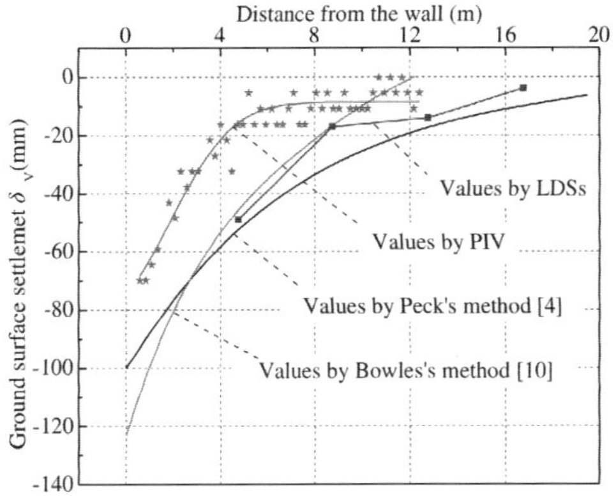

Figure 8. Comparison of δ_V in stage 3 between test results and prediction methods.

IOP Conf. Series: Materials Science and Engineering **652** (2019) 012018 doi:10.1088/1757-899X/652/1/012018

3.4. Relationship between δ_V and δ_H

Horizontal deflection (δ_H) of the retaining wall is closely related to ground surface settlement (δ_V). Once the relationship is determined, δ_V can be predicted by δ_H. Many studies have been carried out in this field [0, 0]. Therefore, this section also discusses the relationship in the case of the cantilever foundation pit engineering in sand ground. As mentioned above, the deformation obtained by PIV method is smaller than that obtained by LDSs because of the inevitable boundary effect of the model box. However, in the context of studying the relationship between δ_V and δ_H, this boundary effect can be eliminated by the dimensionless method.

Figure 9 shows the relationship between the maximum dimensionless horizontal deflection (δ_{Hm}/H_e) of the retaining wall and the maximum dimensionless ground surface settlement (δ_{Vm}/H_e). H_e herein is defined as the excavation depth. As shown in Figure 9, δ_{Vm} is equal to 1.05~1.75 δ_{Hm} for the cantilever foundation pit in sand ground. However, δ_{Vm} is equal to 0.5~0.7 δ_{Hm} for most excavation cases in Taipei [0], and 0.71 ~1.0 δ_{Hm} for those excavation projects in Shanghai [0]. Therefore, it can be seen that the deformation caused by excavation in sand ground is more obvious. The support or anchor structure plays an important role in controlling the deformation of the foundation pit.

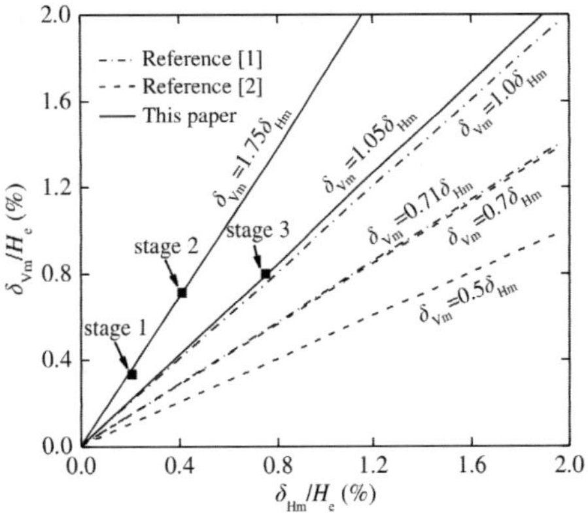

Figure 9. Relationship between δ_H and δ_V by PIV.

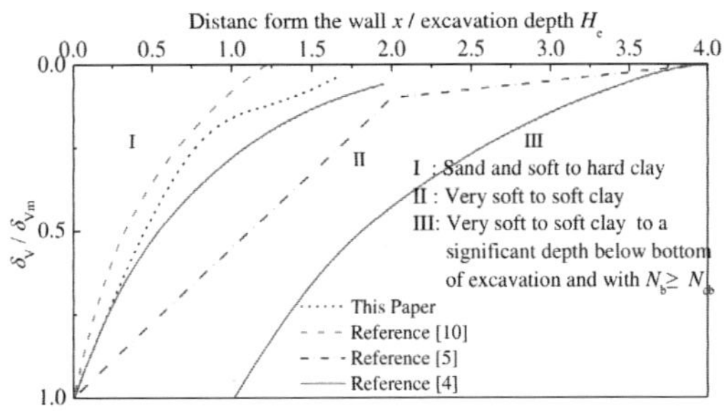

Figure 10. Surface settlement profile.

3.5. Ground surface settlement profile

The magnitude and distribution of the settlement are related to many factors such as construction quality, ground condition, excavation depth, and support system. A theoretical method to predict those result will be complex and inaccurate. Therefore, most of the prediction approaches were obtained based on field measurement. Figure 10 presents the final ground surface settlement profile in this research and some of the empirical methods currently used in engineering practice. It can be seen that the ground surface settlement profile in this study is between the profiles proposed by Bowel and Hsieh&Ou [5, 10]. Besides, it is just in zone I for sand and soft to hard clay, which is proposed by Peck [4].

4. Conclusions

A centrifuge model test was conducted to investigate the excavation-induced deformation characteristics of the foundation pit engineering in sand ground. LDSs and PIV technology were used to obtain the retaining wall deflection and the ground surface settlement. The findings are summarized as follows:

(1) The induced foundation deformation gradually increased during the excavation process. More specifically, the deformation is not obvious in stage 1, but significant in stage 2 and stage 3. The profile of the induced ground surface settlement (δ_V) in this study is a spandrel type, in which the maximum ground surface settlement (δ_{Vm}) occurs very close to the wall. It is consistent with Peck's result. While, the profile of the retaining wall deflection is a straight line with the maximum values at the top.

(2) The deformation of the foundation pit in sand ground is more obvious than those in soft clay ground. The maximum ground surface settlement in stage 3 ($\delta_{\mathrm{Vm-3}}$) is about 1.1 % of the final excavation depth. And δ_{Vm} is equal to 1.05~1.75 times of the maximum wall deflection (δ_{Hm}) for the cantilever foundation pit engineering in sand ground.
(3) The deformation obtained by PIV method is smaller than that obtained by LDSs because of the inevitable boundary effect of the model box. However, PIV is an effective method to get the deformation characteristic in the deep zone.
(4) Further tests in the context of the other soils will be carried out to study the excavation-induced deformation characteristics in the future.

Acknowledgments
This research was sponsored by the Natural Science Foundations of China (grant No. 51479107; 51508310) and the Fund for Science and Technology of Jinan (grant No. 201201145). The authors really appreciate their financial support. All technicians in MOE Key Laboratory of Soft Soils and Geoenvironmental Engineering of Zhejiang University are also acknowledged.

References
[1] Ou C, Hsieh P, and Chiou D, 1993. Characteristics of ground surface settlement during excavation. *Can. Geotech. J.*, **30**(5):758-67.
[2] Liu G and Wang W, 2009. Excavation engineering manual. 2nd ed. Beijing: China Architecture and Building Press.
[3] Roboski J and Finno R, 2016. Distributions of ground movements parallel to deep excavations in clay. *Can. Geotech. J.*, **43**(1): 43-58.
[4] Peck R, 1969. Deep excavation and tunneling in soft ground. Proceedings of the 7th International Conference on Soil Mechanics and Foundation Engineering, State-of-the-Art-Volume, Mexico City, pp. 225-290.
[5] Hsieh P and Ou C, 1998. Shape of ground surface settlement profiles caused by excavation. *Can. Geotech. J.*, **35**(6):1004-17.
[6] Lyndon A and Schofield A, 1972. Centrifuge model test of short term failure in London clay. *Geotechnique*, **20** (4), 440 - 2.
[7] Ong D, Leung C, and Chow Y, 2006. Pile behavior due to excavation-induced soil movement in clay. I: Stable wall. *J. Geotech. Geoenviron.*, **132** (1): 36-44.
[8] Goh A, Teh C, and Wong K, 1997. Analysis of piles subjected to embankment-induced lateral soil movements. *J. Geotech. Geoenviron.*, **123** (9), 792-801.
[9] Li L and Fu Q, 2017. Lateral behavior of composite foundation due to adjacent excavation: centrifuge model test. *China Civil Engineering Journal*, **50**(6): 85-94.
[10] Bowles J, 1988. Foundation analysis and design. 4th ed. McGraw-Hill Book Company, New York.

IOP Conf. Series: Materials Science and Engineering **652** (2019) 012019 doi:10.1088/1757-899X/652/1/012019

Evaluation on low temperature characteristics of SBS/CR modified asphalt binder under different aging conditions

Jie Zhou[1], Xianhua Chen[1,*], Gang Xu[1] and Hanyu Zhang[1]

[1]School of Transportation, Southeast University, 211189 Nanjing, China
*Corresponding author e-mail: chenxh@seu.edu.cn

Abstract. This study discussed the low temperature performance of styrene–butadiene–styrene (SBS)/crumb rubber (CR) modified asphalt under different aging process. The bending beam rheometer (BBR) test was conducted to compare the low temperature performance of SBS and SBS/CR modified asphalt at three different temperatures. The continuous low service temperature of two modified asphalt binders was calculated based on the ASTM method. The simple fractional viscoelastic model was implemented to better describe the creep behaviour of asphalt at low temperatures from the BBR test results. The damping ratio and dissipated energy ratio were also derived from the simple fractional model to assess the low temperature performance of asphalt from the aspect of energy. The analysis results showed that adding CR to SBS modified asphalt was beneficial to reduce the low service temperature and increase the values of damping ratio and dissipated energy ratio before and after the aging process. It has been proved that the low temperature performance of SBS/CR modified asphalt is much better than that of SBS modified asphalt under different aging conditions.

1. Introduction

In cold regions, low temperature cracking is one of the main diseases of asphalt pavement [1]. The anti-crack property of the asphalt pavement is directly related to the low temperature performance of the asphalt binder [2]. So the modification of asphalt by adding modifiers in neat asphalt has become an important means to improve the low temperature performance of asphalt pavement. On the one hand, the styrene–butadiene–styrene (SBS) modified asphalt has become one of the most widely used modified asphalts due to its superior comprehensive performance and mature production technology. However, SBS modified asphalt has insufficient anti-aging properties and high price, which limits its application range [3]. On the other hand, grinding waste rubber tires to obtain crumb rubber (CR) for asphalt modification is an important way to treat waste tires [4]. The low temperature performance and anti-aging performance of CR modified asphalt have been significantly improved compared with neat asphalt [5]. However, the CR modified asphalt has unsatisfied storage stability because of the poor compatibility between CR and asphalt [6]. Due to the shortcomings of both SBS modified asphalt and CR modified asphalt, the SBS and CR modifiers have been applied in the composite modification of neat asphalt [7,8]. The effective use of waste tires and SBS as the modifier for the asphalt is beneficial to protect the environment and save energy.

The bending beam rheometer (BBR) test based on the rheological theory is an effective tool to assess the low temperature performance of asphalt binders. The flexural creep stiffness and creep rate at time 60 s obtained from the BBR test are usually used as the indicators for the low temperature performance of asphalt binders. But there are often contradictions between the two indicators. It is unilateral to evaluate the low temperature performance of SBS/CR composite modified asphalt by creep stiffness or creep rate alone because of the complex composition. The low service temperature, damping ratio, dissipated energy ratio and derivation of creep compliance were also used as the comprehensive indicators to investigate the low temperature performance [9,10]. The rheological models commonly used to describe the viscoelastic behaviour for the asphalt at low temperatures include CAM model, Burgers model and simple fractional viscoelastic model. The simple fractional model has been proved to be an effective and simple tool to predict the creep behaviour of asphalt binders at low temperatures [11].

Considering that the low temperature performance of asphalt has a certain relationship with the aging state of asphalt, but there are still few studies on the SBS/CR modified asphalt in this regard. The objective of this paper is to investigate the low temperature characteristics of SBS/CR modified asphalt under different aging conditions and SBS modified asphalt is selected as the comparative object. The BBR test was conducted for two asphalt binders and low service temperature was calculated according to the ASTM method. Based on the simple fractional model, the damping ratio and dissipated energy ratio were also adopted to compare the low temperature performance of two modified asphalt binders.

2. Experimental program

2.1. Materials

The SBS and SBS/CR modified asphalt binder used in this paper were provided by Jiangsu Baoli International Investment Co. Ltd. in China. The SBS content in the SBS modified was 5.5%. The SBS/CR modified asphalt was produced on the basis of SBS modified asphalt with the addition of liquefied rubber and carbon black.

2.2. Test method

The aging condition considered in this paper included short-term aging and long-term aging, and the corresponding test was Thin-Film Oven Test (TFOT) and accelerated aging of asphalt binder using a Pressurized Aging Vessel (PAV). The samples of two asphalt binders were kept in the thin-film oven for 5 hours at 163 °C. Then the samples were put into the pressurized aging vessel for 20 hours at 90 °C under the pressure of 2.1 MPa.

The BBR test was followed the test methods specified by the “Standard Test Methods of Bitumen and Bituminous Mixtures for Highway Engineering” (JTG E20-2011 in China). The asphalt binders under different aging conditions were used to prepare the prismatic specimens for the BBR test with the dimension of 127 ×6.35×12.7mm. The test temperature was selected as -12 °C, -18 °C and -24 °C. The mid-span displacement data of the specimens were recorded in 0.5s time steps for 240s. Two parallel tests were conducted at different temperatures and the averaged values were calculated as the final results. Then the flexural creep stiffness $S(t)$ can be calculated by the following equation:

$$S(t) = \frac{PL^3}{4bh^3\delta(t)} \tag{1}$$

where P is the constant load applied on the mid-span of the specimen, L is the span length, b is the width of the specimen, h is the height of the specimen, and $\delta(t)$ is the mid-span displacement.

The creep rate $m(t)$ is defined as:

$$m(t) = \left| \frac{d \log S(t)}{d \log(t)} \right| \tag{2}$$

3. Simple fractional viscoelastic model

It has been observed that the fractional viscoelastic model is suitable to characterize the creep behaviour of asphalt [11], as is shown in Figure 1. The simple fractional model requires fewer parameters compared with the widely used Burgers model. The creep compliance $D(t)$ of the simple fractional model is presented as:

$$D(t) = At^{\alpha} \tag{3}$$

in which A and α are the model parameters, and $D(t)=1/S(t)$.

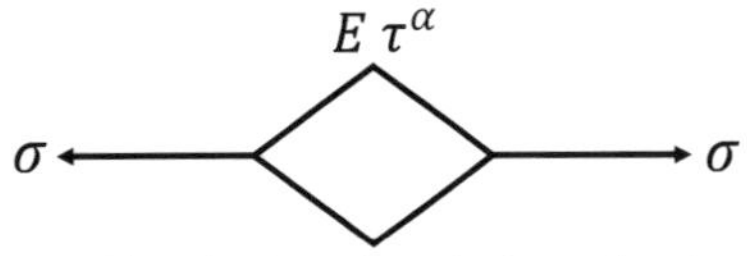

Figure 1. Simple fractional viscoelastic model

4. Results and discussion

4.1. Low service temperature

The low service temperature of two asphalt binders was determined according to the method in ASTM D7643. The logarithmic linear interpolation of $S(t\text{=}60\text{s})$ and linear interpolation of $m(t\text{=}60\text{s})$ at different temperatures were conducted. The continuous low service temperature was the maximum interpolation result when $S(t\text{=}60\text{s})$=300 MPa or $m(t\text{=}60\text{s})$=0.3. The results of low service temperature for two asphalt binders is illustrated in Figure 2.

It is observed from Figure 2 that the low service temperature of SBS modified asphalt decreases slightly after short-term aging while SBS/CR modified asphalt increases after short-term aging. In addition, the low temperature performance of SBS modified asphalt decreases significantly after long-term aging. The low service temperature of SBS modified asphalt increases by 27.9% after long-term aging while that of SBS/CR modified asphalt increases by 8.9%. It can be seen that SBS has poor resistance to long-term aging in terms of low temperature performance. However, this shortcoming has been greatly improved with the addition of CR modifier. Moreover, the low service temperature of SBS/CR modified asphalt before and after aging is lower than that of original SBS modified asphalt. Therefore, the effect of CR on the low temperature performance of SBS asphalt is very obvious.

4.2. Determination of simple fractional model parameters

The parameters of the simple fractional model for two asphalt binders at different temperatures were calculated based on the nonlinear regression method and listed in Table 1. It can be observed that the values of the parameters A and α in the simple fractional model become smaller as the temperature decreases. The coefficient α shows a downtrend after short-term and long-term aging.

According to Equation (3) and the reciprocal relationship between the creep stiffness and creep compliance, the following equation can be obtained:

$$\log\left[S(t)\right] = -\log A - \alpha \log\left(t\right) \tag{4}$$

then the m-value is calculated by simple fractional model based on the definition in the Equation (2):

$$m(t) = \left| \frac{d\log\left[-\log A - \alpha\log\left(t\right)\right]}{d\left[\log(t)\right]} \right| = \alpha \tag{5}$$

Thus it can be seen that the coefficient α has a clear physical meaning because it reflects the creep rate of asphalt binders. The m-value of two asphalt binders at different temperature and aging stage calculated by ASTM method and simple fractional model is compared in Figure 3. It can be seen that the m-value obtained by the two methods has a good linear correlation with an acceptable coefficient of determination. Therefore, it is feasible to predict the creep rate of asphalt binder by the simple fractional model with sufficient accuracy.

Table 1. Parameters of creep compliance based on the simple fractional model

Aging condition	Type of modifier	-12°C			-18°C			-24°C		
		A	α	R^2	A	α	R^2	A	α	R^2
Origin	SBS	0.001083	0.4206	0.9960	0.000765	0.2907	0.9864	0.000704	0.1792	0.9661
	SBS/CR	0.002790	0.5002	0.9982	0.001484	0.4137	0.9959	0.000881	0.2903	0.9936
TFOT	SBS	0.001329	0.3890	0.9972	0.000858	0.2710	0.9942	0.000745	0.1356	0.9747
	SBS/CR	0.003271	0.5005	0.9981	0.001588	0.3716	0.9967	0.000832	0.2529	0.9956
PAV	SBS	0.001264	0.3027	0.9976	0.000889	0.1858	0.9766	0.000620	0.1160	0.9618
	SBS/CR	0.002876	0.4404	0.9984	0.001489	0.3324	0.9974	0.000978	0.2077	0.9860

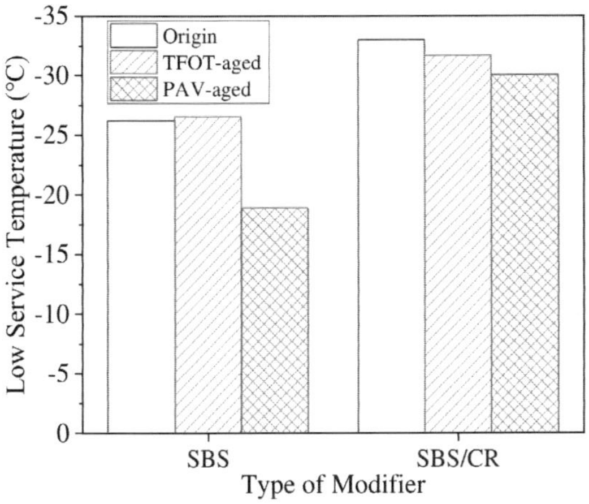

Figure 2. Low service temperature

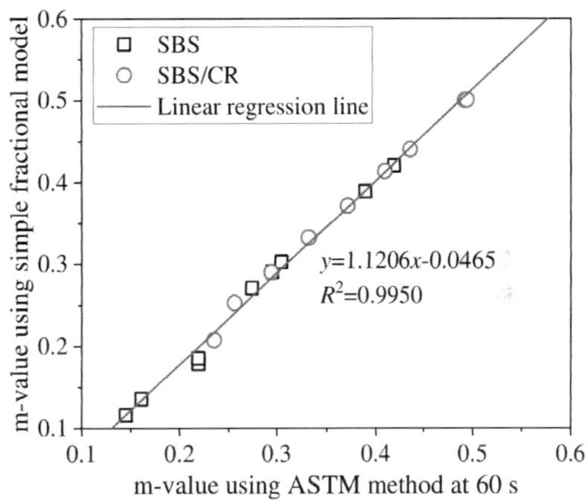

Figure 3. Comparison of m-value

4.3. Damping ratio

The damping ratio describes the ratio of viscous component to elastic component for the viscoelastic materials. For the simple fractional model, the damping ratio is calculated in the following expression:

$$\text{damping ratio} = \tan\left(\frac{\pi\alpha}{2}\right) \tag{6}$$

It can be seen from Equation (6) that the damping ratio calculated by the simple fractional model is directly related to the coefficient α and increases with the increase of α. The damping ratio of two asphalt binders at different temperatures under different aging condition is shown in Figure 4. It can be observed that the damping ratio decreases as the temperature decreases due to the dominant elastic behaviour at low temperatures. In general, the damping ratio of SBS/CR modified asphalt is greater than that of SBS modified asphalt at different temperatures. The greater damping ratio means better performance at low temperatures, which indicates the good anti-crack property of SBS/CR modified asphalt binder at low temperatures.

Table 2. Linear regression results of damping ratio versus temperature

Type	Regression equation	R^2	Crossover temperature(°C)
SBS-origin	y=0.0407x+1.2512	0.9903	-6.2
SBS-TFOT	y=0.0403x+1.1830	0.9999	-4.5
SBS-PAV	y=0.0276x+0.8291	0.9714	6.2
SBS/CR-origin	y=0.0425x+1.5156	0.9989	-12.1
SBS/CR-TFOT	y=0.0485x+1.5669	0.9902	-11.7
SBS/CR-PAV	y=0.0408x+1.3157	0.9996	-7.7

IOP Conf. Series: Materials Science and Engineering **652** (2019) 012019 doi:10.1088/1757-899X/652/1/012019

Then the crossover temperature at which the storage modulus is equal to loss modulus is calculated based on the linear regression method, as is listed in Table 2. It can be seen that the crossover temperature of SBS/CR modified asphalt is much lower than that of SBS modified asphalt. It is indicated that SBS/CR modified asphalt has sufficient viscous components to dissipate energy at low temperatures, which is beneficial to resist low temperature cracks. In addition, the crossover temperature of the long-term aged SBS/CR modified asphalt is at the same level as that of origin SBS modified asphalt. This proves the superiority of CR in terms of anti-aging performance.

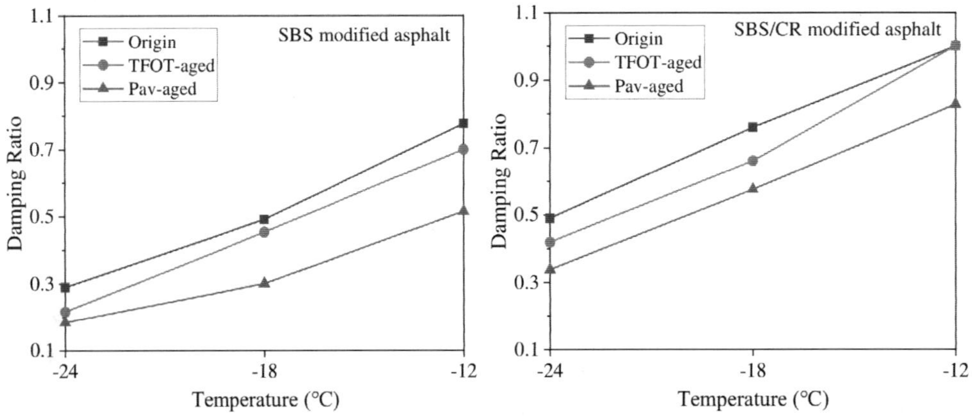

(a) (b)

Figure 4. Damping ratio versus temperature

4.4. Dissipated energy ratio

The dissipated energy ratio(DER) is defined as the ratio of dissipated energy to stored energy. The higher the DER, the better the low temperature performance of the asphalt binder. The DER calculated using the simple fractional model is expressed as:

$$\mathrm{DER} = \frac{2^{\alpha-1}}{1-2^{\alpha-1}} \tag{7}$$

It can be seen from Equation (7) that the DER is independent of time and is only related to the parameter α of the simple fractional model. The DER of two modified asphalt binders before and after aging at different temperatures is illustrated in Figure 5.

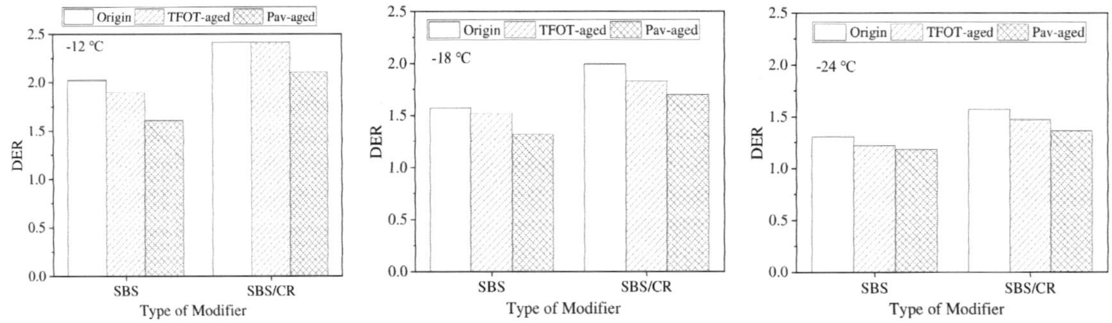

(a) (b) (c)

Figure 5. DER of two asphalt binders under different aging conditions at different temperatures

It is indicated that the DER of the SBS/CR modified asphalt is larger than that of SBS modified asphalt at three temperatures. Therefore, adding CR to SBS modified asphalt can improve the viscosity of asphalt at low temperatures. Moreover, the DER increases with the increase of temperature. As the temperature decreases, the effect of long-term aging on the DER is getting smaller and smaller. In addition, it can be found that the evaluation results of the low temperature performance for the two asphalt binders are the same when using DER and damping ratio [9,10]. The DER has been widely used to evaluate the low temperature performance of asphalt binders. So the damping ratio can be an effective indicator for the evaluation of low temperature performance of modified asphalt binders.

5. Conclusions

In this study, the low temperature performance of SBS modified asphalt and SBS/CR modified asphalt under different aging conditions were compared based on the BBR test. The main conclusions of this study are summarized as follows:

- The low temperature performance of SBS/CR modified asphalt is much better than that of SBS modified asphalt under different aging conditions. The low service temperature of SBS/CR modified asphalt after long-term aging is much lower than that of original SBS modified asphalt.
- The simple fractional viscoelastic model is an effective tool to characterize the creep behaviour of two modified asphalt binders at low temperatures.
- The coefficient α in the simple fractional model has a good linear correlation with the m-value and can be an index to describe the low temperature performance of asphalt binders.
- The two indicators, damping ratio and dissipated energy ratio, also confirm the superiority of low temperature performance of SBS/CR modified asphalt from the aspect of energy.
- The performance of SBS/CR modified asphalt at medium and high temperatures will be further discussed in the future studies.

Acknowledgement

This research was supported by the National Natural Science Foundation of China under grants 51778136.

References

[1] Liu J, Zhao S, Li L, Li P and Saboundjian S 2017 Low temperature cracking analysis of asphalt binders and mixtures *Cold Reg. Sci. Technol.* 141 78–85
[2] Luo X, Gu F, Ling M and Lytton R L 2018 Review of mechanistic-empirical modeling of top-down cracking in asphalt pavements *Constr. Build. Mater.* 191 1053–70
[3] Cuciniello G, Leandri P, Filippi S, Lo Presti D, Losa M and Airey G 2018 Effect of ageing on the morphology and creep and recovery of polymer-modified bitumens *Mater. Struct.* 51
[4] Shu X and Huang B 2014 Recycling of waste tire rubber in asphalt and portland cement concrete: An overview *Constr. Build. Mater.* 67 217–24
[5] Wang S, Cheng D and Xiao F 2017 Recent developments in the application of chemical approaches to rubberized asphalt *Constr. Build. Mater.* 131 101–13
[6] Liang M, Xin X, Fan W, Luo H, Wang X and Xing B 2015 Investigation of the rheological properties and storage stability of CR/SBS modified asphalt *Constr. Build. Mater.* 74 235–40
[7] Dong F, Yu X, Liu S and Wei J 2016 Rheological behaviors and microstructure of SBS/CR composite modified hard asphalt *Constr. Build. Mater.* 115 285–93
[8] Kosma V, Hayrapetyan S, Diamanti E, Dhawale A and Giannelis E P 2018 Bitumen nanocomposites with improved performance *Constr. Build. Mater.* 160 30–8
[9] Liu S, Cao W, Shang S, Qi H and Fang J 2010 Analysis and application of relationships between low-temperature rheological performance parameters of asphalt binders *Constr. Build. Mater.* 24 471–8
[10] Aflaki S and Hajikarimi P 2012 Implementing viscoelastic rheological methods to evaluate low temperature performance of modified asphalt binders *Constr. Build. Mater.* 36 110–8
[11] Hajikarimi P, Aflaki S and Hoseini A S 2013 Implementing fractional viscoelastic model to evaluate low temperature characteristics of crumb rubber and gilsonite modified asphalt binders *Constr. Build. Mater.* 49 682–7

Performance of GFRP bar anchored PBO-FRCM composite on one-way RC slabs under flexure

F Grzymski[1], T Trapko[1] and M Musiał[1]

[1] Faculty of Civil Engineering, Wrocław University of Science and Technology, Pl.Grunwaldzki 11, 50-377 Wroclaw, Poland

Email: filip.grzymski@pwr.edu.pl

Abstract. The use of FRCM (Fabric Reinforced Cementitious Matrix) composites for reinforced concrete elements structural strengthening is gaining increased popularity. This causes the need to better understand the work of the material and to propose an effective way of using it. The mechanical behaviour of FRCM composites related to their ductility and tendency to debond or delaminate causes that its typical use, known from the older FRP (Fibre Reinforced Polymers) technology, is inefficient, that was the reason for the anchoring concept. This paper presents the results of the research based on three slab type elements, two of which were strengthened with one layer of PBO-FRCM composite with GFRP rebar anchorage of the composite mesh ends. The most important mechanical properties of the elements were measured and analysed.

1. Introduction

Strengthening composites are present in civil engineering for around 40 years [1] and have now become an inseparable element in the construction repair situations. The most popular technology of strengthening composites named FRP (Fibre Reinforced Polymers), consisting of the fibres embedded in a polymer matrix, is characterised by good strength-to-weight ratio, adaptability and flexibility of application, and resistance to aggressive and corrosive environment [1]. The main problem with the FRP composites is related to the properties of polymer matrix, which is usually made of epoxy resin. The glass transition temperature (Tg) of the epoxy resin is about +50°C [2], which means that its operation in a fire situation is not effective, because after exceeding the glass transition temperature threshold it starts to become liquid. Additionally, the epoxy resins are expensive and they are characterized by a limited possibility of use in high moisture or cold, and have poor compatibility to the substrate [3].

The majority of the problems related to FRP composites might be reduced or eliminated thanks to the use of FRCM (Fabric Reinforced Cementitious Matrix) composites made of fibrous meshes in mineral mortar. Changing the matrix from epoxy resin to mineral mortar allowed to achieve higher resistance to higher temperatures and better compatibility with concrete substrate [1]. Due to different matrix structure and physical properties the fibrous material needs to be rearranged into loose meshes to allow the mineral mortar to pass through and create a binding. The change of the matrix has a positive impact on many aspects of composite usage, but it changes the mechanics of the material work compared to FRP. Mineral matrix is incapable of penetrating and covering all fibres in the yarns so only external fibres are bonded with the matrix, while internal fibres are subject to telescopic effect inside the yarn as shown in Fig. 1 [4]. FRCM composites usually show bi- or tri-linear behaviour and their failure is generally related to debonding or delamination [4-6]. The slope of the first line of stress-strain curve depends mainly on the properties of mineral matrix, the third line is related to the fibres properties only, while the second line shows the moment of matrix cracking and is an intermediate point of the materials behaviour. Depending on the fibres used different material behaviours and ultimate stresses are obtained as shown in Fig. 2.

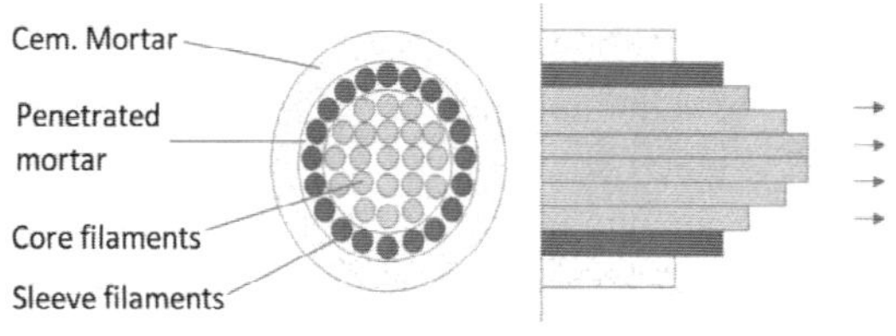

Figure 1. Telescopic effect [4].

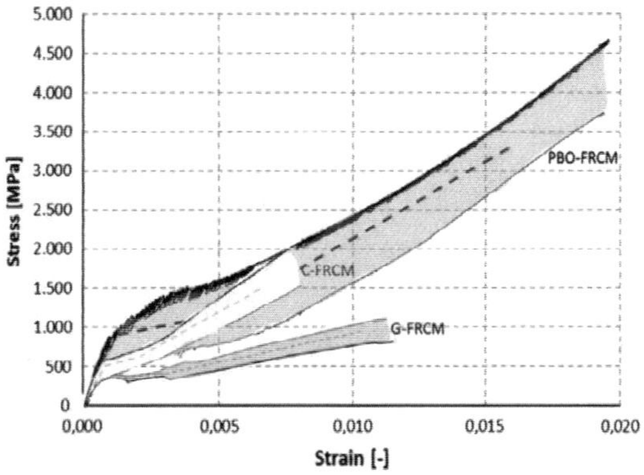

Figure 2. Stress-strain curves of FRCM composites with different fibres used [4].

Such mechanical behaviour of FRCM composites leads to insufficient utilisation of high-strength fibres. To improve the performance of the fibres several ways of anchoring FRCM strengthening were proposed for elements subjected to flexure [7] and shear [8-10]. The anchoring of the fibres allows the higher strains to be obtained and hence improve the load bearing capacity of the composite.

The research on the anchorage of the composites on slab type RC elements under flexure is limited. There is no present research in this field regarding FRCM composites, but the studies on the anchorage FRP strengthening of slab elements show that the use of anchorage can delay the debonding of the composite. It leads to higher strength of the RC slab and allows the element to gain much larger deformation [11, 12]. Despite the limited literature in the field of anchorage of FRCM composites in slab elements the effectiveness of FRCM strengthening of slabs has been confirmed [13-16]. The failure of the FRCM composites on slabs occurs due to slippage of the fibres in the matrix when one layer of mesh is applied or due to delamination of whole composite when more layers of mesh are applied [13]. Similar results were obtained by the other researchers who have compared FRCM and FRP strengthening of slab type elements. Despite the higher strengthening effect of the FRCM system, higher deformations of the elements were observed at the same load level when compared to FRP strengthening [14]. The initial strength of the element, dependent on the reinforcement ratio, has an impact on the strengthening effect. The more steel reinforcement is used in the element, the worse strengthening effect is observed [15]. Another interesting aspect of FRCM is connected with bi-directional meshes that also work in transversal direction of one-way slab element. It might help in gaining more load-bearing capacity and stiffness in strengthened RC slab [16].

The present research results clearly show that FRCM composites are a promising method of reinforced concrete elements strengthening. The benefits of composite materials connected with high thermal resistance make it a very interesting solution. Due to lack of research regarding anchorage of FRCM composites on slab-type elements the authors decided to take this topic and carry out a series of tests to recognize the phenomenon.

2. Experimental research

The research was carried out on the basis of three slab type RC elements under flexure. Two of the elements were strengthened with FRCM composite with anchorage of the mesh ends. One of the elements was left without strengthening as a reference element.

2.1 Test specimens and test stand

The slabs used in the research were 2000 mm long, 1000 mm wide and 150 mm thick. The elements were reinforced with 8 ϕ10 rebars in bottom surface and 8 ϕ8 rebars in top surface. Transverse reinforcement on top and bottom surface was made of ϕ8 rebars at 200 mm spacing. All test specimens were prepared at the precast plant to ensure stability of dimensions and strength. Slabs were subjected to flexure in 4-point bending test with the span length of 1700 mm and the spacing between load forces equal to one third of the span (567 mm). The set-up of the test stand is shown in Fig. 3.

Figure 3. Test stand used in research.

Five LVDTs were used to measure the deflections of the element – two over the supports, two directly under the forces and one in the middle of the span. Strain gauges were placed in three rows on the width of the element to measure strains of main steel reinforcement, top and bottom surface of the concrete, and composite mesh.

2.2 Materials

The elements were made with one batch of concrete and left to reach their full strength. Additional concrete cylindrical elements were made to measure its mechanical properties. The mean value of cylinder compressive strength was 38,8 MPa and the module of concrete elasticity value was 35,2 GPa. Steel reinforcement was made of steel with characteristic yield strength of 500 MPa. The meshes used in FRCM composites were made of PBO (p-Phenylene Benzobis Oxazole) fibres. They are characterised by very high value of tensile strength of 5800 MPa, while the module of its elasticity is 270 GPa, that results in the ultimate strain of about 2% [17].

2.3 Anchorage

To help the PBO fibres in obtaining higher strains the anchorage of the mesh ends has been applied. Grooves with dimensions of 20x20 mm were prepared in test specimens to create a mounting socket. The ends of the mesh were wrapped on a ϕ8 GFRP rebar and firmly pressed into the groove into which the mineral mortar was previously applied. Finally, the entire anchorage and mesh area was covered with a top layer of mineral mortar to complete the FRCM composite formation process.

3. Research results and discussion

All the elements tested failed after reinforcing steel yielding followed by compressed concrete crushing. In the elements with PBO-FRCM strengthening the steel yielding point was the beginning of the increased work of the composite, that was noticeable thanks to many clicking sounds that were heard during the cracking of the matrix and the tension of PBO fibres. The composite failed when the ends of GFRP bars torn out of the groove (Fig. 4), that lead to fabric debonding. No fibres were ruptured during the tests. It suggests that fibres ultimate strain was not obtained.

The ultimate failure force registered in the test of unstrengthened element was 171,4 kN, while the elements with bar-anchored PBO-FRCM strengthening showed the mean ultimate value of 216,2 kN. The composite with proposed way of anchorage gave 26% increase in the load-bearing capacity of the tested slab.

3.1 Load-deflection

The load-bearing capacity increase finds its confirmation in the analysis of the load-deflection curves of the element. Graphs obtained during the study are shown in Fig. 5. Reference element curve is marked black, while the curves referring to two strengthened elements are marked in colour.

Figure 4. GFRP anchoring rebar torn out of the groove.

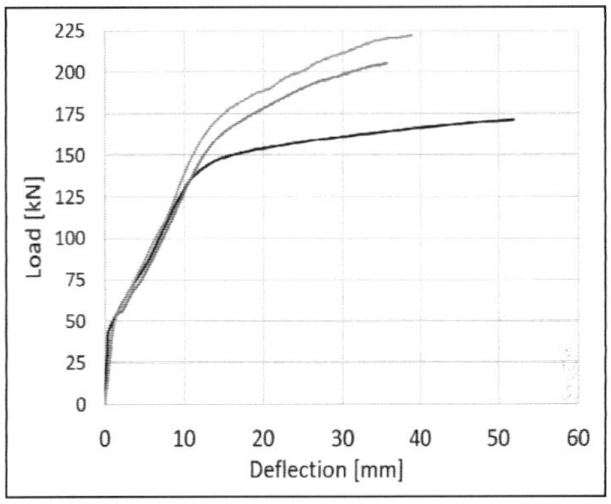

Figure 5. Load-deflection curves obtained in the research.

The behaviour of all the elements is almost identical before reaching the load level of reinforcing steel yielding in reference element. FRCM strengthening does not show any noticeable effect in this range of load forces in terms of load-deflection curves. After exceeding the load of around 140 kN the unstrengthened element starts to deform rapidly, while the elements with FRCM composite attached show stiffness even at higher loads. The slope of the curve in the last phase of the elements work is more steep for the strengthened elements, what indicates the incorporation of external composite stiffness to the stiffness of the whole element. The failure of FRCM strengthened elements occurs at much smaller deflections of the slab (around 40 mm) compared to reference slab (around 55 mm). It may suggest that with better anchorage or fabric adhesion the load-bearing capacity increase can be even greater.

3.2 Strengthening composite and steel strains

The main task for the anchorage is to allow the fabric in FRCM composite to reach higher strains, that will allow them to reveal their full potential. The maximal measured composite strain values of around 10‰ are much below the ultimate value, which may indicate that this type of anchorage does not allow for optimal use of PBO fibers. Mesh strains development comparison on the elements width is presented in Fig. 6. Curve corresponding to the longitudinal axis of the element is marked black while the coloured ones describe the behaviour of the mesh closer to the edges. Strain values were measured until strain gauge failure. It is visible that strains in the middle longitudinal axis of the slab are higher than those closer to the edges. Before concrete cracking the differences are negligibly small, however, the result spread on both sides of the slab (two coloured lines in Fig. 6) may indicate that the mesh strain is mainly dependent on the cracking development rate in the immediate strain gauge surrounding. Another arrangement of the strain gauges could show greater strain values. Reinforced concrete slab strength depends mainly on the reinforcing steel. Application of the strengthening composite may reduce deformation of reinforcing steel, and hence stresses occurring in it. Middle cross-section steel reinforcing strain values of strengthened and unstrengthened element are presented in Fig. 7. The reference element is marked black and the strengthened one marked in colour. Values are compared until the strain value corresponding to the destruction of the first strain gauge. However, it occurred after reinforcing steel yielding, so the results determined to the level shown allow for sufficient analysis.

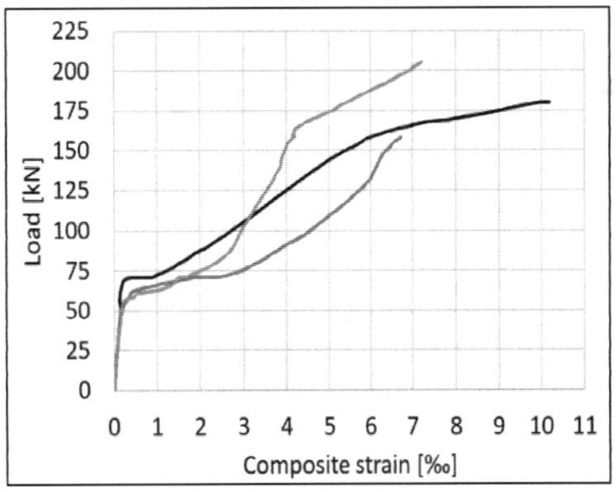

Figure 6. Middle cross-section strains of the composite in the element longitudinal axis and at the edges.

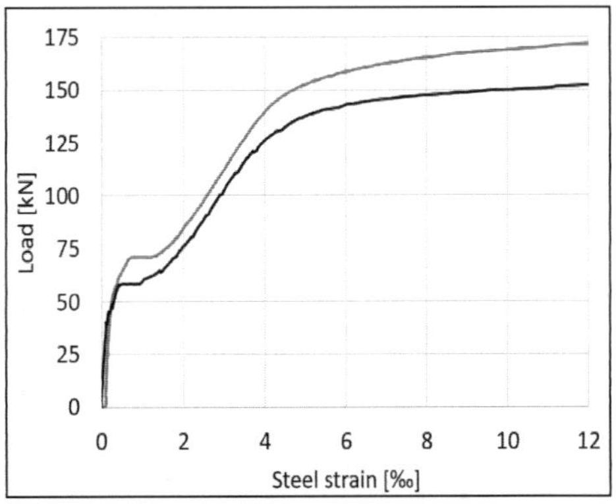

Figure 7. Maximal reinforcing steel strains in strengthened and unstrengthened element.

The diagram clearly shows that steel strains for equal load levels are lower in the case of PBO-FRCM strengthened element. Before concrete cracking the difference is very small, but with further sample loading it becomes more visible. Curves shapes are almost identical in both cases, but they are mutually offset.

4. Conclusions

Externally bonded PBO-FRCM composite showed its effectiveness in one-way RC slab type elements strengthening. The following main conclusions have been drawn from the conducted research:

- Application of one layer of PBO-FRCM composite with bar anchorage have increased the load-bearing capacity of the tested one-way slabs by 26%,
- In addition to increasing the load-bearing capacity of the slab the reduction of deformation after steel yielding was also visible,
- The failure of the composite is initiated by the destruction of the anchorage, which may indicate that the anchorage is a critical element from the element load-bearing capacity point of view,
- The use of FRCM strengthening increases the stiffness of RC slabs after exceeding reinforcing steel yielding point,
- Strain values of around 10‰ and no fibre rupture indicate that the fibre potential has not been fully used,
- PBO-FRCM composite application takes over some of the reinforcing steel loads, thereby reducing its deformation and stress, which leads to increased load-bearing capacity of the element.
- More research in the field of FRCM composites anchorage is still needed in order to study this phenomenon thoroughly. The presented work is still in progress and the aim of the authors is to develop an optimal way of FRCM composites anchorage.

References

[1] O. Awani, T. El-Maaddawy, N. Ismail: Fabric-reinforced cementitious matrix: A promising

IOP Conf. Series: Materials Science and Engineering **652** (2019) 012020 doi:10.1088/1757-899X/652/1/012020

strengthening technique for concrete structures, *Construction and Building Materials* (2017) 132: 94-111
[2] T. Trapko: The effect of high temperature on the performance of CFRP and FRCM confined concrete elements, *Composites Part B* (2013) 54: 138-145
[3] F. Bencardino, C. Carloni, A. Condello, F. Focacci, A. Napoli, R. Realfonzo: Flexural behaviour of RC members strengthened with FRCM: State-of-the-art and predictive formulas, *Composites Part B* (2018) 148: 132-148
[4] F. G. Carozzi, C. Poggi: Mechanical properties and debonding strength of Fabric Reinforced Cementitious Matrix (FRCM) systems for masonry strengthening, *Composites Part B* (2015) 70: 215-230
[5] A. D'Ambrisi, L. Feo, F. Focacci: Bond-slip relations for PBO-FRCM materials externally bonded to concrete, *Composites Part B* (2012) 43: 2938-2949
[6] T. D'Antino, C. Carloni, L.H. Sneed, C. Pellegrino: Matrix-fiber bond behavior in PBO FRCM composites: A fracture mechanics approach, *Engineering Fracture Mechanics* (2014) 117:94-111
[7] T. Trapko, M. Musiał: PBO mesh mobilization via different ways of anchoring PBO-FRCM reinforcements, *Composites Part B* (2017) 118: 67-74
[8] D. Marcinczak, T. Trapko: Experimental research on RC beams strengthened in shear with PBO-FRCM composites, *IOP Conference Series: Materials Science and Engineering* (2018) 365: 042035
[9] D. Marcinczak, T. Trapko, M. Musiał: Shear strengthening of reinforced concrete beams with PBO-FRCM composites with anchorage, *Composites Part B* (2019) 158: 149-161
[10] D. Marcinczak, T. Trapko, M. Musiał: Comparison of RC beams strengthened with PBO-FRCM composite with different types of anchorage, fib 2018 Congresss (Melbourne, Australia 2018)
[11] S. T. Smith, S. Hu, S. J. Kim, R. Seracino: FRP-strengthened RC slabs anchored with FRP anchors, *Engineering Structures* (2011) 33: 1075-1087
[12] S. T. Smith, H. Zhang, Z. Wang: Influence of FRP anchors on the strength and ductility of FRP-strengthened RC slabs, *Construction and Building Materials* (2013) 49: 998-1012
[13] G. Loreto, L. Leardini, D. Arboleda, A. Nanni: Performance of RC slab-type elements strengthened with fabric-reinforced cementitious-matrix composites, *Journal of Composites for Construction* (2014) 18: A4013003
[14] Z. R. Aljazaeri, J. J. Myers, Strengthening of Reinforced Concrete One-way Slabs for Flexure using Composite Materials: Evaluation of Different Composite Materials, Fourth International Conference on Sustainable Construction Materials and Technologies (SCMT4) (Las Vegas, USA 2016)
[15] S. Lee, K. Hong, Y. Yeon, K. Jung: Flexural Behavior of RC Slabs Strengthened in Flexure with Basalt Fabric-Reinforced Cementitious Matrix, *Advances in Materials Science and Engineering* (2018) 2018: 982784
[16] L. Anania, A. Badalà, G. D'Agata: Numerical simulation of tests for the evaluation of the performance of the reinforced concrete slabs strengthening by FRCM, *Curved and Layered Structures* (2016) 3: 63–73
[17] L. Ombres: Concrete confinement with a cement based high strength composite material, *Composite Structures* (2014) 109: 294-304

Study on Performance of Cement Mixing Pile in Marine and Continental Deposition Soft Soil Area

Han Xia[1], Guangyin Du[1,*], Jun Cai[1], Huangsong Pan[1] and Da Qin[2]
[1] Institute of Geotechnical Engineering, Southeast University, 211189 Nanjing, China
[2] Guangxi Communications Investment Group Co., LTD. 530022 Nanning, China

*Email: guangyin@seu.edu.com

Abstract. Soft soil was widely distributed in the coastal area, and the key characters were high water content and low capacity. Cement Mixing Pile (CMP) is an effective method to treat soft soil foundation. While CMP construction quality was difficult to control and it was of different reinforce effect for different geology origin. In this paper, five geology origins have been divided for the littoral area, lagoon area, alluvial area (Yangtse River), alluvial area (Ancient Yellow River) and lacustrine area. CMP, used in 26 expressways which were constructed during the period of 2008 - 2016, has been divided two-part (Upper & Lower part) to test the construction quality of CMP through Standard Penetration Test (SPT), Unconfined Compaction Strength (UCS) and Cement Mixing Pile Quality Designation (CMPQD). Results show that geology origins have great influence on the construction quality of CMP at the same construction maintenance condition. Depth has a significant effect on UCS of the lower part of CMP while the little effect on UCS of the upper part. Construction quality of long CMP was more difficult to control than short CMP. The high quality of the upper part may cause the unqualified lower part evaluated as qualified or good, which may lead to some engineering accidents.

1. Introduction

Jiangsu province located in the eastern coast of China, Yangtse River, and Ancient Yellow River has been passed through here flowing into East China Sea [1-3]. Two of the four largest freshwater lake of China located in the middle and south of this province [4-6]. Geology origin can be divided into five area due to the rivers, lakes, and sea, therefore, geology condition in this area is complicated.

Cement mixing pile was developed in the United States, and China began carrying out experience and developing construction machinery in 1977 [7]. Cement mixing pile plays an important role in the soft foundation treatment of highway in China [8]. Yin [9] introduced large-scale marine reclamation on soft Hong Kong marine deposits. Lorenzo [10] established consolidation model of lime pile and cement mixed pile composite foundation and found that the excess pore pressure of the pile cannot be assumed to be zero at any given time during the consolidation process of a Soil – Cement pile unit cell. Chew [11] reported the relationship between the microstructure and engineering properties of cement-treated marine clay. Yin [12] studied the mechanism of the consolidation deformation and failure of a composite foundation with a deep cement mixing column. Yin [12] used a physical model to test of a steel cylindrical mould composite foundation in order to understand the consolidation behaviour of the composite foundation treated by deeply mixed cement method. Chai [13] put forward a method to predict the consolidation settlement-time curve of floating soil-cement column on improved soft clayey. Pongsivasathit [14] illustrated the interaction behaviour between a floating column and the surrounding soil. As an effective soft soil foundation treatment method, The expressway constructed or under construction in Jiangsu province be using cement mixing pile has more than 200 million meters until 2015, and 25% - 54% of soft soil foundation treatment have been chosen cement mixing pile, it has produced great economic and social benefits in the construction. Cement mixing pile is a hidden project in expressway soft foundation treatment, if the quality problems occur, it will seriously affect the efficiency of soft foundation treatment and stability of subgrade. Therefore, it is necessary to study the impact of new construction technology on testing standards.

In this paper, cement mixing piles applied in 26 expressways in Jiangsu province from 2008 to 2016 were tested including traditional mixing pile and bidirectional mixing pile. The test includes the quality index of the pile, strength, standard penetration, upper, lower score, and comprehensive score. Then according to the regional distribution of soft soil, the testing results of mixing piles are divided, and the detection results and detection parameters of each partition pile have been analyzed.

2. Test Methods

2.1. Standard Penetration Test (SPT)

The free-falling weight method of automatic decoupling was adopted in SPT, reducing the frictional resistance between the guide rod and the hammer. SPT detail operation step and calculate equations of this test followed the test standard [15].

2.2. Unconfined compressive strength (UCS)

Unconfined compressive strength uses unconfined compression test apparatus, the loading rate of 0.2 - 0.5 MPa per second until the specimen was a failure (i.e. soft soil sample can be appropriately reduced the loading rate). The detail operation step and calculate equations of this test followed the test standard [16].

2.3. Cement mixing pile quality designation (CMPQD)

Recording each drill core length and using the following equation to evaluated CMPQD [15]:

$$CMPQD = \frac{\sum l_i}{L} \times 100 \tag{1}$$

Whereas, L is roundtrip meter age; l_i is length of drill core.

3. Analysis methods

In order to analyse the quality of cement mixing pile, 5 areas have been divided according to different sedimentary history, which are the littoral area, lagoon area, alluvial area (Yangtse River), alluvial area (Ancient Yellow River) and lacustrine area (Fig.1).

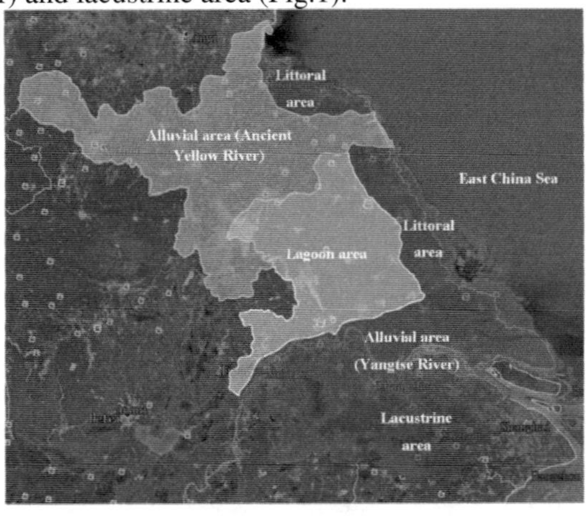

Figure 1. Geological division in Jiangsu

Littoral area soft soil area was located in the eastern coastal Jiangsu Province, the whole terrain was tilted from northwest to southeast, with an average altitude of 3 - 4 m. Lagoon area located in central of Jiangsu Province, belong to Yangtze River - Huai River Plain, A group of lakes and streams distributed in this low-lying region. Jiangsu province in the territory of the Ancient Yellow River 496 km, alluvial area (Ancient Yellow River) including Xuzhou, Suqian, Huai'an and Yancheng cities. Alluvial area (Yangtse River) located in Yangtze River delta plain, including the Yangzhou, Taizhou area. The main area of a lacustrine area are Suchow, Wuxi and Changzhou, this region deposited soft soil thickness was nearly 15 m.

In this construction quality test, set 5 m as the boundary, the CMP was divided into two parts, top surface to 5 m depth as upper part, below 5 m as lower part (see Fig.2), both upper and lower parts are respectively scored. The score of upper part above 75 can be evaluated qualified, and the score of lower part less than 60 can be evaluated not qualified. Then scores of a different part will be evaluated with the following equation:

$$t = (u \times 0.5) + (l \times 0.5) \quad (2)$$

Whereas, t is the total score, u is the upper part score and l is lower part score.

Then, the score can be divided into four grades as great, good, general and not qualified (can be seen as Table 1).

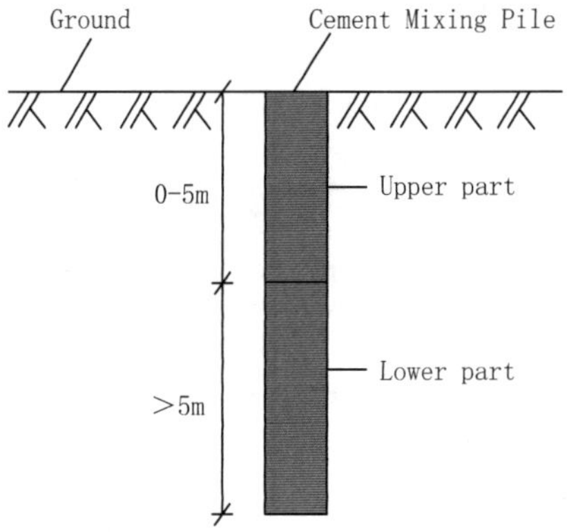

Figure 2. Score boundary of cement mixing pile

Table 1. Standard of drill holes evaluation

Scored	Grade
≥90	Great
80~89	Good
67.5~79	General
<67.5	Not qualified

Different parameters (SPT, UCS and CMPQD) of Cement Mixing Pile have been independently tested to evaluate the qualities of the different part of the cement mixing pile.

4. Results of test in different area

Fig.3 (a) shows the relationship between mean SPT (Standard Penetrate Test) blow counts and depth. With the depth increases, the five areas SPT blow numbers were decreasing. The upper part construction quality was higher than the lower part. That quality of cement mixing in the top 5 m was easier to control, blew 5 m the quality will be decreased sharply.

The relationship between cement mixing pile strength and depth can be seen in Fig.3 (b). Data shows below upper part begin to disperse, and the strength is mainly blowing 1 MPa. For the strength, the trend of the mean value varies little with the depth in each area, and all of them gradually decreases with the increase of depth. For the littoral area and lacustrine area, strength was more stable than

alluvial area (alluvial-Y and alluvial-A) and lagoon area. If simply consider the piles in the same area, the strength means a value of the upper part was more stable, and it was almost a straight line.
Cement mixing pile quality designation (CMPQD) distributed in 0 - 100, the higher score means the better quality. Fig.3 (c) indicated that depth was significantly affecting the score of CMPQD. It was no hard to find that CMPQD increased sharply before former 6 meters and decreased gradually after 6m. The results indicated that the standard of CMP can set 6 m as boundary of upper and lower part.

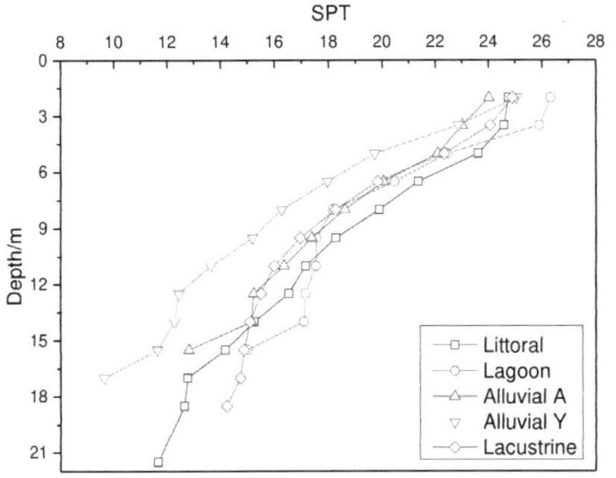

(a) Relationship between mean SPT and depth

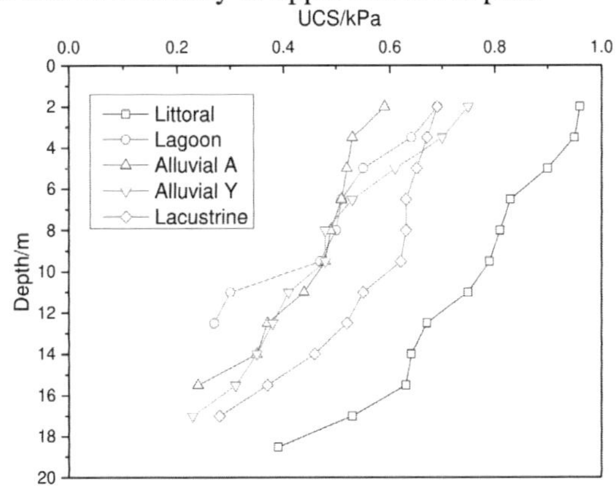

(b) Relationship between mean UCS and depth

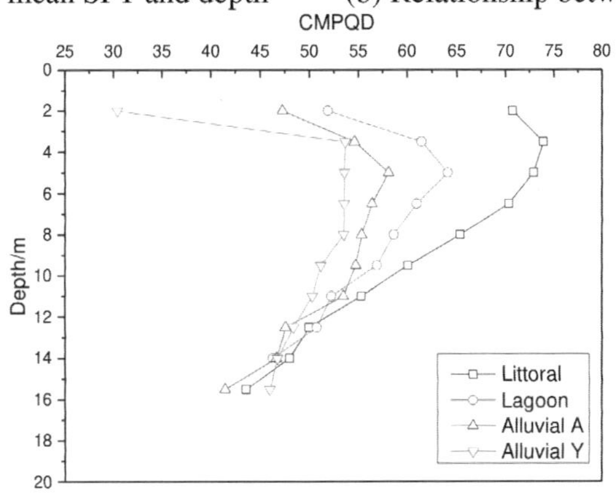

(c) Relationship between mean CMPQD and depth

Figure 3. Test results in different area

5. Evaluation and discussions

In order to compare the quality with each area clearly, proportion diagram has been shown in Fig.4. There have 3 piles evaluated not qualified, in a proportion of 0.14 %; 67 piles evaluated general, in a proportion of 3.68 %; 517 piles evaluated good, in a proportion of 23.77 % and 1588 piles evaluated great, in a proportion of 73.01 %, the qualification rate has reached 99.86 % in the littoral area. No piles are evaluated in unqualified or general, 85 piles are evaluated in good, in a proportion of 7.01 % and 1,127 piles be evaluated in great, in a proportion of 92.99 %, the qualification rate of lagoon area has reached 100 %. Alluvial-A area has 49 piles be evaluated as general, 250 piles are evaluated as good and 570 piles are evaluated as great. Like the lagoon area, the qualification rate also reached 100 %. For the alluvial-Y area, 67.54 % piles were a judge to be great, 22.32 % piles to good, 9.8 % to general and 0.35 % to non-qualified. A lacustrine area including 81.14 % great piles, 14.14 % good piles, 1.64 % general piles and 3.03 % unqualified piles.

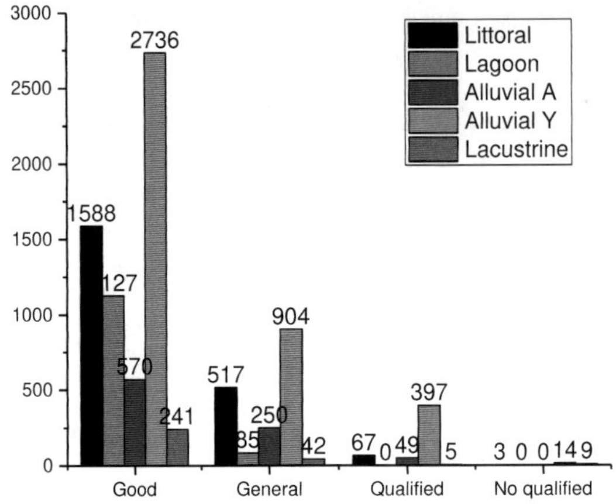

Figure 4. Proportion diagram of different piles

Geology condition can influence CMP strength efficiently, soil deposited in lacustrine and lagoon plain were relatively simple, and the quality of CMP accordingly was higher. Complicated engineering geology of the alluvial area (Yangtse River & Ancient Yellow River) and littoral area because the multiple transgressive and regressive events, the quality was difficult to control. Piles have a significant quality problem in the top due to the initial consolidation but are still be evaluated qualified because of the score of another part. With pile length increasing, the numbers of a pile to be evaluated great and a good proportion are decreasing. The lower part of long CMP can more easily to get the lower score. This can lead the not qualified pile misevaluated as qualified, bring some engineering accident. It is necessary to correct the evaluation score parameters of a standard.

6. Conclusions

Based on deposition history, geology in Jiangsu province can be divided into five soft soil areas, littoral, lagoon, alluvial (Yangtse River), alluvial (Ancient Yellow River) and lacustrine area. Each Cement Mixing Pile in 26 expressways has been grouped into the upper part and a lower part by 5m, then SPT, UCS, and CMPQD results were statistics and analysed to evaluate the influence of quality in each geology engineering condition. The results can be seen as follows:

1. For the same construction maintenance condition, different geology origin affected the quality of Cement Mixing Pile remarkably. CMP is of higher quality in simple geological conditions than in complex conditions.
2. The length of the Cement Mixing Pile only has great influence on the lower part quality and has little influence on the higher part. Thus, it is difficult to control with long cement mixing pile.
3. The upper part's score of the pile is generally higher than the lower, and the upper score has a great contribution to the final score. This may lead the not-qualified-pile misevaluated.
4. Based on the comprehensive evaluation results of the 5 regions, evaluate standards of SPT and UCS should be improved.

Acknowledge

Authors gratefully acknowledge financial support from the National Science Foundation of China (Grant No. 41372308).

References

[1] Yan Y. Channel of the lower reaches of the Ancient Yellow River in the change Qing dynasty prior to the of course at Tongwaxiang [J]. Ancient Yellow River, 1986, (1): 62-65.

[2] Wang K., Ru Y. River and sea demarcation in estuary of Ancient Yellow River: theory and principle [J]. China Water Resources, 2010, 18(9): 6-9.

[3] Chen Z. The deposition, breach, and diversion in the lower Ancient Yellow River and their relationships with human activities during the historical period [J]. Progress in Geography, 2001, 20(1):44-50.
[4] Jian P., Jian D. Studies on historical evolution of Hongze Lake Levee [J]. China Water Resources, 2019, 9(5): 53-56.
[5] Zhou J., Dong Z., Zhe Z. Influence of sea-entering channel phase II project on flood risk in Hongze Lake area [J]. Hydro-Science and Engineering, 2017, (5): 60-66.
[6] Zhu M. Review and prospect of the ten years governance on Lake Tai Basin [J]. Environmental Protection, 2017, (12):34-38.
[7] Liu Z., Liu S., Zhou B., Jing F., Ji F. Geological Genesis And Spatial Variability Characteristic Of The Interactive Marine And Terrestrial Soft Deposit In Yangtze River Estuary [J]. Engineering Geology, 2014, 10(2):102-108.
[8] Zhang D., Fan L., Liu S., Fan L., Deng Y. Laboratory model tests on consolidation mechanism of soft clay improved by deep mixing cement columns [J]. China Journal of Highway and Transport, 2014, 27(12):1-9.
[9] Yin J. Properties and behaviour of Hong Kong marine deposits with different clay contents [J]. Canadian Geotechnical Journal, 1999, (36): 1085-1095.
[10] Lorenzo G. A., Dennes T. B. New consolidation equation for soil–cement pile improved ground Can. Canadian Geotechnical Journal, 2003, (40): 265-275.
[11] Chew S. H., Kamruzzaman A. H. M., Lee F. H. Physicochemical and Engineering Behavior of Cement Treated Clays [J]. Journal of Geotechnical Geoenvironment Engineering, 2004, 130(7): 696-706.
[12] Yin J. H., Fang Z. Physical modelling of consolidation behavior of a composite foundation consisting of a cement-mixed soil column and untreated soft marine clay [J]. Geotechnique, 2006, 56 (1):63-68.
[13] Chai J., Miura N., Kirekawa T., Hino T. Settlement prediction for soft ground improved by columns [J]. Proceedings of the Institution of Civil Engineers, 2010, 12(5):109-119.
[14] Pongsivasathit S., Chai J., Ding W. Consolidation settlement of floating-column-improved soft clayey deposit [J]. Proceedings of the Institution of Civil Engineers, 2013, 166(2):44-58.
[15] Jiangsu Bureau of Transportation Engineering. Inspection specification for cement mixing pile quality of highway engineering [S]. DB32/T 2283-2012, 2012.
[16] Ministry of Transport of the People's Republic of China. Test methods of soils for highway engineering [S]. JTG E40-2007, 2007.

Permeability of soil cement admixed with air foam

Sakulsaksri Chaiyasat[1]

[1]Callege of Industry Technology, King Mongkut's University of Technology North Bangkok, Bangkok, 10800, Thailand

Email: chaiyasat.s@cit.kmutnb.ac.th

Abstract. This research studied the permeability of soil cement mixed with air foam. Three soil samples at different locations were collected. For the investigations, the 20 and 30 percentages of Portland cement by the weight of dry soil and 10 and 20 percentages of air foam by the volume of soil were applied. Method of study consists of the physical properties and permeability tests by the variable head method. The soil samples have been carried out 7, 14, and 28 days of curing time. The test results showed that at 28 days curing time, the permeability of the soil cement samples mixed with air foam are in the range of 5.7 x 10^{-9} cm/sec to 8.8 x 10^{-9} cm/sec. Samples with higher cement content and longer curing time subjected to decrease the permeability. Vice versa the permeability of soil cement increased with the increasing of air foam because the air foam created more pore spaces. The activity of soil has influence in the changing percentages of permeability of soil samples mixed with air foam.

1. Introduction

Nowadays the natural disaster is more seriously cause of the global worming effect such as drought, flooding and variability of the climate. In 2012, the major flooding occurred in Thailand that was serious damage in economics of the whole country. Then it is necessary to apply the resources in the highest efficiency in order to help into the water management.

One of the flooding protection systems which is the low construction cost and usually utilize is the earth dam or embankment. The embankment can be applied by various soils as sand, lateritic and weathering soil. In Bangkok, there is lots of surplus soil from constructions sites. It is very useful to utilize this kind of soils for flooding protection. The application of soil mixing with air foam treated by cement also known as "cement treated soil mixing air foam" is a new alternative of this lightweight material. This material commonly used in Japan, which is called SGM (Super Geo-Material) as material from of mud left from the construction which is low shear strength (Tuchida[1], Tuchida et al.[2]). The cement has hardened and the air is mixed with foam to reduce weight which is the good alternative materials that replace a limited fill material for flooding protection. The surplus soil is normally high water content and difficult to compact. In order to solve these problems, the surplus clay is mixed with Portland cement. Darcy[3] proposed the equation for calculating the velocity of water through the saturated soil equal the multiplication between coefficients of permeability (k) with hydraulic gradient (i). The coefficient of permeability value is one of the important values to design the flooding protection system. There are some properties such as strength, consistency, compressibility, microstructure and permeability of soil treated by lime and cement investigated by the

previous researches [4-7]. Akiije [8] reported the k value of lateritic soil-cement mixture when compared to which of the natural soil has decreased from 10^{-6} cm/sec to 10^{-8} cm/sec. Nguyen and Chai[9] also investigated the soil microstructure using the mercury intrusion porosimetry (MIP) test and scanning electron microscope (SEM) imaging indicates that when the cementation products formed by the pozzolanic reaction fill mainly the intra-aggregate pores, the value of k is comparable for the treated and untreated samples. When the cementation products begin to fill the inter-aggregate pores, the value of k of the treated sample becomes smaller than that of the untreated soil sample.
Although there were some behaviour of cement treated soil from previous research, there are lack knowledge of soil cement admixed with air foam. Then in this research focuses on the permeability of soil cement admixed with air foam. In order to clarify this behaviour the various content of Portland cement and air foam were mixed in soil samples with different curing periods of time.

2. Soil sample and test setup

Three soil samples with different locations were applied in this study. Soil sample 1 was collected from bore pile construction with 2-3 meters depth at Bangkhen district, Bangkok. Soil sample 2 is commercial kaolinite. Soil sample 3 was surplus soil from construction with 1.5 meters depth at Bangsue district, Bangkok.
Their properties are shown in Table 1. The liquid limit of all soil samples are in the rage of 46% to 77%. The clay content is in the rage of 48% to 65%. The sand content is about 13% to 18%. The soil sample 1, 2, and 3 can be classified as CH, CH, and SC by using Unified Soil Classification System. The activity (the ratio of plasticity index to the clay content) of soil samples 1, 2 and 3 are 0.65, 0.90, and 0.66, respectively.
The soil samples were mixed with 20 and 30 percentages of Portland cement per weight of dry soil. The 10 and 20 percentages of air foam per volume of soil were also mixed. The water content of soil samples were mixed at two times of liquid limit. After mixing the specimen were contained in the PVC mould with 5 cm diameter and 10 cm length. The soil specimens were cured with 7, 14, and 28 days in the curing room with constant temperature 25°C before permeability tested. The soil samples were investigated the coefficient of permeability by variable head method (ASTM D5084–03).

Table 1. Properties of soil samples.

	Soil sample		
Properties	Soil 1	Soil 2	Soil 3
Liquid limit	57.70	76.50	45.50
Plastic limit	15.20	25.80	13.70
Sand	12.62	13.22	17.93
Silt	22.38	31.75	34.07
Clay	65.00	55.00	48.00
Specific gravity	2.64	2.67	2.59
USCS	CH	CH	SC
Activity	0.65	0.90	0.66

3. Test results and discussion

From the test results, the various admixtures were plotted against coefficients of permeability at various curing times as shown in Figures 1, 2, and 3. By specifying, the symbol "C" instead of percent Portland cement mixed and "AF" instead of air foam percentage mixed. The coefficient of permeability (k) of soil sample 1 mixed with 20 % of cement and 20 % of air foam (C20-AF20) at curing 7 days is the highest as 1.70 x 10^{-8} cm/sec. For the permeability of C30-AF0 at curing 28 days is the lowest as 4.30 x 10^{-9} cm/sec.

When considering the curing time from figure 1 shows that the coefficient of permeability decreases as the curing time increases. The permeability of C20-AF20 at curing 7, 14, and 28 days are 1.70 x 10-8, 1.40 x 10^{-8}, and 9.20 x 10^{-9} cm/sec, respectively. To considering the effect of Portland cement content on coefficient of permeability, the comparing of cement content at 20 % and 30 % with the same air foam and curing time were investigated. The permeability of cement content at 20 % and 30 % with the 20% air foam (C20-AF20 and C30-AF20) at curing of 28 days are 9.20 x 10^{-9} cm/sec to 7.30 x 10^{-9} cm/sec., respectively. It show that the higher Portland cement content decreases in the permeability of soil specimens. As for the effects of adding air foam to the specimen, it was found that the mixture C30-AF0, C30-AF10 and C30-AF20 by 0, 10 and 20 percent of air foam, respectively, at 28 days of curing were 4.30 x 10^{-9}, 5.70 x 10^{-9} and 7.30 x 10^{-9} cm/sec. The coefficient of permeability tends to increase when increasing the air foam content in the soil specimen.

For soil sample 2, as shown in Figure 2, the permeability coefficient of the mixture 20 percent cement, 20 percent air foam volume, C20-AF20 with 7 days curing, is 1.70 x 10^{-8} cm/sec, which is the highest value for soil sample 2. The permeability of mixture C30-AF0 at 28 days curing is 3.50 x 10^{-9} cm/sec, which is the lowest value for soil sample 2. When considering the curing period from figure 2, it shows that the coefficient of permeability decreases as the curing time increases with the same trend as soil sample 1. The permeability coefficients of the mixture C20-AF20 at the curing of 7, 14, and 28

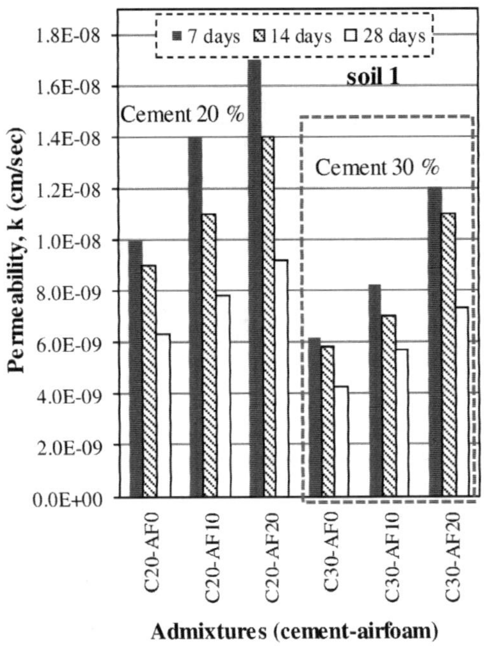

Figure 1. Permeability of soil 1 mixed with various contents of cement and air foam at curing times 7, 14, and 28 days.

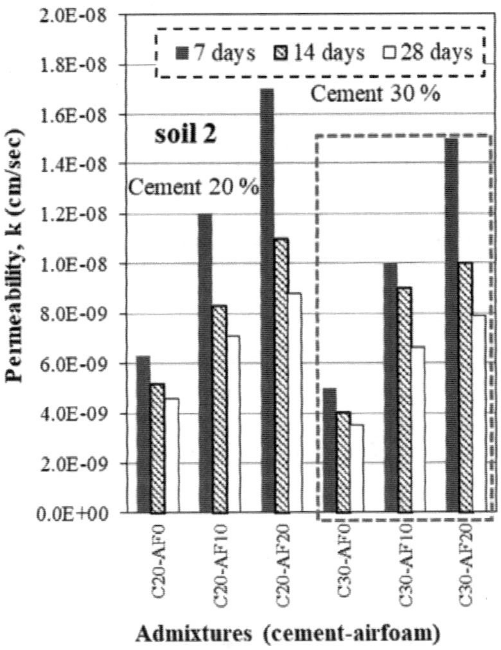

Figure 2. Permeability of soil 2 mixed with various contents of cement and air foam at curing times 7, 14, and 28 days.

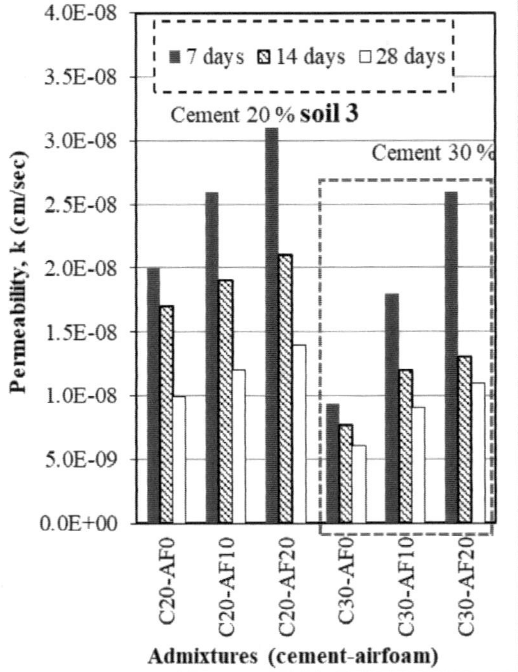

Figure 3. Permeability of soil 3 mixed with various contents of cement and air foam at curing times 7, 14, and 28 days.

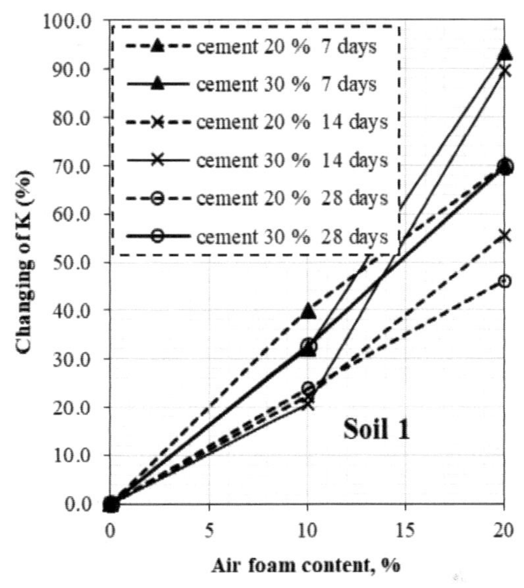

Figure 4. Changing of Permeability of soil 1 mixed with various contents of cement and air foam at curing times 7, 14, and 28 days.

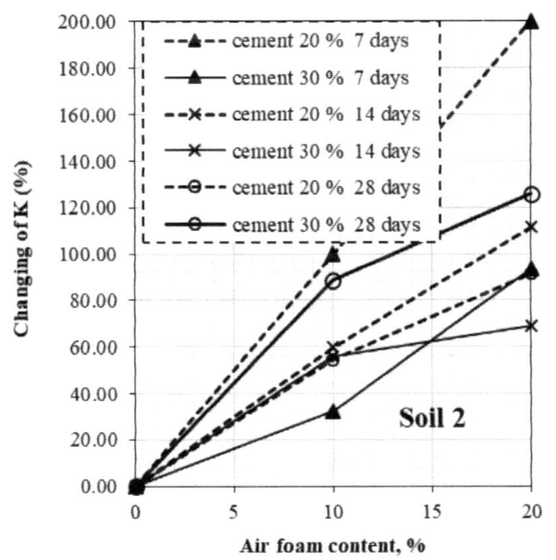

Figure 5. Changing of Permeability of soil 2 mixed with various contents of cement and air foam at curing times 7, 14, and 28 days.

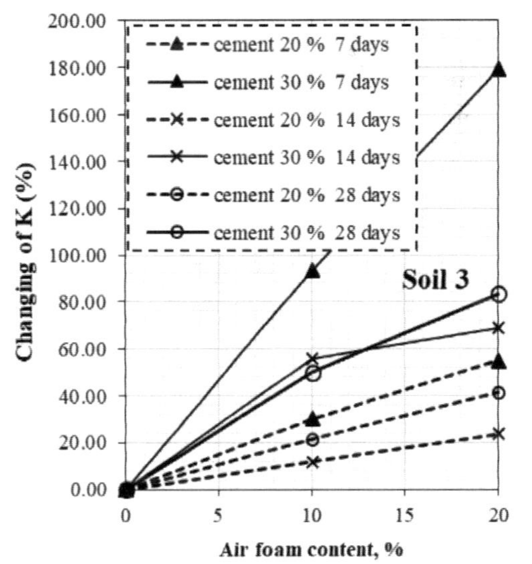

Figure 6. Changing of Permeability of soil 3 mixed with various contents of cement and air foam at curing times 7, 14, and 28 days.

days were 1.70 x 10^{-8}, 1.10 x 10^{-8}, and 8.80 x 10^{-9} cm/sec respectively. It found that the cement content 20% and 30% in the mixtures C20-AF20 and C30-AF20, at the curing time 28 days, the permeability coefficient decreased from 8.80 x 10^{-9} cm/sec to 7.90 x 10^{-9} cm/sec, respectively. The permeability also tends to increase when increasing in air foam content in the soil sample 2. They were found that the mixture C30-AF0, C30-AF10 and C30-AF20 by 0, 10 and 20 percent of air foam, respectively, at 28 days of curing were 3.50 x 10^{-9}, 6.60 x 10^{-9}, and 7.90 x 10^{-9} cm/ sec respectively.

For soil sample 3, as shown in Figure 3, the permeability coefficient of the mixture 20 percent cement, 20 percent air foam volume, C20-AF20 with 7 days curing, is 3.10 x 10^{-8} cm/sec, which is the highest value for soil sample 3. The permeability of mixture C30-AF0 at 28 days curing is 6.00 x 10^{-9} cm/sec, which is the lowest value for soil sample 3. When considering the curing period from Figure 3, it shows that the coefficient of permeability decreases as the curing time increases with the same trend as soil sample 1and soil sample 2. The permeability coefficients of the mixture C20-AF20 at the curing of 7, 14, and 28 days were 3.10 x 10^{-8}, 2.10 x 10^{-8}, and 1.40 x 10^{-8} cm/sec respectively. When considering the Portland cement content on permeability, it found that the comparing the amount of cement at 20% and 30% at the mixtures C20-AF20 and C30AF20, at the curing time 28 days, the permeability coefficient decreased from 8.80 x 10^{-9} cm/sec to 7.90 x 10^{-9} cm/sec, respectively. The permeability also tends to increase when increasing in air foam content in the soil sample 2. They were found that the mixture C30-AF0, C30-AF10 and C30-AF20 by 0, 10 and 20 percent of air foam, respectively, at 28 days of curing were 6.00 x 10^{-9}, 9.00 x 10^{-9}, and 1.10 x 10^{-8} cm/ sec respectively.

From the test results, it found that the coefficient of permeability increased with adding air foam. In order to clarify these mechanisms, the percentage of permeability changing is proposed (Changing of K, %) by comparing the permeability coefficient of the soil cement with air foam (K_F) and the permeability coefficient of soil cement (K_0) as equation 1.

$$\text{Changing of K}(\%) = \left|\frac{K_F - K_0}{K_0}\right| \times 100 \quad (1)$$

Figure 4 shows the percentage changing in permeability of soil sample 1 when the air foam content increases with a curing time of 7, 14, and 28 days. In the case of 20 percent cement, when the air foam was added by 10 and 20 percent at curing 28 days, the percentages of changing permeability are 24 and 46, respectively. For 30 percent cement, when air foam was added to 10 and 20 percent at curing 28 days, the percentages of changing permeability are 33 and 70, respectively.

In the case of soil sample 2 (in figure 5) mixed with 20 percent cement, when the air foam was added to 10 and 20 percent at curing 28 days, the percentages of changing permeability are 54 and 91, respectively. For cement 30 percent, when air foam was added by 10 and 20 percent at curing 28 days, the percentages of changing permeability are 89 and 126, respectively.

For soil sample 3 (in figure 6) mixed with 20 percent cement, when the air foam was added to 10 and 20 percent at curing 28 days, the percentages of changing permeability are 21 and 41, respectively. For cement 30 percent, when air foam was added by 10 and 20 percent at curing 28 days, the percentages of changing permeability are 50 and 83, respectively.

The percentage of changing permeability of soil samples is related to the soil activity (in Table 1)

With the same activity, the percentage of changing permeability of soil samples 1 and 3 are quite the same as while the higher activity in soil 2 showed that the percentage of changing permeability was high in the range of 89 and 126 percent. It showed that the permeability of treated soils depend upon soil properties, cement, air foam, and curing time. The permeability of cement treated clayey sand, SC (soil sample 3) admixed with air foam is higher than which of cement treated highly plastic clay, CH (soil sample 1 and 2) admixed with air foam.

The permeability of treated soils are mainly controlled by the soil types. Because with the high water content as two times liquid limit and high volume of air foam in treated soil, the physico-chemistry in soil has low influence in hydration and pozzolanic reaction which are the reaction between soil and Portland cement.

4. Conclusion

Results of the research showed that it is successful to apply the air foam in the cement treated soil admixed with air foam to alternative materials for flooding protection because the permeability of the admixed material is still adequate for water resistance which are in the range of 5.7 x 10^{-9} cm/sec to 8.8 x 10^{-9} cm/sec at curing time 28 days.

1. The advantages of air foam admixed material are lightweight to reduce pressure on foundation. These materials can also reduce the amount of soil filling by using additional water and can increase the fill volume by using air foam.
2. Samples with higher cement content and longer curing time subjected to decrease the permeability. Vice versa the permeability of soil cement increased with the increasing of air foam because the air foam created more pore spaces. The air foam is more influence in permeability of cement treated clay admixed with air foam than both of cement content and curing time.
3. The activity of soil has influence in the changing percentages of permeability of soil samples mixed with air foam. The high activity soil is subject to increase the changing percentages of permeability of soil samples.
4. The soil properties have also influence in permeability of admixed materials. The permeability of cement treated clayey sand (SC) admixed with air foam is quite higher than which of cement treated highly plastic clay (CH) admixed with air foam.
5. The physico-chemistry of soil has low influence in the high water content as two times liquid limit and high volume of air foam, the mechanisms that controlled the permeability of treated soil are the reaction between Portland cement and soil particles.

Acknowledgement

This research was funded by King Mongkut's University of Technology North Bangkok.
Contract no. KMUTNBGOV-59-10.

References

[1] Tsuchida, T 1995 *Proc. of Int. Symp. On Coastal Ocean Space Utilization COSU'95 Yokohama* 22-31
[2] Tsuchida T, Takeuchi D, Okumura T and Kishida T 1996 *Proc. of Environmental Geotechnics Balkema* 415-420
[3] Darcy H 1856 *the French engineer Henry Philibert Gaspard Darcy* **1803** 58
[4] Otani J, Mukunoki T and Kikuchi Y 2002 *Soil and Foundation* vol 42(3) pp 93-105
[5] Mitchell J, Hooper D and Campanella R 1965 *J. Soil Mech. and Found. Div., ASCE,* vol 91(4) pp 41-65
[6] Locat J, Tremblay H and Leroueil S 1996 *Canadian Geotechnical Journal* vol 33(4) pp 654-669
[7] Al-Mukhtar M, Khattab S and Alcover JF 2012 *Engineering Geology* vol 139 pp 17-27
[8] Akiije I 2014 *International Journal of Scientific & Engineering Research* vol 5(12)
[9] Nguyen D Q and Chai J C 2015 *Canadian Geotechnical Journal* vol 52(9) pp 1-7

IOP Conf. Series: Materials Science and Engineering **652** (2019) 012023 doi:10.1088/1757-899X/652/1/012023

Sparse modeling for the geotechnical observation data

T Kurita

Tokyo Electric Power Services Co., Ltd., Civil Engineering Department, Tokyo, Japan
Email: kurita@tepsco.co.jp

Abstract. In the geotechnical engineering, it is a significant issue to construct the numerical models for simulation analysis based on the observation data. In this study, sparse modeling used as an effective modeling method in the field of machine learning and image processing was applied to the geotechnical engineering and its effectiveness was examined. Aa a case study based on the field observation data, the fused lasso which is a typical method of the sparse modeling was applied to model the velocity structure of the subsurface ground using the PS logging data. As a result of examination, it was found that simplified models can be obtained by increasing the value of the regularization parameter.

1. Introduction

In order to simulate the seismic ground motions, it is necessary to prepare the velocity structure model of the subsurface ground. Since there is arbitrariness in the manually preparation of analytical models from bowling data, more objectively modeling method is required.
Recently, sparse estimation has been utilised as an effective modeling method in the field of machine learning [1] and image processing [2, 3]. In this study, sparse modeling was applied to the geotechnical engineering and examined its effectiveness. Fused lasso which is a representative method of the sparse modeling was applied to model the velocity structure of the ground using PS logging data.

2. Theory

The fused lasso which is a kind of generalized lasso proposed by Tibshirani and Taylor (2011) [4] was referred. The theory of the generalized lasso and the fused lasso is described below.

2.1. The generalized lasso

In this study, following linear regression equation is considered as a basic model.

$$y_i = \beta_1 x_{1,i} + \beta_2 x_{2,i} + \cdots + \beta_m x_{m,i} \qquad (i = 1, 2, \cdots, n) \tag{1}$$

In Eq.(1), y_i is an induced variable, $x_{m,i}$ is an explanatory variable, and β_m is a regression coefficient.

Precede the formulation of generalised lasso, the mathematical notations are defined as follows. $\|\cdot\|_1$ is the L_1 norm. And, $\|\cdot\|_2$ is the L_2 norm. According to Tibshirani and Taylor (2011) [4], the generalized lasso is amounted to the following optimization problem.

$$\hat{\boldsymbol{\beta}} \in \arg\min_{\boldsymbol{\beta}} \frac{1}{2}\|\mathbf{y} - \mathbf{X}\boldsymbol{\beta}\|_2^2 + \lambda\|\mathbf{D}\boldsymbol{\beta}\|_1 \tag{2}$$

Where, $\hat{\boldsymbol{\beta}}$ is a generalized lasso solution vector, $\mathbf{y}$ is a vector of induced variable, $\mathbf{X}$ is a matrix of explanatory variables, $\mathbf{D}$ is a penalty matrix, and $\lambda \geq 0$ is a regularization parameter. The solution

IOP Conf. Series: Materials Science and Engineering **652** (2019) 012023 doi:10.1088/1757-899X/652/1/012023

can be obtained by solving the equivalent Lagrange dual problem. The relationship between the primal and dual solution is as follows.

$$\hat{\boldsymbol{\beta}} = \left(\mathbf{X}^T\mathbf{X}\right)^{-1}\left(\mathbf{X}^T\mathbf{y} - \mathbf{D}^T\hat{\mathbf{u}}\right) \tag{3}$$

Therefore, solving for $\hat{\mathbf{u}}$ in Eq.(3) brings the solution $\hat{\boldsymbol{\beta}}$ in Eq.(2).

2.2. The fused lasso

As a simple example of the generalized lasso of Eq.(2), the 1D fused lasso is formulated as follows.

$$\hat{\boldsymbol{\beta}} = \arg\min_{\boldsymbol{\beta}} \frac{1}{2}\sum_{i=1}^{n}\left(y_i - \beta_i\right)^2 + \lambda\sum_{i=1}^{n-1}\left|\beta_{i+1} - \beta_i\right| \tag{4}$$

Here, Eq.(4) corresponds to setting $\mathbf{X}$ and $\mathbf{D}$ in Eq.(2) to be as follows.

$$\mathbf{X} = \mathbf{I} \tag{5}$$

and

$$\mathbf{D} = \begin{bmatrix} -1 & 1 & 0 & \cdots & 0 & 0 \\ 0 & -1 & 1 & \cdots & 0 & 0 \\ & & & \cdots & & \\ 0 & 0 & 0 & \cdots & -1 & 1 \end{bmatrix} \tag{6}$$

Moreover, y_i indicates an observation data.

3. Results

The sparse modelling was applied to model the ground velocity structure by using the PS logging results of the subsurface ground obtained in the heavily damaged area of the 2016 Kumamoto earthquake.

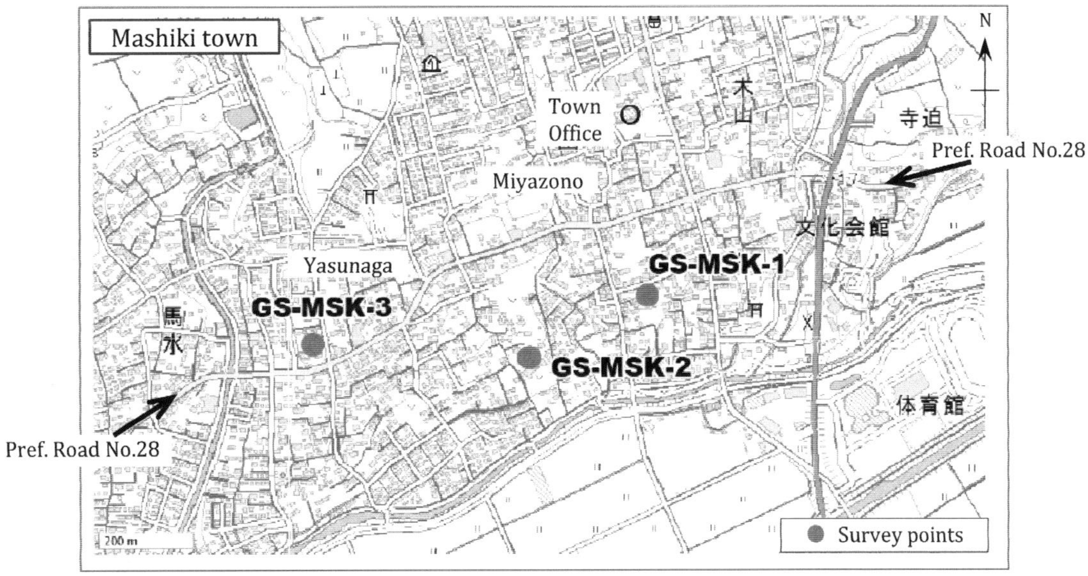

Figure 1. Borehole survey sites in Mashiki town (adapted from Yoshimi *et al.*, 2017 [5])

IOP Conf. Series: Materials Science and Engineering **652** (2019) 012023 doi:10.1088/1757-899X/652/1/012023

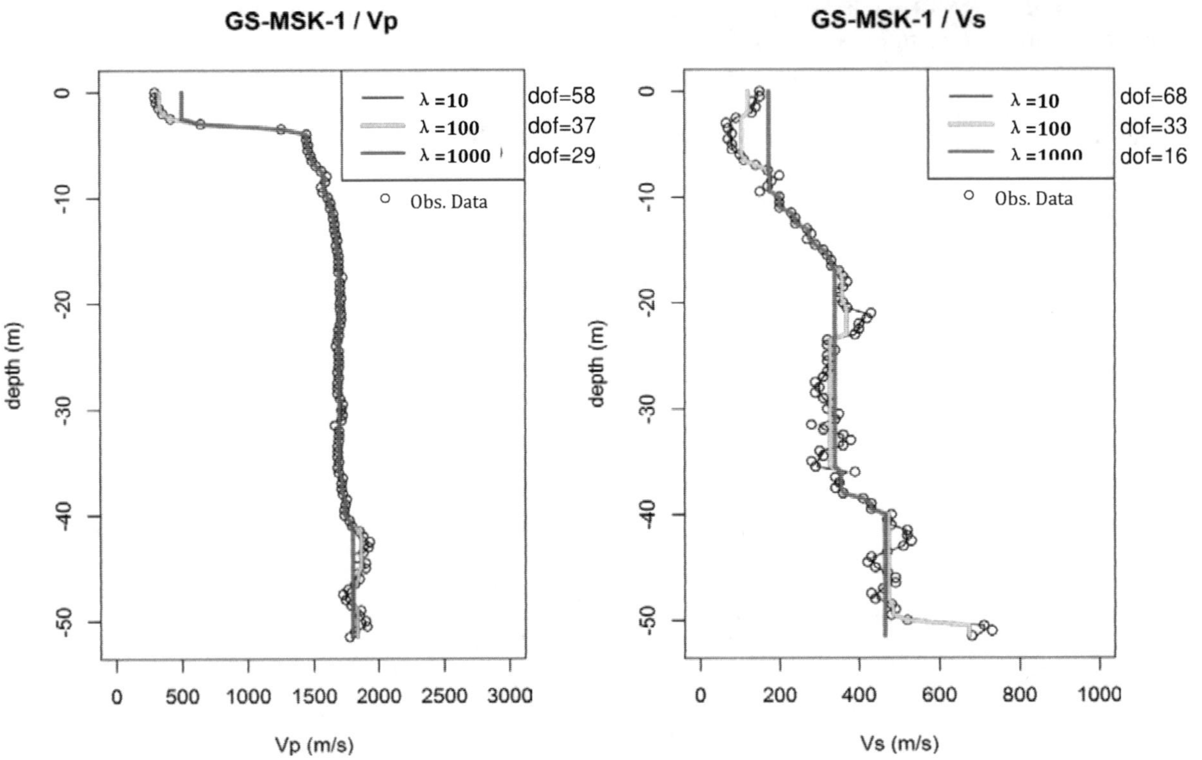

Figure 2. Modeling of the suspension PS logging results at GS-MSK-1 (Number of observed data: 104)

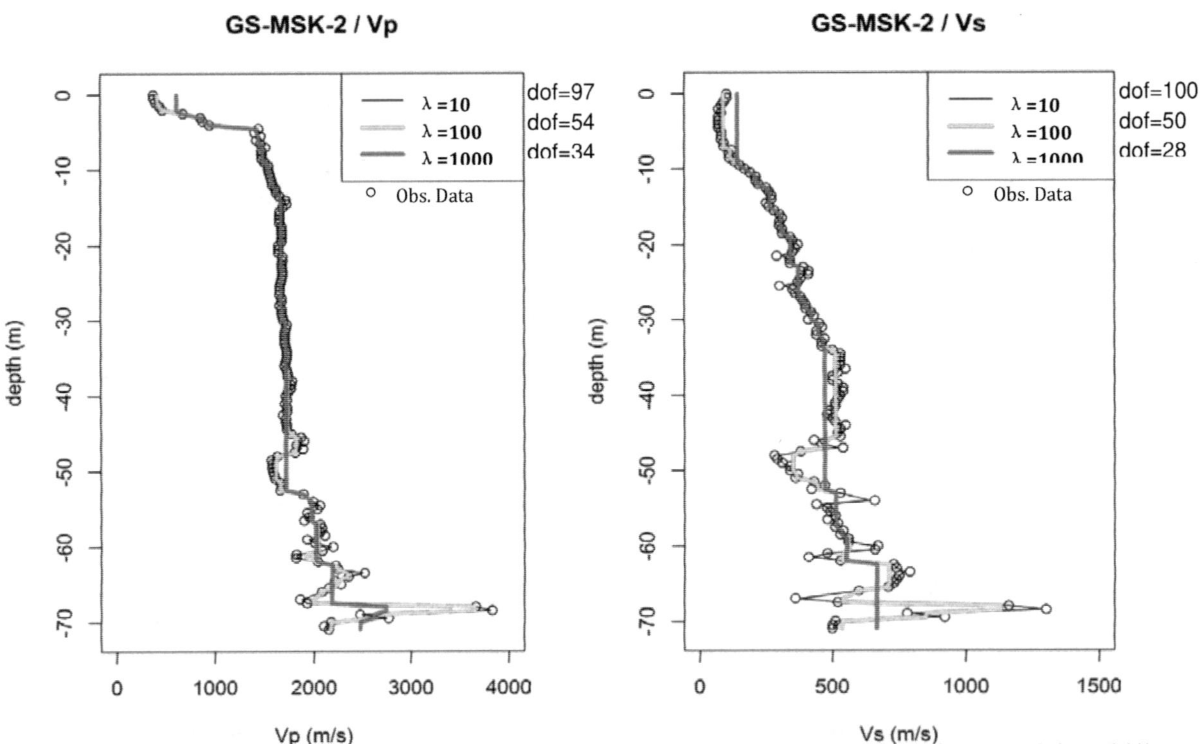

Figure 3. Modeling of the suspension PS logging results at GS-MSK-2 (Number of observed data: 141)

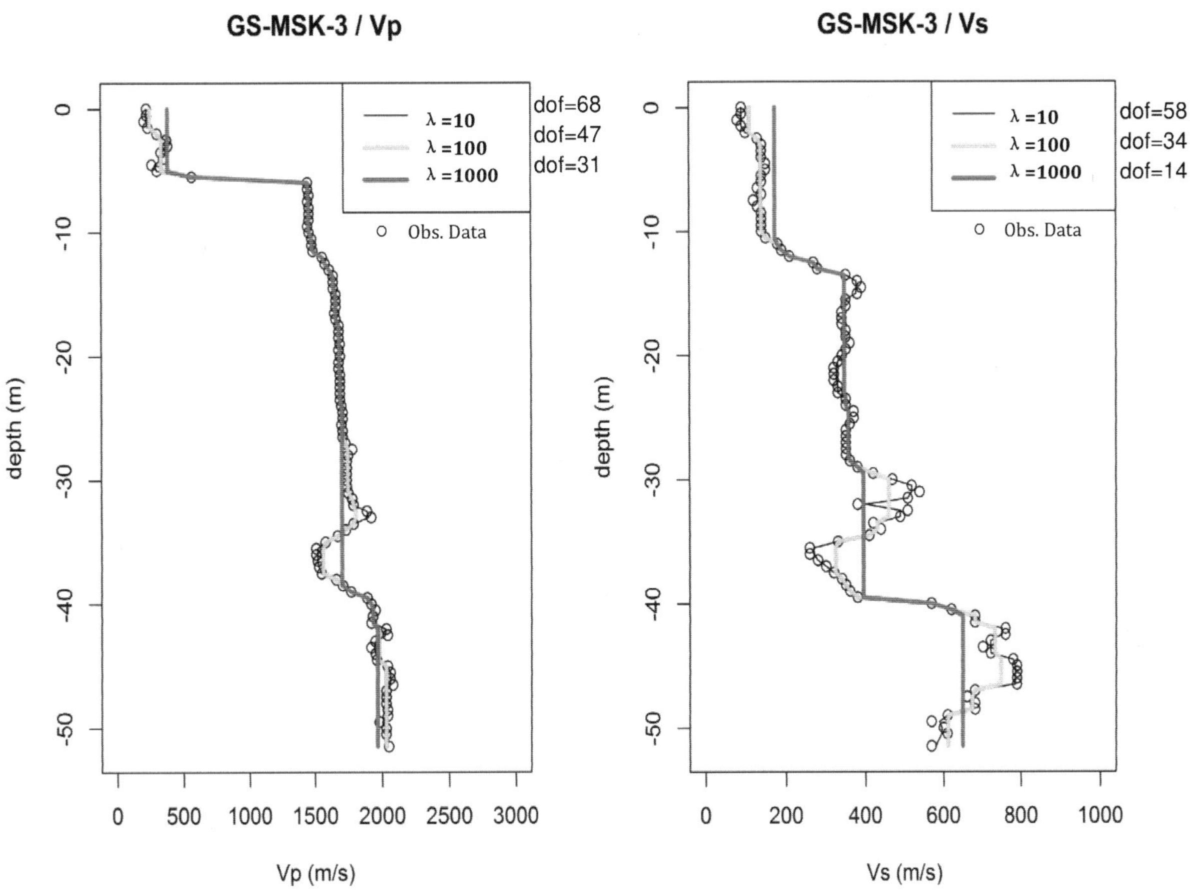

Figure 4. Modeling of the suspension PS logging results at GS-MSK-3 (Number of observed data: 103)

3.1. Outline of the survey data

The PS logging results carried out by Yoshimi et al. (2017, 2016) [5, 6] at three places around Miyazono and Yasunaga areas in Mashiki town, Kumamoto prefecture were used. Miyazono and Yasunaga were heavily damaged areas struck by the 2016 Kumamoto earthquake (Mw 7.0) occurred on April 16. It was a survey conducted to resolve the mechanism of earthquake damage in the area. The ground in the survey area consists of the shallow volcanic ash soil and the deeper pyroclastic flow layer from Mt. Aso. The surveyed sites are shown in Figure 1. Since the suspension type PS logging was carried out, high resolution data was obtained. The step size in the depth direction is 50 cm.
The PS logging data obtained at the three sites are shown Figure 2 to Figure 4. In these figures, open circles indicate observed data.

3.2. The sparse modeling of PS logging data

The sparse modeling of PS logging data is shown in the Figure2 to Figure 4. Each figure shows the fused lasso solutions when the value of the regularization parameters (λ) are changed. The degrees of freedom (dof) of the estimated model are also showed.
Regardless of the surveyed sites, the estimated model follows the fluctuation of the observation data when the value of the regularization parameter is small. On the other hand, with the increasing regularization parameter (λ), it can be seen that the obtained estimation model has many consecutive parts and the inflection points are decreasing. The regularization term of Eq.(4) represents the absolute value of the difference between adjacent model parameters. By minimization of this equation,

estimated values with reduced numbers of discontinuous points have been obtained. Such results are similarly obtained for both P wave velocity (Vp) and S wave velocity (Vs).

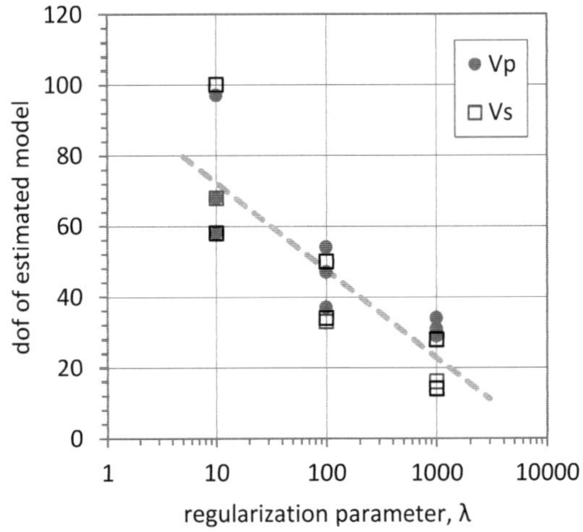

Figure 5. Relationship between the regularization parameter and the dof of estimated model

Figure 5 shows the relationship between the regularization parameter and the dof of estimated model. In this figure, symbols means the results of sparse estimation, and the broken line represents their regression.

4. Conclusions

In this paper, the sparse modeling was applied to the geotechnical observation data and examined its effectiveness. The fused lasso, which is a typical method of sparse modeling, was applied to model the velocity structure of the ground by using the PS logging data. The findings obtained in this research are summarized below.

- ✓ The 1D fused lasso was applied to model the velocity structure of the ground by using the continuous data with large fluctuation. The applicability of this method to the PS logging data was indicated.

- ✓ It was found that simplified models can be obtained by increasing the value of the regularization parameter.

- ✓ A negative correlation was found between the logarithm of the normalization parameter and the dof of the estimation model.

In the future, the applicability of the space modeling to another kind of observation data will be investigated to confirm the validity.

References

[1] Liu B, Wang M, Foroosh H, Tappen M, Penksy M 2015 Sparse Convolutional Neural Networks, *2015 IEEE Conference on Computer Vision and Pattern Recognition*, 806-814

[2] Lusig M, Donoho D, Pauly J M 2007 Sparse MRI: The Application of Compressed Sensing for Rapid MRI Imaging, *Magnetic Resonance in Medicine*, **58**, 1182–1195

[3] Mairal J, Elad M, Sapiro G 2008 Sparse Representation for Color Image Restoration, *IEEE Transactions on Image Proceeding*, **17**, 1, 53-69

[4] Tibshirani R J and Taylor J 2011 The Solution Path of the Generalized Lasso, *The Annals of Statistics*, **39**, 3, 1335-1371

[5] Yoshimi M, Goto H, Hata Y, Yoshida N 2017 Nonlinear site response at the worst hit area of the 2016 Kumamoto earthquakes in the Mashiki town, Kumamoto, Japan, *Document of DPRI Annual Meeting, Disaster Prevention Research Institute, Kyoto University*, A05 *(in Japanese)*

[6] Yoshimi M, Hata Y, Goto H, Hosoya T, Morita S, Tokumaru T 2016 Borehole exploration in heavily damaged area of the 2016 Kumamoto earthquake, Mashiki town, Kumamoto, *Proceedings of the Japanese Society for Active Fault Studies 2016 Annual Meeting*, P17 *(in Japanese)*

IOP Conf. Series: Materials Science and Engineering **652** (2019) 012024 doi:10.1088/1757-899X/652/1/012024

A review of the existing models to evaluate the ultimate bending moment of corroded reinforced concrete beam

H J Wang

School of Civil Engineering, Central South University, 22 Shaoshannan Road, Changsha 410075, China
Email: 174812262@csu.edu.cn

Abstract. Premature failure caused by steel corrosion is one of the most prominent durability problems of the reinforced concrete (RC) beams. Efficient evaluation of the ultimate bending moment of the corroded RC (CRC) beams is important for engineering practice. A great number of researches have been carried out, and various evaluation models have been proposed. However, the advantages and disadvantages of the existing models are still unclear, and less comparison studies among these models have been conducted. Therefore, the objective of this study is to review the existing models and clarify the merits and deficiencies of each model. Six empirical models for evaluating the ultimate bending moment of CRC beams are analyzed. Comparison study among these models is conducted based on 177 sets of experimental data collected from published literatures. According to the comparative analysis results, the advantages and disadvantages of each model are discussed. The comparison results show each existing model can exhibit good performance in evaluating a particular kind of experimental data. However, when the experimental data are out of the applicable range of each model, the existing empirical models may result in significant derivations.

1. Introduction

The durability problem is one of the most common and severe problems of the reinforced concrete (RC) beams, which requires a lot of financial resources and has attracted more and more attention. Among all the causes of durability damage of the RC beams, premature failure of beams caused by steel corrosion is the most prominent disaster. A great number of researches have been carried out to evaluate the residual strength and performance of corroded RC beams. Among all these aspects, the ultimate bending moment of the corroded RC (CRC) beams is focused on in this study.

Many models have been proposed to evaluate the ultimate bending moment of the CRC beams, which can be sorted into two kinds. For one kind, the ultimate bending moment is firstly calculated according to specifications without considering the influence of the corrosion, and then the calculated value is multiplied by a reduction factor corresponding to the influence of the corrosion. Another kind of existing models applies a reduction factor to the steel bars stress to consider the influence of the corrosion, and then the reduced steel bars stress is utilized to calculate the ultimate bending moment according to corresponding specifications.

The evaluated ultimate bending moment of the CRC beams by using each existing model is in good agreement with the experimental data used to propose the particular model. However, the performance of the existing models in evaluating other experimental data is unclear. Furthermore, there is currently less comparative analysis among these empirical models, which means that there may be difficulties in choosing appropriate model in practice. Therefore, the objective of this study is to review the existing models and conduct comparative analysis among these models. This study is organized as follows:

section 2 reviews six existing models; section 3 establishes a database, which consists of 177 sets of experimental data gathered from published literatures; and a comparison analysis among all these existing models based on the established database is conducted in section 4; Finally, the findings of this study is summarized in section 5.

2. Existing models for flexural capacity

In this part, six existing empirical models have been selected for calculating the ultimate bending moments of CRC beams. At the same time, a brief introduction to the calculation process of each model is given as follows.

2.1 Zhang et al.'s model [1]

In Zhang *et al.*'s model, the deformation coordination between longitudinal reinforcement and concrete after corrosion is considered to be no longer established [1], which means that the plane section assumption is no longer valid after corrosion. This model defines the strain relationship $\phi(\eta)$ between the concrete and the steel bar under the action of the ultimate bending moment. The effect of corrosion on the yield strength f_y' of steel bars is considered. The calculation formula for the height of the concrete compression zone of the beam x with different corrosion rates η_s under the ultimate bending moment is given. Considering the influence of corrosion, the test beams are divided into two types of failure modes. According to different types of failure modes, the formula for calculating the ultimate bending moment of the beam with different failure modes is given:

$$\begin{cases} M_u = (1-\eta_s)A_s f_y'(h_0 - \frac{x}{2}), & \varepsilon_s \geq \frac{f_y'}{E_y} \\ M_u = f_{cd} bx(h_0 - \frac{x}{2}), & \varepsilon_s < \frac{f_y'}{E_y} \end{cases} \tag{1}$$

where ε_s is the stress of tensile steel bars in limit state; E_y is elastic modulus of steel; A_s is section area of steel; f_{cd} is the experimental value of concrete axial compressive strength; h_0 is effective section height of beam.

2.2 Xu's model [2]

Xu [2] points out that corrosion will cause the bond performance between steel and concrete to degrade, and the strain coordination relationship between the two will change. It is first proposed that the strain relationship between steel and concrete after corrosion is no longer satisfied with the plain section assumption. Through the finite element analysis and experimental research, the relationship of the strain ratio between concrete and steel bar $m(\eta)$ under the action of ultimate bending moment is defined. Based on the bending experiment and nonlinear finite element analysis of the test beams, the calculation model of the ultimate bending moment of CRC beams is established by using the modified limit equilibrium theory:

$$M_u = T(h_0 - \frac{T}{2\alpha f_c b}) \tag{2}$$

where T is the tensile force of steel; α is the ratio of the stress value of the equivalent rectangular stress diagram to the concrete axial compressive strength f_c.

2.3 Sun's model [3]

This model firstly studies and analyzes beam members with perfect bond and the unbonded beam members. Based on this, the causes and mechanisms of the reduction of RC beams' flexural capacity caused by corrosion are analyzed. Sun's model considers the influence of corrosion, and defines the section reinforcement index of the corroded beams β_{0c}. The linear interpolation method is used to construct the relationship between the section reinforcement index β_{0c} of the corroded beam members and the reinforcement coefficient α_{sc} of the steel bar in the two extreme states of the perfect bonded and unbonded beams. The ultimate bending moment of CRC beam members is calculated:

$$\alpha_1 f_c bx = \alpha_{sc} f_{yc} A_{sc} \tag{3}$$

$$M_u = \alpha_1 f_c bx(h_0 - \frac{x}{2}) = \alpha_{sc} f_{yc} A_{sc}(h_0 - \frac{x}{2}) \tag{4}$$

where f_{yc} is the actual yield strength of longitudinal reinforcement after corrosion; A_{sc} is the cross-sectional area of steel after corrosion.

2.4 Azad et al.'s model [4]

The corrosion activity index I_{CCORT} is considered to be the most critical factor affecting the ultimate bending moment of CRC beams according to the experimental results. This model calculates the ultimate bending moment of CRC beams in two steps. In the first step, only the influence of corrosion on the section of the steel bar A_s' is considered, and the ultimate bending moment of the CRC beams is calculated according to the existing specifications:

$$M_{u1} = f_y A_s'(h_0 - \frac{x}{2}) \tag{5}$$

In the second step, considering the influence of corrosion on the bond strength between steel and concrete, a collaborative work coefficient β is fitted through the experiment and multiplied by the result calculated in the first step to obtain the final calculation result:

$$M_{u2} = \beta M_{u1} \tag{6}$$

2.5 Azad et al.'s modified model [5]

This model is a new modified model proposed by Azad *et al.* [5] considering the size effect of the test beam section. The modified model is consistent with the previously proposed model calculation, and is also divided into two steps. The difference is that the new modified model taking the influence of the cross-section dimensions of the test beam into account. And a new collaborative work coefficient C_f considering the influence of the cross-section dimensions is proposed.

2.6 Torres-Acosta et al.'s model [6]

This model points out that the maximum corrosion depth of steel bars caused by corrosion PIT_{MAX} is the most important parameter affecting the ultimate bending moment of CRC beams. By the regression analysis of the experimental results, this model gives the relationship between the average corrosion depth x_{AVER} and the maximum corrosion depth PIT_{MAX}. The relationship between the maximum corrosion depth of the corroded steel bar and the residual flexural capacity of the test beams RLC_{COR} is fitted:

$$RLC_{COR} = -0.8596 \cdot (PIT_{MAX} / r_0) + 0.9707 \tag{7}$$

where r_0 is radius of steel.

3. Experimental data

3.1 Establishment of experimental database

In this section, experimental data of 177 sets of CRC beams are collected from published literatures. All the specimens in the database are loaded by single concentrated loads or symmetrical two-point loads, which includes: 17 sets of Jin and Zhao [7], 24 sets of Hui *et al.* [8], 12 sets of Cao *et al.* [9], 20 sets of Xia *et al.* [10], 60 sets of Azad *et al.*[4, 5], 8 sets of Shang [11], 16 sets of Rodriguez *et al.* [12], 7 sets of Chen [13], 13 sets of Zhang [14].

3.2 Conversion of concrete strength

When calculating the ultimate bending moment of CRC beams, the strength of concrete is required. However, different testing standards are used to define the compressive strength of concrete for different experimental data. For comparative analysis, different concrete strengths require uniform standards. Eurocode 2 proposes the conversion relationship between $f_{cyl,150}$ (cylinder compressive strength of concrete with specimens of 150mm×300mm), $f_{cu,150}$ (cube compressive strength of concrete with specimens of 150mm×150mm×150mm) and f_t (axis tensile strength of concrete) as Table 1. GB50010-2002 gives the conversion relationship between $f_{cu,150}$ and f_{ck} (prism compressive strength of concrete with specimens of 150mm×150mm×300mm) as Table 2.

3.3 Conversion of corrosion ratio

In the published literatures, different methods are used to characterize the corrosion degree of the steel bars, such as the weight loss ratio (η_{wt}) and the section loss ratio (η_{sn}). According to the experimental investigation, the conversion relationship between the section loss ratio and the weight loss ratio of corroded steel bars can be expressed as:

$$\eta_{sn} = \begin{cases} 0.013 + 0.987\eta_{wt}, & \eta_{wt} < 10\% \\ 0.061 + 0.939\eta_{wt}, & 10\% \le \eta_{wt} < 20\% \\ 0.129 + 0.871\eta_{wt}, & 20\% \le \eta_{wt} < 30\% \\ 0.199 + 0.801\eta_{wt}, & 30\% \le \eta_{wt} \le 40\% \end{cases} \tag{8}$$

where η_{wt} and η_{sn} are the weight loss ratio and the section loss ratio of longitudinal steel bars, respectively. According to experimental investigation, the calculation of η_{wt} and η_{sn} can be expressed according to Eqs. (9) and (10), respectively:

$$\eta_{wt} = (m_0 - m)/m_0 \times 100\% \tag{9}$$

$$\eta_{sn} = (A_0 - A_{\min})/A_0 \times 100\% = (d_0^2 - d_{\min}^2)/d_0^2 \times 100\% = \left[d_0^2 - (d_0 - X)^2\right]/d_0^2 \times 100\% \tag{10}$$

4. Comparison among different models

By using different models, the ultimate bending moments can be evaluated based on the experimental condition, and the results are plotted in Figs. 1(a)-(f). The abscissa of the graph represents the weight loss ratio of the longitudinal reinforcement, and the ordinate represents the ratio between the test value of ultimate bending moment of the CRC beams and the model's predicted value ($M_{fx,exp}/M_{fx,cal}$). The calculation results of each model are summarized as follows:

(a) From the calculation results of Zhang *et al.* [1], it can be seen that 77.4% of the 177 experimental data are more than 1, that is, the experimental value of the ultimate bending moment is larger than the predicted value of the model, so it can be seen that the calculation result of this model is conservative. It can also be seen from the Fig. 1(a) that when the weight loss ratio of the steel (η_{wt}) is between 0 and 10%, the calculation result is closer to 1, which means that the prediction result is more accurate. When the weight loss ratio of the longitudinal reinforcement η_{wt} is large, the prediction error of this model will increase. This model has a mean value of 1.118 and a standard deviation of 0.246 for the 177 sets of data in the experimental database. The maximum value of the calculation result is 2.221 and the minimum value is 0.564.

(b) It can be seen from the model calculation results of Xu [2] that more than 87% of the data results are greater than 1, and only a small part of the calculation results are less than 1, so the calculation result of this model is conservative. At the same time, a clear trend can be found from the Fig. 1(b) that as the weight loss ratio of the longitudinal reinforcement increasing, the calculation result will also increase. That means the greater the corrosion rate of the longitudinal reinforcement, the larger the calculation error of this model and the more conservative of this model's calculation result. The mean value of the model calculation is 1.250, and the standard deviation is 0.310. The maximum value of the calculation result is 2.682, and the minimum value is 0.597.

(c) Sun's [3] model has better calculation results in existing empirical models. In the results of 177 sets of experimental data, the average value of the model is 0.998 and the standard deviation is 0.208. Moreover, compared with other models, the calculation result of Sun's model is more accurate when the weight loss ratio of steel bars is large. However, according to the experimental data of the model calculation results, it is known that nearly 50% of the data results are less than 1. In other words, this model overestimates the ultimate bending moment of CRC beams and the calculation result may be unsafe for engineering practice.

(d) It can be seen from the model calculation results of Azad *et al.* [4] that about 92% of the experimental data calculation results greater than 1. The calculation results of this model are conservative and underestimate the ultimate bending moment of CRC beams. What's more, when the test beam specimens come from the test beam members made by this model itself, the model's calculation results will be more accurate. But when the test beam specimens come from other literatures, the calculation error of this model is relatively large indicting that this model has poor universality when calculating the ultimate bending moment of CRC beams with the values of the mean 1.394 and the values of the standard deviation around 0.375.

(e) It can be seen from the calculation results that the modified model of Azad *et al.* [5] is better than the previous model proposed by himself. The mean of the calculated results is reduced to 1.294 and the standard deviation is reduced to 0.329. However, the modified model still has the problem that the calculation results are conservative and the generality of the model calculation is poor with the values of maximum is 2.979 and the values of minimum is 0.722 of the result.

(f) As can be seen from the Fig. 1(f), more than 90% of the calculation results are greater than 1 in the Torres-Acosta *et al.*'s model [6]. The calculation results of the model are conservative and underestimate the ultimate bending moment of CRC beams. It can also be seen from the Fig. 1(f) that as the weight loss ratio of the steel increases, the calculation error increases continuously where weight loss ratio of steel is greater than 20% and the calculation result of the model is too large. The mean of this model's result is 1.509 and the value of the standard deviation is 0.917.

Table 1. Conversion relationship between $f_{cyl,150}$, $f_{cu,150}$ and f_t.

$f_{cyl,150}$ (MPa)	12	16	20	25	30	35	40	45	50	55	60	70	80	90
$f_{cu,150}$ (MPa)	15	20	25	30	37	45	50	55	60	67	75	85	95	105
f_t (MPa)	1.6	1.9	2.2	2.6	2.9	3.2	3.5	3.8	4.1	4.2	4.4	4.6	4.8	5.0

Table 2. Conversion relationship between $f_{cu,150}$ and f_{ck}.

$f_{cu,150}$	15	20	25	30	35	40	45	50	55	60	65	70	75	80
f_{ck} (MPa)	10.0	13.4	16.7	20.1	23.4	26.8	29.6	32.4	35.5	38.5	41.5	44.5	47.4	50.2

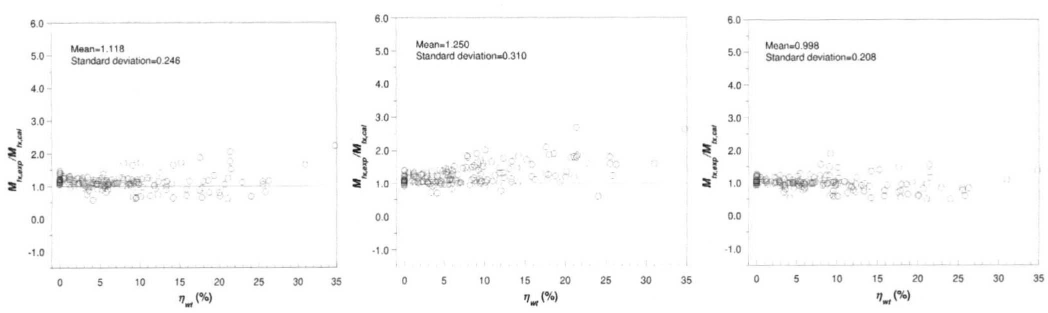

(a) Zhang *et al.*'s model [1] (b) Xu's model [2] (c) Sun's model [3]

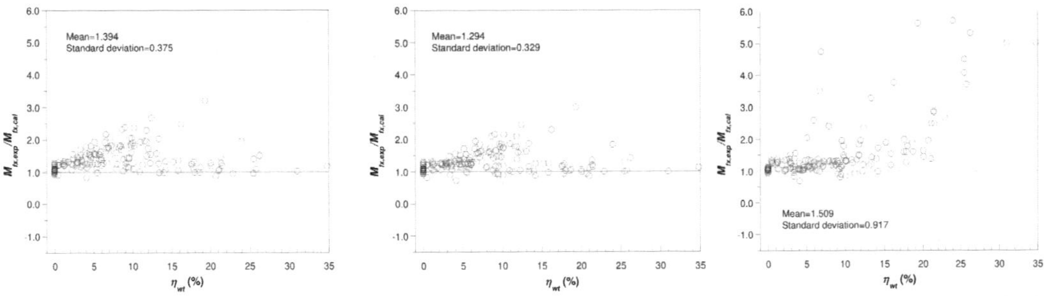

(d) Azad *et al.*'s model [4] (e) Azad *et al.*'s modified model [5] (f) Torres-Acosta *et al.*'s model [6]

Figure 1. The relationship between the ratios ($M_{fx,exp}/M_{fx,cal}$) and weight loss ratio of the longitudinal reinforcement (η_{wt}) of existing models.

Table 3. Comparisons on CRC beam ultimate bending moment from tests and predictions of existing models.

Literature	The ratios of test results to the predictions of existing models and new model				
	Mean	Standard deviation	Maximum	Minimum	Maximum-Minimum
Zhang *et al.*[1]	1.118	0.246	2.221	0.564	1.657
Xu [2]	1.250	0.310	2.682	0.597	2.085
Sun [3]	0.998	0.208	1.895	0.497	1.398
Azad *et al.* [4]	1.394	0.375	3.197	0.824	2.373
Azad *et al.*'s modified [5]	1.294	0.329	2.979	0.722	2.257
Torres-Acosta *et al.* [6]	1.509	0.917	5.501	0.683	4.818

In summary, in the six empirical models for calculating the ultimate bending moment of corroded reinforced concrete beams, five models [1, 2, 4, 5, 6] underestimate the ultimate bending moment of CRC beams, and the other model [3] overestimates the ultimate bending moment of CRC beams. The mean, standard deviation, maximum, minimum, and the difference between maximum and minimum of the ratio of all 177 test data are also listed in Table 3.

5. Conclusion

(1) According to the results of established database of flexural tests, five of the six empirical models [1, 2, 4, 5, 6] underestimate the ultimate bending moment of CRC beams with around 77.4%-92% lower than the experimental results. The model of Sun [3] provides the overestimated results, which is insecure in engineering practice.

(2) The comparison results show existing empirical models exhibit good agreement in calculating the residual flexural capacity of CRC beams, especially when the calculated experimental data come from the experiment of the corresponding empirical model itself. But when the experimental data involved are from other experiments and the corrosion rate of the steel involved varies within a large range, the existing empirical models will have a certain deviation in the calculation.

(3) Further research needs to be carried out. Researchers should know under what circumstances which model is more accurate to calculate the ultimate bending moment of CRC beams. And accurate and more universal models need to be proposed.

Acknowledgement

The authors acknowledge the research supports from the National Natural Science Foundation of China (51820105014 and 51738001) as well as the research funds from Australian Research Council (DE150101751 and IH150100006), Australia. The authors are also grateful for the financial supports of the National Engineering Laboratory for High-speed Railway Construction (Central South University), China (HSR2017001) and the State Key Laboratory of Silicate Materials for Architectures (Wuhan University of Technology), China (SYSJJ2018-01).

IOP Conf. Series: Materials Science and Engineering **652** (2019) 012024 doi:10.1088/1757-899X/652/1/012024

References

[1] Zhang J.R., Zhang K.B., Peng H., Gui C 2009 *Chin. J. Highw. Transport.* **22**(3), 45-51 (in Chinese).

[2] Xu S.H. 2003 *The deteriorated models and durability evaluation of reinforced concrete structure*, Xi'an University of Architecture and Technology (in Chinese).

[3] Sun B 2006 *Structural performance degrading and seismic evaluation of existing reinforced concrete structures*, Xi'an University of Architecture and Technology (in Chinese).

[4] Azad A.K., Ahmad S., Azher S.A. 2007 *ACI. Mater. J.* **104**(1), 40-47

[5] Azad A.K., Ahmad S., AL-Gohi B.H.A 2010 *Mag. Concrete. Res.* **62**(6), 405-414

[6] Torres-Acosta A.A., Navarro-Gutierreza S., Ter'an-Guill'en J. 2007 *Eng. Struct.* **29**(6), 1145-1152

[7] Jin W.L., Zhao Y.X. 2001 *Industrial Construction* **31**(5), 9-11 (in Chinese).

[8] Hui Y.L., Li R., Lin Z.S.1997 *Industrial Construction* **27**(6), 14-18 (in Chinese).

[9] Cao F.B., Wang C.X., Liu L.G., Xin Y.D., Li J.H., Tian Z.G.2015 *Building Structure*, **45**(10), 49-55 (in Chinese).

[10] Xia. J., Jin W.L., Li L.Y.2012 *Mag. Concrete. Res.* **64**(6), 471-485

[11] Shang D.F.2005 *Study on flexural behavior of corroded reinforced concrete beams*, Tongji University(in Chinese)

[12] Rodriguez J., Ortega M., Casal J.1997 *Constr. Build. Mater.* **11**(4), 239-245

[13] Chen J.2013 *Journal of Disaster Prevention and Mitigation Engineering* **33**, 83-87 (in Chinese).

[14] Zhang Z.2010 *Experimental study of bending behaviour of corroded reinforced concrete beams*, Shanghai Jiao Tong University (in Chinese)

IOP Conf. Series: Materials Science and Engineering **652** (2019) 012025 doi:10.1088/1757-899X/652/1/012025

Analysis of corrosion performance of rebar in Red mud blended concrete

U Raghu Babu[1], **B Kondraivendhan**[1,2]

[1] Applied Mechanics Department, SV National Institute of Technology, Surat 395007, India.
[2]Corresponding Author, Email: kondraivendhan78@gmail.com

Abstract. In this paper, electrochemical technique was employed to evaluate the performance of Red mud (RM) on the reinforcement corrosion. This paper also reports the effect of chloride, sulphate and the combined chloride-sulphate solutions on the corrosivity of rebar embedded in Ordinary Portland Cement (OPC) and RM blended concrete (95% OPC & 5% RM) slab specimens. The corrosion performance was evaluated by Linear polarisation resistance sweep test on the reinforced slab specimens made with mix water contaminated with 5% NaCl, 2% $MgSO_4$ and the combination of both salts. The corrosion performance was monitored for every 30 days over a period of 180 days, in terms of corrosion rate values. In order to investigate the overall performance of experimental test results they were analysed as per the specifications given by ASTM G16-13. The analysis of the corrosion data includes the descriptive statistics of the Normal, Weibull lognormal, and Smallest Extreme Value probability distribution functions (Pdfs) and the test of fit significance by the Anderson-Darling (AD) goodness of fit statistics. In addition, the analysis of variance was also carried out to determine the influence of each variable on the corrosion data. Further, the analysis of variance is also carried out to evaluate the influence of cementitious material and type of salt on corrosion data.

1. Introduction

Red Mud (RM) is a solid industrial waste obtained from Aluminium industries which is considered as a serious environmental issue due to its alkaline nature (pH ranges from 10.5 to 12.5) and chemical composition. The disposal of RM is also a huge challenge in Aluminium industries due to the need of vast landfills. Throughout the world 120 tons of RM is produced annually. In India, annually significant quantities of RM are stored in areas adjoining the Alumina plants. Due to inappropriate disposal of RM in large quantities, the soil and ground water resources can be contaminated severely. Therefore the utilisation of RM in construction industry needs to be explored to minimize the environmental as well as disposal problems.

From the experimental studies, Belén Díaz et al. [1] explored that the diffusivity values of chlorides and carbon dioxide through RM blended cement paste were significantly reduced. The high Aluminium content of RM able to trap chlorides to produce Friedel's salt is pointed as responsible for the reduction in chloride diffusivity. On the other hand, the retention in the CO2 penetration is assisted by the larger amount of C-S-H gel developed in the RM-mixtures along with the alkaline nature of the residue. Ribeiro et al. [2] investigated the possible use of RM in corrosion resistance of rebar by monitoring chloride diffusion, corrosion potential and electrical resistivity study. The results showed that the addition of red mud is beneficial to concrete, reducing its diffusion coefficients and corrosion potential and increasing its electrical resistivity. Cabeza et al.[3] studied the effectiveness of RM as chloride induced corrosion inhibitor of rebar in mortar specimens and also in the solutions made with red mud and alkaline solutions having the same pH 12. The results show that RM is able to maintain steel passivity in presence of high chloride concentration. Manfroi et al. [4] investigated the pozzolanic property of RM, microstructure and the mineralogy of pastes produced with up to 15% of RM in substitution of OPC. The results have shown that the most appropriate substitution is up to 5% of RM calcined at 600°C.

Several studies have been done on the properties of RM waste and its utilizations for different applications. Nikbin et al. [5]showed that Portland cement can be replaced by as high as 25% of RM. Senff et al.[6]used RM as much as 50% cement replacement in mortar and showed that increasing the RM content decreased the compressive and tensile strengths. As earlier research works suggested the possibility of using RM as the partial replacement of cement for effective corrosion resistance, the present work is aimed to investigate the performance of RM blended concrete in chloride, sulphate and combined chloride-sulphate admixed concrete.

2. Experimental Work

2.1. Materials and specimens

Reinforced concrete slab specimens were prepared with six concrete mixes. The major test variables include cementitious material type and admixed salt type. The plain reinforced concrete specimens were cast with Ordinary Portland cement (OPC) 53 grade conforming to IS: 12269-2013[7] and the blended concrete specimens were prepared with the replacement of cement with Red Mud at 5% by weight of OPC cement. The oxide composition of the cement and blended materials used in the current study were shown in Table 1.The diameter of the steel reinforcing bar was 10 mm. Coarse aggregates of size 20mm and 10mm maximum size of aggregate (MSA) were chosen in the ratio of 1.5:1 respectively as optimum packing. The specific gravity and water absorption of coarse aggregate were 2.8, and 2.25 % respectively, and the corresponding values for fine aggregate were 2.7, and 0.75%. Locally available river sand was used as fine aggregate. Laboratory tap water was used for the preparation of concrete mixtures with w/cm ratio 0.51. Analytical reagent grade chemicals were added to the concretes. The 5% sodium chloride and 2% of magnesium sulphate were added in mix water as the source of chloride and sulphate ions.

Reinforced concrete slab specimens with size 320mm x 320mm x 52mm were prepared with a centrally embedded Thermo Mechanically Treated (TMT) steel reinforcement of diameter 10mm. The steel specimens were cleaned and prepared as per the recommendations of ASTM G109-99a.The reinforcing steel was prepared such that the central portion of length 250 mm remains bare to expose to corrosive environment as shown in Fig. 1.

2.2. Corrosion monitoring

Corrosion monitoring instrument (make: ACM, Gill AC guard serial no. 1824-sequencer) was used to monitor the corrosion performance of rebar in concrete slab specimens. Linear Polarization Resistance (LPR) technique was adopted to determine the corrosion current density values of the concrete slab specimens. This LPR sweep test was conducted with IR compensation, using the guard ring arrangement. The slab specimens were polarized to 20mV from the corrosion potential at a sweep rate of 0.1 mV/second.

3. Results of statistical fitting analyses of corrosion data

Figs.2-4 depicts the I_{corr} values of rebar in OPC and RM blended concrete specimens admixed with NaCl, $MgSO_4$ and NaCl plus $MgSO_4$ respectively. According to ASTM G16-13 [8] statistical evaluation is a necessary step in the analysis of results regarding corrosion performance from any procedure which provides quantitative information. In order to evaluate the overall performance of the variables namely cementitious material type, salt type on corrosion of rebar, the different probability distribution functions (pdf's) were adopted. The descriptive statistics for the measured corrosion data were measured using Normal, Lognormal, Weibull and Smallest Extreme Value pdfs as per ASTM G16-13 [8].

Results of statistical analyses of measured corrosion data from each OPC and RM blended concrete specimen contaminated with chlorides, sulphates and combination of both were shown in Fig.5. In order to choose the best fitting distribution, the results of Anderson-Darling (AD) goodness-of-fit (Gof) test [9] of the corrosion test data were also reported in Fig.6. AD p-value with significance level (α) 0.05 is used to choose which distribution fits corrosion data well [9]. A p-value greater than or equal to 0.05 suggests that the modelled distribution significantly drawn from the fitting distribution otherwise not accepted. In Fig. 6 linear plot of significance level (α) 0.05 was included for directly indicating the suitable statistical distribution which fits the corrosion data well. This figure showed that all the I_{corr} data distributed like Weibull pdf while no group of I_{corr} data was distributed like Smallest Extreme Value distribution. It is also observed that Normal distribution is not a good fit for the corrosion data of the plain concrete admixed with chloride and composite solutions as well as the RM blended concretes admixed with sulphates. The p-value less than 0.05 of the lognormal distribution of RM blended concrete admixed with sulphate and composite solutions suggest that the distribution is not a good fit.

Table 1. Chemical composition of cementitious materials.

Chemical compound	OPC	RM
Al_2O_3	5.32	16.15
Fe_2O_3	4.23	54.8
SiO_2	20.65	6.3
TiO_2	--	3.7
MgO	1.13	--
CaO	64.12	1.98
Na_2O	--	3.55
SO_3	2.16	--
Others	--	1.72
LOI	2.39	11.8
Specific gravity	3.13	3.1

Figure 1. Experimental setup for corrosion monitoring of slab specimen

From the Fig.6, it is observed that the P-Value for Weibull distribution of all concrete specimens with different admixed solutions was more than the significant value 0.05, which means that all the corrosion current density values were drawn from the Weibull distribution .The overall performance of OPC and RM blended concrete specimens was evaluated from the use of this selected Weibull distribution as shown in Fig.7. These identifications of descriptive statistics for detailing the corrosion data highlights the importance of the use of goodness-of-fit test for determining the probabilistic distribution fitting followed by the corrosion data.

4. Analysis of Variance

Analysis of Variance (ANOVA) was carried out for the experimental I_{corr} data as the response and the two levels (OPC, RM) of cementitious material type and three (Chloride, sulphate and combination of both the salts) levels of salt type as discrete variables. The ANOVA was adopted to determine the statistical significant difference between the type of cementitious material and slat type.

IOP Conf. Series: Materials Science and Engineering **652** (2019) 012025 doi:10.1088/1757-899X/652/1/012025

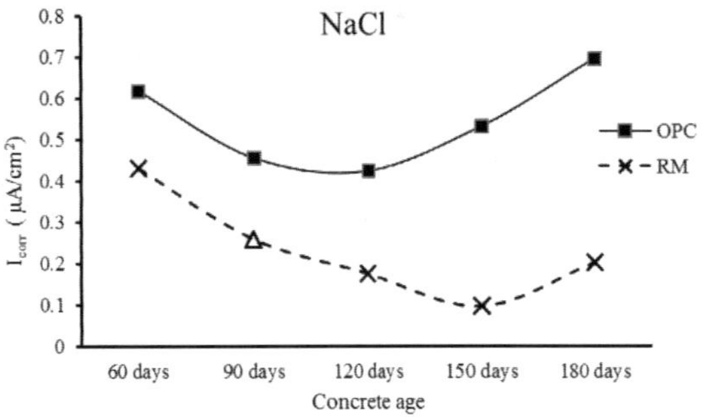

Figure 2. Experimental test results of NaCl admixed concrete

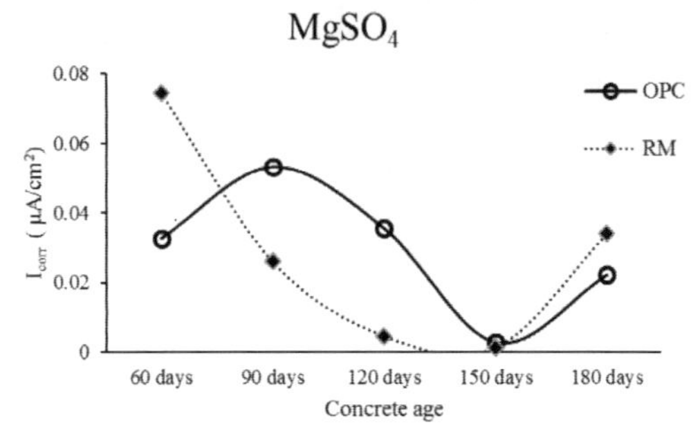

Figure 3. Experimental test results of $MgSO_4$ admixed concrete

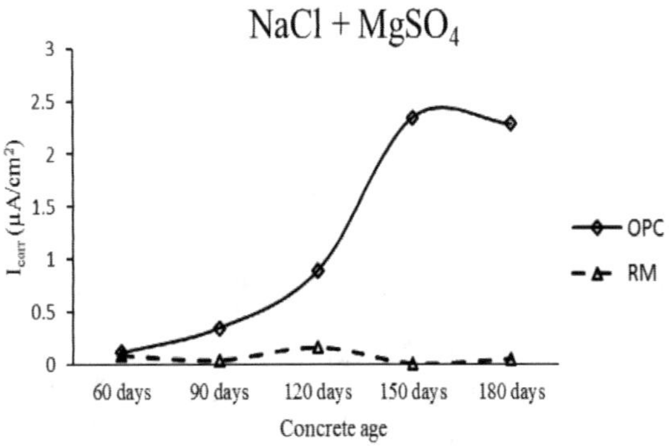

Figure 4. Experimental test results of NaCl plus $MgSO_4$ admixed concrete

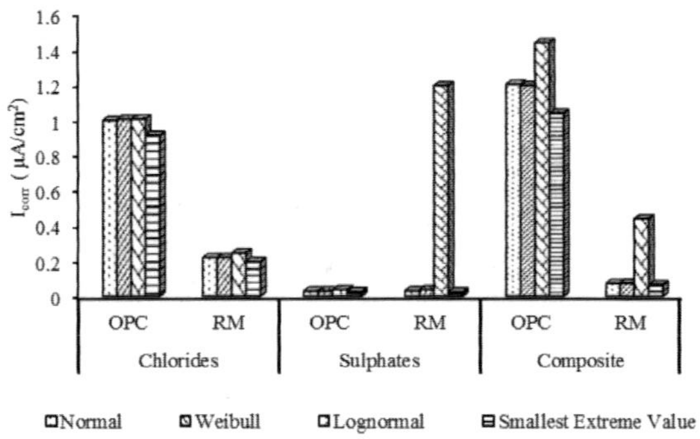

Figure 5. Analysed results of I_{corr} data of OPC and RM concretes admixed with chloride, sulphate and composite solutions.

The obtained I_{corr} values at various levels of cementitious material type(CM), salt type are arranged in tabular form. After that the sum of the squares (SS) for factor and residual error were calculated. Then the mean square (MS) of the factor is calculated by dividing their corresponding sum of squares by associated degree of freedom (DF). Then the effect of individual factors is evaluated by testing the hypothesis of equality of variances. In analysis of variance the most important statistics is the P-value, which used to determine whether the level of means are significantly different from each other about the hypothesis with 95% confident level (α=0.05). The hypothesis for the present study is given below[10]:
P value $\leq$ 0.05: The level means are significantly different from each other
P value > 0.05: The level means are not significantly different.

While observing the effect of salt type on corrosion of rebar from the Fig.6, it is observed that the corrosion due to sulphate ions is very less as compared to chlorides. As well as it is also observed that the presence of sulphate ions with chlorides enhances the chloride induced corrosion in OPC concrete but not in RM blended concrete. The replacement of cement with 5% of RM enhanced the corrosion resistance of rebar in concretes contaminated with chlorides, sulphates and combination of both the salts.

Table 2. ANOVA results for Salt type variable

ource	DF	SS	MS	F	P
Salt	2	4.532	2.2662	3.39	0.041
Error	56	37.392	0.6677		
Total	58	41.924			

Table 3. Results of ANOVA for I_{corr} values of NaCl contaminated concrete

Source	DF	SS	MS	F	P
CM	1	2.858	2.857	6.85	0.018
Error	17	7.096	0.4174		
Total	18	9.958			

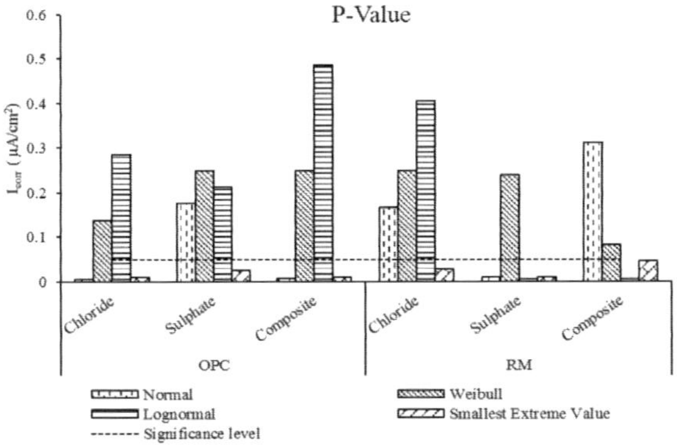

Figure 6. The plots of Anderson Darling goodness-of-fit test results of corrosion data

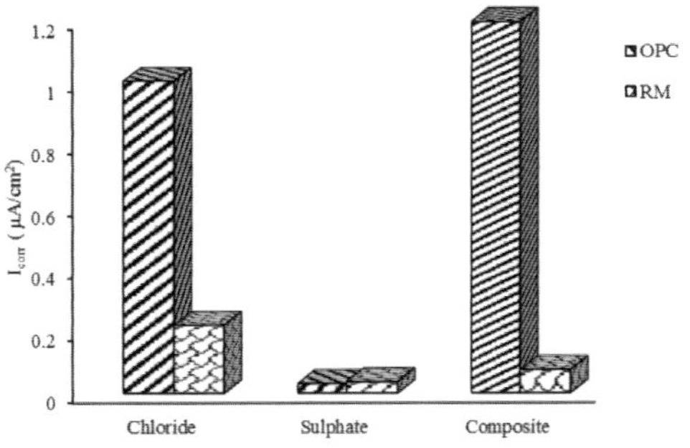

Figure7. Overall performance of OPC and RM blended concrete

Table 5. Results of ANOVA for I_{corr} values of NaCl plus $MgSO_4$ contaminated concrete

Source	DF	SS	MS	F	P
CM	1	6.337	6.337	5.41	0.032
Error	16	21.080	1.171		
Total	17	27.417			

Table 4. Results of ANOVA for I_{corr} values of $MgSO_4$ contaminated concrete

Source	DF	SS	MS	F	P
CM	1	0.0020	0.002	1.87	0.190
Error	17	0.0171	0.001		
Total	18	0.0191			

5. Conclusions

From the present experimental study the following conclusions are drawn:

1. The incorporation of 5% RM in concrete is beneficial to enhance the corrosion resistance of rebar embedded in concrete.
2. The corrosion due to sulphates is negligible as compared to chloride induced corrosion in both plain and RM blended concrete.
3. The presence of sulphates enhances the chloride induced corrosion in normal OPC concrete but the opposite trend was observed in RM blended concrete.

References

[1] B. Díaz, L. Freire, X.R. Nóvoa, M.C. Pérez 2005 Chloride and CO_2 transport in cement paste containing red mud, *Cem. Concr. Compos.* **62** 178–186.

[2] D. V. Ribeiro, J.A. Labrincha, M.R. Morelli 2012 Effect of the addition of red mud on the corrosion parameters of reinforced concrete, *Cem. Concr.* Res. **42** 124–133.

[3] M. Cabeza, A. Collazo, X. Novoa, M. Pérez 2003 Red mud as a corrosion inhibitor for reinforced concrete, *J. Corros. Sci.* **6** 1–14.

[4] E.P. Manfroi, M. Cheriaf, J.C. Rocha 2014 Microstructure, mineralogy evaluation of cementitious composites produced with red mud waste, *Constr. Build. Mater.* **67** 29–36.

[5] I.M. Nikbin, Aliaghazadeh, Sh Charkhtab, A. 2018 Environmental impacts and mechanical properties of lightweight concrete containing bauxite residue, *J. Clean. Prod.* **172** 2683–94.

[6] L. Senff, D. Hotza, J.A. Labrincha, Effect of red mud addition on the rheological behaviour and on hardened state characteristics of cement mortars 2011 *Constr. Build. Mater.* **25** 163–170.

IOP Conf. Series: Materials Science and Engineering **652** (2019) 012025 doi:10.1088/1757-899X/652/1/012025

[7] BIS 12269 2013, Ordinary Portland cement, 53 Grade-Specification (First Revision).
[8] ASTM G16-13 2013 Standard Guide for Applying Statistics to Analysis of Corrosion Data.
[9] G.S.M. Kumar Molugaram Rao, G.S. RAO, *Statistical Techniques for Transportation Engineering*, BSP Books Pvt Ltd. Published by Elsevier Inc.

The Effect of Dynamic Vulcanization on the Mechanical and Thermal Properties of Stereocomplex PLLA/PDLA/rubber Blends

S Phattarateera[1,] and C Pattamaprom[1,*]

[1]Departments of Chemical Engineering, School of Engineering, Thammasat University, Pathumthani 12120, Thailand
*Corresponding author, Email: cattalee@engr.tu.ac.th

Abstract. This study investigated the influence of dynamic vulcanization of sterecomplex PLA/PDLA (ST) containing various types of rubbers by using dicumyl peroxide (DCP) as a vulcanizing agent. In this work, natural rubber (NR), isoprene rubber (IR), silicone rubber (SI), acrylic rubber (AR), thermoplastic copolyester (TPE) and thermoplastic polyurethane (TPU) were selected. Here, 5 wt% of PDLA, 15 wt% of rubber and 2 wt% DCP as an initiator were melt-blended with PLA in an internal mixer at 180 0C and 50 rpm. It was found that the addition of DCP in (NR and IR) led to a significantly increase in the impact strength and % elongation at break. On the other hand, for the rubbers that do not have double bonds in the main chain (SI, AR, TPE and TPU), the addition of DCP led to a decrease in impact strength and % elongation at break compare to the case without DCP due to the lack of double bonds to react with DCP. For the effect of DCP on the rubber dispersion, it was found that the addition of DCP on the NR and IR could improve the interface compatibility of the rubber and PLA leading to smaller average rubber domain sizes than without DCP. For other types of rubbers, the average rubber domain sizes were bigger with poorer interface compatibility than without DCP. For the effect of DCP on the crystallinity, it was found that the addition of DCP could increase the crystal density and could consequently reduce the crystal sizes of the ST/rubber blends.

1. Introduction

PLA has been used in a broad range of medical applications, including drug delivery systems and tissue engineering [1-5]. However, the major deficiencies of PLA that limited its wider range of applications are low thermal resistance and high brittleness. Many researches have attempted to improve the thermal properties of PLA. A few studies have focused on the addition of inorganic fillers which act as heterogeneous nucleation sites for PLA crystals. Urayama et al. (2003) [6] found that talc was the most effective nucleating agent for PLLA. Even though the addition of a nucleating agent could enhance the crystallization rate of PLA, it decreased its toughness. One of the interesting methods to induce crystallization of PLA without introducing alien inorganic components is by using stereocomplex crystals formed by the enantiomeric blends of PLLA and PDLA. Ikada et al. (1987) [7] reported that the 50:50 blend of PLLA and PDLA produces a stereocomplex with the melting temperature (T_m) around 230 ^{0}C, which is 50 ^{0}C higher than the melting points of its homocrystal counterparts. Yamane et al. (2003) [8] studied the effect of PDLA on the thermal property of PLLA. They found that a small amount of PDLA could significantly enhance the crystallization of PLLA by forming stereocomplex crystallites in the PLLA matrix. The stereocomplex structure with small amount of PDLA (1-5 wt%) could improve the heat tolerance, tensile strength and Young's modulus of PLA. For the improvement in toughness of PLA, citrate ester, triacetin, poly(ethylene glycol) (PEG) and other small molecules have been reported as efficient plasticizers for PLA [9-13]. However, blends of PLA with plasticizers caused a blooming problem when used at high concentration. The elastomers such as commercial natural and synthetic rubbers, as well as thermoplastic elastomers are better choices for improving the toughness of PLA because it is extremely tough and easy to process without a blooming problem. Bitinis et al. (2011) found that the optimal NR content to improve the brittleness of PLA was 10 wt%, where the tensile tests of the samples showed a drastic increase in the average elongation at break from 5% to 200% [14]. Kowalczyk and Piorkowska (2012) researched mechanisms of plastic deformation in biodegradable polylactide/poly(1,4-cis-isoprene) blends and found that incorporation of only 5 wt% of the poly(1,4-cis-isoprene) rubber could significantly increase the strain at break by 3 fold and impact strength by 1.81 fold, respectively [15]. Petchwattana and Covavisaruch, (2014) studied the toughness improvement of PLA/rubber wood sawdust composites by using acrylic core-shell rubber (CSR). They found that 5% of CSR could improve the impact strength of the composite by more than five times with improved tensile elongation at break [16]. Rui-lei et al. (2014) studied the improvement in toughness of PLA by melt blending it with bio-based polyurethane elastomer (TPU). They found that 15 wt% of TPU had the highest improvement in impact strength from 4 kJ/m^2 to 9 kJ/m^2 [17].

In our group, we found that the improvement in toughness and impact strength by adding rubbers came with the sacrifice of lower thermal resistance. It seems that less attention has been paid to resolve both problems simultaneously. Because most rubbers and polymer were not naturally compatible and they required one or more steps to make them compatible, the dynamic vulcanization using peroxide as crosslink agent is a simple technique that has been widely applied. The free radicals obtained from decomposed peroxides could promote chain scission, branching, crosslinking or any combination of the three, which affect the mechanical and thermal properties of polymers. For decades, dynamic vulcanized of various polymer/rubber systems [18-29] have already been investigated.They suggested that dynamic vulcanization could lead to fine morphologies and better interfacial interaction [30-36]. For PLA, Rytlewski et al. (2011) [37] studied the influence of DCP content on thermal and mechanical properties of PLA. They found that the only 0.2 wt% of DCP led to higher % crystallinity and adding of DCP (above 0.4 wt%) caused crosslinking of PLA and led to the decrease in molecular weight due to decomposition and degradation processes. Moreover, the high content of DCP led to a significant decrease in the tensile strength and impact strength. Daosheng et al. (2014) [38] studied the use of 1.5 wt% DCP in PLA/NR blend at the PLA:NR ratio of 60:40. They found that DCP led to an increase in impact strength from 5 kJ/m^2 to 40 kJ/m^2 compared to neat PLA and an increase in % elongation at break from 30% to 120%. Moreover, the addition of DCP could increase the percent crystallinity of PLA/NR from 9.71% to 15.44% compared to neat PLA. Nevertheless, there is only one research article reported on the use of DCP in the PLA/PDLA stereocomplex system. Wan-jie et al. (2017) [39] investigated the effect of dynamic vulcanization using DCP in the PLLA/PDLA/ bioelastomer (BE)) blends. They found that the addition of DCP (0.1 wt%) in PLLA/BE enhanced the impact strength from 50 J/m to 600 J/m and they also found that the addition of PDLA (only10wt%) in the blend(70%PLLA/20%BE10%PDLA) had the best impact strength and excellent heat resistance.

In this paper, the effect of dynamic vulcanization by using dicumyl peroxide (DCP) as a cross-linking agent has been applied to various rubber-toughened stereocomplex PLA/PDLA systems. The rubbers used in this study included natural rubber (NR), isoprene rubber (IR), acrylic rubber (AR), acrylic core-shell rubber (CSR), thermoplastic polyester (TPE) and thermoplastic polyurethane (TPU). The effect of DCP on the crystallization, morphology, and mechanical properties of the blends has been systematically investigated

IOP Conf. Series: Materials Science and Engineering **652** (2019) 012026 doi:10.1088/1757-899X/652/1/012026

2. Materials and Methods

2.1 Materials

Polylactic acid, PLA (4043D) used in this study was produced by NatureWorks Co. Ltd., USA. Poly (D-lactide) (PDLA) grade D100H with M_n = 40-45 kg/mol was purchased from Sulzer ChemTech Co. Ltd. PLA and PDLA were dried at 70°C for 24 hours in a hot air oven before use. Polyisoprene (IR) grade IR2200 and acrylic rubber (AR) grade ACRON 801 were obtained from Innovation Group Co. Ltd., Thailand. The thermoplastic polyester (TPE) grade ARNITEL EM 400 and thermoplastic polyurethane (TPU) grade WHT-1195IC were purchased from Ableone Engineering Co. Ltd., Thailand. Liquid silicone rubber (SI) grade RTV S815 was purchased from Super-Silicone and Resin Art Co. Ltd., Thailand.

2.2 Preparation of the rubber-toughened PLA/PDLA blends

The melt blending of PLA/PDLA with a rubber was carried out in an internal mixer at 180 °C with the rotor speed of 50 rpm. The ratio of PLA to PDLA was 95:5 by weight and the rubber content was fixed at 15 wt%. The amount of DCP was fixed at 2.0% by weight of the rubber. The mixing began by feeding PLA and PDLA into the internal mixer, which was set at 180 °C. The specified amount of rubber and DCP were added at 2 and 5 minutes, respectively, while the total mixing time was 12 minutes. The blend sample was compression molded at 190 °C for 15 minutes before further characterization. Here, the compound of PLA with PDLA is denoted as ST and the blend of ST with xx rubbers is denoted as ST/xx. In addition, the blend of ST/xx with DCP is called ST/xxD.

2.3 Characterization

2.3.1 X-ray diffraction (XRD). XRD is primarily used for distinguishing between amorphous and crystalline portions of a material and for quantifying the percent crystallinity of a sample. Wide-angle X-ray diffraction (WAXD) measurements of a compression-molded specimen were carried out with a DX-1000 X-ray diffractometer (Dandong Fanyuan Instrument Co. LTD, China) using a Cu Kα radiation source (λ = 0.154 056 nm, 40 kV, 25 mA) in the scanning angle range of 2θ = 5°-40° at a scan speed of 3^0/min. The percent crystallinity of the homocrystals (%X_{hc}) and stereocomplex crystals (%X_{sc}) were calculated using the following equations:

$$X_{hc}\,(\%) = \frac{Ihc}{Ihc+Isc+Iamor}\,x100 \qquad (1)$$

$$X_{hc}\,(\%) = \frac{Isc}{Ihc+Isc+Iamor}\,x100 \qquad (2)$$

where Ihc is the integrated intensity of homocrystals, Isc is the integrated intensity for the stereocomplex crystallites and Iamor is the integrated intensity of the amorphous portion.

2.3.2 Differential scanning calorimetry (DSC). The thermal properties of PLA and ST/rubber blends were investigated by a METTLER TOLEDO (DSC 3) differential scanning calorimetry (DSC) at the heating rate of 10 °C/min from 50 to 230 °C.

2.3.3 Mechanical properties. Tensile tests were performed according to ASTM D 638 type I on an Instron machine Model 5969, Instron Engineering corporation USA, at 25°C, with the cross-head speed of 5 mm/min. Dumbbell-shaped specimens were cut from the compression-molded sheet prepared at the pressure of 1500 psi and 190 °C. At least five specimens of each sample were tested to determine the mean value of the results. The notched Izod impact tests were measured according to ASTM D256 by the Rasil impactor using the pendulum energy of 2.75 J at 25 °C. For this measurement, the samples were cut, notched, and kept in desiccators overnight before the tests.

2.3.4 Morphology of the blends by SEM. The size and dispersion of rubber domains in the ST matrices were observed on the freeze-fractured surfaces of the samples using a scanning electronic microscope (SEM) (Camtrad, JEOL MX2000 operated at 5-15 keV). The fractured surfaces after impact tests of selected samples were also observed by SEM to investigate the fracture behavior of the blends. All samples were coated with gold before SEM analysis.

3. Results and Discussion

3.1 Characteristics of ST and ST/rubber blends

In this part, the stereocomplex formation of PLA/PDLA was confirmed by XRD and DSC .The XRD profile of PLA and ST in Figure 1 illustrated that, even though the PDLA content was only 5 wt%, the stereocomplex crystals were successfully formed .The characteristic of ST peaks were observed at the 2ϴ of 12.0°, 20.9°, and 24.0°, which were assigned to the (110), (300)/(030), and (220) planes of stereocomplex crystals [40]. On the other hand, the characteristic peaks of PLA homocrystals were at the 2ϴ of 14.8°, 16.9°, 19.0°, and 22.5° corresponding to the α or β-form of PLA homocrystals[16]. The amorphous portion of the blend was indicated by the underlying broad peak. From the DSC pattern in Figure 2, the characteristic of sterecomplex crystals was indicated by the high melting point peak at around 220 °C, which could pose a potentially improved heat tolerance of the sample.

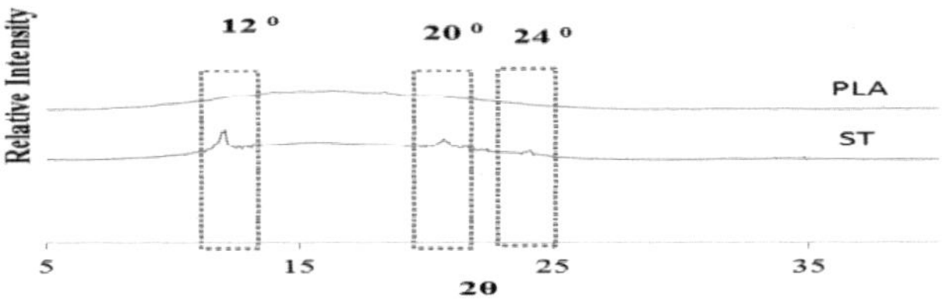

Figure 1. Wide-angle X-ray scattering profiles of PLA and ST

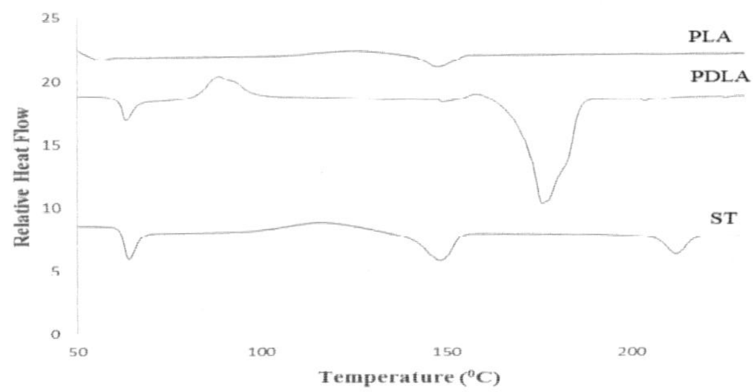

Figure 2. DSC heating curves of PLA, PDLA and ST

3.2 The effect of DCP on mechanical properties

In this section, the effect of dynamic vulcanization was investigated on rubber-toughened stereocomplex PLA/PDLA systems by using 2.0 wt %of DCP as a radical initiator. The rubber blends with PLA were categorized by the double bonds in rubber main chains. NR and IR were classified as unsaturated rubbers, whereas SI, AR, TPE and TPU were classified as saturated rubbers. The effect of DCP on the impact strength, % elongation at break and tensile strength of ST/rubber blends are demonstrated in Figures 3-4, respectively. It can be seen that the addition of DCP in PLA or ST led to decreased impact strengths, which was from 27 J/m to 22 J/m for neat PLA and from 31 J/m to 28 J/m for ST. This was probably because high DCP content could cause a chain scission reaction in PLA chains [17]. For the ST/rubber blends, it was found that DCP could significantly improve the interfacial adhesion between unsaturated rubbers (NR) and (IR) and PLA matrix as observed by smaller rubber domain sizes (Figure 5 and 6) leading to improved impact strength, tensile strength and % elongation at break of the blends (Figures 3-4). This was probably because DCP could easily react with the double bonds in the rubber chains and thus provide improved linkage between the rubbers and PLA .However, for the blends of ST with saturated rubbers (SI, AR, TPE, and TPU), the addition of DCP led to significant decreases in the impact strength, tensile strength and % elongation at break with increased rubber domain sizes. This finding is consistent with a previous work, which reported that the reaction of DCP in miscible blends tended to induce phase separation, where the dispersed domain sizes increased from nanometer range to sub micrometer range [41]. However, in case of PLA/TPU, Shi et al. (2014) [42] reported that the addition of DCP at the lower content of 0.2 wt% could improve impact strength by nearly 3 times of the unmodified PLA/TPU blend with a decrease in rubber size.

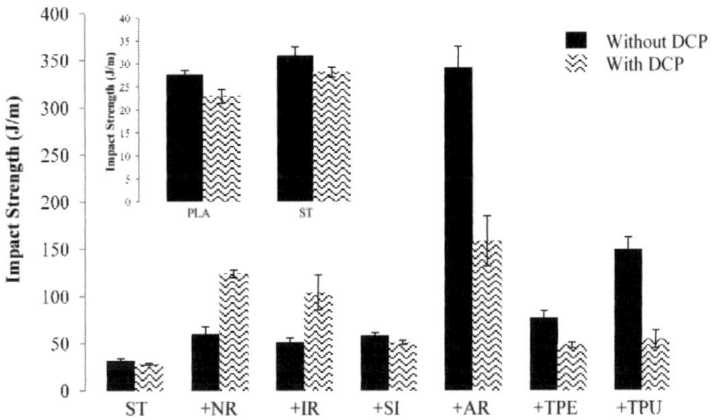

Figure 3. The impact strength of the ST/rubber blends without and with DCP.

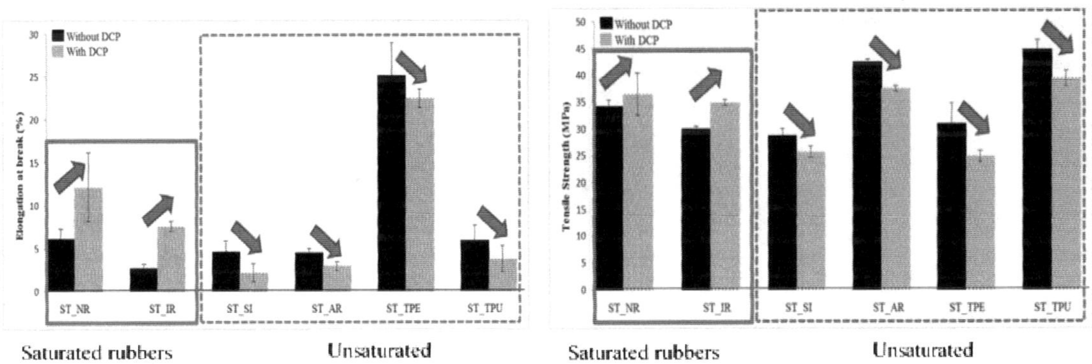

Figure 4. % Elongation at break and Tensile strength of ST/rubber blends without and with DCP .

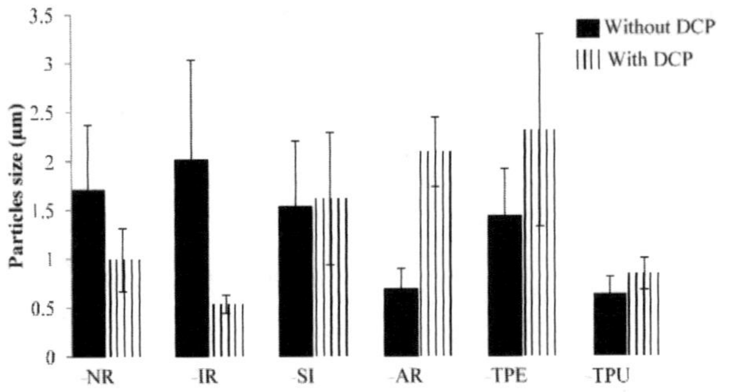

Figure 5. the average rubber domain sizes of ST/rubber blends without and with DCP

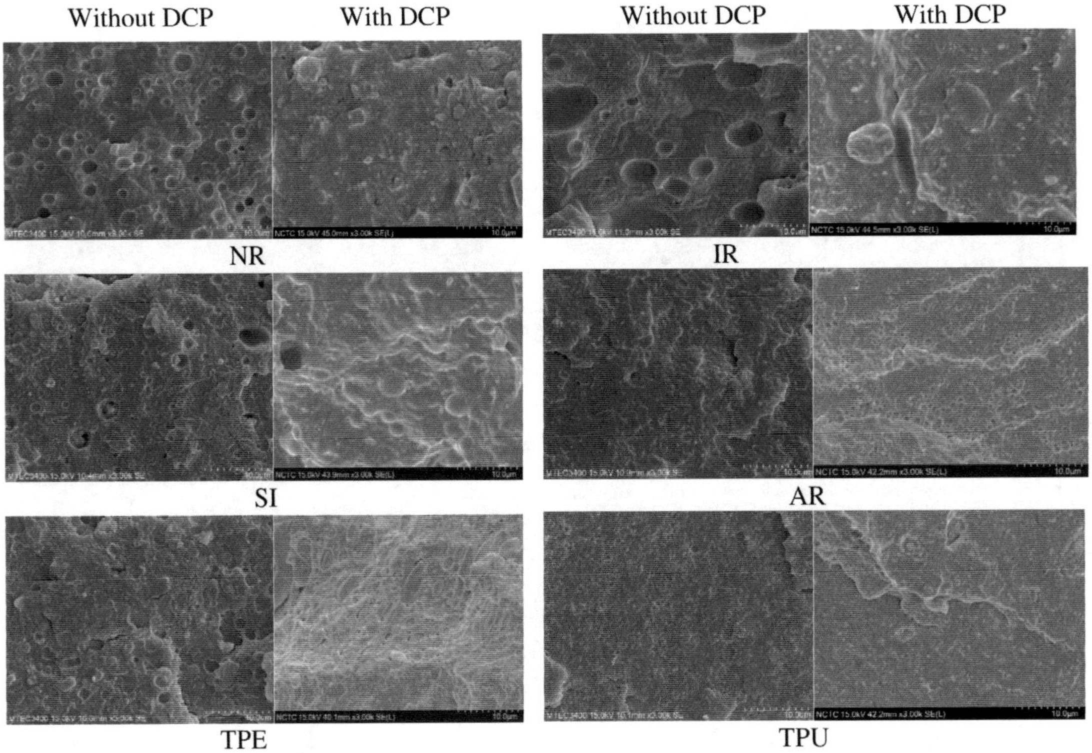

Figure 6. SEM micrographs of the freeze-surfaces without and with DCP

3.3 Effect of DCP on thermal properties

The crystallization behaviors of ST/rubber blends with DCP were compared to those without DCP by XRD technique. The XRD profiles of the blends are not shown and the percent crystallinities of the blends are summarized in Table 1. For neat PLA, it was found that crystalline peaks could not be detected from XRD (Figure 1). This is probably because the PLA sample used in this work was processed the same way with the PLA/rubber blends without annealing to control the similarity in their process conditions and cooling pattern. It was found that the addition of DCP into neat PLA, ST, and all ST/rubber blends could lead to significant increases in the percent crystallinities of both of homo and stereocomplex crystallize. In particular, the percent crystallinity of ST with DCP increased by 62.7%. With the presence of unsaturated rubbers, the percent crystallinity of ST/NR and ST/IR increased by 12.8% and 47.4%, respectively. Whereas, for the blends with saturated rubbers, the percent crystallinity of ST/SI, ST/TPE and ST/TPU increased by 18.5%, 32.8% and 56.1%, respectively. For ST/AR, the percent crystallinity did not increase much with the presence of DCP as its original value was already high. These results indicated that the DCP could contribute to improvement in the crystallization of PLA phase, possibly due to the enhanced chain mobility or improved heterogeneous nucleation effect [43]. In addition, according to Scherrer equation [44], it was found that the relative intensity of the blends with DCP was higher than the blends without DCP indicating that the crystal sizes of ST/rubber blends with DCP were smaller than those without DCP.

Table 1. The percent crystallinity calculated from XRD profile

Formulate	**%X_c (Homo)**		**%X_{sc} (Stereocomplex)**	
	Without DCP	**With DCP**	**Without DCP**	**With DCP**
ST	N/A	N/A	4.13	6.72
ST_NR	N/A	1.11	3.98	4.49
ST_IR	N/A	2.09	3.86	5.69
ST_SI	N/A	3.07	4.64	5.50
ST_AR	N/A	2.97	4.89	4.92
ST_TPE	N/A	2.69	4.03	5.35
ST_TPU	N/A	2.96	3.49	5.45

* N/A = the crystals were not detectable.

4. Conclusions

The dynamic vulcanization of stereocomplex PLA/PDLA in the presence of various rubbers was successfully prepared in an internal mixer by using 2 wt% DCP as a radical initiator. From the results, the addition of DCP in unsaturated rubber blends led to a significant increase in the impact strength, tensile strength and % elongation at break with smaller rubber domain sizes. This is probably because the addition of DCP into unsaturated rubbers could improve phase interaction and crosslink the rubber. On the other hand, for saturated rubber blends, the addition of DCP led to decreases in impact strength, tensile strength and % elongation at break with increased rubber domain sizes. On the other hand, saturated rubbers already had excellent phase interaction without DCP as indicated by originally very small rubber domain sizes. The addition of DCP led to poorer phase interactions leading to bigger average rubber domain sizes. For crystallization properties, the addition of DCP in ST/rubber blends could increase the % crystallinity and reduce the crystallite size of both homo and stereocomplex crystals in the blends.

References

[1] K.S. Anderson, K.M. Schreck, M.A. Hillmyer 2008 *Polym. Rev.* **48** 85-108
[2] H. Liu, J. Zhang 2011 *J. Polym. Sci. Part B-Polym. Phys.* **49** 1051-1083
[3] H. Urayama, T. Kanamori, K. Fukushima, Y. Kimura 2003 *J. Polym.* **44** 5635–41
[4] Y. Ikada, K. Jamshidi, S. Hyon and H. Tsuji 1987 *J. Macromol.* **20** 904-905
[5] H. Yamane, K. Sasai 2003 *J. Polym.* **44** 2569–75
[6] Z. Kulinski and E. Piorkowska 2005 *J. Polym.* **46** 10290-10300
[7] Z. Kulinski, E. Piorkowska, A. Galeski and R. Masirek 2006, *J. polym.***47** 7178-7188
[8] O. Martin and L. Avérous 2001 *J. Polym.* **42** 6209-6219
[9] N. Ljungberg and B. Wessle´n 2003 *J. Polym.* **44** 7679-7688

[10] H. Fatima , M. R. Jean-, A. Frédéric , D. Philippe , T. Valérie and R. David 2011 *Eur. Polym. J.* **47** 2134-2144
[11] D. Garlotta 2011 *J. Polym. Environ* **9** 63-84
[12] R.M. Rasal, A.V. Janorkar, D.E. Hirt 2010 *Prog. Polym.* Sci. **35** 338-356
[13] O. Martin, L. Averous 2001 *Polym* **42** 6209-6219
[14] N. Bitinis, R. Verdejo, P. Cassagnau and M. Lopez-Manchado 2011 *Mater. Chem. Phys.* **129**, 823-831
[15] M. Kowalczyk, E. Piorkowska 2011 *J. Appl. Polym. Sci.* **124** 4579-4589
[16] N. Petchwattana, S. Covavisaruch 2014 *Bionic Engineering.* **11** 630-637
[17] Y. Rui-lei, L.-s. Zhang, Y.-h. Feng, R.-y. Zhang and J. Zhu 2014 *J. Polym. Sci.* **32** 1099-1110
[18] F.Liao, A.Su, T. Hsu 1994 *Polym* 2579-2586
[19] A. R. Payne 1962 *Appl. Polym. Sci.* **6**, 57-63
[20] R. Mancke, J.Ferry 1968 *Transactions of the Society of Rheology* **12** 335
[21] B. Na, Q. Zhang, Q. Fu, G. Zhang, K. Shen 2002 *Polym* **43** 7367-7376
[22] A. Jha, A. Bhowmick 1997 *Rubber Chem. Technol* **70** 798-814 (1997)
[23] S. George, K. Ramamurthy, J. Anand, G. Groeninckx, K. Varughese, S. Thomas 1999 *Polym* **40** 4325-4344
[24] Y. Coran Aubert 1995 *Rubber Chem. Technol* **68** 351-375
[25] H. Ismail Salmah, M. Nasir 2001 *Polym. Test.* **20** 819-823
[26] W. Si, W. Yuan, Y. Li, Y. Chen, J. Zeng 2018 *Polym. Test.* **65**, 249-255
[27] Y.K. Chen, C.H. Xu, L.M. Cao, Y.P. Wang, L.M. Fang 2013 *J Phys Chem B* **117** 7819-7825
[28] Y. He, T. Zhao, Y. Li, M. Wang, J. Zeng 2017 *Polym. Test.* **59** 470-477
[29] D. Yuan, K. Chen, C. Xu, Z. Chen, Y. Chen 2014 *Carbohydr. Polym* **113** 438-445
[30] P. Ezzati, I. Ghasemi, M Karrabi 2008 *Ira Poly J.* **17** 265–272
[31] W. Brostow, T. Datashvili, KP. Hackenberg 2010 *Poly Compos* **12** 1678–1691
[32] R. Rajesh Babu, K. Naskar 2011 *Adv Polym Sci.* **239** 219–248
[33] Y. Wang, C. Zhang, Y. Ren, T. Ding, D. Yuan, Y. Chen 2018 *Polym Adv Technol* 1–8
[34] H.Z. Liu, F. Chen, B. Liu, G. Estep, J.W. Zhang 2010 *Macromole* **43**, 6058-6066
[35] H.W. Liu, W.J. Song, F. Chen, L. Guo, J.W. Zhang 2011 *Macromole* **44** 1513-1522
[36] Y.K. Chen, D.S. Yuan, C. H 2014 *ACS Appl Mater Interfaces* **6** 3811-3816
[37] P. Rytlewski, M. Z˙enkiewicz, R. Malinowski 2011 *Intern. Polymer Processing XXVI* **5**
[38] D. Yuan, K. Chen, C. Xu, Z. Chen, Y. Chen 2014 *Carbohydr. Polym.* **113** 438–445
[39] W.-J. Si, X.-P. An, J.-B. Zeng, Y.-K. Chen and Y.-Z. Wang 2017 *Science China Materials* 1-15
[40] X.-F. Wei, R.-Y. Bao, Z.-Q. Cao, W. Yang, B.-H. Xie and M.-B. Yang 2014 *Macromole* 1439-1448
[41] R. Rajesh Babu and K. Naska 2010 *Advanced Rubber Composites* 219-247
[42] U. Basuli, T. K. Chaki, K. Naskar 2008 *J. Appl. Polym. Sci.* **108** 1079–1085
[43] B. G. Soares, D. M. Santos, A. S. Sirqueira 2008 *eXPRESS Polymer Letters* 602–613
[44] M. Shi, D. Zhang, J. Zhu, Y. Shi, J. Sun and Y. Ji, J. 2018 *Thermoplast Compos. Mater* 1–14
[45] Y. Shi, X. Du, J. Yang, T. Huang, N. Zhang, Y. Wang, G. Yuan, and C. Zhang 2014 *Polymer-Plastics Technology and Engineering* 53, 1344–1353
[46] G.-C. Liu, Y.-S. He, J.-B. Zeng, Q.-T. Li and Y.-Z. Wang 2014 *Biomacromolecules* 4260-4271
[47] B. Ingham, M.F. Toney 2014 *Structure, Processing and Properties* 3-38

Structural and seismic performance of RC columns with L-shaped hoop details

Y H Choi[1], Y K Lee[1], Y J Chun[1], H S Kim[1] and Y S Shin[1, 2]

[1] Architectural and Urban Systems Engineering Department, Ewha Womans University 03760 Seoul, South Korea
[2] Corresponding author, Email: shinys@ewha.ac.kr

Abstract. This paper aims at developing L-shaped hoop detail and evaluating structural and seismic performance of reinforced concrete columns when the developed hoop detail is applied. L-shaped hoop is introduced in this study to save construction time period by improving constructability. RC columns are fabricated having test parameters of concrete strength (24MPa and 35MPa) and hoop details (general hoop and L-shaped hoop). Axial and seismic loading tests are performed, and the test results of maximum column strength and load-strain curve of the column are similar regardless of hoop types. In addition, seismic test results show that hysteresis loops and energy dissipation capacity of the column specimens using two different hoop types (general hoop and L-shaped hoop) are similar.

1. Introduction

Korean Building Code (KBC 2016[1]) recommends details of seismic hoop as shown in figure1 to increase seismic performance with 135° anchorage hooks. However, it is hard to place such design hoops at the top of longitudinal bar and pull them down to fixed position. Considering that a height of one story is about 4-5m, such hoop design needs to be improved.

The previous studies with alternative hoop details have been reported by Kim G Y [2], Kim C G et al. [3-5], Eom T S [6-7], Kim D H [8], and Ko S H [9]. Kim G Y suggests two kinds (C-shaped hoop and L-shaped hoop) of hoop as alternatives. Both hoops show similar axial and seismic performance in their test, but it is reported that C-shaped hoop is sometimes loosened when the cover of the concrete is failed. Therefore, this study choose L-shaped hoop (figure 2) for the evolution of axial and seismic performance.

The L-shaped hoop used in the study is prefabricated in L shape and has 135° anchorage hooks at both ends. Because it is relatively easy for workers to open the tie of the hoop momentarilly as long as the diameter is 10mm, the L-shaped hoop can be placed at designated location directly without being pulled down from the top of the longitudinal bar. Thus, the L-shaped hoop can improve the constructability and shorten construction period.

This study aims at evaluating structural and seismic performance of reinforced concrete columns by carrying out axial and seismic tests on the columns applied with the developed hoop detail.

Figure 1. General hoop.

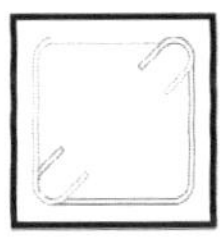

Figure 2. L-shaped hoop.

2. Test program

The experimental parameters are hoop detail (general and L-shaped type) and designed concrete compressive strength, fc (24MPa and 35MPa). Considering these parameters, total eight specimens (4 for axial loading test specimens, and 4 for seismic test specimens) are fabricated as tabulated in Table 1. [10-13]

Table 1. Details of test specimens.

Specimen	f_c	Hoop type
A^a (C^b)24-08	24MPa	General type
A^a (C^b)35-08	35MPa	
LA^a (C^b)24-08	24MPa	L-shaped type
LA^a (C^b)35-08	35MPa	

[a] A: Axial loading test specimen
[b] C: Seismic test specimen

2.1 Test specimens

There are two types of test specimens, which are for axial loading test and for seismic test. Axial loading test specimens are fabricated in straight rectangular shape and seismic test specimens are fabricated in reversed T shape.

Figure 3 shows details of specimens for axial loading tests. The cross section is 400x400mm2 and the height of the test specimen is 1400mm. Thickness of 5mm steel plates are welded on both ends of specimen to prevent eccentricity during loading. In order to prevent excessive cracking in the region close to the loading point, three layers of carbon fiber sheet is attached to the ends of the specimens at a width of 250mm. Figure 4 shows details of specimens for seismic test. Test specimens are composed with column section at upper part, and foundation section at the bottom. Cross section of column is 400x400mm^2 and that of foundation is 1400x800mm^2. The height of the column is 1500mm and the total height of specimen is 2000mm.

For longitudinal bars and hoops of all the specimens, D22 and D10 are used respectively, which means that the nominal diameter of longitudinal bar is 22mm and that of hoop is 10mm. Concrete mix proportion and strength are tabulated in table 2 and mechanical properties of longitudinal bars and hoops are tabulated in table 3. The nominal yield strength of longitudinal bar is 500MPa and the nominal yield strength of hoop is 400MPa. However, in our study, strength measured from tensile strength tests are found as 504.92MPa and 467.41MPa for longitudinal bars and hoops, respectively.

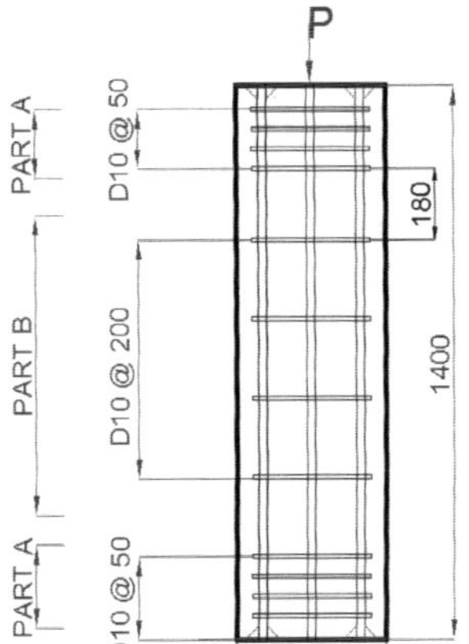

Figure 3. Axial loading test specimen.

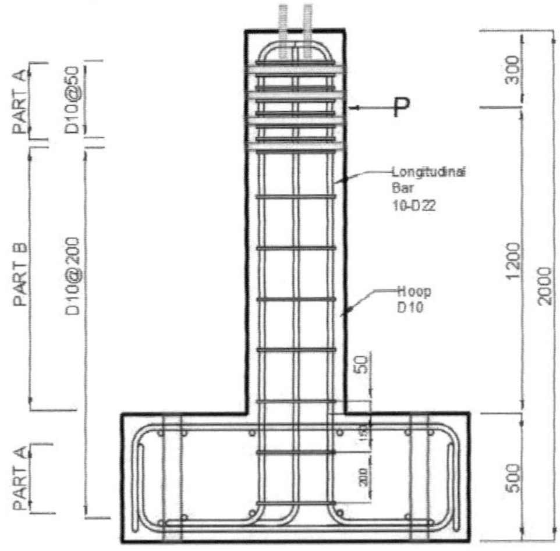

Figure 4. Seismic loading test specimen.

Table 2. Concrete mix proportion and strength.

f_c (MPa)	f_{ck} (MPa)	Mixed proportion			
		Max. aggregate size (mm)	Slump (mm)	W/C (%)	Fine aggregate (%)
24	33.50	25	150	49.7	49.5
35	43.60	25	150	39.5	47.5

Table 3. Steel material property.

Bar type	f_y (MPa)	ε_y (mm/mm)	E_s (GPa)
Longitudinal bar	504.92	0.0028	179.90
Hoop	467.41	0.0026	179.74

2.2 Test setup and loading plan

For the axial loading test, the load is applied using a 1000 ton capacity hydraulic machine until the column is failed as shown in figure 5. Strain gauges are attached to longitudinal bars and hoops.

Figure 6 illustrates the test setup for the seismic loading test. Axial load (N) is applied to the columns by actuator placed on the top of the column. During seismic loading, axial load N, sustained to be around 10% of the axial strength of the column. As the axial load is sustained, the cyclic lateral loading is applied in a form of controlling the displacement at a height of 1200mm from the column base. The total of 11 steps of seismic loading are planned and the same loading magnitude is repeated for three times at each step as shown in figure 7. After the each step, the load is increased about 1.25~1.5 times to the drift ratio of previous step, according to ACI 374.1 [14]. Strain gauges are attached at longitudinal bars and hoop sides parallel to the lateral force direction in plastic hinge section, where in the range of 400mm from the column bottom.

Figure 5. Axial loading test setup.

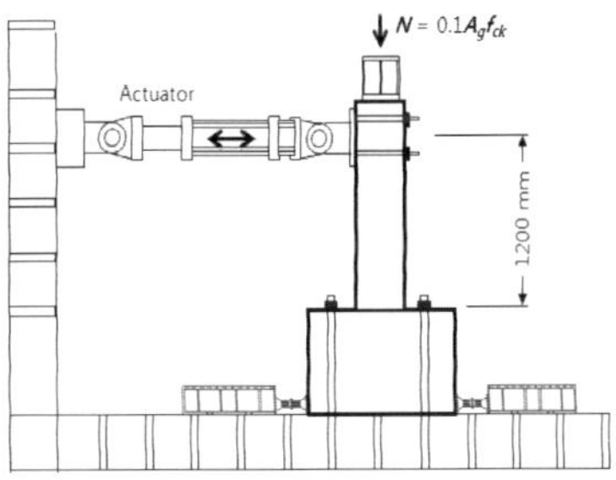

Figure 6. Seismic test setup.

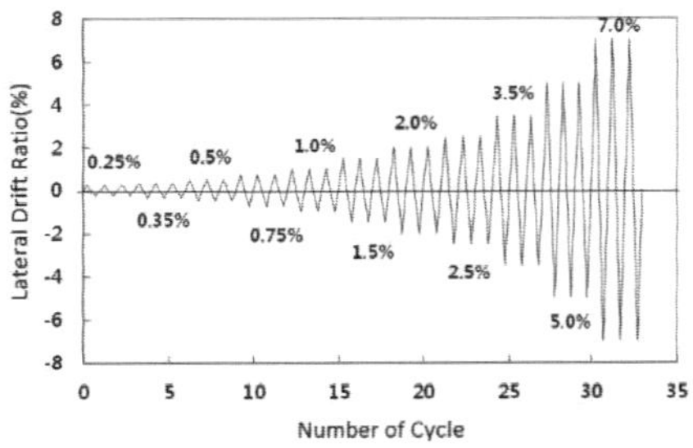

Figure 7. Seismic loading plan.

3. Test results

3.1 Axial loading test results

3.1.1 Maximum column strength

Based on the results obtained from the axial force tests, P_{test} is defined as failure load at the time of failure. For the case of columns having concrete strength of 24MPa (A24-08 and LA24-08), the difference of maximum loads between the columns with the general hoop and L-shaped hoop is only 0.1%. For the case of columns having concrete strength of 35MPa (A35-08 and LA35-08), the difference of maximum load between the columns with the general hoop and L-shaped hoop is 4.4%. Therefore, the test specimens show similar axial strengths regardless of general or L-shape hoop types. To compare the test result with the theoretical value, column strength (P_{theo}) is calculated according to the concrete design code from KBC 2016 [1] as shown in Equation (1). In the equation, A_g and A_{st} denote cross sectional areas of the column and longitudinal bars, respectively. Compressive strength measured 28 days after the curing is generally regarded as standard compressive strength (f_{ck}) and the yield strength of longitudinal bar obtained by steel tensile strength test (f_y in table 3) is used.

$$P_{theo} = 0.85f_{ck}\,(A_g - A_{st}) + f_yA_{st} \tag{1}$$

As listed in Table 4, theoretical values, P_{theo}, are very close to the values of test results, P_{test} , which shows that not only L-shaped hoop functions similar to general hoop but also axial strength of the column having L-shaped hoop is predictable using the classical design code. The maximum difference of axial strength between the P_{theo} and P_{test} is about 3.3%.

Table 4. Comparison of test result and expectation.

Specimen	P_{theo}(kN)	P_{test}(kN)	P_{test} / P_{theo}
A24-08	6031.55	6200.426	1.028
LA24-08		6191.509	1.027
A35-08	7378.56	7298.728	0.989
LA35-08		7655.720	1.033

3.1.2 Load-strain curve of hoop

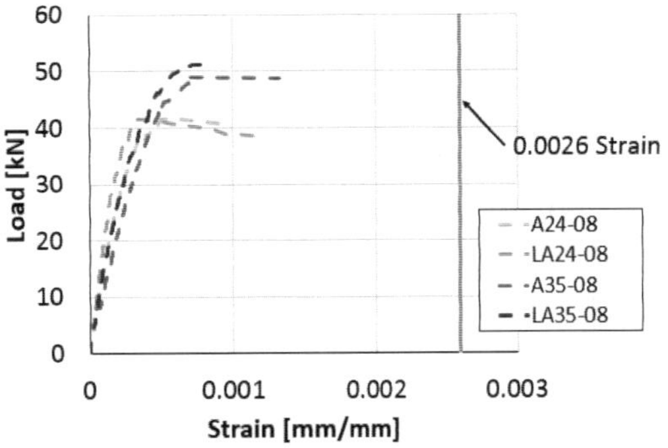

Figure 8. Load-strain curve of mean of 4 sides of hoops.

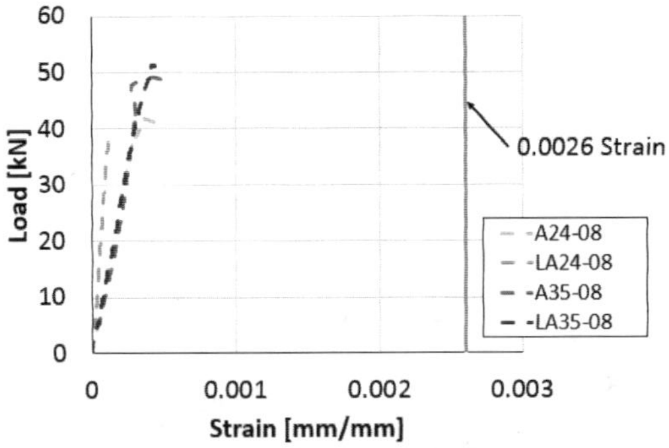

Figure 9. Load-strain curve of hook side of hoops.

The relationships between axial load and strains obtained from the axial loading tests are illustrated in figure 8 and figure 9. Strain values are measured from the gauges attached to the hoops. For the analysis, averaged values of strains obtained from different locations of hoops are used. The solid line of 0.0026 of strain represents the point where D10 bar is yielded.

From figure 8 and figure 9, it can be seen that the graphs show increase of hoop strain as the column is axially loaded and the axial load-strain relationships of general hoop and L-shaped hoop are very similar to each other. In addition, strains of hoops do not reach 0.0026, which means hoops are not yielded until the specimen is failed.

Figure 9 shows axial load versus strain obtained from a gauge attached to the hook of the hoop. The graph shows that the strains of hook are smaller than those measured from other locations of hoops and they are not yielded as well. This means that the hooks are fully tied until the specimens fail regardless of hoop type.

3.2 Seismic test results

3.2.1 Hysteresis loops of columns

Seismic loading is terminated when the specimens show severe damage at the plastic hinge area. At the moment, all the specimens show similar failure mode, such that flexural and diagonal shear cracks occur in plastic hinge area of the specimens.

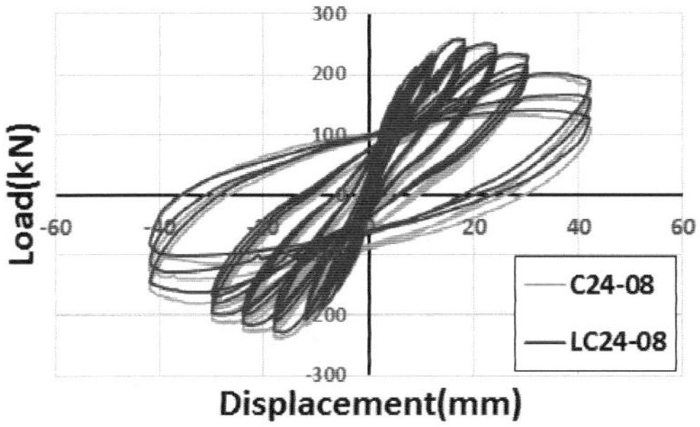

Figure 10. Hysteresis loops of specimens having compressive strength 24MPa.

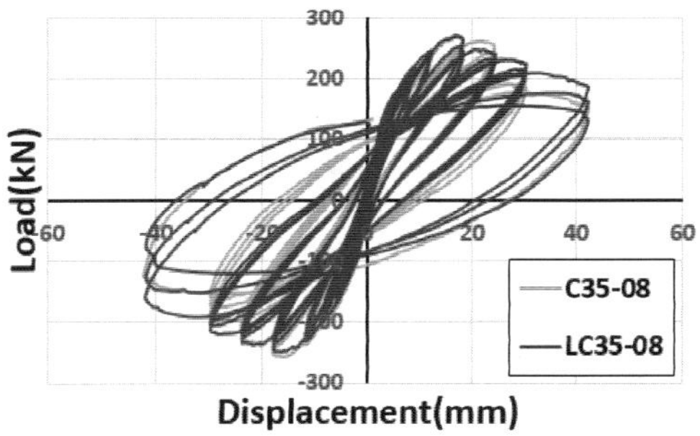

Figure 11. Hysteresis loops of specimens having compressive strength 35MPa.

The hysteresis loops obtained from the seismic loading test are illustrated in figure 10 and figure 11. From the graphs, the hysteresis behaviors of the specimens using the L-shaped hoop and general hoop are similar to each other in terms of stiffness and maximum shear load. For all the specimens, shear force reaches its maximum value at the 6th loading step (drift ratio of 1.5%) regardless of hoop types. After the 6th loading step, maximum shear force decreases steadily until the final loading step, which means the specimens show ductile behaviors.

3.2.2 Energy dissipation

Energy dissipation analysis is important as it represents energy absorption capacity of columns under seismic loading. Accumulated energy dissipation is obtained from the area inside the hysteresis loops shown in figure 10 and figure 11.

As tabulated in table 5, the energy dissipation capacity of specimens using the general hoop or L-shaped hoop have almost the same energy dissipation capacity until 6th loading step (drift ratio of 1.5%), where the maximum shear force occurs. In the 6th loading step, the accumulated energy dissipation of C24-08 and LC24-08 are very similar with 1.01% difference. For the C35-08 and LC35-08, difference of accumulated energy dissipation in 6th loading step is 2.03%.

At the 7th and 8th loading step, the energy dissipation capacity is similar regardless of hoop types for the columns having concrete strength of 24MPa. However in the last loading step (drift ratio of 3.5%), the energy dissipation capacity of the C24-08 is 11.5% higher than that of LC24-08.

For the columns having concrete strength of 35MPa, the energy dissipation capacity of C35-08 is higher than LC35-08. However, in the last loading step, due to the yielding of the longitudinal bar, C35-08 cannot withstand the axial force sufficiently and the cumulated energy dissipation decreases. On the other hand, in the case of the specimens using the L-shaped hoop, accumulated energy dissipation steadily increase with the loading cycle.

Table 5. Accumulated energy dissipation.

Drift ratio (%)	Loading step	Accumulated energy dissipation (kN·mm)			
		C24-08	LC24-08	C35-08	LC35-08
0.25	1	0.527	0.507	0.550	0.550
0.35	2	0.712	0.755	0.725	0.850
0.5	3	1.156	1.308	1.216	1.316
0.75	4	2.183	2.459	2.303	2.377
1	5	3.728	3.619	3.387	3.742
1.5	6	7.447	7.372	7.948	8.109
2	7	11.976	10.973	14.049	12.083
2.5	8	18.656	17.346	23.440	19.165
3.5	9	39.205	34.662	16.350	42.939

3.2.3 Yield step of hoop.

In this study, strains of the hoops at 50mm and 250mm away from the bottom of the column are focused because the earliest yielding step occurs in those area compared to other locations.

All hoops are yielded after the 6th loading step where the maximum shear forces occur. In addition, all hoops are yielded after the yielding of the longitudinal bars. Hooks are yielded later than 8th loading step which means that the hooks have been tied with no significant deformation.

Table 6 shows yield step of hoop. For the hoops located on the first layer (1H1 and 1H2), it can be seen that the general hoops are yielded at the same or later steps than the L-shaped hoops are yielded. For the hoops located on the second layer (2H1 and 2H2), yield steps of general hoops and L-shaped hoops are similar.

Table 6. Yield step of hoop.

Specimen	1H1	1H2	2H1	2H2	Gauge location
C24-08	x[a]	9	8	7	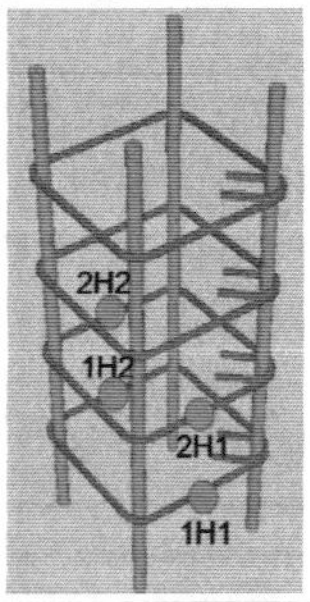
LC24-08	7	7	8	x[a]	
C35-08	x	x[a]	8	9	
LC35-08	-[b]	x[a]	9	9	

[a] x : Not yielded
[b] - : Not measured

3.2.4 Failure mode

Figure 12~15 show crack patterns of the tested specimens after the test. There is no significant difference in failure mode among the tested specimens regardless of hoop types. For all specimens, flexural and diagonal shear cracks occurred in the plastic hinge zone. In addition, the concrete cover in the plastic hinge zone was gradually spalled off as cracks at the bottom of the column are developed with load.

Between the columns having concrete strength of 24MPa (C24-08 and LC24-08), concrete failure at the bottom of the column and diagonal shear cracks are more obvious in C24-08 than LC24-08. On the other hand, the crack patterns of specimens are similar between the columns having concrete strength of 35MPa (C35-08 and LC35-08). Thus, one can say that there is no significand difference in failure modes due to different hoop types.

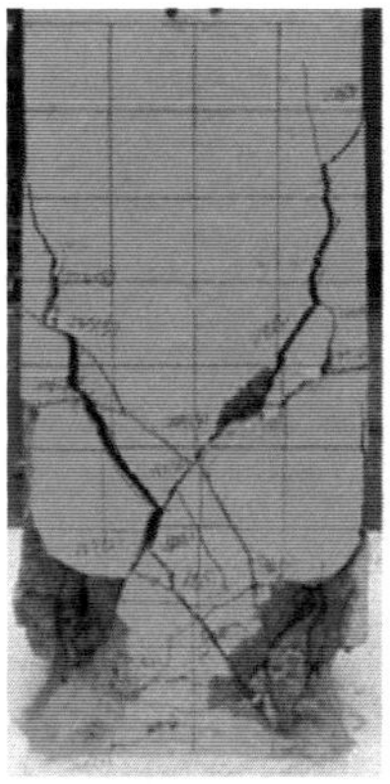

Figure 12. Failure mode of C24-08.

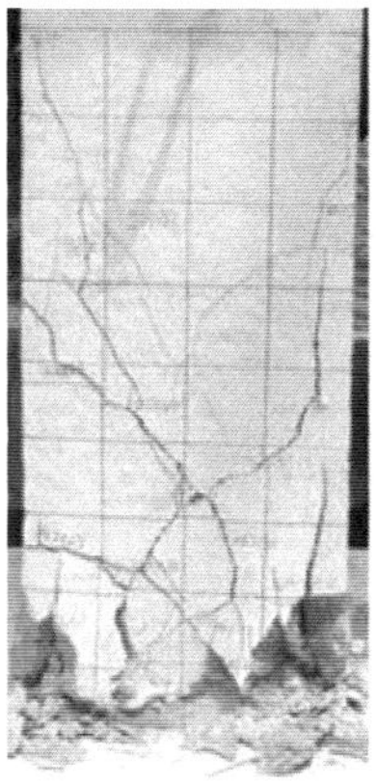

Figure 13. Failure mode of LC24-08.

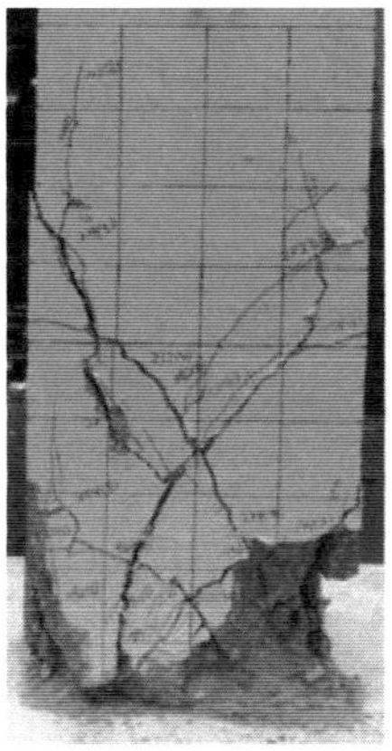

Figure 14. Failure mode of C35-08.

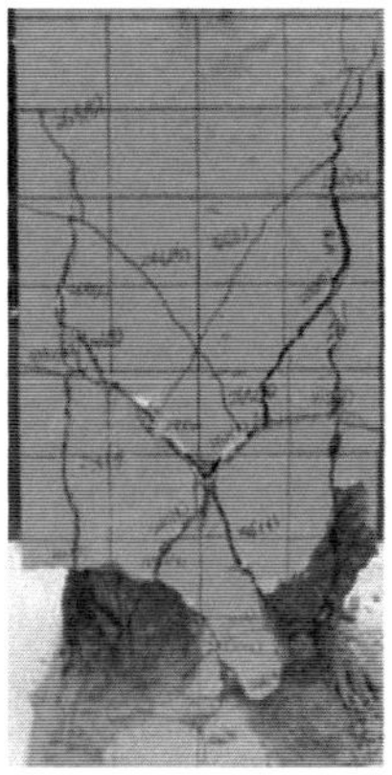

Figure 15. Failure mode of LC35-08.

4. Summary and conclusion

- The axial strengths of the columns using the general hoops or the L-shaped hoops are similar with 4.4% of maximum difference.
- The hysteresis loops of the specimens with the same designed concrete compressive strength show almost same behavior between the specimens using general hoop and L-shaped hoop. All specimens show maximum shear force at 6th loading step (drift ratio of 1.5%) and shows ductile behavior afterwards.
- The energy dissipation capacity of all the specimens is very similar to each other until 6th loading step (maximum shear load). Only for the specimen of C35-08, energy dissipation capacity is relatively small due to the earlier failure than that of specimen LC35-08.
- The hoops located in plastic hinge section are yielded after the yielding of longitudinal bar. Yield step of general hoop and L-shaped hoop is similar, which confirms that seismic performance of general hoop and L-shaped hoop is similar.
- For the failure mode of all the tested specimens, diagonal shear crack is dominant and cross-sectional area of the concrete at the bottom of the column is gradually reduced. The crack patterns of the all specimens are similar to each other, which means there is no significant difference in failure mode among the specimens having different hoop types.

Acknowledgments

This research was supported by TK Engineering (tkeng2015@naver.com). The authors are grateful for their supports.

References

[1] Korea Ministry of Land, Infrastructure and Transport, *Korean Building Code* (2016)
[2] Kim G Y, Structural performance of RC columns with separate type hoops, (Ewha Womans University, Master degree of thesis, 2017)
[3] Lee C G, Alternative reinforcement details for RC moment resisting frames with limited ductility demands, (Seoul National University, Doctoral degree of thesis, 2017)
[4] Lee C G, Park H G, Eom T S and Kim T W, Alternative seismic hoop details for reinforced concrete columns in low rise buildings, 2015 *The proceeding of ASEM15*
[5] Kim C G, Park H G, ASCE M and Eom T S, Cyclic load test and shear strength degradation model for columns with limited ductility tie details, 2019 *Jounal of Structural Engineering*, **145(2)**
[6] Eom T S, Kang S M, Park H G, Choi T W and Jin J M, Cyclic loading test for reinforced columns with continuous rectangular and polygonal hoops, 2014 *Engineering Structures*, **67**, 39-49
[7] Eom T S, Choi T W, Park H G, Kang S M and Jin J M, Seismic tests on reinforced concrete columns confined with continuous hoops, 2013 *J. Architectural Institute of Korea Structure&Construction*, **29(2)** , 29-37
[8] Kim D H, Kim J C and Choi K K, 2016 *J. Architectural Institute of Korea Structure&Construction*, **32(1)** , 35-44
[9] Ko S H, Seismic performance of octagonal flared RC columns using oblong hoops, 2015 *J. Korea Institute for structural maintenance and inspection*, **19**, 1-9
[10] Abdullah J A, Sumei Z and Jiepeng L, Shear strength and behavior of tubed reinforced and steel reinforced concrete(TRC and TSRC) short columns, 2010 *Thin-Walled Structures*, **48**, 191-99
[11] Li Y A, Huang Y T and Hwang S J, Seismic response of reinforced concrete short columns failed in shear, 2014 *ACI Structural Journal*, **111**
[12] Zhou X H and Liu J P, Seismic behavior and shear strength of tubed RC short columns, 2010 J. *Constructional Steel Research*, **66**, 385-97
[13] Youssf O, ElGawady M A and Mills J E, Experimental investigation of crumb rubber concrete columns under seismic loading, 2015 *The Institution of Structural Engineers*, **3**, 13-27
[14] ACI Committee 374, *Acceptance Criteria for Moment Frames Based on Structural Testing and Commentary* (2005)

IOP Conf. Series: Materials Science and Engineering **652** (2019) 012028 doi:10.1088/1757-899X/652/1/012028

Effect of freeze-thaw cycling on fatigue behaviour in concrete

Andrew J. Boyd[1,2] and Andrea Leone[1]

[1] Department of Civil Engineering, McGill University, Montreal, QC, H3A 0C3, Canada
[2] Author to whom any correspondence should be addressed.

Email: andrew.boyd@mcgill.ca; andrea.leone@mail.mcgill.ca

Abstract. Freeze-thaw damage is a deterioration mechanism caused when saturated concrete is exposed to temperature variations that cycle above and below its freezing point. The expansion of water in the concrete pore space due to freezing can lead to internal pressures that induce cracking, thus expediting the deterioration process through freeze-thaw cycles. The purpose of this investigation was is to evaluate the effect of freeze-thaw deterioration on the tensile fatigue life of air-entrained concrete at early stages of deterioration. Concrete cylinders were prepared with 0.45 and 0.65 water to cement ratios (W/C), then cured for 28 days before being subjected to the freeze-thaw cycling. After 0, 25 and 50 freeze-thaw cycles, the pressure tension (PT) test was used to induce tensile fatigue loading cycles until failure. The PT was capable of determining the decrease in fatigue life, represented as the number of cycles to failure, of the concretes when subjected to pure cyclical tensile loading after having first been exposed to freezing and thawing cycles. The results indicated that even though the ultimate static tensile strength of the specimens did not vary significantly due to the freeze-thaw cycles, the residual fatigue properties were degraded. UPV monitoring was able to determine the increase in internal damage, as the UPV of the specimens decreased with continued freeze-thaw cycling. It was shown that freeze-thaw deterioration reduces the ability of concrete to withstand cyclic loading, even at early stages prior to a decrease in static tensile strength.

1. Introduction

Almost one third of concrete cast in Nordic countries undergoes freeze-thaw conditions due to the temperature fluctuation above and below the freezing point of the pore water solution within the concrete [1]. In these regions, freeze-thaw damage can cause severe deterioration of the infrastructure, which in turn leads to a significant economic impact, as repairs and preventative measures are required in order to ensure public safety [2].

Freeze-thaw damage is an attrition process, beginning at the surface causing separation or scaling until complete collapse. The concrete loses stiffness and strength while irreversible expansion is induced by the repeated freezing and thawing cycles. The two properties of concrete mixtures that control the damage caused by freezing and thawing cycles: water to cement ration (W/C) and entrained air content [1]. Air entrainment improves the performance of concrete against freeze-thaw damage, but it does not completely stop the deterioration [3].

Most structures will also be exposed to cyclic loading of some kind throughout their service life. Cyclic loading could include traffic loads, temperature changes, wind gusts and vibrations due to machinery [4]. Such load reversals can lead to internal microcracks, which in turn result in increased strain and a change in the mechanical properties of concrete [5]. Damage caused by repeated loading is defined by fatigue.

Understanding the effects of cyclic loading and fatigue failure is essential for both safety and economic reasons. The combination of freeze-thaw damage and cyclic loading can lead to the premature deterioration of materials. This deterioration can be categorized as a degradation in the durability of concrete due to an increase in transport properties resulting from fatigue cracks [6].

The objective of this research was to determine whether freeze-thaw cycling could have a significant effect on tensile fatigue life of air-entrained concrete. Two different W/C concrete mixtures were cast and subsequently subjected to freeze-thaw cycling in accordance with ASTM C666. The pressure tension (PT) apparatus was then used as a destructive test for cyclic loading of the concrete specimens in pure axial tension in order to determine the residual fatigue life of the concrete.

The Pressure Tension (PT) test was chosen as the evaluation mechanism due to its past success in detecting tensile stress induced damage at much earlier ages, or earlier stages of damage, than other destructive techniques [7-9]. The PT device has been used successfully for several years in static tensile testing of concrete, but recent modifications of adding a second servo-vale controller provide the equipment with the capability of a controlled unloading rate, allowing full cyclic tensile loading for the first time.

This method applies compressed air over the curved surface of a cylinder, while the flat ends are left exposed to regular atmospheric pressure [10]. The pore water contained within the sample comes to a hydrostatic equilibrium with the applied gas pressure [11]. Since the applied gas pressure is applied only along the curved surface a net tensile stress field develops parallel to the axis of the cylinder [12]. This indirect tensile effect will act to apply tension at every point within the cylinder at a magnitude equal to the applied stress, resulting in a tension failure crack that propagates across the specimen as the specimen is pushed apart from the inside.

The PT method, because its mode of action applies stress from within the sample as opposed to an external application (Fig. 1), is more sensitive to expansive damage such as that associated with F/T damage [7-9]. Any microstructural cracks that are present in a test specimen will be 'pushed apart' at the stress and, thus, this test will indicate deterioration as a loss of tensile strength not otherwise detectable using more common methods [9].

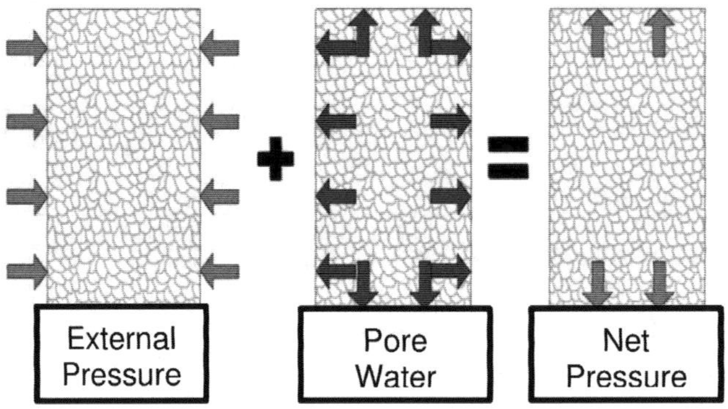

Figure 1. Pressure Tension Mechanism

2. Materials and Methods

Two mixture designs with different W/C were used to determine the deterioration effects of freeze-thaw cycles, as shown in Table 1. For each of these mixtures, 36 concrete cylinders were cast with dimensions of 100 mm x 200 mm, as required by ASTM C 666. Two different W/C's were chosen in order to compare concrete of different strengths under the same exposure conditions. An air entrainment admixture was added to both mixtures to achieve a target air content of 7% ± 0.5%.

After demolding at 24 hours, the specimens were cured for a 28-day period in a saturated limewater bath prior to the initiation of freeze-thaw cycling. The 72 specimens were split into three sample

groups of 24 specimens that were used for 0, 25 and 50 freeze-thaw cycles, respectively. Each group of specimens contained 12 - 0.45 and 12 - 0.65 W/C specimens.
After the 28-day curing process, each concrete cylinder was placed into an individual container conforming to the dimensional specifications laid out in ASTM C 666, specifically the maximum clearance between the edge of the container and the sample being less than 3 mm. The containers were filled with tap water to a depth of 3 mm above the specimen surface. The water was replaced during the thawing portion of each cycle. In order to bring the temperature below freezing, two commercially available upright freezers were used. The freezers brought the specimens' interior temperature down to a minimum of -23 ± 2 °C. Thermocouples embedded in additional concrete specimens were used to confirm the temperature range.

Table 1. Mixture designs.

	0.45 W/C	0.65 W/C
Water (kg)	22.27	22.51
Cement (kg)	48.51	33.58
Coarse Aggregate (kg)	99.36	99.36
Fine Aggregate (kg)	50.63	58.10
Air Entrainment Agent (L)	0.053	0.948
Air Content (%)	6.5	6.9

The 72 specimens were split into three sample groups of 24 specimens that were used for 0, 25 and 50 freeze-thaw cycles, respectively. Each group of specimens contained 12 - 0.45 and 12 - 0.65 W/C specimens. After the 28-day curing process, each concrete cylinder was placed into an individual container conforming to the dimensional specifications laid out in ASTM C 666, specifically the maximum clearance between the edge of the container and the sample being less than 3 mm. The containers were filled with tap water to a depth of 3 mm above the specimen surface. The water was replaced during the thawing portion of each cycle. In order to bring the temperature below freezing, two commercially available upright freezers were used. The freezers brought the specimens' interior temperature down to a minimum of -23 ± 2 °C. Thermocouples embedded in additional concrete specimens were used to confirm the temperature range.
A custom-made water tub with a water heater and submersible recirculation pump were used to thaw the specimens. This ensured that the specimens always remained at 100 % saturation and prevented any significant temperature variations between samples. The thawing process was stopped at the 2-hour mark (±15 minutes), when the interior temperature was between 5 and 7 °C. The samples were then placed back into the freezer for the next freezing cycle.
The 12 specimens in each group, from each W/C, were then subdivided based on the tensile strength level at which they were to be tested. These included the control specimens tested to failure under static loading (Ft), specimens loaded to 60% of the control strength (0.6 Ft) and specimens loaded to 80% of the control strength (0.8 Ft). The distribution of specimens can be seen in Table 2.

Table 2. Specimen groups.

	0 Cycles	25 Cycles	50 Cycles
		0.45 W/C	
F_t	4	4	4
0.6 F_t	4	4	4
0.8 F_t	4	4	4
		0.65 W/C	
F_t	4	4	4
0.6 F_t	4	4	4
0.8 F_t	4	4	4

IOP Conf. Series: Materials Science and Engineering **652** (2019) 012028 doi:10.1088/1757-899X/652/1/012028

3. Results and Discussion

Figure 2 shows the ultimate tensile strength (Ft) of the two mixtures at the respective freeze-thaw cycles. As expected, there was little variation in tensile strength over the course of the experiment. This was a direct result of the inclusion of air entrainment in both concrete mixtures. The increased mean tensile strength at the 25-cycle point can be attributed to continued hydration, followed immediately by the very early stages of deterioration.

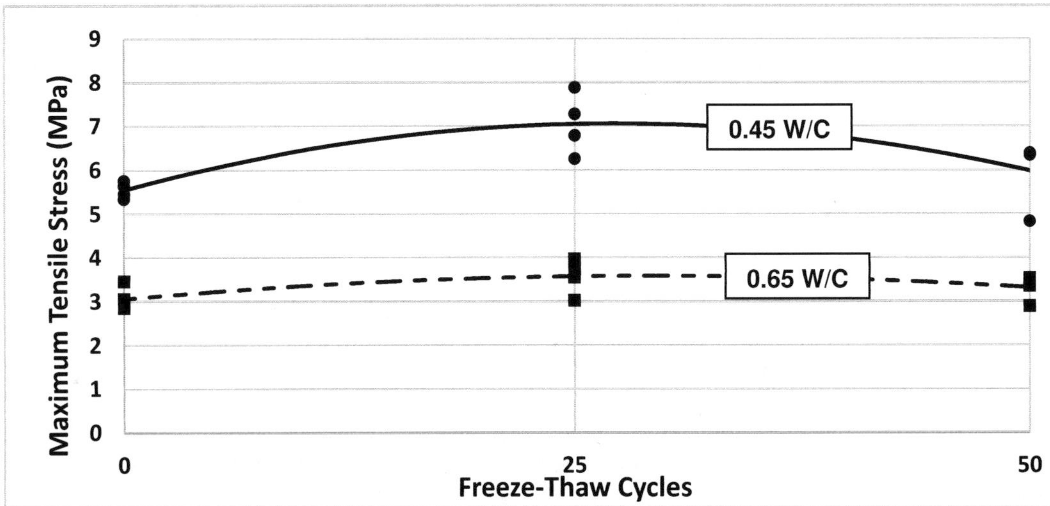

Figure 2. Ultimate tensile strength (static).

Neither of these changes was found to be statistically significant. Even though the static tensile strength was unaffected, there was a large reduction in residual fatigue life, indicated by a decrease in the number of cycles to failure as freeze-thaw cycling progressed. This reduction in was found for both the 0.45 and 0.65 W/C specimens, as shown in Figures 3 and 4.

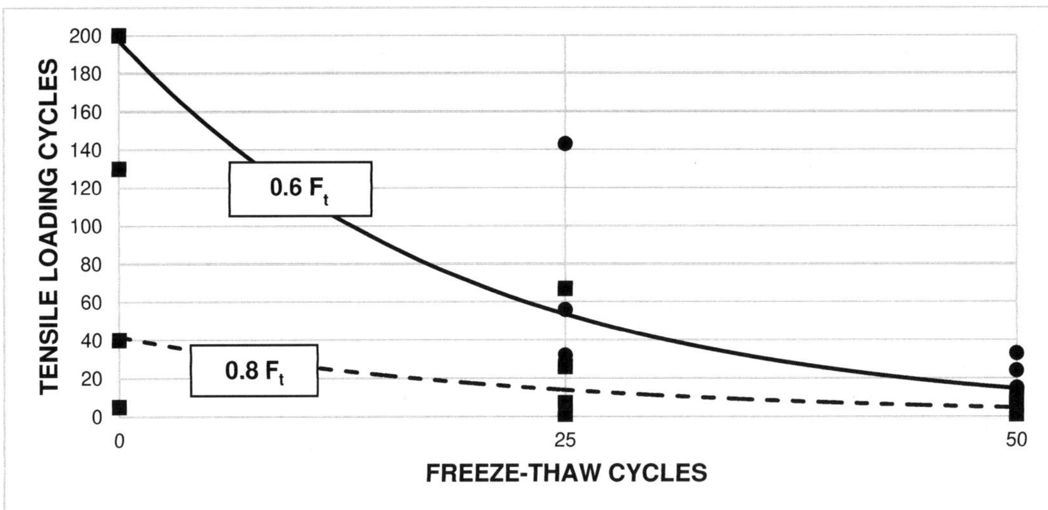

Figure 3. Fatigue behaviour (0.45 W/C).

It should be emphasised that these reductions in fatigue life were not accompanied by any reduction in tensile strength before cyclic loading, as shown earlier. The decrease in fatigue life was caused by the internal damage of the specimens due to freezing and thawing cycles. The cracks that developed due to freezing acted as initiation points for further propagation and damage under cyclic tensile loading. Under cyclic tensile stress, these cracks propagated and caused weakening at the bonds between the aggregates and concrete. Failure then occurred through the weakest and most damaged plane in the concrete specimen.

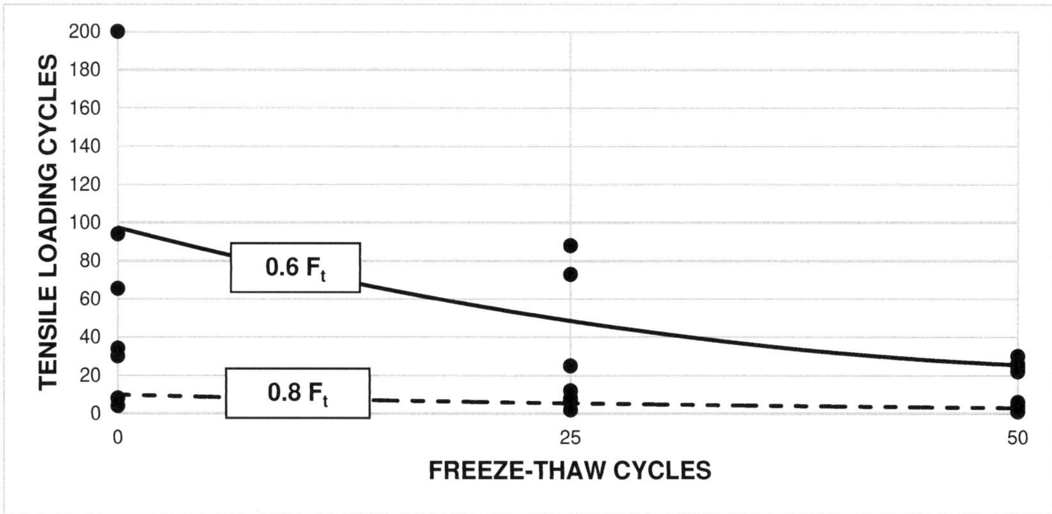

Figure 4. Fatigue behaviour (0.65 W/C).

Cyclic loading at the higher stress level (0.8 Ft) caused the specimens of both mixtures to fail much sooner than when testing the specimens at the lower cyclic stress level (0.6 Ft). There was a fairly large variation in the number of cycles to failure at any given F/T/ cycle due to the infinite number of possible crack orientations that can occur within concrete. If the orientation of cracks is perpendicular to the loading plane, then the specimens are affected more by the loading and will have a decreased fatigue life as the crack opens and propagates with each loading cycle. This variation is typical in fatigue testing of brittle materials.

Interestingly, the average percentage decrease in fatigue life for specimens tested at 0.6 Ft correlated very well with the percentage decrease in specimens tested at 0.8 Ft for both mixtures. Essentially, as the freeze-thaw cycles increased the fatigue life of the specimens decreased by the same proportion, whether the specimens were tested at 0.6 or 0.8 Ft.

At 25 freeze-thaw cycles, the fatigue life of the 0.65 W/C specimens decreased on average to 49.81 and 50.00 % of the average control fatigue life when tested at 0.6 or 0.8 Ft, respectively. At 50 freeze-thaw cycles, the fatigue life of the 0.65 W/C specimens decreased an average of 26.19 and 25.93% of the average control fatigue life when tested at either 0.6 or 0.8 Ft, respectively. For the 0.45 W/C mixture, the drop were 31.18 / 26.93 % at 25 cycles and 9.11 / 7.73 % at 50 cycles, for the 0.6 and 0.8 Ft loading levels, respectively. This comparison is included in Table 3.

Table 3. Reduction in cycles to failure (% of F_t loading cycles to failure).

		F/T Cycles		
		0	25	50
0.45 W/C	0.6 F_t	100	31.18	9.11
	0.8 F_t	100	26.93	7.73
0.65 W/C	0.6 F_t	100	49.81	26.19
	0.8 F_t	100	50.00	25.93

4. Conclusions

The findings determined that freeze-thaw cycles affect the fatigue life of concrete when loaded in pure axial tension but had little effect on the ultimate tensile strength after 50 freeze-thaw cycles. This demonstrated that the microcracks caused by freezing and thawing cycles affect the fatigue life of the concrete even at very early stages of deterioration since they act as initiation points for further propagation, however; they do not yet affect the ultimate tensile load of the concrete at such an early stage.

It was also demonstrated that the pressure tension machine, an apparatus unique to this research program, is fully capable of testing concrete cyclically in tension. Such a loading configuration is extremely difficult to achieve with any other equipment currently available. This success will lead to more detailed investigations into the importance of cyclic behavior of concrete with respect to its durability properties.

This paper represents the initial investigation into the effect of deterioration on fatigue performance of concrete structures. The results clearly indicate that this is a potentially critical issue, since the dramatic reduction in resistance to cyclic loading has significant implications toward cracking in concrete and thus long-term durability. Future research will delve further into these effects, not only due to for freeze-thaw deterioration, but to address other durability mechanisms as well. Footnotes should be avoided whenever possible. If required they should be used only for brief notes that do not fit conveniently into the text.

References

[1] Penttala, V 2006 Surface and internal deterioration of concrete due to saline and non-saline freeze–thaw loads. *Cement and Concrete Research* **36**(5) 921-928

[2] Shang H S, and Yi T H 2013 Freeze-thaw durability of air-entrained concrete *Scientific World Journal* 2013

[3] Chatterji S 2003 Freezing of air-entrained cement-based materials and specific actions of air-entraining agents *Cement and Concrete Composites* **25**(7) 759-765

[4] Kessler-Kramer C, Mechtcherine V and Müller H S 2004 Testing and modeling the behavior of concrete under cyclic tensile loading 5th International Conference on Fracture Mechanics of Concrete Structures Vail

[5] Lee M K and Barr B I G 2004 An overview of the fatigue behaviour of plain and fibre reinforced concrete *Cement and Concrete Composites* **26**(4) 299-305

[6] Li W, Sun W and Jiang J 2011 Damage of concrete experiencing flexural fatigue load and closed freeze/thaw cycles simultaneously *Construction and Building Materials* **25**(5) 2604-2610

[7] Boyd A J and Mindess S 2004 The use of tension testing to investigate the effect of W/C ratio and cement type on the resistance of concrete to sulfate attack *Cement and Concrete Research* **34**(3) 373-377

[8] Boyd A J, Bremner T W, Holm T A and Boyd S R 1998 Tensile Testing to Evaluate the Effect of Alkali-Aggregate Reaction in Concrete T192-2, IRC National Research Council, Ottawa, Canada

[9] Komar A J K and A J Boyd 2017 Evaluating freeze-thaw deterioration with tensile strength 2nd International Conference on Civil Engineering and Materials Science, Seoul, South Korea, 6 pp

[10] Clayton N and Grimer F 1979 The Diphase Concept, with Particular Reference to Concrete Developments in Concrete Technology, Elsevier, Waterford, UK, 283-317

[11] Terzaghi K and Peck R 1967 Soil Mechanics in Engineering Practice Wiley, New York

[12] Uno T, Fujikake K, Mindess S and Xu H 2010 The Nitrogen Gas Tension Test of Concrete, Part 1: Effect of Boundary Conditions and Axial Strain Response Materials and Structures. 44 p 857-864

IOP Conf. Series: Materials Science and Engineering **652** (2019) 012029 doi:10.1088/1757-899X/652/1/012029

Analysis of Geometry Effect on Laterally-loaded Large Diameter Bored Pile Deflection in Soft Soil

K Tamara[1] and T Ilyas[2]

[1]Student, Department of Civil Engineering, Universitas Indonesia, Depok 16424, Indonesia
[2]Professor, Department of Civil Engineering, Universitas Indonesia, Depok 16424, Indonesia

Email: kaishatamara@gmail.com; tilyas@eng.ui.ac.id

Abstract. A bored pile is one of the foundation type widely used in construction because of its design flexibility. For a laterally-loaded bored pile, additional pile and pile cap play an important role in influencing the lateral capacity. Large diameter bored pile with a 3.5 m diameter then proposed to substitute group piles in order to simplify the construction method aiming the same capacity. This study examines laterally-loaded large diameter bored pile deflection on soft soil as parametric studies with different pile length of 7 m, 14 m, and 21 m. Thick soft soil layer interpreted through soil testing data at Ancol, North Jakarta with the domination of silt. A thorough analysis of this system is modelled using a three-dimensional numerical method, PLAXIS 3D Foundation using push-over analysis that the deflection of laterally-loaded foundation reaches its tolerable deflection according to Indonesian National Standardization (SNI).

1. Introduction

Lateral capacity of bored pile plays important role when it comes to deep foundation in soft soil. Soft soil with cohesive structure contained of soil particle bond which then quantified as shear strength. According to [1] cohesive soil has variation of shear strength based on the direction of principal stress, which then affect soil resistance. The differences in between using group pile and single pile is that the response of group pile affected by nonlinear interaction between pile-soil-pile [2]. Large diameter bored pile then proposed to substitute group pile in order to simplify construction method with the same lateral capacity. There are many studies related to large diameter bored pile under lateral load on cohesion-less soil which tend to gain soil resistance from its cohesion [3-6]. This study aims to examine laterally-loaded large diameter bored pile deflection on soft silty soil in order to comprehend its behavior since there are limited researches available for evaluating large diameter bored pile behavior on soft soil as one of the typical residual soil found in Jakarta.

2. Literature Review

Mono-pile with massive diameter typically used to resist bending moment, also relatively easier constructed [4] compared to group pile. Typical diameter used for offshore structure has slenderness ratio (L/D) at about 5-7 thus influencing pile behavior as rigid piles. Slenderness ratio of soil structure then affecting soil stiffness response for soil interaction with pile increased compared to flexible pile [5]. Laterally-loaded pile performance is predicted with two soil stiffness idealization, constant and linearly proportional to depth [7]. There are many models proposed to analytically predict this inconsistency and variation between models and existing data might be substantial [8]. Force system for laterally-loaded bored pile is very complex and in three-dimensional space. These forces dominated by passive soil resistance. But, there are pile tip shear resistance, lateral flow, vertical side resistance, and axial resistance component at pile tip also influencing lateral behavior. At shallow depth, soil surrounding single pile might move vertically or laterally [9].

3. Research Methodology

This study conducted with numerical approach using PLAXIS 3D Foundation. Soil parameters selected from N-SPT interpreted through soil testing data at Ancol, North Jakarta and laboratory experiments of five soil sample for various depth taken from deep boring at site. Ground water level defined from field investigation at 1.4 m. Two points chosen at 540 m span with relatively planar elevation and one point located at midpoint. In numerical simulation, soil layers defined using three Borehole[1] to simulate each location's stratification. This study established soil stratification into 4 soil type; silt, silty clay, sandy silt, and sand tabulated in Table 1.

Table 1. Soil stratification.

Soil Type	Depth [m]	N-SPT
Silt-1	0.00 – 2.20	9
Sand	2.20 – 12.50	5
Silty Clay	12.50 – 20.50	20
Silt-2	20.50 – 28.00	60
Sandy Silt	28.00 – 36.20	24
Silt-3	36.20 – 40.50	60

This study using correlation from N-SPT to soil parameter calculated in PLAXIS such as cohesion, friction angle, stiffness, permeability, unit weight, and Poisson's ratio. In general, soil differentiated to two constitutive models based on their analysis type. Drained materials are simulated with Mohr Coloumb model while undrained materials are simulated with Soft Soil Creep model. Silt-dominated soft soils are analyzed in undrained condition thus its stiffness parameters defined by Cc, Cs, and Cα. For Soft Soil Creep model, failure behavior is defined by cohesion and friction angle as Mohr Coloumb model does. Other than that, this study using laboratory experiment result on index and engineering properties to verify soil parameter correlation. Table 2 shows parameter used in PLAXIS for each soil type.

Table 2. Soil parameter.

Soil Type	**Stiffness**			**Failure Behavior**	
Drained	**E [kN/m^2]**		**v**	**c [kN/m^2]**	**φ [°]**
Sand	55000		0.35	11	38
Sandy Silt	12000		0.27	10.5	32.53
Undrained	**Cc**	**Cs**	**Cα**	**c [kN/m^2]**	**φ [°]**
Silt-1	0.6	0.02	0.036	2.93	8.25
Silt-2	0.77	0.07	0.046	15.21	26.32
Silt-3	0.395	0.038	0.023	39.24	31.33
Silty Clay	0.39	0.08	0.021	12.32	17.78

As for loading, lateral load applied in pile load test are modelled for validation of pile modelling. While on large diameter bored pile, lateral load applied on 1 m depth and increased until pile reach maximum allowed lateral deflection (push-over analysis). Pile interface is defined surrounding pile

[1] Borehole (in PLAXIS) is defined to determine soil thickness, soil type, and ground water level

model in PLAXIS to simulate interaction between soil and concrete by R-inter value. Calculation steps are divided in 4 phases; initial phase, piling, excavation, and loading.

3.1. Validation

Numerical modelling validated for both soil geometry and pile modelling. Soil geometry validated from its stratification in meshing step compared to actual soil stratification data. For pile modelling validation, pile lateral displacement compared to actual lateral displacement from lateral loading test data. Pile lateral displacement then assessed both for its magnitude and direction as Fig. 1 showed. Pile modelling validation confirmed with such a slight deviation due to modelling process in between model and actual lateral displacement tabulated in Table 3.

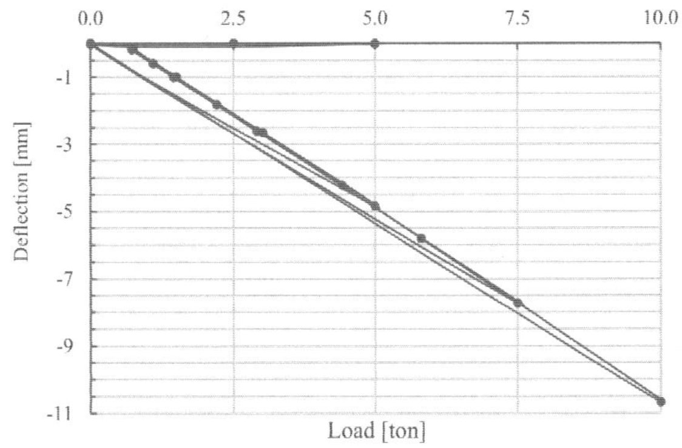

Figure 1. Load-Deflection curve for pile modelling validation

Table 3. Actual and modeling pile lateral displacement comparison.

Load [ton]	Load [kN]	Lateral Displacement [mm]	
		Loading Test	PLAXIS
5	50	2.44	4.8
7.5	75	4.34	7.7
10	100	10.54	10.66
Final Displacement Deviation			1.138%

4. Result and Discussion

Depth effect on bored pile slenderness ratio (L/D) is selected to be 2, 4 and 6 to understand the influence especially through pile deflection. For each pile length, taken 7 m, 14 m, and 21 m, pile lateral capacity at maximum deflection allowed by [10] increase in non-linear function. As Fig. 2 showed, it is known that pile behavior tends to increase head deflection as more load applied. Other than pile length, pile lateral capacity influenced by surrounding soil resistance. Non-linear lateral capacity increase then happens since 7 m-length pile sits on fill material (sand), 14m-length pile sits on silty clay, while 21 m-length pile sits on silt. From this model, it is known that the most increment lateral capacity reach at the 14 m-length piles with 55.26% incremental. Pile deflection significantly increases at 500 kN lateral load for each pile, then after 5 mm deflection each pile capacity differs with gap due to different pile length.

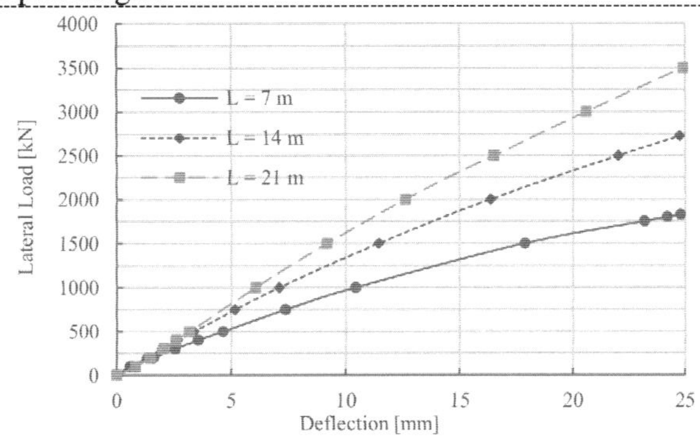

Figure 2. Lateral Load-Deflection curve for large diameter bored pile at different pile length.

It is known that variation of pile geometry influencing pile deflection along the length. As for 7 m-length piles, deflection tends to be influenced by rigid behavior generating tip displacement for 1.54 mm. In contrary, 14 m and 21 m-length pile deflection at pile head and tip are shifting to different direction indicating influence of soft soil on its tip. This condition resulted into a higher probability of pile rotation since its deflection probably reach point of rotation. Since load-deflection behavior of surrounding soil around the pile is non-linear, response of the pile subjected to lateral load along pile depth is non-linear. Other than that, soil strength around the upper part of pile mobilized and generating load transfer to greater depth. This outcome then developing moment increase more rapidly than the load on the same depth. For soft soil dominated by silt, this model confirmed that thick fill material (6 m layer of sand) can influence the whole pile behavior as the 14 m-length pile sits on silty clay and underlying silt material but generating the most significant lateral capacity increment. The calculated value from numerical modelling agree well with previous behavior in which bored pile bending moment rapidly increase rather than applied lateral load.

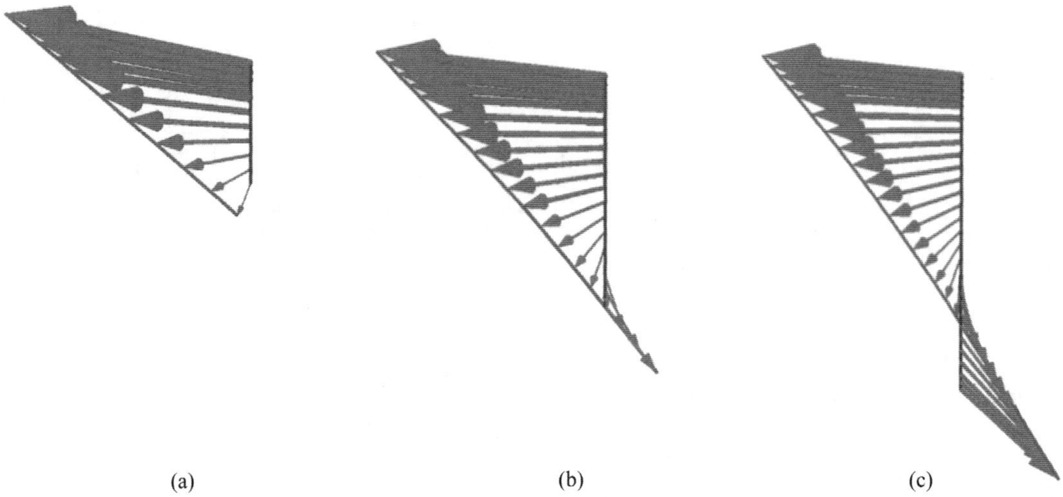

(a) (b) (c)

Figure 3. Three-dimensional pile deflection at maximum load: (a) L = 7 m; (b) L = 14 m; (c) L = 21 m

To understand pile deflection response against lateral load, this study using three-dimensional model. From Fig. 3, the stiffness of pile is not sufficient to fix the pile head against rotation. Soft soil layer then generating large bending moment at pile tip on opposite direction. As seen above, 7 m-length piles deflect on the same direction while the rest can rotate at some point. This response explains the presence of lateral flow surrounding bored pile as one of the component of soil-pile lateral resistance.

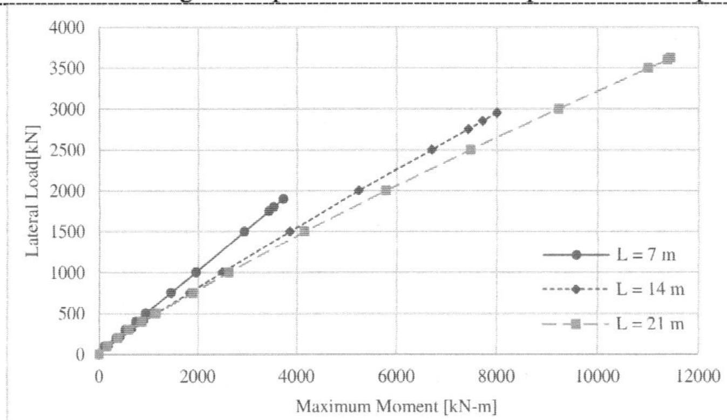

Figure 4. Comparison of maximum moment increment for various pile length

As seen in Fig. 4, it is known for the same load (1000 kN, and 1500 kN) that 7 m-length large diameter bored pile generating the least maximum bending moment amongst all pile length. While 14 m and 21 m-length piles, influenced by its soft soil behavior, generating identical bending moment. Maximum moment occurred at shallow depth for all pile length. This behavior explains lateral load transfer to greater depth. 7-m length pile maximum bending moment occurs at half span (3.67 m), causing lateral tip displacement which then verify its very rigid behavior. 14 m and 21 m-length pile maximum bending moment occurs at one-third span, causing point of rotation of pile which then verify its rigid behavior also started to lessen, now influenced by bending.

5. Conclusion

Analysis of depth effect on laterally-loaded large diameter bored pile deflection is presented using three-dimensional numerical model (PLAXIS). The model is adjusted by comparison to lateral load test. Various pile length is analyzed to understand large diameter bored pile response on form of pile deflection and bending moment. The following conclusions can be drawn:

1. The effect on pile length gradually changing pile influencing behavior. As the pile lengthen, slenderness ratio increases, then large diameter bored pile will response with bending influence.
2. Non-linear behavior for pile capacity occurred since soil layer changes into soft soil at greater depth. The most significant pile capacity increment is achieved at 14 m-length pile.
3. It is known that due to rigid behavior, tip displacement occurred at 7 m-length pile. While for the rest pile length, tip deflection response occurred in opposite direction to pile head deflection. This notes that soil strength around the upper part of pile mobilized and generating load transfer to greater depth.

This study explains behavior of laterally-loaded large diameter bored pile on soft silty soil as it provides a simple representation of pile response in terms of deflection on various pile length. It is proposed to be a thorough analysis on other loading and resistance component in lateral soil-pile system such as the influence of tip shear and axial resistance.

Acknowledgement

The authors wish to gratefully acknowledge the financial support received through the Indexed International Publication Grant for Student Final Project (PITTA) No. PENG-1/UN2.R3.1/PPM.00/2019, which is funded by Universitas Indonesia.

References

[1] Zhang Y, Andersen K H and Tedesco G 2016 *Mar. Struct* **50** 260-275
[2] Ng C W, Zhang L and Nip D C 2001 *J. Geotech. Geoenviron. Eng* **127** 658-669
[3] Achmus M and Kuo Y S 2009 *Comput. Geotech* **36** 725-735
[4] Alderlieste E A 2011 *Experimental Modelling of Lateral Loads on Large Diameter Mono-Pile Foundations in Sand* (Delft: Delft University of Technology)
[5] Ashford S A and Juirnarongrit T 2003 *J. Geotech. Geoenviron. Eng* **129** 234-242
[6] Arshad M and O'Kelly B C 2016 *Int. J. Geomech* **17** 04016040
[7] Elson W K 1984 *Design of Laterally-loaded Piles* (London: Construction Industry Research and Information Association)
[8] Chen Y J and Kulhawy F H 1994 *Case History Evaluation of The Behaviour of Drilled Shafs Under Axial and Lateral Loading* (California: Electric Power Research Institute)
[9] Reese L, Cox W and Coop F 1974 *6th Offshore Technology Conf.* **2** 473-483
[10] Standar Nasional Indonesia, *Persyaratan Perancangan Geoteknik*, SNI 8640:201

Dynamic compliances of rigid foundation on layered poroelastic soils

Suraparb Keawsawasvong[1], Teerapong Senjuntichai[1,*], Rawiphas Plangmal[1]

[1]Applied Mechanics and Structures Research Unit, Department of Civil Engineering, Faculty of Engineering, Chulalongkorn University, Bangkok 10330

Email: Teerapong.S@chula.ac.th

Abstract. Dynamic interaction between a rigid foundation and layered poroelastic soils is studied in this paper. The foundation is subjected to time-harmonic vertical horizontal and moment loading, and bonded to a layered poroelastic half-space. The interaction problem is formulated by employing a discretization technique based on the rigid body displacement boundary conditions at the foundation-soil contact surface and the displacement influence functions of a layered poroelastic half-space. The discretization method used in this paper yields a flexibility equation system for the contact traction at the foundation-soil contact area. Comparison with existing solution for a rigid square foundation on an elastic half-space is presented to verify the accuracy of the present scheme. Selected numerical results for compliances of rigid square foundations are presented to portray the influence of poroelastic effects on the dynamic interaction problem.

1. Introduction

The classical problem concerned with dynamic interaction between foundations and supporting soils has a rich history in geomechanics due to its close relevance to various practical problems in geotechnical engineering such as the analysis and design of pavement, mat and raft foundations, anchor plates and theoretical modelling of some in-situ tests. In the past, several researchers have employed a variety of techniques to study dynamic interaction between rigid foundations and elastic soils. For example, Wong and Luco [1] presented time-harmonic response of a rigid rectangular foundation resting on an elastic half-space under vertical, rocking and horizontal motions. Vibrations of rigid rectangular foundations bonded to an elastic medium subjected to excitations from horizontal waves were considered by Wong and Luco [2]. In addition, Shahi and Noorzad [3] presented dynamic impedances of rigid foundations of arbitrary shape resting on a semi-infinite elastic medium.
Geomaterials often consist of two phases, i.e. solid and voids filled with water, and they are commonly known as poroelastic materials. These materials are considered to be more suitable representation of soils and rocks than ideal elastic materials. The theory of elastic wave propagations in a poroelastic medium was presented by Biot [4, 5], and it has been widely employed by several researchers to study a variety of practical problems related to poroelastic materials including dynamic interaction between foundations and poroelastic soils (e.g., Halpern and Christiano [6], Kassir and Xu [7], Rajapakse and Senjuntichai [8], Zeng and Rajapakse [9], Senjuntichai and Kaewjuea [10], Senjuntichai et al. [11]).
In this paper, dynamic compliances of a square foundation bonded to the surface of layered poroelastic soils are determined. The foundation is assumed to be massless and rigid, and subjected to time-harmonic vertical, horizontal, and moment loading. The supporting soils under consideration consist

of a poroelastic layer perfectly bonded to an underlying poroelastic half-space, and are governed by Biot's poroelastodynamics theory. A computer program based on the proposed solution scheme has been developed, and its accuracy is confirmed by comparing with existing solutions. Selected numerical results are presented to demonstrate the influence of poroelastic effects on dynamic compliances of rigid square foundation on different layered poroelastic soils.

2. Problem description

Consider a layered poroelastic half-space with a Cartesian coordinate system as shown in Fig. 1. Let u_i and w_i denote the average displacement of the solid matrix and the fluid displacement relative to the solid matrix in the i-direction ($i = x,y,z$) respectively. The constitutive relation of a homogeneous poroelastic medium can then be expressed according to Biot [12] by using the standard indicial notation as,

$$\sigma_{ij} = 2\mu\varepsilon_{ij} + \lambda\delta_{ij}\varepsilon_{kk} - \alpha\delta_{ij}p\,, \quad i, j = x, y, z \tag{1a}$$

$$p = -M\left(\alpha\varepsilon_{kk} + w_{i,i}\right) \tag{1b}$$

In the above equations, σ_{ij} is the total stress component of the bulk material; ε_{ij} is the strain component, which is related to the displacement u_i as in ideal elasticity; μ and λ are Lame' constants of the bulk material; δ_{ij} is the Kronecker delta; and p is the excess pore fluid pressure (suction is considered negative). In addition, α and M are Biot's parameters accounting for compressibility of the two-phased material.

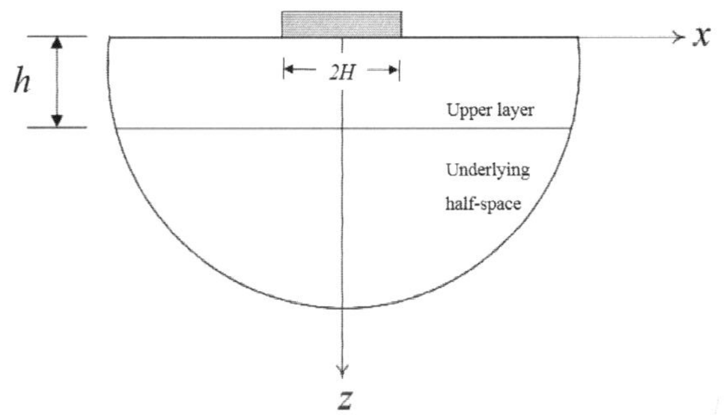

Figure 1. A square foundation on a layered poroelastic half-space.

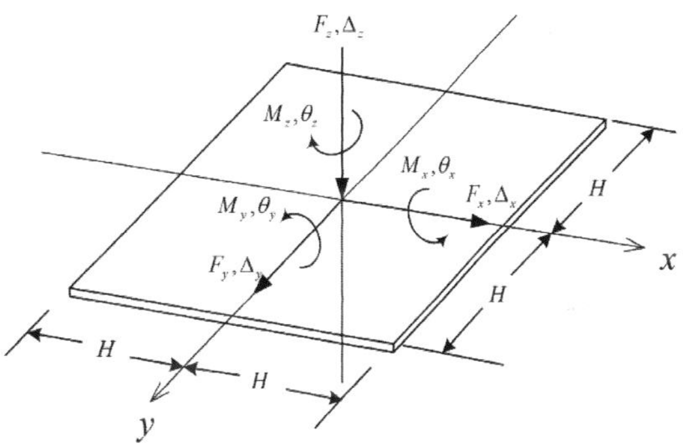

Figure. 2. Applied time-harmonic loading on a square foundation.

A square foundation perfectly bonded to a layered poroelastic half-space with fully permeable contact surface is shown in Fig 1. The foundation is assumed to be massless and rigid, and subjected to time-harmonic vertical, horizontal, and moment loading with the factor of $e^{i\omega t}$, where ω is the frequency of excitation. The amplitudes of loads and displacements, which are also assumed to be time-harmonic with the factor of $e^{i\omega t}$, are shown in Fig. 2. The discretization technique is employed in the formulation of this dynamic interaction problem. The proposed technique requires a set of influence functions, which are the displacements on the surface of a layered poroelastic half-space under uniform surface traction of unit intensity applied in the i-direction ($i = x,y,z$). These influence functions are determined in the Fourier transform space by employing an exact stiffness matrix scheme presented by Senjuntichai et al. [13] and the details of this scheme are omitted here for brevity. The obtained influence functions are then employed in the formulation of dynamic interaction between a rigid rectangular foundation and a layered poroelastic half-space outlined in the next section.

IOP Conf. Series: Materials Science and Engineering **652** (2019) 012030 doi:10.1088/1757-899X/652/1/012030

3. Formulation of interaction problem

Consider a rigid square foundation of size $2H \times 2H$ on the surface of a layered poroelastic half-space illustrated in Fig. 1 subjected to the time-harmonic loading as shown in Fig. 2. In addition, the contact condition between the foundation and the supporting medium is assumed to be perfectly bonded and fully permeable. Under this condition, the traction in the *i*-direction ($i = x,y,z$), denoted by T_i, on the contact area is unknown. The displacements of rigid foundation U_i ($i = x,y,z$) can be expressed in terms of three translation and three rotation components as,

$$U_x = \Delta_x - \theta_z y, U_y = \Delta_y - \theta_z x, U_z = \Delta_z + \theta_x y - \theta_y x \tag{2}$$

On $|x| \le H$ and $|y| \le H$. In addition, Δ_i represents the displacement amplitude in the *i*-direction $(i = x, y, z)$ at the centre of the foundation, and θ_i denotes the amplitude of the rotation about the *i* axis $(i = x, y, z)$ respectively. These equations can be represented by the following matrix equation:

$$\boldsymbol{U} = \Omega \Delta \tag{3}$$

where $\boldsymbol{U} = \{U_x, U_y, U_z\}^{\mathrm{T}}, \Delta = \{\Delta_x, \Delta_y, \Delta_z, \theta_x, \theta_y, \theta_z\}^{\mathrm{T}}$ and Ω is a 3×6 matrix, defined as,

$$\Omega = \begin{bmatrix} 1 & 0 & 0 & 0 & 0 & -y \\ 0 & 1 & 0 & 0 & 0 & x \\ 0 & 0 & 1 & y & -x & 0 \end{bmatrix} \tag{4}$$

Let A be the square contact area between the foundation and the layered half-space. The equilibrium equations of forces applied to the rigid foundation shown in Fig. 2 can then be expressed as,

$$F_i = -\int_A T_i dA \quad (i = x, y, z), M_x = -\int_A T_z y dA \tag{5a}$$

$$M_y = -\int_A T_z x dA, M_z = -\int_A [T_y x - T_x y] dA \tag{5b}$$

In the present study, the discretization technique [1] is adopted to solve for the unknown contact traction T_i ($i = x,y,z$). The contact area is divided into a total number of N_e square elements, in which $N_e = N_x \times N_y$, and N_x and N_y represent the number of elements in the x and y directions respectively. Nodal points are located at the centre of each discretized element. It is assumed that the unknown contact traction T_i (i=x,y,z) is constant within each element. The foundation displacement U_i ($i = x,y,z$) at the discretized element with the nodal point at (x_k, y_l, 0) can then be expressed as,

$$U_i(x_k, y_l, 0) = \sum_{n=1}^{N_y} \sum_{m=1}^{N_x} G_{ij}(x_k, y_l; x_m, y_n) T_j(x_m, y_n) \tag{6}$$

where $G_{ij}(x_k, y_l; x_m, y_n)$ denote the displacement in the *i*-direction ($i = x,y,z$) at the nodal point with the coordinate (x_k, y_l, 0) of the layered half-space due to uniform traction of unit intensity applied in the *j*-direction ($j = x,y,z$) over the discretized element with the nodal location at (x_m, y_n, 0). The influence function G_{ij} are obtained from the exact stiffness matrix scheme [13]. Eq. (6) can then be expressed in the following matrix form:

$$\boldsymbol{GT} = \boldsymbol{U} \tag{7}$$

where the vector $\boldsymbol{U}$ is defined in Eq. (3), and the elements of the matrices $\boldsymbol{G}$ and $\boldsymbol{T}$ correspond to G_{ij} and the contact traction at the discretized elements of the contact area respectively. Finally, the dynamic response of a rigid square foundation is characterized by the following relationship:

$$\Delta = \boldsymbol{CF} \tag{8}$$

where

$$\boldsymbol{C} = \begin{bmatrix} C_h & 0 & 0 & 0 & C_{hm} & 0 \\ 0 & C_h & 0 & -C_{hm} & 0 & 0 \\ 0 & 0 & C_v & 0 & 0 & 0 \\ 0 & -C_{mh} & 0 & C_m & 0 & 0 \\ C_{mh} & 0 & 0 & 0 & C_m & 0 \\ 0 & 0 & 0 & 0 & 0 & C_t \end{bmatrix} \tag{9}$$

The matrix $\boldsymbol{C}$ given by Eq. (9) is a 6 × 6 symmetric complex frequency-dependent compliance matrix and its elements are defined as follows: $C_v = \mu^{(1)} H\Delta_z / F_z$ for vertical compliance; $C_h = \mu^{(1)} H\Delta_x / F_x$ for horizontal compliance; $C_m = \mu^{(1)} H^3\theta_y / M_y$ for rocking compliance; $C_{hm} = \mu^{(1)} H^2\Delta_x / M_y$ for coupling compliance; $C_t = \mu^{(1)} H^3\theta_z / M_z$ for torsional compliance.

4. Numerical results and discussion

In this study, an adaptive numerical quadrature scheme with a 21-point Gauss–Kronrod rule is employed to evaluate the semi-infinite integral for the inversion of Fourier integral transform. A non-dimensional frequency $\delta = \omega H\sqrt{\rho^{(1)} / \mu^{(1)}}$ is used in the numerical results presented in this paper, where $\rho^{(1)}$ and $\mu^{(1)}$ are mass density and shear modulus of the upper layer respectively. Fig. 3 shows the comparison of dynamic compliances of a rigid square foundation ($2H \times 2H$) resting on a homogenous elastic half-space with Poisson's ratio of 0.33 between the present study with $N_e = 64$ elements ($N_x = N_y = 8$) and the solution by Wong and Luco [2]. It is evident from Fig. 3 that all compliances obtained from the present study agree very closely with the existing solutions [2].

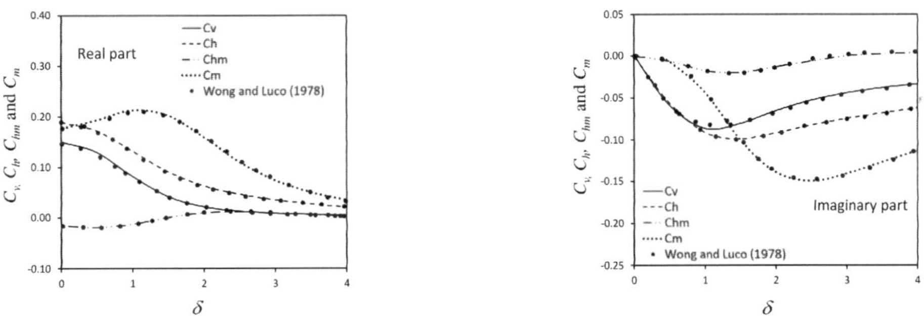

Figure 3. Comparison of compliances of a square foundation on an elastic half-space

The dynamic interaction between a rigid square foundation of size $2H \times 2H$ and a layered poroelastic half-space is investigated in the remaining of this paper. Non-dimensional dynamic compliances are presented for the square foundation bonded to a layered poroelastic half-space consisting of an upper layer of finite thickness $h = H$ and an underlying half-space. Three systems are considered in this set of numerical results, namely System A, System B and System C. The material properties for both upper layer and underlying half-space of all three systems are given as follows: $\lambda = \mu = 2 \times 10^8$ N/m^2; $M = 24.4 \times 10^8$ N/m^2; $\rho = 2 \times 10^3$ N/m^2; $\rho_f = 1.06 \times 10^3$ N/m^2; $m = 2.2 \times 10^3$ N/m^2; and $\alpha = 0.97$, where ρ and ρ_f are the mass densities of the bulk material and the pore fluid respectively; and m is a density-like parameter that depends on ρ_f and the geometry of the pores. In addition, the parameter b, which takes into account the internal friction due to the relative motion between the solid matrix and the pore fluid, is set to be different in the upper layers of the three systems. This parameter is found to have the most significant influence on compliances of rigid foundation in poroelastic media when compared to other poroelastic material parameters [9]. The values of b in the upper layers of System A,

IOP Conf. Series: Materials Science and Engineering **652** (2019) 012030 doi:10.1088/1757-899X/652/1/012030

System B and System C are equal to 6.32×10^{2} N/m^2, 1.45×10^{6} N/m^2, and 6.32×10^{6} N/m^2 respectively, whereas the values of b for the underlying half-spaces of all three systems are identical, and they are equal to 1.45×10^{6} N/m^2. Thus, System B is a homogeneous poroelastic half-space.

Figs. 4 and 5 display the vertical (C_v) and horizontal (C_h) compliances of the rigid square foundation resting on the three systems. The rocking (C_m) and torsional (C_t) compliances of the square footing are shown in Figs. 6 and 7 respectively. In addition, the coupling compliance (C_{hm}) is presented in Fig. 8. Numerical results shown in Figs. 4 to 8 indicate that all compliance components of the surface square footing vary smoothly with δ over the frequency range considered in the numerical study. The real part of C_v and C_h decreases with frequency over the range $0 < \delta < 4$ whereas those of C_m, C_t, and C_{hm} increase with increasing δ at low frequencies. In addition, the imaginary parts of all compliance components increase with the frequency in the low frequency range, and remain negative over the range $0 < \delta < 4$.

Comparison of the compliances for Systems A, B and C indicates substantial difference in the obtained values. Note that the difference between the three systems under consideration is in the value of b in the upper layer. The parameter b is related to the internal friction between solid matrix and pore fluid, and it is inversely proportional to permeability. The material with lower value of b is then more permeable with the one with higher b. Numerical results presented in Figs. 4 to 8 reveal that both real and imaginary parts of all compliance components in System A are higher than the corresponding values in System B and System C indicating that the system with more permeable upper layer, i.e. System A, is less stiff and less damped than the other two systems.

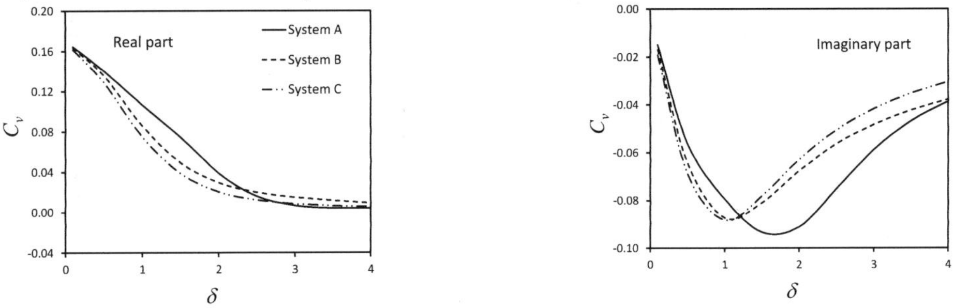

Figure 4. Vertical compliances of a square foundation on different poroelastic soils

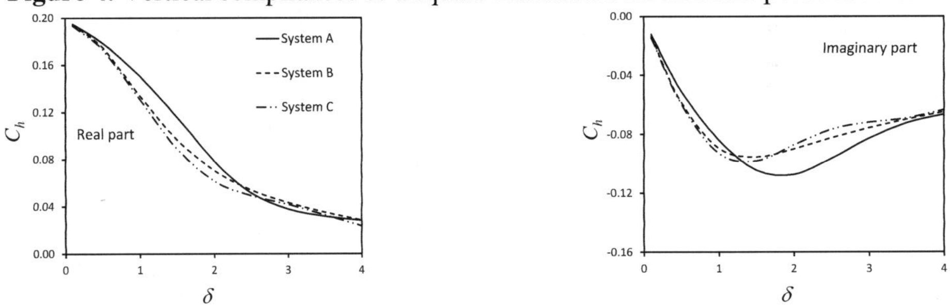

Figure 5. Horizontal compliances of a square foundation on different poroelastic soils

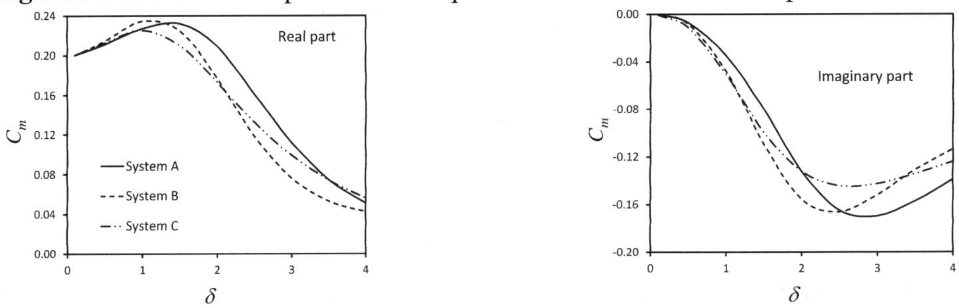

Figure 6. Rocking compliances of a square foundation on different poroelastic soils

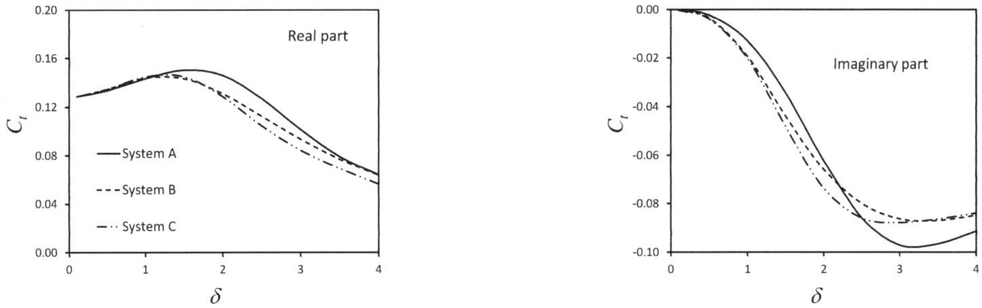

Figure 7. Torsional compliances of a square foundation on different poroelastic soils

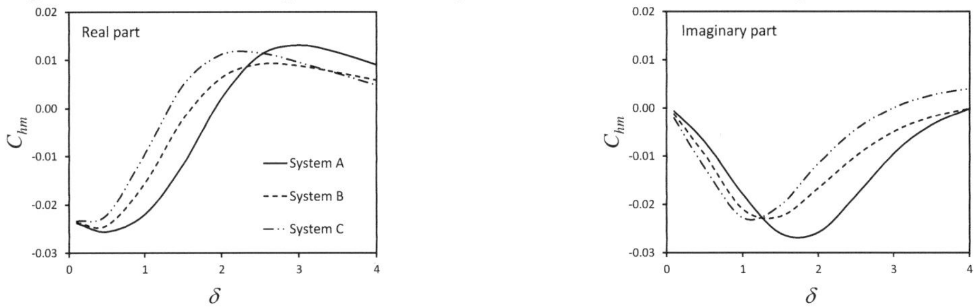

Figure 8. Coupling compliances of a square foundation on different poroelastic soils

5. Conclusions

The dynamic interaction between a rigid foundation and a layered poroelastic medium is investigated in this paper by employing a semi-analytical discretization technique and the exact stiffness matrix scheme. The rigid foundation is subjected to the time-harmonic vertical, horizontal and moment loading, and the contact surface between the foundation and the layered medium is assumed to be perfectly bonded and fully permeable. The accuracy of the present solution is confirmed by comparing with an existing solution for compliances of a rigid foundation on an elastic half-space. Numerical results presented in this paper indicate that the compliances of rigid square foundations are significantly influenced by the parameter b, which takes into account the internal friction of the layered medium due to the relative motion between the solid matrix and the pore fluid. The frequency of excitation also shows a significant influence on the foundation response. The numerical solution scheme developed in this paper can be extended to investigate more complicated problems involving foundations with other shapes and foundations with small flexural rigidity.

References

[1] H. Wong, J.E. Luco, Earthquake Eng. Struct. Dyn. 4, 579-587. (1976)
[2] H. Wong, J.E. Luco, Earthquake Eng. Struct. Dyn. 6, 3-16. (1978)
[3] R. Shahi, A. Noorzad, Int. J. Geomech. 11, 391-398. (2011).
[4] M.A. Biot, J. Acoust. Soc. Am. 28, 168-179 (1956)
[5] M.A. Biot, J. Appl. Phys. 334, 1482-1498 (1962)
[6] M.R. Halpern, P. Christiano, Earthquake Eng. Struct. Dyn. 14, 439-454 (1986)
[7] M.K. Kassir, J. Xu, Int. J. Solid Struct. 24, 915-936 (1988)
[8] R.K.N.D. Rajapakse, T. Senjuntichai, Int. J. Numer. Analyt. Meth. Geomech. 19, 587-614 (1995)
[9] X. Zeng, R.K.N.D. Rajapakse, Int. J. Numer. Analyt. Meth. Geomech. 23, 2075–2095 (1999)
[10] T. Senjuntichai, W. Kaewjuea, J. Mech. Mater. Struct. 3, 1885-1901. (2008)
[11] T. Senjuntichai, S. Keawsawasvong, R. Plangmal, Comput. Geotech. 100, 121-134. (2018)
[12] M.A. Biot, J. Appl. Phys. 122, 155-164 (1941)
[13] T. Senjuntichai, S. Keawsawasvong, R. Plangmal, Mar. Georesour. Geotec. (2019) doi: 10.1080/1064119X.2018.1446200

IOP Conf. Series: Materials Science and Engineering **652** (2019) 012031 doi:10.1088/1757-899X/652/1/012031

Effect of time dependent corrosion rate on residual capacity of corroded RC beam

M Kioumarsi[1], MH Baghban[2], S Imperatore[3]

[1] Department of Civil Engineering and Energy Technology, Oslo Metropolitan University, Oslo, Norway
[2] Department of Manufacturing and Civil Engineering, Norwegian University of Science and Technology, Gjøvik, Norway
[3] Department of Civil Engineering, Niccolò Cusano University, Rome, Italy
Email: mahdi.kioumarsi@oslomet.no; mohammad.baghban@ntnu.no; stefania.imperatore@unicusano.it

Abstract. Corrosion is the most commonly severe defect in reinforced concrete (RC) structures and it mainly causes reduced rebar cross-section. The corrosion rate is the determining parameter of the progress of corrosion-induced damage. The present work is focused on the application of finite element (FE) analysis, using DIANA, to predict the residual capacity of reinforced concrete beam with different degree of corrosion. To this aim, existing experimental data is used to develop a FE model of corroded beams and to investigate their behaviour. This model is validated to predict the impact of time and current rate of corrosion on the residual load carrying capacity of corroded RC beams.

1. Introduction

Corrosion of steel rebar embedded in reinforced concrete (RC) members is one of the major causes of deterioration of these structures [1-3]. Rebar corrosion is a chemical process, which has physical consequences such as decreased ultimate strength and serviceability of structures [1]. The infrastructures today are therefore facing a new challenge that governments, industry and research need to address. The early research in the field of reinforcement steel corrosion, focused on the causes and mechanisms through which steel transforms to rust [4]. However, in recent years, the research focus have moved towards more practical oriented topics. These topics include the assessment of risk, reliability and residual load carrying capacity of corroded concrete elements [2, 4-14].

Early investigations have shown that corrosion is a chemical process that stems from the attack to reinforced concrete by compounds present in the environment. Ingress of chloride ions into concrete affect rebars, causing depassivation of steel and beginning of the corrosion problems [15]. The loss of steel cross section is one of the main physical effects of corrosion [16]. Furthermore, concrete cracking due to stresses exerted by corrosion products as well as loss of bond between concrete and rebars are the other consequences of corrosion of the reinforcement [3, 4, 10, 17]. As a result, serviceability and ultimate strength of reinforced concrete elements may be reduced [2, 7, 10, 18, 19]. Collecting experimental data of corroded concrete elements is challenging, as it means having to analyse reinforcement from old structures [20], or undertake long laboratory experiments with induced corrosion [1, 5, 21]. Thus, much effort has been directed to modelling the corrosion and developing simulation tools that enable investigating residual capacity of corroded concrete elements [10, 22-24].

Accelerated corrosion test is another way to assess corroded RC members. However, in the accelerated experimental test, corrosion rate is one of the most important input parameters, which could affect residual capacity of the RC structures [25].

In this paper, existing experimental data by Rodriguez et al. [21] is used to develop a finite element (FE) model of corroded beams and to investigate their mechanical behaviour. This model is validated and used to predict the impact of time and current rate of corrosion on the load carrying capacity of corroded beams.

2. Finite element simulation of corroded beams

2.1 Experimental setup

The beams for numerical simulations were selected from experimental tests by Rodriguez et al. [21]. Thirty-one beams were tested to investigate the effect of different variables such as details of reinforcement, and a range of corrosion percentages on the corroded beams. In the current study, three of the beams were selected for the numerical simulation. An un-corroded reference beam (no.1), slightly corroded beam (no.2) and severely corroded beam (no.3) were addressed to mainly account for the effect of reinforcement corrosion on RC beams loaded in a four-point bending. Geometry dimensions and rebar arrangement of the selected simply supported beams are shown in Figure 1. The tensile, compressive and stirrups rebars with diameter of 10mm, 8mm and 6mm are constructed in the tested beams, respectively.

The compressive strength of the reference beam is reduced to 34 MPa (from 50 MPa) by adding calcium chloride solvent (3 percent by weight of cement) to accelerate the corrosion process in the test data from [21]. The beams were cured for 28 days in wet conditions. By this way, an accelerated corrosion procedure was established to achieve the required level of corrosion. The reinforcement was subjected to a constant corrosion current rate of 1 μA/cm2. However, due to the presence of stirrups, the corrosion is not distributed uniformly along the bar [21]. In present study, the authors are intending to test the residual capacity of beam for different corrosion rates of 1, 2, 3 and 4 μA/cm2. For this purpose, the corroded beam no. 2 was studied in the deterioration interval time of 50 years. More details are presented in Section 3.

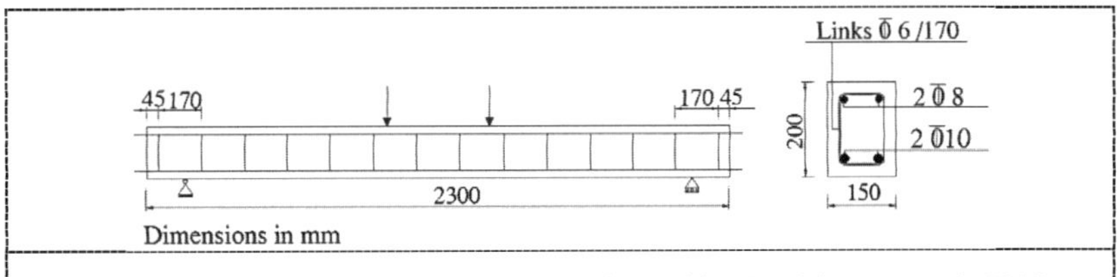

Figure 1. Geometry of beams investigated by Rodriguez et al. [21].

2.2 Finite Element Analyses

The FE method is a suitable numerical technique to investigate the global behaviour of the corroded reinforced concrete structures and the impact of corrosion on the bond strength between the corroded reinforcement and the surrounding concrete [16]. Many research studies were carried out by using different commercial FE programs or in-house developed codes to assess the corrosion impact on the concrete structure [3, 4, 7, 16, 26]. In this study, two-dimensional nonlinear FE analysis were carried out using DIANA. Due to the symmetry condition, only half part of the geometry is simulated. An axial displacement with a constant rate is imposed to the loading point of the beam. To solve the nonlinear problem, the Newton-Raphson (NR) approach was applied in the solver. It should be mentioned that the uniform corrosion assumption is taken into account in the current model in which the initial diameter of the reinforced bars is gradually reduced due to the uniform corrosion attack. The residual cross section area of the tensile reinforcement, which is induced by the uniform corrosion, is calculated by:

$$A_{res} = \frac{\pi.\varphi_R^2}{4} = \frac{\pi.(\varphi_0 - \alpha x)^2}{4} \quad (1)$$

where φ_0 is the initial (nominal) diameter, φ_R is the residual bar diameter, α is coefficient depending on the type of attack and x is corrosion penetration.

2.2.1. Material properties of concrete and steel. The concrete was modelled using four-node quadrilateral plane stress elements Q8MEM. In addition, a constitutive model based on nonlinear fracture mechanics using a smeared rotating crack model based on total strain was used to model the concrete. The constitutive model for concrete in tension is shown in Figure 2.

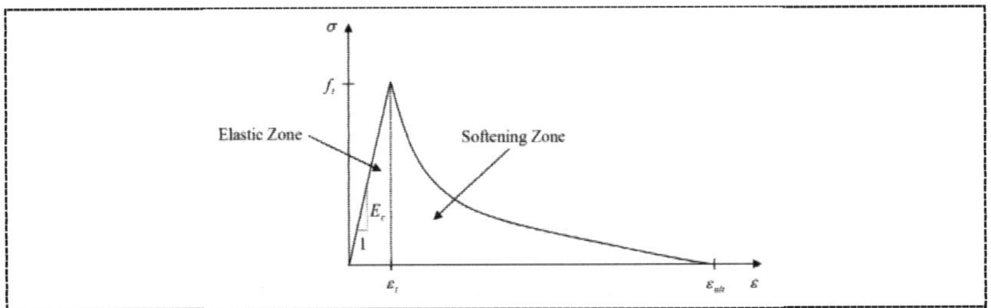

Figure 2. Stress-strain curves for concrete in tension.

The Summary of material properties used in numerical simulation of concrete beams are illustrated in Table 1. It should be noted that the compressive strength of the corroded beams were only about 70% of the strength of the reference [21].

Table 1. Material properties of concrete.

Concrete Properties	Symbol	Reference beam (no.1)	Corroded beams (no.2 & 3)
Compressive strength	f_{cc}	50 MPa	34 MPa
Tensile strength	f_{ct}	3.5 MPa	2.8 MPa
Young's modulus	E_c	32200 MPa	28000 MPa
Tensile fracture energy	G_F	0.12 N/m	0.09 N/m
Crack band width	h	10 mm	10 mm
Poisson's ratio	ϑ	0.2	0.2

The reinforcement was modelled by two-node straight truss elements L2TRU. The constitutive behaviour of the reinforcing steel was assumed to follow that of a standard elastic-ideally plastic with hardening. The yield strength f_{sy}, the ultimate strength f_{su} and the elastic modulus E_s for the steel reinforcement are given in Table 2.

Table 2. Steel strength for corroded and non-corroded models, data from [21].

Type of bar	Strength (MPa)	Ref. beam (no. 1)	Corroded beam (no. 2)	Corroded beam (no. 3)
Stirrups (ϕ6)	Yield, Ultimate	626, 760	506, 610	451, 550
Compressive (ϕ8)	Yield, Ultimate	615, 673	615, 673	615, 673
Tensile (ϕ10)	Yield, Ultimate	575, 655	569, 648	552, 626

While the uniform corrosion may considerably changes the mechanical properties of reinforcement [1], reduction of yield and ultimate strength of corroded rebar is calculated based on the following equations [1]:

$$f_y^D = (1 - \alpha_y(\frac{A_{cor}}{A_0})100)f_{y0} \quad (2)$$

$$f_u^D = (1 - \alpha_u(\frac{A_{cor}}{A_0})100)f_{u0} \quad (3)$$

where f^D and f_0 are the residual and initial strength respectively, A_{cor} is the pit area, A_0 is the cross-sectional area of the non-corroded rebar, α_y = 0.005, and α_u = 0.005 [1]. As a result of uniform corrosion, the rebar diameter is reduced. The sum of uniform corrosion area was subtracted from the original steel cross-section.

2.2.2. Bond slip model. The mechanical behaviour of reinforced concrete structures are mainly controlled by the bond strength and bond-slip between the concrete and reinforcing steel. This behaviour is drastically affected by corrosion in the studied structures. The relative displacements of the reinforcement bars should be measured to investigate the bond slip between the reinforcement steel and the concrete [20]. For this purpose, the CEB-FIP 2010 model code is slightly modified here to describe the impact of corrosion on bond between the reinforcement steel and the concrete. The bond slip behaviour was changed based on the amount of the uniform corrosion and kept constant along the tensile rebar [1]. Moreover, the bond strength is calculated using Equation 4 [27], which describes the effect of concrete and steel stirrup confinement on bond strength.

$$U_{max}^{D} = R\left[0.55 + 0.24\left(\frac{c}{d_0}\right)\right]\sqrt{f_c} + 0.191\left(\frac{A_{sw}f_{yt}}{s_s d_0}\right) \tag{4}$$

In which U_{max}^{D} is the reduced bond strength, c is the thickness of concrete cover, d_0 is the diameter of the anchored rebar, f_c is the concrete compressive strength, A_{sw} and f_{yt} are the stirrups cross-section and the stirrup yield strength respectively, S_S is the stirrup spacing, R is a factor which is function of the amount and rate of corrosion [27]. Bond slip behaviour of simulated beams with/without corrosion are indicated in Figure 3.

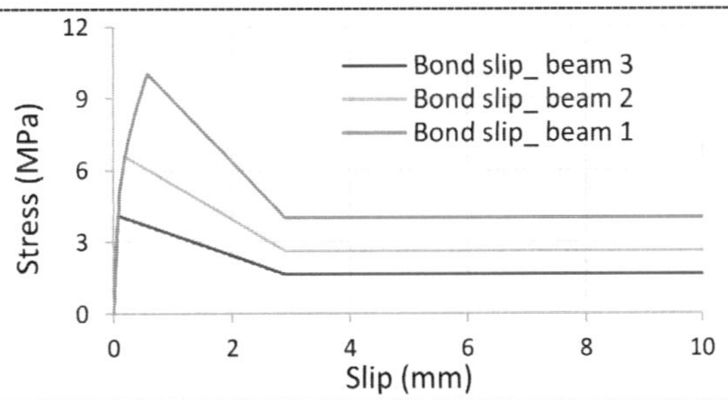

Figure 3. Bond slip behaviour of the reference beam (no.1) and the corroded beams (no.2 and 3).

3. Results and discussion

3.1 Comparison between FE results and the experimental data

The load-displacements at the mid-span of the three different simulated beams were compared with the three corresponding experimental results (see Figure 4). The results of numerical solution are in agreement with the test data. In the experimental investigation, the reference beam no.1 had a low ratio of tensile reinforcement and failed due to yielding of tensile reinforcement bars. The same failure type was observed in the FE simulation. In addition, a good correlation is obtained between the two curves in the yield loads as well as the ultimate loads. Similar to beam no.1, the simulation results for slightly corroded beam no.2 and severely corroded beam no.3 are in alignment with the experimental results. However, a higher difference between experimental and numerical results was observed in the severely corroded beam no.3 compared to slightly corroded beam no.2. Moreover, before the first crack, in all of the simulated beams, the initial stiffness is higher than the experimental results. This difference might be due to micro cracking in the experimental condition, which can result in lower initial stiffness compared to numerical analysis. Based on results from experimental and numerical investigations, notable reduction is clear on the load carrying capacity of corroded beam by increasing the level of corrosion. Since steel bar corrosion affects serviceability as well as ultimate capacity, both the Serviceability Limit State (SLS) and the Ultimate Limit State (ULS) should be taken into account. Serviceability limit state is defined based on deflection L/500 (L is length of the beam), according to ACI code.

In FE analyses, corrosion leads to reduction of SLS load up to 7% for beam no.2. This percentage for beam no.3 is 38%. As it can be seen in Figure 4, obtained SLS loads from the FE analyses of beams no. 1 and 2 are in good agreement with the experimental specimens results, while a deviation between FE and experimental results for severely corroded beam (no.3) for SLS load is seen (around 20%). Long corroding period for beam no.3 (compared to no.2) may have caused stiffness reduction and increased crack width, which could be the reason of this deviation.

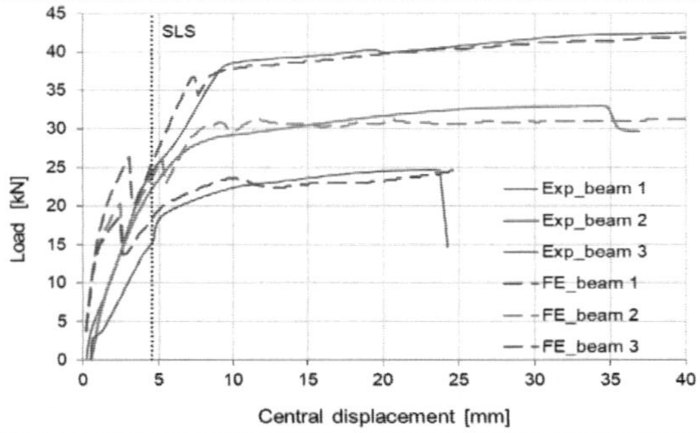

Figure 4. Load-displacement curve of the reference beam (no.1) and the corroded beams (no.2 and 3), after [28].

3.2 Current rate and time of corrosion

Notable reduction of the ultimate capacity is directly associated with the level of corrosion in the reinforced concrete structure. In fact, the level of corrosion is a time dependent quantity, which significantly increases by the evolution of time. Another key quantity, which induces severe deterioration in RC structures, is corrosion rate, which mainly corresponds to loss of rebar cross section area per unit of time. A model for the time dependent loss of steel cross-sectional area as a function of corrosion rate can be derived using the relation for mass loss of reinforcing steel based on Faraday's low, which is described in details by Baingo et al. [26].

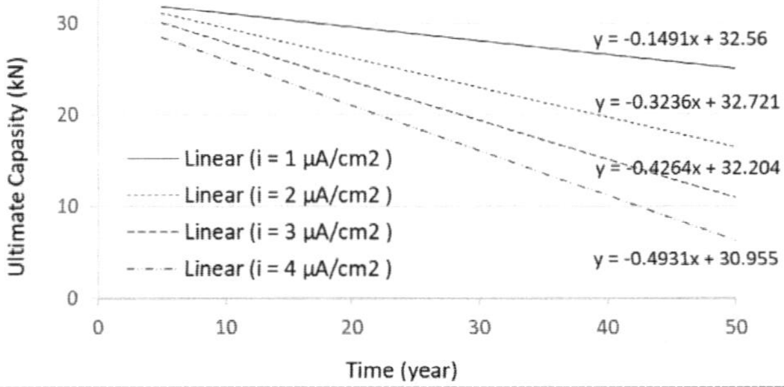

Figure 5. Trendlines of ultimate capacity vs time for the corroded RC beam with different current rates of corrosion: a) i =1 $\mu A/cm^2$, b) i =2 $\mu A/cm^2$, c) i =3 $\mu A/cm^2$ and d) i =4 $\mu A/cm^2$.

A corrosion rate can be expressed as a current density (current per unit surface area), a rate of weight loss or a rate of section loss, which induces the reduction of structural safety and reliability. In this study, the corrosion rate of 1 μA/cm^2 is implemented in the analyses and it is increased up to 4μA/cm^2. The beam with the moderate corrosion (beam no.2) is addressed for this investigation and the ultimate capacity is studied as a function of evolution time and current corrosion rate. The results of numerical studies represent the impact of moderate to high corrosion rate on the residual capacity of reinforced concrete structure during the service lifetime of 50 years. As illustrated in the Figure 5, by increasing the time of corrosion, ultimate capacity will decrease and the failure occurs at earlier stage of the corrosion. By increasing the current rate from 1 to 4 μA/cm^2, the 5-year-old corroded beams, show 4% capacity reduction. However, this reduction is about 65% for the 50-year-old corroded beams. The general behaviours of capacity reduction are derived by the linear equations presented in Figure 5. It is indicated that up to 80% of loading capacity is lost at higher current rates of corrosion (4 μA/cm^2), while at lower rate (1 μA/cm^2) residual capacity is decreased up to 20% compare to the no corrosion case.

4. Conclusions

The objective of the present work was to develop a nonlinear FE analysis to simulate the mechanical behaviour of reinforced concrete structures while affected by the time-dependent corrosion rate over the interval time of 50 years. This validated FE models are implemented to predict the impact of time and current rate of corrosion on the load carrying capacity of corroded beams. Results show that in the 5-year-old corroded beam, increasing current rates of corrosion has less effect on capacity reduction in comparison with the 50-year-old corroded beam. By increasing the time of corrosion form 5 to 50 years, ultimate capacity will decrease with a linear trend. For the beam with corrosion rate of 4μA/cm^2, loading capacity is reduced up to 80%, while at 1 μA/cm^2, residual capacity is decreased up to 20% compare to the no corrosion case.

Acknowledgments

This paper is part of the research convention "Effect of the corrosion on the steel-to-concrete interaction" between Niccolò Cusano University and Oslo Metropolitan University. We would like to show our gratitude to the Oslo Metropolitan University for the financial support of this project.

References

[1]. J. Cabrera, Deterioration of concrete due to reinforcement steel corrosion. Cement and concrete composites, 18(1): p. 47-59, (1996).
[2]. M.M. Kioumarsi, M.A. Hendriks, M.R. Geiker, Quantification of the interference of localised corrosion on adjacent reinforcement bars in a concrete beam in bending. Nordic Concrete Research (NCR), 49: p. 39-57, (2014).
[3]. I. Sæther, B. Sand, FEM simulations of reinforced concrete beams attacked by corrosion. ACI Structural Journal, 109(2): p. 15-31, (2012).
[4]. K. Zandi Hanjari, P. Kettil, K. Lundgren, Analysis of mechanical behavior of corroded reinforced concrete structures. ACI Structural Journal, 108(5): p. 532-541, (2011).
[5]. M.S. Darmawan, M.G. Stewart, Spatial time-dependent reliability analysis of corroding pretensioned prestressed concrete bridge girders. Structural Safety, 29(1): p. 16-31, (2007).
[6]. S. Hooshmandi, B. Kioumarsi, M. Kioumarsi, M. H. Baghban. Application of response surface method (RSM) on sensitivity analysis of reinforced concrete bridge pier wall. Nordic Concrete Research (NCR), Proceedings of the XXIII Nordic Concrete Research Symposium. p. 303-306, (2017).
[7]. A.N. Kallias, M.I. Rafiq, Finite element investigation of the structural response of corroded RC beams. Engineering Structures, 32(9): p. 2984-2994, (2010).
[8]. M. Kioumarsi, G. Markeset, S. Hooshmandi, Effect of pit distance on failure probability of a corroded RC beam. Procedia Engineering, 171: p. 526-533, (2017).

[9]. M.M. Kioumarsi, M.A. Hendriks, M. Geiker, Effect of mesh alignment on simulated interference of localised corrosion on adjacent reinforcement rebars. Proceedings of the XXII Nordic Concrete Research Symposium, Reykjavik, Iceland 2014. 20. p. 431-434, (2014).
[10]. M.M. Kioumarsi, M.A. Hendriks, J. Kohler, M. Geiker, The effect of interference of corrosion pits on the failure probability of a reinforced concrete beam. Engineering Structures, 114: p. 113-121, (2016).
[11]. M.G. Stewart, Mechanical behaviour of pitting corrosion of flexural and shear reinforcement and its effect on structural reliability of corroding RC beams. Structural safety, 31(1): p. 19-30, (2009).
[12]. M.G. Stewart, Q. Suo, Extent of spatially variable corrosion damage as an indicator of strength and time-dependent reliability of RC beams. Engineering Structures, 31(1): p. 198-207, (2009).
[13]. D.V. Val, R.E. Melchers, Reliability of deteriorating RC slab bridges. Journal of structural engineering, 123(12): p. 1638-1644, (1997).
[14]. M. Hajmohammadian Baghban, M. Ghazaei, Seismic performance of reinforced concrete frame for the risk assessment process. Proceedings of the XXIII Nordic Concrete Research Symposium, (2017).
[15]. D.V. Val, P.A. Trapper, Probabilistic evaluation of initiation time of chloride-induced corrosion. Reliability Engineering & System Safety, 93(3): p. 364-372, (2008).
[16]. J. Li, S. Glenn, M. Kioumarsi, Assessment of Existing RC Structures by Virtue of Nonlinear FEM: Possibilities and Challenges, Nordic Concrete Research. 57, (2017).
[17]. S. Imperatore, A. Leonardi, Z. Rinaldi. Mechanical behaviour of corroded rebars in reinforced concrete elements. Mechanics, Models and Methods in Civil Engineering. Springer, Berlin, Heidelberg, p. 207-220, (2012).
[18]. S. Imperatore, A. Leonardi, Z. Rinaldi. Strength decay of RC sections for chloride attack. International Journal of Structural Integrity, 7(2), 194-212, (2016).
[19]. S. Coccia, S. Imperatore, Z. Rinaldi. Influence of corrosion on the bond strength of steel rebars in concrete. Materials and structures, 49(1-2), 537-551, (2016).
[20]. R. Palsson, M.S. Mirza, Mechanical response of corroded steel reinforcement of abandoned concrete bridge. Structural Journal, 99(2): p. 157-162, (2002).
[21]. J. Rodriguez, L. Ortega, J. Casal, Load carrying capacity of concrete structures with corroded reinforcement. Construction and building materials, 11(4): p. 239-248, (1997).
[22]. E. Gestsdóttir, T. Gudmundsson, Bond Behaviour of Naturally Corroded Reinforcement in Concrete Stuctures, Experimental and Numerical Study, (2012).
[23]. J. Li, G. Markeset, M. Kioumarsi, Nonlinear FEM Simulation of Structural Performance of Corroded RC Columns subjected to Axial Compression. Nordic Concrete Research . 57, (2017).
[24]. G. Lin, Y. Liu, Z. Xiang, Numerical modeling for predicting service life of reinforced concrete structures exposed to chloride environments. Cement and concrete composites, 32(8): p. 571-579, (2010).
[25]. M. Otieno, H. Beushausen, M. Alexander, Prediction of corrosion rate in RC structures-A critical review, Modelling of Corroding Concrete Structures, Springer, Dordrecht, 15-37, (2011).
[26]. D. Baingo, A Framework for Stochastic Finite Element Analysis of Reinforced Concrete Beams Affected by Reinforcement Corrosion, PhD thesis, Université d'Ottawa/University of Ottawa, (2012).
[27]. T. El Maaddawy, K. Soudki, T. Topper, Analytical model to predict nonlinear flexural behavior of corroded reinforced concrete beams. ACI Structural Journal, 102(4): p. 550-559, (2005).
[28]. M.M. Kioumarsi, M.A. Hendriks, M. Geiker, Interference of localised corrosion in adjacent reinforcement bar of a beam in bending, in *Concrete innovation conference_CIC*. Oslo, Norway, (2014).

IOP Conf. Series: Materials Science and Engineering **652** (2019) 012032 doi:10.1088/1757-899X/652/1/012032

Lateral displacement capacity of reinforced concrete elements damaged by corrosion

S Imperatore[1], M Kioumarsi[2]

[1] Department of Civil Engineering, Niccolò Cusano University, Rome, Italy
[2] Department of Civil Engineering and Energy Technology, Oslo Metropolitan University, Oslo, Norway
Email: stefania.imperatore@unicusano.it; mahdi.kioumarsi@oslomet.no

Abstract. The aim of this work is to evaluate the influence of corrosion on the lateral displacement capacity and on the global ductility of reinforced concrete (RC) elements, using finite element (FE) analysis. The corrosive phenomenon is simulated accounting for its consequences on the reinforcing bars (variation of the steel geometry and mechanical properties), on the concrete (cracking) and at the interface (variation of the bond-slip behaviour). The parametric analysis, using numerical models, highlights how corrosion significantly changes the structural performance of the element in terms of deformability.

1. Introduction

The corrosion is the most important degradation phenomenon in reinforced concrete (RC) structures. Although several researchers studied the degradation phenomena due to corrosion effect, more efforts are needed to optimize the costs of maintenance and restoration of the deteriorated structures [1]-[4]. In a RC element, the macroscopic effects of the corrosion are the concrete cover cracking, splitting and the reduction of the reinforcement cross-section. Rebar cross-section reduction can be uniform or localized depending on the corrosion typology: in the case of carbonation, uniform corrosion usually occurs; elements deteriorated by chloride attack, instead, are typically characterized by pitting corrosion. In any case, the degraded element experiences a drastic reduction in strength and ductility, as well as a consequent expected variation of the rotational capacity, dissipation capacity and failure mode. All these aspects are also related to a change in the bond-slip behaviour after corrosion.
In order to evaluate the safety of the structure and to optimize any rehabilitation measurements, the influence of the corrosion, affecting the structural performance of a RC element, should be adequately investigated. At international level, the issue has been tackled in both the scientific and regulatory levels. Since the beginning of the century, the International Federation for Structural Concrete (fib) produced the Service Life Model Code in 2006 and the new Model Code in 2010. In both the cases, the specific aspects related to the reinforcement corrosion are described and the criteria for assessing the service life of degraded structural elements are defined.
In the scientific field, technical literature deals with the topic mainly from the static point of view. In this context, a wide state of the art is available. The reduction of the steel cross-section is accompanied by a mechanical properties decay both in the cyclic and monotonic behaviours, both in tension and in compression [5]-[11]. At the same time, as a result of corrosion, the steel cross-section reduction is accompanied by the formation of corrosion products, cracking and spalling of the concrete cover and changing the bond-slip behaviour [12]-[18]. Only taking into account the degradation of the mechanical properties of reinforcement, the sectional capacity of any RC element appears significantly

reduced in the time [19]-[26]. This aspect, may significantly affect the seismic behaviour of the structures [27]. Obviously, the decay may become more significant if also bond and cracking are considered.
In the present work, basing on the numerical FE models, a parametric analysis is performed in order to investigate the lateral displacement capacity of RC elements damaged by corrosion.

2. Decay laws to account for the corrosion: bond-slip behaviour and reinforcements

The reinforcement cross-section reduction is the first consequence of corrosion degradation. Since the production process of the steel rebars induces a multilayer microstructures able to satisfy the regulatory mechanical performances, the corrosive attack also changes in the reinforcement constitutive law. The variation of the mechanical characteristics of reinforcing bars can be statistically represented with a linear regression of the strength and an exponential regression of the ultimate strain [11].
With specific reference to the uniform corrosion, the degradation laws are analytically expressed by the following formulations, which define the mechanical features for a given percentage of corrosion.

$$\sigma_{y,ad} = 1\text{-}0.0143453 \cdot M_{loss}\ [\%] \tag{1}$$

$$\sigma_{uad} = 1\text{-}0.0125301 \cdot M_{loss}\ [\%] \tag{2}$$

$$\varepsilon_{u,ad} = e^{-0.0204983 \cdot Mloss\ [\%]} \tag{3}$$

where $\sigma_{y,ad}$ and $\sigma_{u,ad}$ are respectively the yielding and ultimate nominal strength of the corroded reinforcement, adimensionalized respect to the initial un-corroded nominal strength; $\varepsilon_{u,ad}$ is the fracture strain of the corroded reinforcement, adimensionalized respect to the un-corroded one; M_{loss} is the percentage of mass loss due to the corrosion degradation.

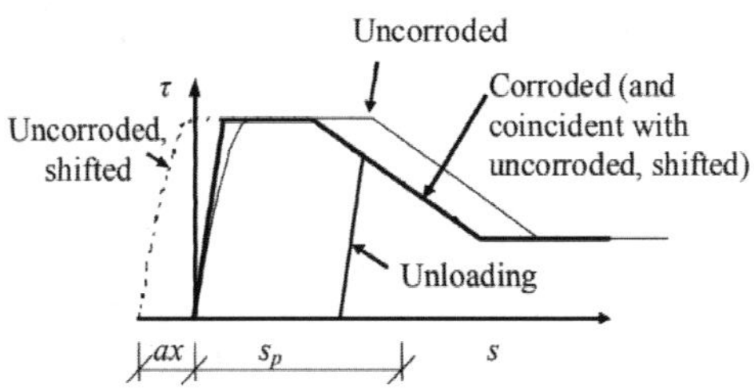

Figure 1. Local bond-slip law model before and after corrosion, from [28].

Another aspect that significantly compromised by corrosion is the bond-slip performance. Due to the electro-chemical reactions, iron oxides form at the steel-concrete interface and exert a radial pressure on the concrete cover and modify the bond condition. Typically, three different stages occur:

- In the so-called pre-cracking, an improvement of the bond strength is detected. In fact, the corrosion products exert a pressure able to promote the concrete confinement.
- In the second stage, cracking stage, a crack forms and then the bond capacity returns to be similar to the initial un-corroded specimen.
- In the third stage, defined of post-cracking, the cracks spread in the concrete cover and enlarge. The confinement effect is loosen and a sharp reduction of the bond conditions are observed.

In this paper, to consider the bond-slip behaviour due to the corrosion in FE model, the plasticity model proposed in [28] has been adopted. In model is shown in Figure 1, where τ represent the bond strength, s the slip of the reinforcement, s_p the plastic slip of the un-corroded bar, and ax is a parameter introduced to account for the corrosion effects. According to model, the degradation is accounted for shifting the CEB-FIP bond-slip formulation [29] along the slip axis of a quantity ax, where a is a constant parameter fixed equal to 8.1 and x is the actual corrosion penetration.

IOP Conf. Series: Materials Science and Engineering **652** (2019) 012032 doi:10.1088/1757-899X/652/1/012032

3. Parametric analyses

Effects of corrosion on the lateral displacement capacity of RC elements is evaluated using FE analysis, performed by ATENA 2D software. In particular, a wall-plate joint has been considered to be studied, see Figure 2. The wall-plate joint has been modelled using the plane stress idealization and has been designed to evaluate the bond-slip capacity of anchoring bars according to [30], and it is here modified to account for the corrosion effects. Aim of this study is, in fact, to understand how the bond-slip capacity of corroded reinforcement affects the lateral displacement capacity of the element.

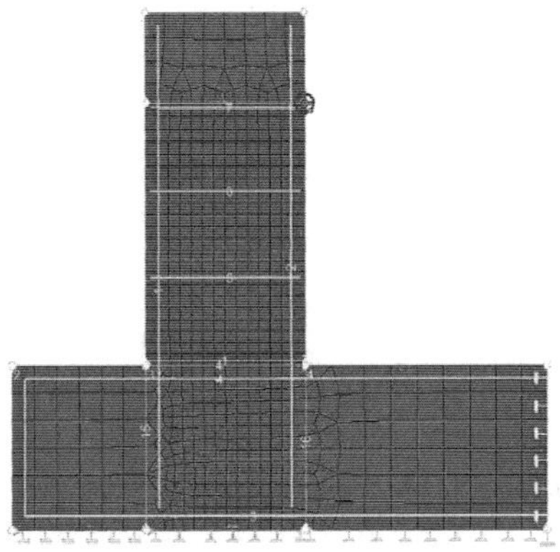

Figure 2. FE model adopted for the parametric analysis.

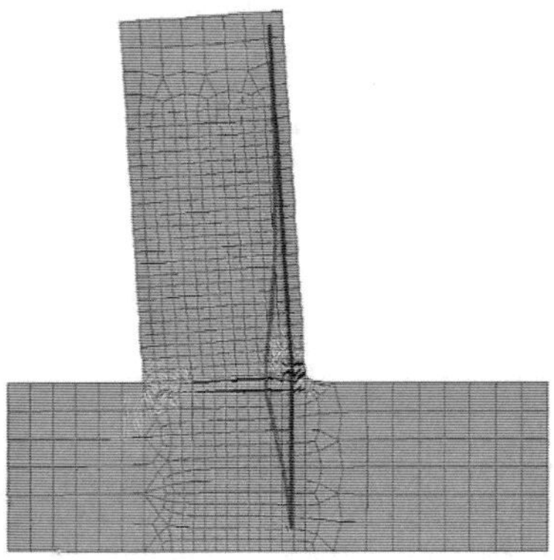

Figure 3. Collapse configuration and stress distribution in the un-corroded element.

The wall has 30 cm depth and is characterized by a height of 120cm; the wall consists of seven bars of 12mm diameter on each side and presents a cover of 30mm, the stirrups have a diameter of 8mm and a spacing of 30cm. The plate has a width of 120cm and a depth of 35cm, the transversal reinforcement consists in 10mm diameter stirrups with a spacing of 20cm; the lower reinforcements consist of seven bars of 12mm diameter, while the upper reinforcements are 13 bars of 16mm diameter. In this way, the failure of the element is localized on the wall-plate interface. The concrete has a cylindrical characteristic strength of 25MPa and is modelled with a nonlinear constitutive law both in tension and in compression. The reinforcements are characterized by a yielding characteristic strength of 450MPa and are modelled with a trilinear constitutive law. The loading scheme involves in an eventual axial load of 3500kN (to investigate the full structural capacity) and in a horizontal displacement, monotonically increasing, and acting 20cm under the top of the element. The collapse configuration of the un-corroded reference element, with the stress distribution on the tensed reinforcement, is represented in Figure 3.

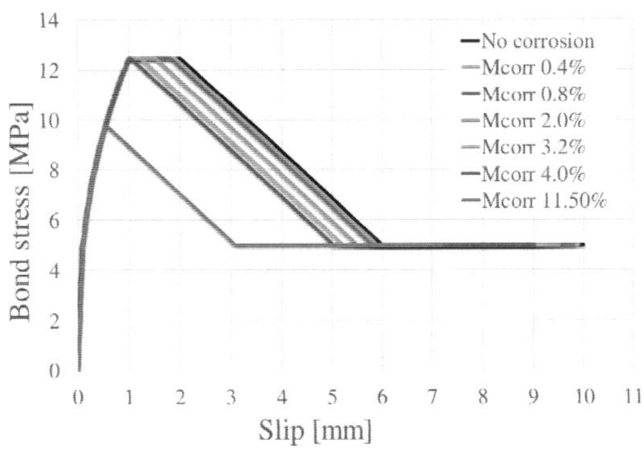

Figure 4. Bond-slip curves employed in the numerical simulation, according to [28].

To investigate the structural degradation, when the corrosion does not still cause a significant reduction in the reinforcement mechanical characteristics (i.e., the below the regulatory limits), the parametric analysis has been carried out considering corrosion levels up to the 11.5% in terms of mass loss. Moreover, the case of the uniform corrosion is chosen, since it seems strongly affects the bond-slip behaviour. The bond-slip curves considered in the parametric analysis are reported in Figure 4.

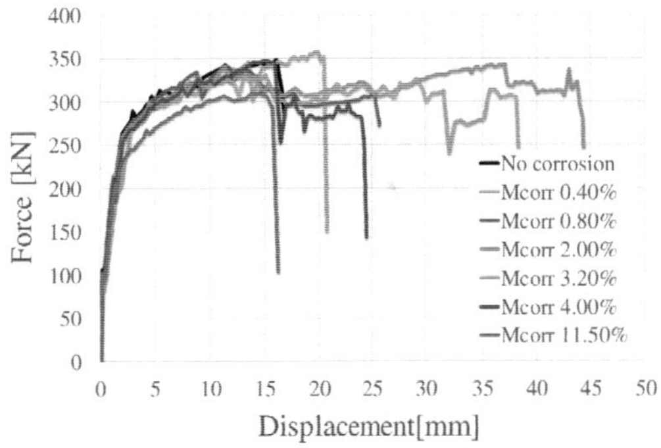

Figure 5. Pushover curves for different corrosion levels: pure bending.

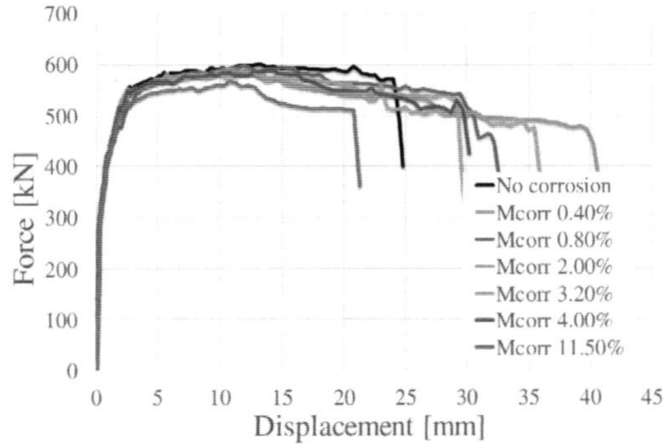

Figure 6. Pushover curves for different corrosion levels: flexure and axial load.

The obtained results are represented in Figure 5 in the case of the pure bending, while the axial load is also considered in Figure 6. Whether or not the axial load is considered, the lateral displacement capacity of the element it is significantly influenced by corrosion. In particular, for a mass loss up to the 2%, the global ductility increases from 7.15 to 19.35 if the pure bending action is considered, while it varies from 7.87 to 13.02 if an axial load of 3500kN is superimposed. Instead, corrosion levels higher that the 2% induce a progressive reduction of the lateral displacement capacity, which returns similar to that obtained for the un-corroded case in both the load cases, when the 11.5% mass loss is considered. Moreover, it is worth to note that the latter corrosion level is the only issue characterized also by a slight strength reduction, due to the greater impact of the mechanical properties degradation of reinforcement.

4. Conclusions

In this paper, a wall-plate joint into the in-plane-stress idealization has been considered in order to evaluate as the bond-slip capacity of corroded reinforcement affects the lateral displacement capacity of the element. The variation of the bond-slip law due to the corrosion has been modelled according to [28]. Moreover, a decay of the reinforcement mechanical properties for the considered degradation level has been introduced. Two different load cases are analysed: the pure bending and the combined compressive and bending stress, the latter considers an axial load level comparable to the point of maximum in the ultimate strength domain. The effects of corrosion on the lateral displacement capacity is evaluated analysing the response of the element in terms of pushover curves, crack pattern, stresses distribution. For sake of brevity, only the load-displacement curves are shown in the present work. The following conclusions can be drawn:

The obtained results clearly remark the influence of corrosion on the lateral displacement capacity of a RC element. In both the load cases the global ductility significantly increases for corrosion level up the 2%. It is worth noting that this degradation level can be achieved in few years from the casting of the reinforced concrete element, depending on the environmental aggressiveness. Higher corrosion levels are instead characterized by a progressive reduction of the lateral displacement capacity. This effect is related only to the variation in the bond-slip law. In fact, for the considered corrosion levels, the reinforcement mechanical properties decay appears negligible and then does not affect the response of the element. In this optic, the effects of the influence of corrosion on the steel-to-concrete interaction and on the reinforcement can be disregarded and the key factor of the study become the

influence of the bond-slip behavior for degradation levels up to 10% in mass loss. Moreover, the obtained results show that the lateral displacement capacity is lower than the initial un-corroded accompanied by a slight strength reduction for an 11.5% corrosion level in the both considered load cases. This means that the most common corrosion levels, characterized by mass losses greater 10%, modify the structural behaviour in terms of both strength and lateral displacement capacity. Consequently, both the degradation of mechanical properties of reinforcement and the influence of the corrosion on the bond-slip behaviour should be considered in the numerical simulation of deteriorated structures. It is worth observing that corrosion modify also the crack pattern, due to the progressive reduction of the bond strength capacity. In particular, a reduction of the number of crack and an increase of their width is detected by increasing the corrosion level.
In conclusion, the corrosion can significantly influence the performance of a reinforced concrete element in terms of strength and, above all, of ductility. The correct simulation of the degradation, then, appears the key point to effectively catch the performance modifications and collapse configurations.

Acknowledgments
This paper is part of the research convention "Effect of the corrosion on the steel-to-concrete interaction" between Niccolò Cusano University and Oslo Metropolitan University. We would like to show our gratitude to the Oslo Metropolitan University for the financial support of this project.

References
[1] Zanini M. A., Faleschini F., Pellegrino C. *Cost analysis for maintenance and seismic retrofit of existing bridges*. Structure and Infrastructure Engineering, 12(11), 1411-1427, (2016).
[2] Zanini M. A., Faleschini F., Pellegrino C. *Probabilistic seismic risk forecasting of aging bridge networks.* Engineering Structures, 136, 219-232, (2017).
[3] Bossio A., Lignola G. P., Fabbrocino F., Monetta T., Prota A., Bellucci F., Manfredi G. *Nondestructive assessment of corrosion of reinforcing bars through surface concrete cracks.* Structural Concrete, 18(1), 104-117, (2017).
[4] Bossio A., Bellucci F. *Environmental degradation of reinforced concrete structures.* Corrosion Reviews, 37(1), 1-2, (2019).
[5] Almusallam A. A. *Effect of degree of corrosion on the properties of reinforcing steel bars.* Construction and Building Materials, 15(8), 361-368, (2001).
[6] Apostolopoulos C. A., Papadopoulos M. P. *Tensile and low cycle fatigue behavior of corroded reinforcing steel bars S400.* Construction and Building Materials, 21(4), 855-864, (2007).
[7] Lee H. S., Cho Y. S. *Evaluation of the mechanical properties of steel reinforcement embedded in concrete specimen as a function of the degree of reinforcement corrosion.* International journal of fracture, 157(1-2), 81-88, (2009).
[8] Apostolopoulos C. A., Demis S., Papadakis V. G. *Chloride-induced corrosion of steel reinforcement–Mechanical performance and pit depth analysis.* Construction and Building Materials, 38, 139-146, (2013).
[9] Kashani M., Crewe A. J., Alexander N. A. *Nonlinear stress–strain behaviour of corrosion-damaged reinforcing bars including inelastic buckling*. Engineering Structures, 48, 417-429, (2013a).
[10] Kashani M., Crewe A. J., Alexander N. A. *Nonlinear cyclic response of corrosion-damaged reinforcing bars with the effect of buckling*. Construction and Building Materials, 41, 388-400, (2013b).
[11] Imperatore S., Rinaldi Z., Drago C. *Degradation relationships for the mechanical properties of corroded steel rebars*. Construction and Building Materials, 148, 219-230, (2017).
[12] Al-Sulaimani G. J., Kaleemullah M., Basunbul I. A. *Influence of corrosion and cracking on bond behavior and strength of reinforced concrete members*. Structural Journal, 87(2), 220-231, (1990).

[13] Andrade C., Alonso C., Molina F. J. *Cover cracking as a function of bar corrosion: Part I-Experimental test.* Materials and structures, 26(8), 453-464, (1993).
[14] Cabrera J. G. *Deterioration of concrete due to reinforcement steel corrosion.* Cement and concrete composites, 18(1), 47-59, (1996).
[15] Coronelli D. *Corrosion cracking and bond strength modelling for corroded bars in reinforced concrete.* Structural Journal, 99(3), 267-276, (2002).
[16] Coccia S., Imperatore S., Rinaldi Z. *Influence of corrosion on the bond strength of steel rebars in concrete.* Materials and structures, 49(1-2), 537-551, (2016).
[17] Bossio A., Monetta T., Bellucci F., Lignola G. P., Prota A. *Modelling of concrete cracking due to corrosion process of reinforcement bars.* Cement and Concrete Research, 71, 78-92, (2015).
[18] Bossio A., Lignola G. P., Prota A. *An overview of assessment and retrofit of corroded reinforced concrete structures.* Procedia Structural Integrity, 11, 394-401, (2018).
[19] Vu K. A. T., Stewart M. G. *Structural reliability of concrete bridges including improved chloride-induced corrosion models.* Structural safety, 22(4), 313-333, (2000).
[20] Zandi Hanjari K., Kettil P., Lundgren K. *Analysis of mechanical behavior of corroded reinforced concrete structures.* ACI Structural Journal, 108(5), 532-541, (2011).
[21] Imperatore S., Leonardi A., Rinaldi Z. *Mechanical behaviour of corroded rebars in reinforced concrete elements.* In: Mechanics, Models and Methods in Civil Engineering, pp. 207-220, Springer, Berlin, Heidelberg, (2012).
[22] Imperatore S., Leonardi A., Rinaldi Z. *Strength decay of RC sections for chloride attack.* International Journal of Structural Integrity, 7(2), 194-212, (2016).
[23] Kioumarsi M., Markeset G., Hooshmandi S. *Effect of pit distance on failure probability of a corroded RC beam.* Procedia Engineering, 171, 526-533, (2017).
[24] Kioumarsi M., Hendriks M.A., Geiker M. *Failure probability of a corroded beam with interference effect of localised corrosion.* Nordic Concrete, 53, p.39, (2015).
[25] Kioumarsi M.M., Hendriks M.A., Geiker M.R. *Quantification of the interference of localised corrosion on adjacent reinforcement bars in a concrete beam in bending.* Nordic Concrete Research (NCR), 49: p. 39-57, (2014).
[26] Kioumarsi M.M., Hendriks M.A., Kohler J., Geiker M. *The effect of interference of corrosion pits on the failure probability of a reinforced concrete beam.* Engineering Structures, 114: p. 113-121, (2016).
[27] Bossio A., Fabbrocino F., Monetta T., Lignola G. P., Prota A., Manfredi G., Bellucci F. *Corrosion effects on seismic capacity of reinforced concrete structures.* Corrosion Reviews, 37(1), 45-56, (2019).
[28] Lundgren K., Kettil P., Zandi Hanjari K., Schlune H., Roman A. S. S. *Analytical model for the bond-slip behaviour of corroded ribbed reinforcement.* Structure and Infrastructure Engineering, 8(2), 157-169, (2012).
[29] CEB Bulletin d'Information 213/214. CEB-FIP Model Code 1990, Lausanne, Switzerland, (1993).
[30] Kabele P., Cervenka V., Cervenka J. *Atena program documentation Part 3-1, Example manual* ATENA Engineering. Prague (Czech Republic): Cervenka Consulting, (2010).

IOP Conf. Series: Materials Science and Engineering **652** (2019) 012033 doi:10.1088/1757-899X/652/1/012033

Material image segmentation based on feature similarity and non-spherical clustering

Huan Wang

School of Computer Engineering and Science, Shanghai University, Shanghai, 200444, China

Email: wanghuanshucs@163.com

Abstract. The microstructure of the material determines its physical and chemical properties. Although image segmentation and target extraction algorithms have advanced rapidly in recent years, few segmentation algorithms for material images have emerged. By analyzing the characteristics of material images, A new image segmentation method based on feature similarity and non-spherical clustering is proposed in this paper. First, the central pixels of the region are selected according to the feature similarity, and secondly, the attribution of the remaining pixels or the non-region central points are determined by the non-spherical clustering method. In this paper, ceramic images are used as experimental materials. The results show that the proposed method achieves satisfactory segmentation effect and provides some help for subsequent image-based material analysis.

1. Introduction

The internal structure of the material, also known as the microstructure, stores the genes of the material and all its physical, chemical and other important properties. Obtaining the properties of materials by studying the structural information in the microscopic images of the cross-section of materials has always been an important research direction in the field of materials science[1, 2]. Although the importance of material microstructures is well known, automatic segmentation of microstructures in material microscopic images is still very challenging. As the first and most important part of acquiring image information, the quality of segmentation directly determines the accuracy and precision of subsequent information acquisition.

The difficulty of material image segmentation mainly comes from the characteristics of the material itself and its imaging method. Scanning electron microscopy (SEM) is often used in materials science, which can provide high resolution, up to 200,000 times. In this situation, the microstructure of material image can be observed, but the target area of material cross-section in high resolution image usually contains complex textures, or even different phases. Besides, the grinding process of imaging sample can increase noise, artificially. In comparison, natural images do not have such problems. This leads to two problems. The first one, the commonly used natural image segmentation methods can't be effectively applied to material image segmentation. The second one, the intersection of different regions in material image contains a lot of noise, which is difficult to segment. Therefore, how to extract effective features according to the characteristics of the target area in the material image and design the corresponding segmentation algorithm is an urgent and valuable problem.

2. Related work

Image segmentation has always been a popular research direction in image processing field. According to the complexity of the features used, it can be divided into low-level semantic segmentation, middle-level semantic segmentation and high-level semantic segmentation[3-5]. At present, there is no segmentation algorithm designed for the characteristics of material images, and most of the existing researches are based on low-level semantic segmentation with single pixel features[6], such as watershed[7] and Markov[8] model based methods. In addition, interactive segmentation methods have also developed in recent years. For example, Zhang et al.[9] used threshold binarization method combined with mathematical morphology operation to segment the SEM image of compact rock, and calculated the pore size distribution (PSD) of samples. Sourov et al. [10] used interactive segmentation method combined with binarization to segment SEM images of the catalyst layer cross-section in a polymer electrolyte membrane fuel cell. Ruth et al. [11] used simple manual threshold selection based on gray histogram to segment the fiber SEM image. Zhao et al.[12] proposed a spectral clustering segmentation algorithm which combines LBP features and gray-scale features, which is applied to ceramic image segmentation. N.Vyas et al.[13] combined fast random forest with interactive segmentation to segment the surface SEM images of complex biomaterials covered with biofilm. L.Drumetz [14] proposes a semi-automatic segmentation method based on SVM and binary partition trees (BPT), which can be used for segmentation of gel material SEM images.

In general, the shortcoming of the above method is that the area information of the image is not used, and only the single pixel feature is processed, so that the segmentation result is sensitive to noise, the segmentation is inaccurate, images that with complex textures cannot be effectively segmented.

3. Algorithm

In order to solve the above problems, a new material image segmentation method based on feature similarity and non-spherical clustering is proposed in this paper. By analyzing the characteristics of the material microscopic image, the pixels of the image are divided into two categories: the regional central pixels and the non-region central pixels. Among them, the regional central is relatively flat, and the feature similarity between pixels is relatively high, which can be selected by constructing similarity measure function and setting threshold. On the contrary, the feature difference of the non-region central pixels is large, to ensure the connectivity of the pixel class, the class to which it belongs is discriminated by non-spherical clustering. The algorithm is described in detail below.

3.1. Selection of Regional Central pixels. In the previous region segmentation methods, seed point selection is usually random, and then transits to flat region by iteration. This method has two shortcomings: First, iteration requires a large amount of computation and takes a long time. Second, the number of seed points has nothing to do with the number of regions, which easily leads to over-segmentation. Aiming at this problem, this paper proposes a region central pixel set extraction algorithm based on feature similarity.

3.1.1. Construction of feature vectors. As mentioned above, the texture of material images is complex and there are abundant changes even within the same region. As a result, the difference discrimination method based on single-pixel feature, which is often used in natural images, will be affected by a lot of noise and can not accurately distinguish the category of pixels. Therefore, in this paper, the statistical feature of gray histogram in the 5×5 region around a pixel are taken as the feature of the pixel. Assuming that there are L gray levels in the region, and f represents the frequency of the level, then the mean value of the gray in the region can be expressed by m, and the n-order statistical moments of the region can be expressed by u, that is:

$$\mathrm{m}=\sum_{g=1}^{L} r_g f(r_g) \tag{1}$$

$$u_n = \sum_{g=1}^{L} (r_g - m)^n f(r_g) \tag{2}$$

In this paper, mean, second-order statistical moment, third-order statistical moment and fourth-order statistical moment are used as the characteristics of pixel. That is：

$$\mathbf{V}=(m, u_2, u_3, u_4) \tag{3}$$

Due to considering the information of the surrounding area, the construction of the feature vector can effectively reduce the impact of complex texture in the material microscopic image.

3.1.2. Similarity measure function. As mentioned above, relative to the edge, the inner region is relatively flat, and the feature similarity between the pixels is large. Assuming the feature vector of the pixel $P_{(i,j)}$ is $V_{(i,j)}$, the feature similarity measure function $diff_{(i,j)}$ of the pixel $P_{(i,j)}$ is defined as follows:

$$diff_{(i,j)} = \max_{(x,y)} \left\| \mathbf{V}_{(i,j)} - \mathbf{V}_{(x,y)} \right\|^2 \tag{4}$$

Where $V_{(m,n)}$ is the feature vectors of point $P_{(m,n)}$, $P_{(m,n)}$ satisfies the chessboard distance from point $P_{(i,j)}$ less to T_{dist}, which is set to 5 in this paper. That is to say, firstly, all the pixels whose chessboard destance to $P_{(i,j)}$ is less than 5 constitute the neighborhood; secondly, in this neighborhood, the feature vectors of each pixel are calculated according to the method of 3.1.1; finally, the maximum Euclidean distance between the feature vectors of each pixel and all the pixels in its neighborhood is selected as the similarty. It can be seen that the smaller the value of the function, the higher the degree of similarity within the neighborhood, the more likely it is to be the central pixel of the region. On the contrary, the larger the value of the function, the lower the degree of similarity within the neighborhood, and the more inclined the edge is the non-regional central pixel.
Let the similarity threshold be T_{diff}, which is set to 50 in this paper, then:

$$type_{P_{(i,j)}} = \begin{cases} 1 \ , & diff_{(i,j)} \le T_{diff} \\ 0 \ , & others \end{cases} \tag{5}$$

If $type_{P_{(i,j)}}$ equals 1, then the pixel $P_{(i,j)}$ is the regional central pixel, otherwise it is the non-regional central pixel.

3.2. Non-spherical clustering of non-regional central pixels
For non-regional central pixels, the existing methods mostly use spherical clustering to determine their categories, that is, to calculate the similarity between the pixel and all the class central to determine their attribution. However, this method can not guarantee the connectivity of the pixel categories. Inspired by[15], this paper proposes a non-spherical clustering based discriminant method: each non-regional central pixel should be classified into one group with the most similar feature in its eight-connected neighborhoods. That is:

$$label\ [P_{(i,j)}] = label\ [P_{(x',y')}] \tag{6}$$

$$s.t.\ \left\| \mathbf{V}_{(i,j)} - \mathbf{V}_{(x',y')} \right\|^2 = \min_{(x,y)} \left\| \mathbf{V}_{(i,j)} - \mathbf{V}_{(x,y)} \right\|^2 \tag{7}$$

The specific flow chart is shown in Figure 1.

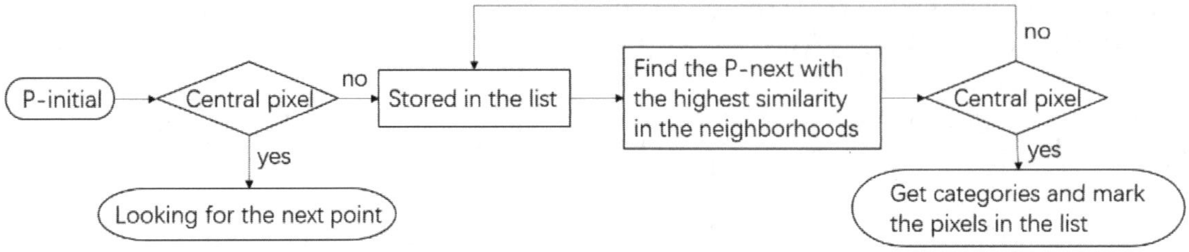

Figure 1. Clustering process of non-regional central pixels

Firstly, a non-regional central pixel *P*-initial in the image is taken and stored in the list. In its eight-connected neighborhoods, the pixel *P*-next which is most similar to *P*-initial is obtained. If *P*-next is the central pixel of a class, *P*-initial is classified as that class; otherwise, *P*-next is also stored in the linked list, and the pixel with the smallest feature difference from *P*-next in its eight-connected neighborhoods is searched until it is connected to the regional central pixel. Then all the non-regional central pixels in the list are classified into this class. The non-spherical clustering process of non-regional central pixels is completed by traversing the image.

4. Experiment

In order to verify the effectiveness of the proposed segmentation algorithm, the ZrB2-SiC ceramic images are selected as experimental materials, as shown in Figure 2. It can be seen that, firstly, there are areas with obvious contrast and areas with low contrast in the image; secondly, there are complex textures in the same area; thirdly, there are obvious artificial polishing marks in the image, which are the typical characteristics of material images.

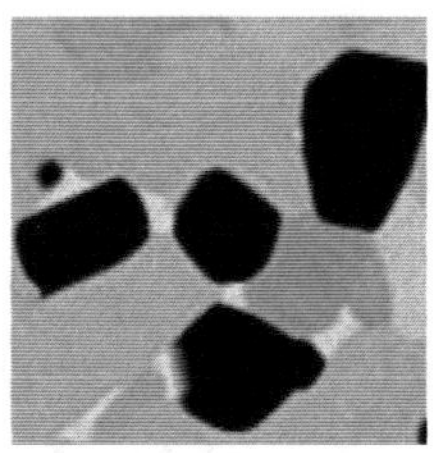

Figure 2. ZrB2-SiC ceramic image

Firstly, we use Canny operators with different thresholds to extract the edges of ceramic images. The results are shown in Figure 3. It can be find that with the increase of high threshold, Canny operator can extract more obvious edges. With the increase of low threshold, trivial edges can be significantly reduced. However, no matter how to adjust the parameters, the complex texture of the material image makes it difficult for Canny operator to extract the closed region, and there are still trivial edges. In addition, it can be found that many important edges Canny operators have not been extracted, such as the dark region in the upper left corner of the original image.

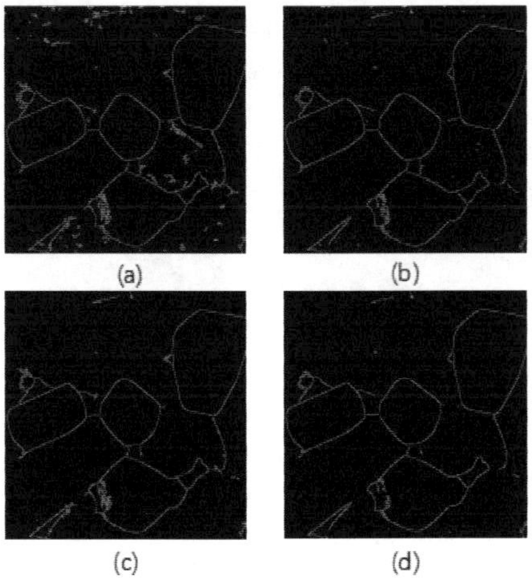

Figure 3. Edge Extraction Using Canny Operator. (a) LT=50, HT=150. (b) LT=50, HT=200. (c) LT=100, HT=150. (d) LT=100, HT=200.

Figure 4 shows the segmentation results of the algorithm proposed in this paper. Where Figure 4(a) is the result obtained by 3.1, the white pixels belong to the regional central and the black pixels belong to the non-regional central. We can see that although there are complex textures in the material image, because the local region similarity is taken into account in the design process of the algorithm, the regional central points maintain good connectivity. Even hand-polished traces rarely cause fracture in the same area. Figure 4(b) is the result of non-spherical clustering. It can be found that the segmentation accuracy has been significantly improved. It can effectively segment the connected regions with different degrees of difference. Although it has not been pre-processed by image smoothing, the algorithm in this paper has strong resistance to complex texture and strong noise. There are also some shortcomings in this algorithm. For example, a small number of unclassified black pixels appear at the boundary of the region, after analysis, self-locking phenomenon appears during non-spherical clustering, that is to say, the most similar pixel of pixel A, in its neighborhood, is pixel B; while the most similar pixel of pixel B in its neighborhood is pixel A, too, resulting in two pixels can not be classified. Besides, a few of details are not accurately segmented, possibly because the neighborhood scale is too large when acquiring features. The above problems can be solved by some mathematical morphological methods, but in the most of other areas, the algorithm proposed in this paper has achieved satisfactory segment results.

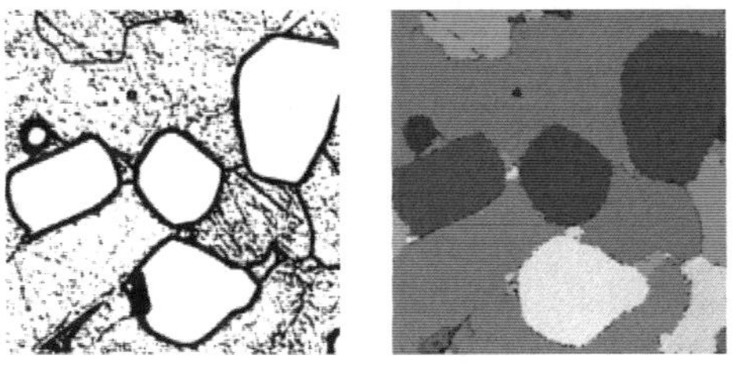

Figure 4. Experimental Results of the Proposed Algorithm. (a) is the regional center point. (b) is the result of non-spherical clustering.

5. Conclusion

Aiming at the characteristics of material image, a new image segmentation method is proposed in this paper. First, we construct a feature vector and similarity measure function to select the regional central pixels. Secondly, we use non-spherical clustering to classify the non-region central pixels. The experimental results show that the proposed method can effectively reduce the influence of complex texture of material images, and can effectively segment various contrast regions. Due to self-locking, a small number of boundary pixels cannot be classified, which is a challenge. So how to build more effective local features and eliminate self-locking will be the focus of our future work.

Acknowledgments
The experimental image used in this article is from Dr. Dongli Hu.

References
[1] Hu Q, Cai Q, He L, Zhao X, Shi R and Ye T 2017 Novel method to determine the image segmentation threshold during the quantitative test on meso-structure of geo-material *Journal of Wuhan University of Technology-Mater. Sci. Ed.* **32** 1408-1412
[2] Carrera D, Manganini F, Boracchi G and Lanzarone E 2017 Defect detection in SEM images of nanofibrous materials *IEEE Transactions on Industrial Informatics* **13** 551-561
[3] Watson M and Marshall M 2017 A Novel Image Segmentation Approach for Microstructure Modelling *Coatings* **7** 166
[4] Li Y, Qi H, Dai J, Ji X and Wei Y 2017 Fully convolutional instance-aware semantic segmentation. In: *Proceedings of the IEEE Conference on Computer Vision and Pattern Recognition,* pp 2359-2367
[5] Dai J, He K, Li Y, Ren S and Sun J 2016 Instance-sensitive fully convolutional networks. In: *European Conference on Computer Vision*: Springer) pp 534-549
[6] Sanyal P, Bhattacharya U and Bandyopadhyay S K 2008 Analysis of SEM images of stomata of different tomato cultivars based on morphological features. In: *2008 Second Asia International Conference on Modelling & Simulation (AMS)*: IEEE) pp 890-894
[7] Jiang L-j and Zhu Y 2007 Region segmentation for SEM image based on watershed transform and concave spot *JOURNAL-EAST CHINA UNIVERSITY OF SCIENCE AND TECHNOLOGY* **33** 861
[8] Zhu Y, Zuo T and Wang Y 2009 SEM Microscopic Image Segmentation Based on Markov Field Models. In: *2009 Fifth International Conference on Image and Graphics*: IEEE) pp 177-782
[9] Zhang Y, Jin S, Wang Y and Wang Y 2015 Characterization of the pore size distribution with SEM images processing for the tight rock. In: *2015 IEEE International Conference on Information and Automation*: IEEE) pp 653-656
[10] Ghosh S, Ohashi H, Tabata H, Hashimasa Y and Yamaguchi T 2015 Microstructural pore analysis of the catalyst layer in a polymer electrolyte membrane fuel cell: a combination of resin pore-filling and FIB/SEM *International Journal of Hydrogen Energy* **40** 15663-15671
[11] Hinrichs R, Frank P R O and Vasconcellos M 2017 Short range shooting distance estimation using variable pressure SEM images of the surroundings of bullet holes in textiles *Forensic science international* **272** 28-36
[12] Zhao Z, Ding G T, Fan M L, Zhang H R, Wang L and Chen L 2016 Material Image Segmentation Combined LBP Texture and Local Gray Level Feature *Computer Technology & Development*
[13] Vyas N, Sammons R, Addison O, Dehghani H and Walmsley A 2016 A quantitative method to measure biofilm removal efficiency from complex biomaterial surfaces using SEM and image analysis *Scientific reports* **6** 32694
[14] Drumetz L, Mura M D, Meulenyzer S, Lombard S and Chanussot J 2015 Semiautomatic classification of cementitious materials using scanning electron microscope images *Journal of Electronic Imaging* **24** 061109
[15] Rodriguez A and Laio A 2014 Clustering by fast search and find of density peaks *Science* **344** 1492-1496

IOP Conf. Series: Materials Science and Engineering **652** (2019) 012034 doi:10.1088/1757-899X/652/1/012034

Development of new wood-concrete connectors

A. Khelil, C. Kiniagi and R. Boissière

IJL, UMRS CNRS 7198, Université de Lorraine, France
Email: abdel.khelil@univ-lorraine.fr

Abstract. The application of timber-concrete composites (TCC) connections in structures has increased over the past decades. The mechanical behaviour of these connections is largely influenced by the stiffness of the joints, both with regard to deformations and strength. This paper presents laboratory experiments and numerical simulations using ABAQUS performed on TCC with a new connector that has been developed in our laboratory. Collapse tests of Shear (push-out) tests, four-point bending tests on beams with different configuration and spacing are carried out within the serviceability load range to verify the slip modulus of connections which were derived from the push-out tests. Although the new connection is applied on notched timber, both experiments and 3D FEM simulations showed that the new connector exhibits stronger stiffness and has a significantly higher peak force performance.in constructing both.

1. Introduction

The assembly of Timber-concrete composite (TCC) structures efficiently optimizes the structural performance of buildings and bridges by technically improving the strength and stiffness of the system as a whole thereby reducing construction cost of buildings and bridges, as well retrofitting and strengthening of existing structures [1, 2, 3]. Timber has to its lower density in compared to reinforced concrete, therefore this decreases the total weight of these hybrid structures hence having several advantages over reinforced concrete floors which include; better efficiency in terms of load applied increase in the rigidity which improves the seismic performance of the overall structure. Compared to concrete structures, TCC significantly reduce CO2 emissions through a sequestration process because Timber can act as a carbon store depending on its end of life treatment [4]. So generally the concrete slab increases the overall stiffness and reduces the structure's vibrations, while the timber beam provides the resistance and improves the environmental impact and the aesthetic of the structures. This structural assembly is achieved by the union of a timber joist to a reinforced concrete slab through connectors, whereby the concrete slab and timber joist mainly resist compression and tension forces, respectively, and the connection provides a mechanism for the transfer of shear. The bond mechanical behaviour has a direct influence on the most important mechanical properties of the composite system, namely load-carrying capacity, stiffness and ultimate deformation capacity. Among the bond properties, the ones with the greatest influence on the mechanical behaviour of the composite system are the load-carrying capacity, stiffness and ductility. Experiments carried out by numerous researches cover various local and global aspects of this behaviour. Local aspects include the joints and connections which are effected by the connection type and the timber interlayers strength, ductility and stiffness of TCC beams [5], application of different types of concrete namely light-weight (LW) and normal-weight (NW) concrete [6] while global aspects are such as the various timber types in TCC beams which have been studied. [7]

IOP Conf. Series: Materials Science and Engineering **652** (2019) 012034 doi:10.1088/1757-899X/652/1/012034

2. Concept of a new composite connector

Over the past decades, there has been development of various new connectors and the analytical models for TCC beams which include discrete elements such as various types of dowels [8], coach screws, SFS-screws [9] [10], nails [11], connections with rebar [12] stud [3], sleeved connectors to mention but a few

The new connector (fig.1) adopted is of steel that is used in notched timber but also has screws which are drilled into the timber. Therefore, it takes the advantages of screwed and notched timber connections. The screws are parallel to the grain of timber and also to the direction of the shear forces. It has a stud of 70 mm that gets embedded into the concrete after to achieve maximum liaison between the steel and concrete

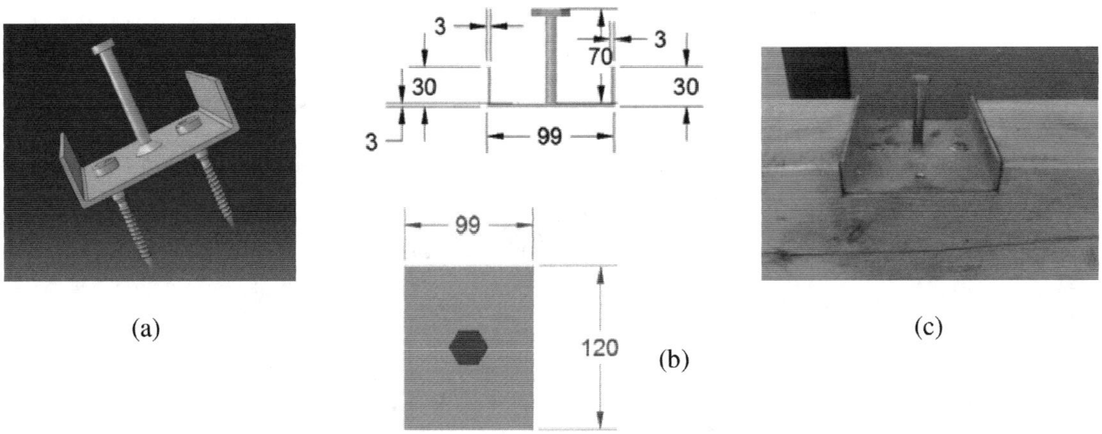

Figure 1. (a) Conception of new connector, (b) Dimensions and corresponding (c) image of the new connector

3. Experimental programs

Push-out or shear tests always precede beam collapse tests in order to obtain important information of the mechanical properties of the connection. Strength, stiffness and post-peak behaviour of a connector are usually investigated by conducting symmetrical or asymmetrical shear tests. These are referred to push out tests, since they involve the "pushing out" of one of the components constituting the test specimens. Fig 2 shows the principal setup of a symmetrical push-out test, where the horizontal lines symbolize the shear connections.

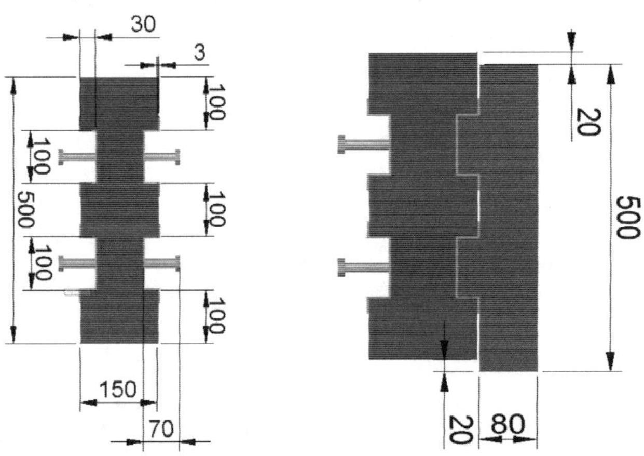

Figure 2. Side, and top views of the setup configuration of the push-out test of new connector, Specimen NC2

The aim of the experiment was to determine the sliding modulus (Kser, Ku) and the maximum force with shear failure (Fmax) in order to be able to correctly size the concrete wood floor with the new connectors. The slip modulus of a connection system is got from calculating two ratios: the instantaneous slip modulus (Kser) and the instantaneous slip modulus for ultimate limit states (Ku), corresponds to the initial levels of loading obtained by the slope of the secant line through the beginning of the load–slip curve and the point corresponding to 40% of the rupture load. According to EUROCODE 5 [13], instantaneous slip modulus of a connection for ultimate limit states (Ku) should be calculated according to: Ku=2/3 Kser. The connectors were then sorted in relation to their stiffness or slip modulus: nails, screws, and dowels are the most flexible, whereas notches cut in the timber and continuous connectors glued to the timber are the most rigid.

Following the tests on the specimens of the new connector, the following results were obtained (Table 1).

Table 1. Mechanical properties of Specimens NC1 and NC2 with new connection

Mechanical property	Fmax(kN)	Ki(kN/mm), Initial slip modulus	Kser(kN/mm), Slip modulus at SLS	ku(kN/mm), Slip modulus at ULS	Smax(mm), Maximum slip	v0.1(mm), 10% secant shear modulus	vi(mm), Initial slip	vi mod(mm), Modified Initial slip	v24 (mm), slip for a load of 0.4 Fest in the load branch until breakage.	v0.6, Slip at 0.6 Fmax	v0.8(mm), Slip at 0.8 Fmax
Specimen NC1	109.99	33.84	36.66	24.44	2.52	0.4	1.3	1.20	1.46	1.60	1.34
Specimen NC2	240.70	57.65	82.06	54.70	4.82	0.79	1.67	1.17	1.70	2.17	1.64

The first pushout test of the new variant connector failed due to the splitting of the horizontal fibres of the timber element with some concrete crack failure on the adjacent side with the timber. The second variant failure generally because of cracking of the concrete around the internal areas of the connection. The crack propagation preceded downwards and outwards the inner assembly to the outsides producing more fissures in the concrete and the general assembly. The contact between the timber member and the concrete blocks was also reduced due to the sliding motion during charging hence reducing the friction between the composites therefore larger displacements. After removing with concrete to observe the state of the connector, little or no deformation of the connector and its screws were being observed.

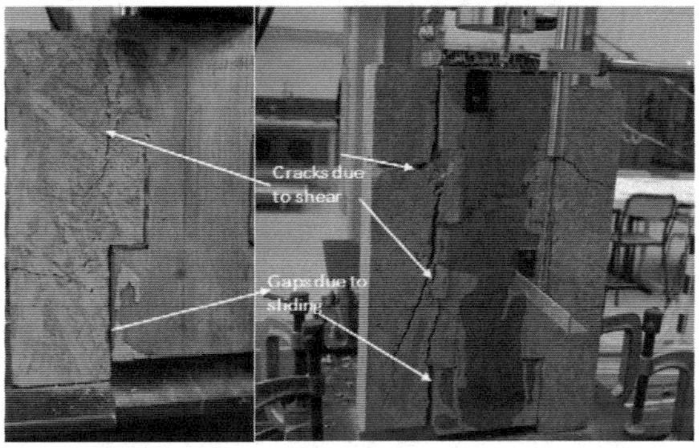

Figure 3. Failure of new connector pushout of variants 1 and 2 showing cracks and development of gaps between the assemblies

4. Finite element simulation

The objective of the FEM simulation was to predict the mechanical behaviour of the timber-concrete joints under the two different types of connection. According to the standard practice, the load carrying capacity of a single fastener, per shear plane, loaded laterally, can be obtained according to the timber density, the fastener diameter and the geometry of the connection. This is method is the standard practice and is based on Eurocode 5 part 1 [13].

4.1. Materials modelling

The constitutive law of steel was assumed to be isotropic as well as the material yield criterion. The hardening rule was also assumed to be isotropic. The steel behaviour was simulated using a tri-linear behaviour corresponding to, respectively: initial elastic phase, hardening phase and yielding as presented. The connectors and the screws were assumed as isotropic elasto-plastic material with the following properties: E = 200,000 MPa, ν=0.35 and σy=450MPa. The concrete material was modelled using the concrete damage plasticity model (CDPM) supported by the FEM package ABAQUS [14]. The model assumes that the uniaxial tensile and compressive response of concrete is characterized by damaged plasticity. It requires Concrete compression hardening, concrete tension stiffening, and isotropic elasticity.

4.2. Boundary conditions and meshing

Two types of external boundary conditions were considered: those due to the support and those due to symmetry considerations. In the model, the supports were simulated considering prescribed values of zero displacements for the three translations in all the nodes that correspond to support areas in the shear tests (the nodes in the timber elements base). Using this approach, a rigid connection was assumed between the specimen and the supports and at the same time no horizontal displacement was allowed in the bottom of the timber block. The loads were introduced by increasing controlled displacements applied on all the nodes in the top of the concrete element, as done in the laboratory tests. In ABAQUS, the material modelling of timber is complicated possibly because of natural imperfections like distortions in the grains or knots that can influence its structural behavior. In terms of elastic behavior, modelling the timber was first considered as isotropic material (E= 11000MPa and v =0.3), but the results were not satisfactory and it was therefore modelled as an orthotropic material. The relations between the elasticity modulus in the various directions and between elasticity moduli and moduli of distortion were kept constant as given in EN 338 [15] for softwoods as given in Equation 1 and Equation 2 This was assumed as a first step, to be checked later with parametric simulations.

$$E0 = Exx = 30Ezz = 30Eyy \qquad (1)$$

$$Gxy = Gzx = \frac{Exx+Eyy}{2} \times \frac{1}{16}; Gyz = \frac{Ezz+Eyy}{2} \times \frac{1}{16} \qquad (2)$$

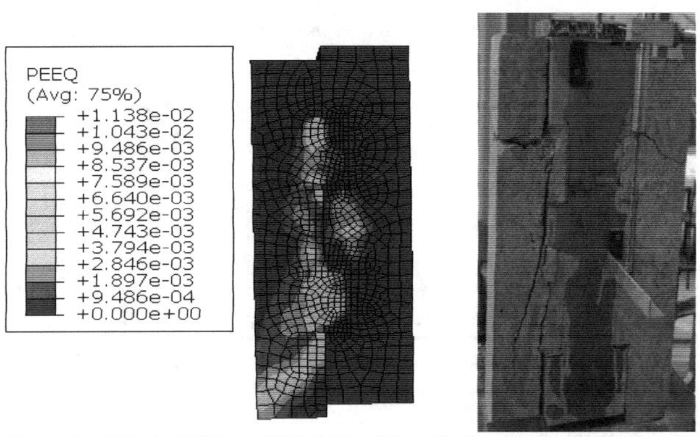

Figure 4. Pushout Failure in ABAQUS at 208 kN of load showing PEEQ values, displacement and stress, the experimental test collapsed at 240kN

The figure 4 shows the comparison between laboratory tests and FEM numerical simulations. The global load slip manner obtained in both tests shows similarity qualitatively. However, there was overestimation of the maximum load and the initial stiffness in ABAQUS. This could probably be explained by the perfect linear elastic behavior assumed for timber until the maximum strength is reached. The overestimation of the ultimate load carrying capacity of the joint was probably caused by the assumption of a yielding strength equal to the embedding strength of timber.

5. Conclusion

The new connector takes advantages of resisting more shear forces compared to the Tecnaria connector namely of the notched connection system, a head stud and screws. Further improvement of its design can be made for easy installation on structures especially when casting concrete. In the pushout tests, the notched connection was rigid enough that failure was due to crack propagation in the concrete members.

The system transferred the forces laterally downwards forcing cracking and high stress forces in the areas below the notches attached to the connector. The model used in ABAQUS shows that it has the capacity to describe the behavior of timber–concrete composite joints having screws and notches connections.

References

[1] A. Buchanan, B. Deam, M. Fragiacomo, S. Pampanin and A. Palermo, *Multi-Storey, Prestressed Timber Buildings in New Zealand*," vol. **18**, no. 2, pp. 166-173, 28 March 2008.

[2] D. Yeoh, M. Fragiacomo, M. D. Franceschi and a. K. H. Boon, "State of the Art on Timber-Concrete Composite Structures: Literature Review," *Journal of Structural Engineering,* vol. 137, 1 October 2011.

[3] S. C. Auclair, L. Sorelli and A. Salenikovich, A new composite connector for timber-concrete composite structures, 2016 *Construction and Building Materials*, vol. **112,**

[4] J. N. Rodrigues, A. M. P. G. Dias and P. Providencia, Timber-Concrete Composite Bridges: State-of-the-Art Review, *Bioresources,* vol. **8**, no. 4, .

[5] L. Deam, M. Fragiacomo and A. H. Buchanan, Connections for composite con-crete slab and LVL flooring systems 2008 *Materials and Structures*, vol. **41**, no. 3.

[6] L.F.Jorge, J.Schänzlin, S.M.R.Lopes, H.Cruz and U.Kuhlmann,Time-dependent be-haviour of timber lightweight concrete composite floors 2010 *Engineering Structures*, vol. **32**,

[7] D. Yeoh, M. Fragiacomo and B. Deamc, Experimental behaviour of LVL–concrete composite floor beams at strength limit, 2011 *Engineering Structures*, vol. **33**, no. 9,

[8] M.R.LeBorgne and R.M.Gutkowski, Effects of various admixtures and shear keys in wood–concrete composite beams, 2010 *Construction and Building Materials*, vol. **24**, no. 9,

[9] M. Fragiacomo, Experimental behaviour of a full-scale timber-concrete composite floor with mechanical connectors, *Materials and Structures,* vol. **45**, no. 11,

[10] L.F.Jorge, J.Schänzlin, S.M.R.Lopes, H.Cruz and U.Kuhlmann, Time-dependent behaviour of timber lightweight concrete composite floors, 2010 Engineering Structures, vol. **32**, no. 12.

[11] D.D Djoubissie, A. Messana, E. Fournely and A. Bouchaïr, Experimental study of the mechanical behavior of timber-concrete shear connections with threaded reinforcing bars, 2018 *Engineering Structures*, vol. **172**, no. 1.

[12] E.P.Carvalho and E. V. M. Carrasco, Influence of test specimen on experimental characterization of timber–concrete composite joints, 2010 *Construction and Building Materials*, vol. **24**, no. 8.

[13] EN. 1995-1-1, Eurocode 5: design of timber structures – part 1.1: general rules and rules for buildings. 2004 European committee for standardization, Brussels.

[14] SIMULIA,"Abaqus Documentation," Dassault Systèmes, 2016. [Online]. Available: http://130.149.89.49:2080/v2016/index.html.

[15] C. EN 338, Structural timber-strength classes. 1995

Effect of fiber reinforcement on tensile strength and flexibility of corn starch-based bioplastic

Jethoo A.S.
Department of Civil Engineering, Malaviya National Institute of Technology, Jaipur, India
Email: asjethoo.ce@mnit.ac.in

Abstract. Plastics are a vital asset for humanity, often providing functionality that cannot be easily or economically replaced by other materials. Bioplastics, made from biological materials represent an effective way of keeping the huge advantages of conventional plastics but mitigating their disadvantages. They can be prepared from plant, animal or bacterial sources. However, the majority of today's available bioplastics is produced from plants rich in carbohydrate such as grains or sugarcane. They not only fulfill the current demands of the modern population but also provide sustainable development; as they are eco-friendly. This paper describes the effect of fiber reinforcement on mechanical properties of starch-based bioplastics. The results show that the strength of bioplastics can be increased by using various natural fibers such as cotton, jute, hair and wool.

1. Introduction

During the course of development, plastics have become an important part of our life. From waking up in the morning to going to sleep in the night, we use plastics in some or the other form. Each year over 140 million tons of petroleum-based polymers are produced worldwide and used in the production of such plastics as polyethylene, polystyrene, polyvinyl chloride, polyurethane, and others [1]. These plastics are highly resistant to micro-organism and other degradation forces because of their chemical and physical properties. This leads to environmental and land shortage problems for solid waste management [2].

The growing global trend in material research towards sustainability and materials from renewable sources has increased research efforts for development of alternative plastic materials and composites. Biopolymers and fiber-based bio-composites are one focus area of these new plastic materials. According to European Bioplastics, a plastic material is defined as a bioplastic if it is either bio-based, biodegradable, or features both properties. The feedstocks for bioplastics come from a variety of biological sources. Starch based bioplastics are usually made from wheat, corn, rice, potatoes, barley, and sorghum. Cellulose based bioplastics are commonly made from wood. Natural oils from soy, palm, and other plants have also been used to make bioplastics. Other bioplastics rely on bacteria to supply raw materials [3].

Bioplastics are bio-based and do not have any hazardous effect on environment on disposal. They release 2/3 times less green house gases than conventional plastics & thus, cause less air pollution and less global warming [4]. During decomposition also, they don't release any toxic chemicals & by-products. In fact, their decomposition releases CO2, water and biomass which can be absorbed back into the earth. Moreover, it is much easier to recycle bioplastics.

Industry for bioplastics is in its infancy but is developing rapidly. According to European Bioplastics, the worldwide production capacity for bioplastics will increase from around 1.2 million tonnes in 2011

IOP Conf. Series: Materials Science and Engineering **652** (2019) 012035 doi:10.1088/1757-899X/652/1/012035

to approximately 5.8 million tonnes by 2016. The high increase in production capacity is also an indication of the growing demand for these sustainable materials [5].

Currently, bioplastics are used for applications such as films, bottle, food service items, bags, packaging and mulching. However, this trend is expected to change in the coming years, with innovations and upgradation in quality being targeted to more demanding end-use segments.

New bioplastics are poised to meet the expanding market demand with usage in products such as mobile phones, cameras, medical devices, electronics, as well as automotive parts. With improved crop yield and efficient farming, the demand of crops for bioplastics manufacturing can be achieved without any extra demand for agriculture land and food supply. Although, bioplastics are prepared from a variety of sources such as plant, animal or bacterial sources but mainly food crops are used. Starch based bioplastics make up the majority of the bioplastics market.

The bioplastic prepared from starch is known as thermoplastic starch (TPS). The transformation of granular starch into TPS is influenced by the processing conditions such as temperature and plasticizer content. Glycerol and water are basically used as plasticizers in TPS materials [6]. The amount of glycerol influences the flexibility and tensile strength of bioplastics prepared from corn starch.

In fiber reinforced thermoplastics, the fibers serve as reinforcement by giving strength and stiffness to the structure while the plastic matrix serves as the adhesive to hold the fibers in place so that suitable structural composites can be made [7]. This study was carried out with the objective to find out the effect of fiber reinforcement on tensile strength of TPS.

2. Experiment

The experiment was carried out by taking four types of bio-composites using natural fibers, namely, cotton (BC), jute (JC), hair (HC), and wool (WC) as a reinforcement, and thermoplastic starch as a biodegradable polymer matrix was prepared. The Figure 1 to 4 showing respectively.

3. Materials

Materials used for the experiment was a biopolymer (corn starch), glycerol, de-ionized water, 5% acetic acid solution, natural fibers like cotton, jute, hair and wool.

4. Methodology

Solution prepared by 5g of commercial starch was mixed with 30ml of water, 2ml of glycerol and 2ml of acetic acid. The solutions were added with varying amount of different types of natural fibers (2%, 4%, 6%, 8% and 10%). The solution was then heated on a hot plate at 150 0 C for 15 minutes with continuous stirring using glass rod. When the solution became viscous and its transparency was lost, heating was stopped. After the solution turned clear again, it was removed from the hot plate and was spread on aluminum sheet for drying at room temperature. Each sample was weighed every day and when their weight became constant they were tested for tensile strength.

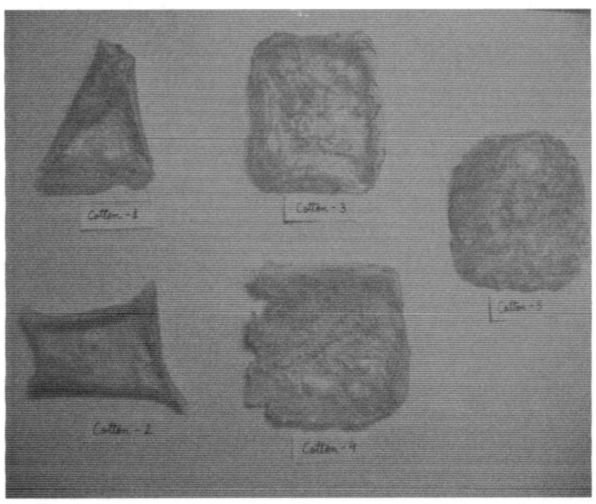

Figure 1.Cotton fiber based bioplastics

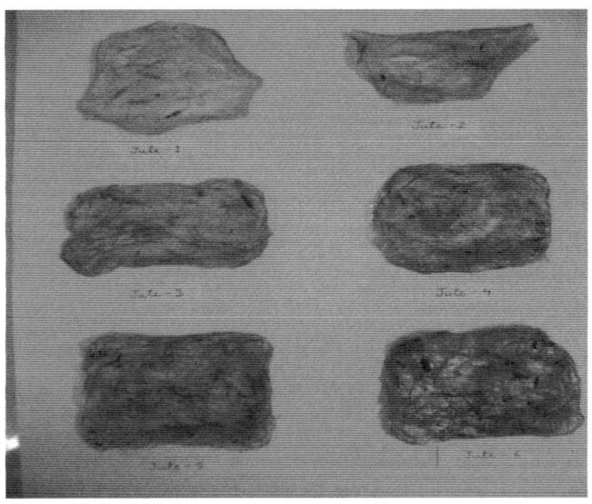

Figure 2.Jute fiber based bioplastic

Figure 3. Hair fiber based bioplastics

Figure 4. Wool fiber based bioplastics

The average thickness of each bioplastic sample, reinforced with different types of natural fibers was calculated. Then, each sample was cut in size of (6cm x 1.5cm) and loaded on tensiometer with fixed specifications of load range, extension range & speed. The tensile strength of each sample was determined and corresponding graphs were obtained.

5. Results and conclusions

The total 20 bioplastic specimen samples, reinforced with natural fibers like wool, cotton, hair & jute, were prepared and tested for their mechanical strength. The thickness of different samples is presented and average thickness is also mentioned in Table 1.

Normally, starch based bioplastics are slightly hard & brittle in nature, they break easily when folded. The brittleness is removed by using adequate amount of plasticizer as excess amount of plasticizer decreases the tensile strength of bioplastics.

The tensile strength of bioplastics depends on the amount of fiber reinforced in it and the strength increases on increasing the amount of fibers up to a certain limit but beyond that it decreases which may be due to decrease in the binding material i.e. biopolymer. The tensile strength of various bioplastics with different type of natural fibers and varying percentage of fibers is presented in Table 2.

Table 1: Thickness of different bioplastic samples

S. No.	Samples	Thickness (mm)			Average thickness(mm)
		(i)	(ii)	(iii)	
1	BC1	1.041	0.965	1.066	1.024
2	BC2	1.092	1.066	1.168	1.108
3	BC3	1.219	0.584	0.723	0.842
4	BC4	1.066	0.863	0.698	0.531
5	BC5	0.711	0.711	0.825	0.749
6	BH1	0.800	0.914	0.736	0.816
7	BH2	0.660	0.685	0.685	0.676
8	BH3	0.647	0.660	0.660	0.655
9	BH4	0.774	0.812	0.749	0.778
10	BH5	0.685	0.660	0.698	0.681
11	BJ1	0.787	0.812	0.749	0.782
12	BJ2	0.762	0.965	0.762	0.829
13	BJ3	0.965	1.016	0.787	0.922

14	BJ4	1.168	1.016	1.143	1.109
15	BJ5	0.889	1.054	1.066	1.000
16	BW1	0.914	1.066	1.092	1.024
17	BW2	0.924	0.863	0.914	0.900
18	BW3	0.965	1.803	0.762	1.170
19	BW4	0.889	0.889	0.914	0.897
20	BW5	0.838	0.863	0.787	0.829

Table 2: Tensile Strength of various Bioplastics

S.No.	%Of Natural Fibers	Bioplastics reinforced with different types of natural fibers							
		Cotton Bioplastics		Wool Bioplastics		Jute Bioplastics		Hair Bioplastics	
		Sample	Tensile Strength (Kg/cm^2)	Sample	Tensile Strength (Kg/cm^2)	Sample	Tensile Strength (Kg/cm^2)	Sample	Tensile Strength (Kg/cm^2)
1.	2	BC 1	21.576	BW 1	24.401	BJ 1	13.694	BH 1	14.581
2.	4	BC 2	15.336	BW 2	22.096	BJ 2	11.685	BH 2	28.408
3.	6	BC 3	21.597	BW 3	23.677	BJ 3	16.039	BH 3	23.351
4.	8	BC 4	36.168	BW 4	17.620	BJ 4	22.984	BH 4	28.398
5.	10	BC 5	33.354	BW 5	41.002	BJ 5	15.805	BH 5	27.450
Average Tensile Strength			**25.6062**		**25.7592**		**16.0414**		**24.4376**

Strength of fiber reinforced bioplastics also depends on the distribution of fibers throughout the sample. Thus, proper mixing and distribution of natural fibers is required during sample preparation of bioplastics. Apart from this, the property of tensile strength is also governed by the properties of fiber like, mechanical strength, thickness, etc. being reinforced in the bioplastics.

The Comparison of tensile strength of bioplastics having different types of natural fibers graphically shown in Figure 5.

Jute fibers have the highest mechanical strength but samples made from jute fibers showed lower average tensile strength. This may be because jute fibers are much thicker than other natural fibers resulting in poor mixing and distribution of jute fibers within the bioplastic samples.

Although cotton, wool and hair have low mechanical strength but their samples showed higher average tensile strength than jute. This may be because cotton, wool & hair are fine, thereby allowing proper mixing and uniform distribution of natural fibers within the samples.

Bioplastics reinforced with wool showed the highest tensile strength among all the samples because of the same reason.

The tensile strength of bioplastics reinforced with wool fiber showed the maximum average tensile strength of 25.7592 Kg/cm^2 as compared to bioplastics reinforced with cotton, hair and jute fibers. This may be due to proper mixing and uniform distribution of natural fibers within the bioplastic samples.

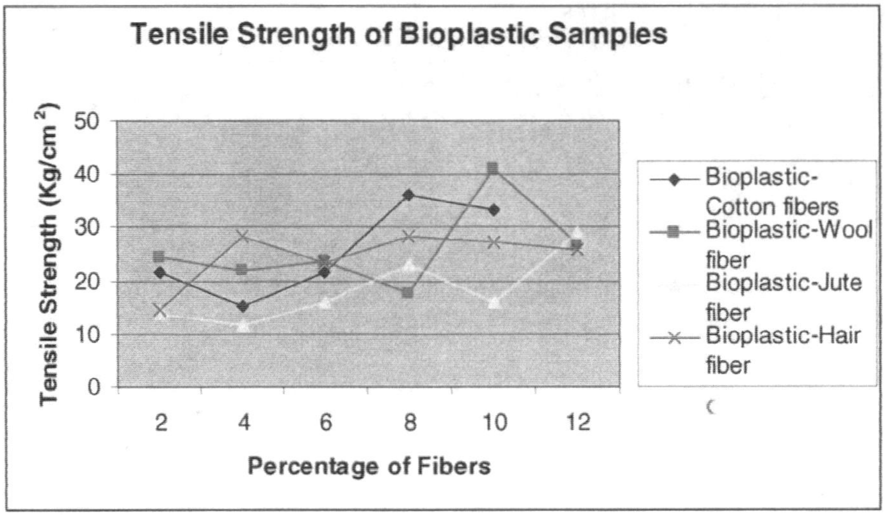

Figure 5: Comparison of tensile strength of bioplastics having different types of natural fibers

This work has shown that bio-composites on the base of thermoplastic starch reinforced with natural fibers, both of which are renewable materials, can be a potential alternative for conventional plastics, which adversely affect the environment and also require ample amount of precious fossil fuel for its manufacturing. The major limitation of present biopolymers is that their production is often more energy intensive than petro-plastic production and cost more than petroleum-based plastics but renewable resource based bioplastics are currently being developed and need to be researched more to overcome the performance limitations.

References

[1] Barry E. DiGregorio, 'Biobased Performance Bioplastic: Mirel', Journal of Chemistry & Biology, Vol. 16, Issue 1, 2009, pp 1-2.
[2] Shah A.A., Hasan F, Hameed A, Ahmed S,'Biological degradation of plastics: a comprehensive review', Biotechnol Adv., Vol. 26, Issue 3, 2008, pp 246–265.
[3] Momani B, 'Assessment of the Impacts of Bioplastics: Energy Usage, Fossil Fuel Usage, Pollution, Health Effects, Effects on the Food Supply, and Economic Effects Compared to Petroleum Based Plastics', Worcester Polytechnic Institute, Worcester, Massachusetts, United States, 2009, pp1-58.
[4] Reddy R.L, Reddy V.S, Gupta G.A, 'Study of Bioplastics As Green & Sustainable Alternative to Plastics', International Journal of Emerging Technology and Advanced Engineering, 2013, Vol 3, pp. 82-89.
[5] 'Bioplastics industry overview guide', The Society of the Plastics Industry, 2012
[6] Jeroen J.G. van Soest, Remko C. Bezemer, Dick de Wit, Johannes F.G. Vliegenthart: 'Influence of glycerol on the melting of potato starch', Industrial Crops and Products, Vol. 5, 1996, pp. 1-9.
[7] S. Kuciel, P. Kuźniar, A. Liber-Kneć: 'Polymer biocomposites with renewable sources', Archives of Foundry Engineering, Vol. 10, 2010, pp. 53-56.

IOP Conf. Series: Materials Science and Engineering **652** (2019) 012036 doi:10.1088/1757-899X/652/1/012036

Properties of Kenaf Fibers/Epoxy Biocomposites: Flexural Strength and Impact Strength

Hendra Suherman[1, a], Edri Azwar[1], Duskiardi[1], Yovial[1], Edi Septe[1]

[1]Department of Mechanical Engineering, Universitas Bung Hatta, 25133 Padang, West Sumatera-Indonesia

[a] hendras@bunghatta.ac.id

Abstract. Kenaf is a non-woody plant, which has high potential fiber to be used as polymer matrix reinforcement. Different compositions of kenaf fiber as bio reinforcing material at different fabrication process parameters will produce different biocomposite properties. This study uses kenaf fibers in the composition of 10, 20, and 30 wt.% as reinforcing material. Kenaf fiber and epoxy resin were mixed based on predetermined compositions, stirred using a mechanical stirrer at 200 rpm for 10 minutes. Then the mixture is poured into a mold and placed on a hot press machine. Pressure of 50 kg/mm^2 for 30 minutes at 150^{0}C is used to produce kenaf fiber/epoxy biocomposite. The result showed that the increasing of kenaf fiber content in the matrix is succeeded to increase the flexural and impact strength of the biocomposite kenaf fiber/epoxy produced. The highest flexural strength (136 MPa) and highest impact strength (180 KJ /m^2) were obtained at a composition of 30 wt.% Kenaf fiber.

1. Introduction

Kenaf (Hibiscuss cannabinus L.) from the Malvaceae family, is a non-wood plant whose fiber has good potential to be used as a reinforcing or filler material in the polymer matrix. Kenaf grow fast about four to five months after sowing the seeds, and are able to reach height of more than 3 m with a base diameter of 3 to 5 cm [1]. Kenaf can grow well under various weather conditions. Different polymer composite manufacturing techniques can be applied based on the type of matrix used. The appropriate formation process can turn the material into a new material with the desired properties. The injection, extrusion, or compression molding, and hot pressing were still the main manufacturing processes until now [2]. In general, the processing method for biodegradable polymer materials such as polylactide (PLA), is similar to that developed for conventional composite materials. According to Barkoula et al. [3] , composites produced through injection molding show a lower impact strength compare to composite produce by compression molding.

Fabrication of natural fibers with PP is one of the most suitable routes for developing natural-synthetic polymer composites [4]. Several techniques are carried out to ensure proper matrix fusion and good fiber impregnation. Zampaloni et al. [5] in his research, focused on the use of kenaf fibers as reinforcing materials with polypropylene (PP) as a matrix to obtain biocomposite with a good fiber distribution. The first experiment used a 130 mm kenaf sandwich between two layers of PP sheets,

IOP Conf. Series: Materials Science and Engineering **652** (2019) 012036 doi:10.1088/1757-899X/652/1/012036

indicating poor distribution of kenaf fiber. To improve the uneven kenaf distribution, 20 mm long chopped kenaf were mixed with PP matrix using a mechanical stirrer. This method produces a circular pattern, causes clamping and voids, making it more difficult to achieve even distribution of fibers. Finally, the sieving method of powdered polypropylene microfin (to be the layered shifting of a powder) is used (20μm compared to the original PP size of 400μm) using cut-off kenaf fibers. This method has been proven to be able to produce better fiber distribution. Meanwhile, Lee et al. [6] used a hot-pressing technique to evaluate the effect of kenaf fiber length orientation and laminate properties. The study results indicate that mechanical properties increase with the increasing of fiber content at parallel orientation to the direction of load. The study reported that the highest tensile strength was obtained at 50% kenaf content using horizontal fiber orientation. Long and continuous fibers are easy to orient and process compared to short fibers. This is because short fibers are more difficult to control and to determine the right process orientation [7]. Hao et al. [8] produced PP kenaf material with a ratio of 50/50 using carding and the needle-punching technique. The results showed that the mobility of liquid polypropylene will increase at high temperatures. This greatly determines the fiber bonds formed in the matrix. The results also show that 6 mm is too thick for a sample, because 1 minute duration of processing time is not enough for heat transfer from surface to inside of the sample.

This study were carried out to find out the effect of length variations (1 cm and 3 cm) and composition based on weight percent (wt.%) of kenaf fiber to the flexural strength and impact strength of the biocomposites produced

2. Materials and method

2.1 Materials

The reinforcing fiber used is kenaf fiber at average fiber diameter of 30 μm, obtained from The National Kenaf and Tobacco Board, Malaysia. The binder (matrix) used is epoxy resin type 635 thin. The ratio of epoxy resin and hardener is 4: 1, according to the manufacturer. While the viscosity of the epoxy resin used is 6 Poise.

2.2 Fabrication processes of kenaf fiber/epoxy biocomposites

The fabrication process of kenaf /epoxy biocomposites were consists of two stages. In stage 1, epoxy resin and hardener with a ratio of 4:1 were stirred using a mechanical stirring machine (RW-IKA) with a rotation speed of 200 rpm for 10 minutes. In stage 2, kenaf fibers based on predetermined compositions was also stirred at the same rotation speed and time, which is 200 rpm for 10 minutes. The mixture of epoxy resin and kenaf fiber is then poured into the mold and placed on a hot pressing machine with a pressure of 50 kg/mm^2, at a temperature of 150 ^{0}C, for 30 minutes.

2.3 Characterization

The characterization performed on kenaf fiber/ epoxy biocomposite are flexural strength and impact strength tests. The flexural strength testing was carried out using the ASTM D 790 standard while the impact strength uses the ASTM E 23 Standard. Each composition of kenaf fiber/epoxy biocomposites were tested 3 times for each test. Scanning electron microscopic (SEM) Hitachi S-3400 N was used to see the kenaf fiber dispersion in the matrix (epoxy resin) used.

3. Results and discussion

3.1 Variation of kenaf fiber on flexural strength

The variation of kenaf fiber addition to the flexural strength of kenaf fiber/epoxy biocomposites are shown in Figure 1. Kenaf fiber addition as reinforcing material showed an increase in flexural strength for both fiber lengths used (1 cm and 3 cm). The highest flexural strength of kenaf fiber/epoxy biocomposite was obtained at 1 cm fiber length at the highest kenaf fiber composition (30 wt.%), ie 130 MPa. In the same composition, kenaf fibers which have a length of 3 cm produce the lower flexural strength (125 MPa). This is caused of longer fibers found not to be well dispersed in the epoxy matrix [6-10].

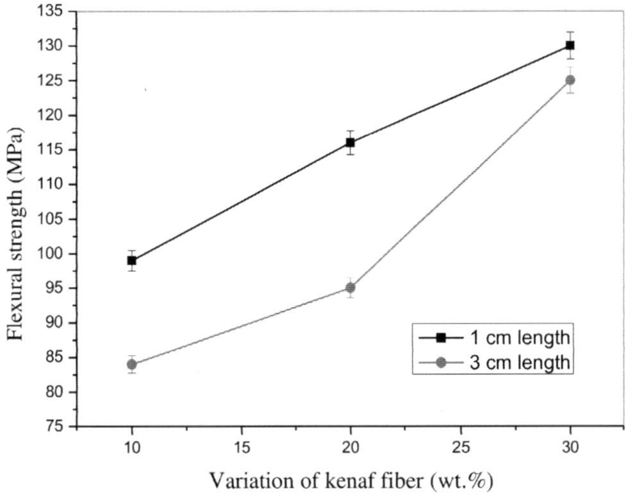

Figure 1. Variation of kenaf fiber on the flexural strength of kenaf fiber/epoxy biocomposites

3.2 Variation of kenaf fiber on impact strength

The effect of variations of kenaf fiber addition (10 wt.%, 20 wt.% and 30 wt.%) and variations of kenaf fiber length (1 cm and 3 cm) on the impact strength of kenaf fiber/epoxy biocomposites are shown in Fig. 2.

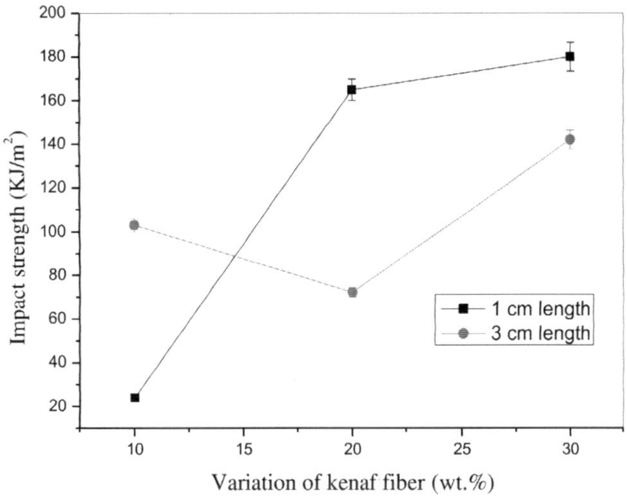

Figure 2. Variation of kenaf fiber on the impact strength of kenaf fiber/epoxy biocomposites

Overall, kenaf fiber addition from 10, 20, to 30 wt.% with variation of kenaf fiber length of 1 and 3cm were successfully increased the impact strength of kenaf fiber/epoxy biocomposite. The highest impact strength were obtained at a composition of 30 wt.% for both fiber lengths (1 cm and 3 cm), namely 180 KJ/m^2 and 140 KJ/m^2. Significant differences in the impact strength of biocomposite kenaf fiber/ epoxy obtained at the same composition (30 wt.%), caused of 1 cm long kenaf fibers were more evenly dispersed in the matrix (epoxy resin). This condition causes a perfect bond between kenaf fiber as reinforcement and the matrix [8, 9]. Kenaf fiber length of 3 cm look more agglomerate in the matrix, thus reducing the impact strength obtained [11-15].

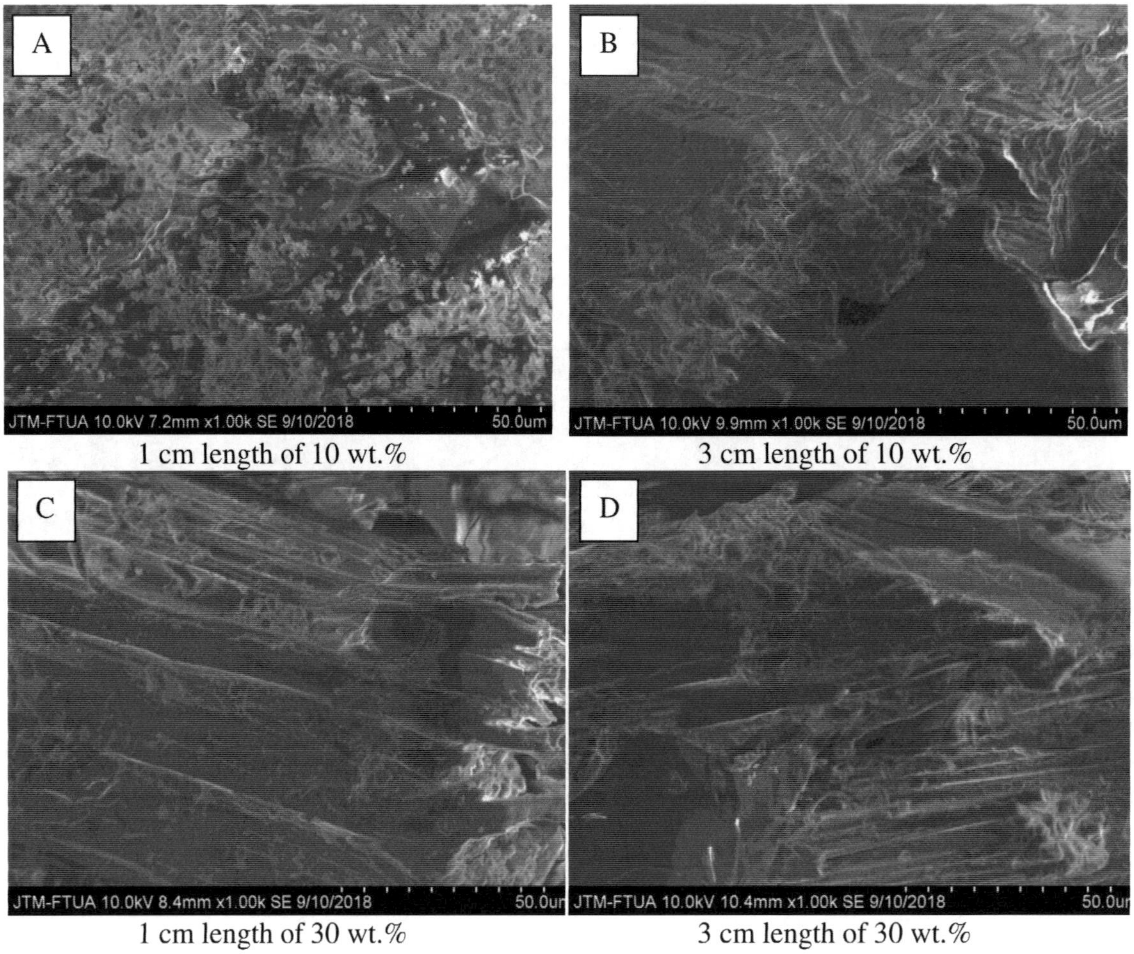

Figure 3. Scanning electron microscopic (SEM) images of facture surface of kenaf fiber/epoxy biocomposite

Figure 3 shows SEM image from the fracture surface of kenaf fiber /epoxy biocomposite produced. Fig. 3a and 3c shows that the fiber length of 1 cm in composition of 10 wt.% and 30 wt.% are well dispersed in the matrix (epoxy resin). This condition produces a good bond between epoxy resin as a binder and kenaf fiber as reinforcement so that increase the flexural and impact strength obtained [9] . In contrary, Fig. 3 (b) and 3 (d) shows that 3 cm long kenaf fiber in composition of 10 wt.% and 30 wt.% tends to be agglomerates and show some voids, resulting in lower flexural strength and impact strength [10].

4. Summary

Research on flexural and impact strength of kenaf fiber/epoxy biocomposite by varying the length and content of kenaf fiber as reinforcing material was completed. The results obtained can be summarized as follows:

1. The addition of kenaf fiber as reinforcement material (10 wt.%, 20 wt.% and 30 wt.%) with a fiber length of 1 cm and 3 cm managed to increase the flexural strength and impact strength of kenaf fiber/epoxy biocomposite.
2. The highest flexural strength and impact strength were obtained at fiber length of 1 cm and highest kenaf fiber content of 30 wt.% that is 130 MPa and 180 KJ /m^2.
3. Kenaf fiber length of 1 cm is dispersed properly for all compositions compared than 3 cm length.

5.Acknowledgment

The author wishes to thank the Ministry of Research and Technology through applied of excellent research of university (PTUPT) with grant no. 023-34/LPPM-Research/Hatta/II-2018, so that this research can be carried out well.

6. Reference

[1] H. M. Akil, M. F. Omar, A. A. Mazuki, M, Safiee, S. Ishak, and A. Bakar: Mater. Des Vol 32 (2011), p. 4107
[2] U.B. Emeka, A.B. Sulong, N. Muhamad, Z. Sajuri, and F. Salleh: Jurnal Kejuruteraan, Vol. 29(2018), p. 49
[3] N. M. Barkoula, S.K. Garkhail, and T. Peijs: Ind. Crops Prod. Vol. 31(2010), p. 34
[4] Q.T. Shubhra, A. Alam, and M. Quaiyyum: J. Thermoplas. Compos. Vol. 26(2011), p. 362
[5] M. Zampaloni, F. Pourboghrat, S.A. Yankovich, B.N. Rodgers, J. Moore, L.T. Drzal, and M. Misra: Compos. Part A Appl. Scie. Manuf. Vol. 38 (2007), p. 1569
[6] S. Lee, S. Shi, L. Groom, and Y. Xue: Polym. Composite Vol. 31 (2010), p. 15
[7] N.A. Mohd Radzuan, A.B. Sulong, and M. Rao Somalu: Sains Malays Vol. 45 (2016), p. 1913
[8] A. Hao, H. Zhao, W. Jiang, L. Yuan, and J.Y. Chen: J. Polym. Environ. Vol. 20 (2012), p. 959
[9] N. Nasia, M. Khairuddin, A, Mansur, M.A. N. ATIQAH: Adv. Mat. Res. Vol 1134 (2015), p. 1
[10] B.H. Lee, H.S. Kim, S. Lee, H.J. Kim, and J.R. Dorgan: Compos. Sci. Technol. Vol. 69 (2009), p. 2573
[11] R. Mahjoub, J.M. Yatim, A.R.M. Sam, M. Raftari: Mater. Des. Vol. 64 (2014), p. 640
[12] V. Fiore, G. Di Bella, and A. Valenza: Compo Part B Vol. 68 (2015), p 14
[13] K.L. Pickering, M.G. AruanEfendy, and T.M.Le: Compos Part A Appl. Scie. Manuf. Vol. 83 (2016), p. 98
[14] M. H. Zamri, H. Md. Akil, Z.A. M. Ishak, A. A. Bakar: Polym. composite Vol 36 (2015), p 1224
[15] M.R.Sanjay P, Madhu, M.Jawaid, P. Senthamaraikannan, S.Senthil, and S.Pradeep: J. Cleaner Prod. Vol. 172 (2018), p. 566

IOP Conf. Series: Materials Science and Engineering **652** (2019) 012037 doi:10.1088/1757-899X/652/1/012037

Research of Concrete Durability in Compressed Elements with Different Types of Confinement Reinforcements

Anatoly L. Krishan[1,a], Elvira P. Chernyshova[1,b] and Vladislav E. Chernyshov[2,c]

[1]Nosov Magnitogorsk State Technical University, 38, Lenin St., Magnitogorsk, 455000, Russia
[2]Saint-Petersburg Mining University, 21 line, 2, Vasilievsky Island, 199106, Saint-Petersburg, Russia
[a]kris_al@mail.ru, [b]ch-elvira@bk.ru, [c]ch_vlad99@mail.ru

Abstract. The paper presents experimental research of short center compressed elements with circular cross section in different types of concrete confinement reinforcements, namely by introducing glass fiber-reinforced shell, steel spiral and by using these two types together was conducted. The results of the conducted experiments show the positive influence of both outer glass fiber-reinforced shell and spiral reinforcements on the durability of such elements. The highest degree of durability was achieved when two types of confinement reinforcements were used simultaneously. The durability of the concrete core of the samples made from class C40 concrete increased by 3.25 times and equaled 135 MPa. In case of the initial class C80 concrete the effect of confinement reinforcements was considerably smaller but nevertheless it was high enough. The durability of the concrete core increased by 2 times. The introduction of two types of confinement reinforcements considerably improved the deformability of compressed elements. The maximum recorded values of contraction longitudinal strain of the samples made of high endurance concrete were about 1%, and in case of concrete class C40 they were 2.4%. Such high level of deformability in compressed elements will help to effectively use high durability longitudinal reinforcements with the compression yield strength up to 2000 MPa and higher in them.

1. Introduction

Over the last years compressed elements made of concrete in concrete filled glass fiber-reinforced tube (CFGFRT) have become very commonly used in construction [1-9]. The increased interest in them is generated due to the considerable economy of labour and non-human resources in actual use. In compressed CFGFRT elements the outer shell is simultaneously the stay-in-place formwork and the longitudinal and confinement reinforcements. Confinement reinforcements in sufficient amount can significantly increase the durability and deformability. Due to these qualities these constructions are highly effective.

Wider use of CFGFRT elements is limited due to the relatively high cost of glass fiber-reinforced tubes. Today the most feasible way to use them is in constructions that require increased corrosion durability qualities. For example, these can be pile foundations, various shore facilities at the sea shore, bridge pylons, etc.

In order to lower the cost of CFGFRT elements, the size of the outer glass fiber-reinforced shell (thickness of the wall and the diameter of the tube) must be reduced while keeping the necessary load-bearing capacity by using high-durability concrete and (or) additional reinforcement of the concrete core. The analysis of possible variants of the concrete core reinforcement shows that spiral reinforcements prove to be the most effective [10-12].

The purpose of this research is to compare the durability of compressed short CFGFRT elements made of concrete of different durability and various types of confinement reinforcements.

2. Methods and results of the experimental research

The experiments tested the durability of short centre compressed lab samples with circular cross section. The ratio between the length of the samples and their diameter was 4. 5 series of test elements were made and tested. Each series consisted of 3 sample twins. The tested samples had the following parameters.

The CT series samples were made of concrete with the outer shell that consisted of a glass fiber-reinforced tube with the diameter of 113 mm and the wall thickness of 6.1 mm.

CR series was represented by reinforced concrete elements with the diameter of the cross section of 100 mm. The test samples of this series had confinement reinforcements in the form of a spiral. Wire ∅5 Bp500 was used as the spiral reinforcements of the frame that was wound with a 30 mm spacing around four longitudinal rods ∅6 A500C. The diameter of the spiral was 90 mm. The compressional yield strength of the reinforcement wire was σ_{ys}=548 MPa, and that of the rod reinforcements was σ_{ys}=552 MPa.

The CRT series samples were made of concrete in a glass fiber-reinforced tube with the diameter of 113 mm and the wall thickness of 6.1 mm. This said, the concrete core had reinforcements similar to the samples of the CR series. The concrete with compression durability class C40 was used in all the three series.

The samples in the other two series was made of concrete class C80. The HCT series samples, except for the concrete, are the exact copy of the CT series constructions, and the HCRT series samples were similar to the construction of the relative CRT series.

A glass fiber-reinforced tube used as the outer shell had the following characteristics:

- elasticity model in case of axial compression E_{pc}=10200 MPa;
- elasticity model in case of circular stretching E_{pt}=29500 MPa;
- durability in case of axial stretching f_{lp}=265 MPa;
- durability in case of circular stretching f_{op}= 303 MPa;
- poisson ratio υ_{po}=0.21;
- specific weight γ_p= 18.5 kN/m^3.

In order to get more objective data when comparing durability of the samples of the tested constructions made of concrete mix of one batch, the samples of different series that had the relevant class of concrete, were made simultaneously.

After the manufacturing all the tested samples were kept at the temperature of about 20 ℃ during 28 days. The tests were conducted using a 500-ton hydraulic press by applying short-term compression pressure. It was applied to the whole cross-section of the constructions. The standard method specified in GOST 8829 was used.

3. Results of the experimental research

The main results of the experiments are represented in Table 1. Besides the durability of the initial concrete f_c and collapse load $N_u^{\exp}$, the durability of the volumetrically compressed concrete core f_{cc}, the coefficient of the concrete durability increase $\alpha_c = f_{cc}/f_c$ and threshold contraction axial deformation of the tested samples $\varepsilon_u^{\exp}$ are represented here. The value of the volumetrically compressed concrete core durability f_{cc} was calculated using the following formula

$$f_{cc} = \frac{N_u^{\exp} - N_s}{A - A_s}, \tag{1}$$

where N_s and A_s are the threshold compressing force in the rods of the longitudinal reinforcements and the area of their cross-section (they are taken into account if there is a spiral frame); A is the area of the cross-section of the test sample.

Table 1. Main Test Results for Laboratory Specimens

Set, specimen	f_c, [MPa]	$N_u^{\exp}$, [kN]	f_{cc}, [MPa]	α_c	$\varepsilon_u^{\exp}$, [%]
CT-1	41.8	990	98.7	2.36	1.50
CT-2	40.9	1020	101.7	2.49	1.55
CT-3	42.6	1080	107.7	2.53	1.38
CR-1	41.8	590	68.2	1.63	0.93
CR-2	40.9	590	68.2	1.67	1.11
CR-3	42.6	620	72.0	1.69	0.82
CRT-1	41.8	1410	135.9	3.25	2.57
CRT-2	40.9	1380	132.8	3.25	2.22
CRT-3	42.6	1440	138.9	3.26	2.40
HCT-1	80.5	1283	121.1	1.50	0.67
HCT-2	82.4	1300	122.8	1.49	0.72
HCT-3	81.7	1300	122.8	1.50	0.77
HCRT-1	80.5	1653	160.4	1.99	1.18
HCRT-2	82.4	1693	164.4	2.00	0.82
HCRT-3	81.7	1733	168.4	2.06	0.90

The analysis of the results shows that the type of confinement reinforcements influenced the durability of the concrete core and the tested samples the most. Thus, with class C40 concrete if only spiral reinforcements were used, the concrete durability increased on average by 1.66 times, the glass fiber-reinforced shell increased it by 2.46 times and when both types of confinement reinforcements were used, it rose by 3.25 times. The CRT series samples proved to be stronger than CT series samples by about 35%. The same rise in durability was achieved by using spiral reinforcements of the samples made of high-durability concrete in a glass fiber-reinforced shell.

The comparison of the CRT and HCRT series samples as well as these of the CT and HCT series shows that when the durability of the initial concrete increased by two times, the durability of the samples rose slower, by 20% and 26% correspondingly. This is explained by the fact that high-durability concrete samples in their limit state with the same confinement reinforcements as the regular durability concrete have a lower value of relative side pressure σ_{cr}/f_c. According to the published research data [13], the value of relative side pressure influences directly the rise in the durability of volumetrically compressed concrete.

Maximum recorded values of longitudinal deformations $\varepsilon_u^{\exp}$ of the tested samples essentially depended on the intensity of the confinement reinforcements and the class of the initial concrete. Minimum values of those deformations were recorded in the samples made of high-durability concrete of HCT series. Their value on average was 0.72% that is three times higher than in similar deformations of monoaxially compressed concrete. Deformability of similar samples made of class C40 concrete was significantly higher. For instance, the values of deformations $\varepsilon_u^{\exp}$ in CT series samples on average reached 1.48%. The highest deformability was delivered by the CRT series samples in glass fiber-reinforced shell with spiral reinforcements of concrete. Their average value $\varepsilon_u^{\exp}$ was 2.4%. High level of deformability of compressed elements helps to effectively use high-durability longitudinal reinforcements in them.

The characteristics of the damage of tested samples were different. Reinforced concrete elements with spiral reinforcements showed almost complete destruction of the protective layer (Figure 1,a) before the load limit was reached. When the load was further increased, the concrete core was smashed inside the spiral.

The destruction of CT and HCT series samples was fragile. Initially, (right after the outer shell bulged) parts of the concrete core were moved. After that the shell was breached (Figure 1,b). Spiral reinforcements in CRT and HCRT series samples slightly changed the characteristics of the destruction by making it more flexible. However, visually it looked the same as in the CT and HCT series samples.

a) b)

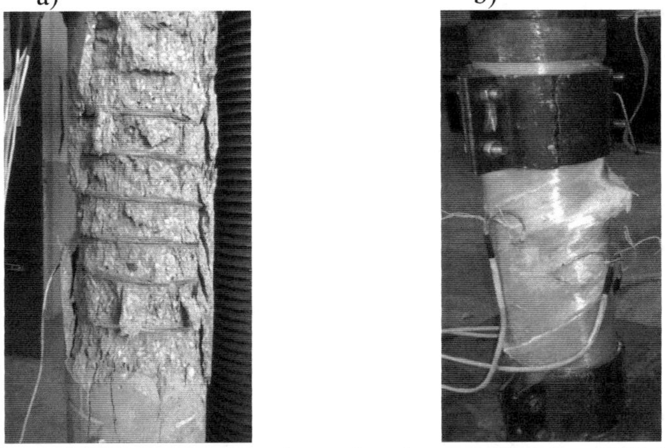

Figure 1. Fracture pattern of samples of series CR (a) and CT (b)

4. Conclusions

The results of the conducted experiments show the positive influence of outer glass fiber-reinforced shell and spiral reinforcements on the durability of short compressed CFGFRT elements. High load performance of such samples can be explained by the significant durability of the concrete that has two types of confinement reinforcements at the same time. Thanks to those reinforcements the durability of the concrete core of the samples made from class C40 concrete increased by 3.25 times and equaled 135 MPa. In case of the initial class C80 concrete the effect of confinement reinforcements was considerably smaller but nevertheless it was high enough. The durability of the concrete core increased by 2 times.

What is more, the introduction of two types of confinement reinforcements considerably improved the deformability of compressed elements. The maximum recorded values of contraction longitudinal strain of the samples made of high endurance concrete were about 1%, and in case of class C40 concrete they were 2.4%. Such high level of deformability in compressed elements will help to effectively use high durability longitudinal reinforcements with the compression yield strength up to 2000 MPa and higher in them. The relative tests are scheduled in the foreseeable future.

5. Acknowledgments

This article was prepared from the results of the implementation of a scientific project within the state task of the Ministry of Education and Science of the Russian Federation No. 7.3379.2017/4.6.

6. References

[1] J.F. Berthet, E. Ferrier and P. Hamelin: Compressive behavior of concrete externally confined by composite jackets Part A: Experimental study, Construction and Building Materials, Vol. 19, No. 3 (2005), pp. 223-232.

[2] A.Z. Fam and S.H. Rizkalla: Behavior of axially loaded concrete-filled circular fiber-reinforced polymer tubes, ACI Struct. J., 98 (3) (2001), pp. 280-289.

[3] R. Jayasooriya, D.P. Thambiratnam and N.J. Perera: Blast response and safety evaluation of a composite column for use as key element in structural systems, Engineering Structures, Vol. 61, No. 1 (2014), pp. 31-43.

[4] A.M. Fattah: Behaviour of concrete columns under various confinement effects: A dissertation doctor of philosophy, Kanzas, USA: Kanzas State University (2012), 399 p.

[5] R. Sudhakar and Dr. P. Partheeban: Strengthening of RCC Column Using Glass Fibre Reinforced Polymer (GFRP), International Journal of Applied Engineering Research, Vol. 12, No. 14 (2017), pp. 4478-4483.

[6] B. Drishya and A.K. Rajesh: Study of Glass Fibre Wrapped Concrete Column under Axial Compression, International Journal of Engineering Research & Technology (IJERT), Vol. 5 Is. 09 (2016), pp. 454-458.

[7] M.N.S. Hadi, Q.S. Khan and M.N. Sheikh: Axial and flexural behavior of unreinforced and FRP bar reinforced circular concrete filled FRP tube columns, Construction and Building Materials, 122 (2016), pp. 43-53.

[8] S. Karthiyaini and S. Nagan: Behaviour of geopolymer concrete circular column using glass fiber reinforced polymer, Indian Journal of Engineering and Materials Sciences (IJEMS), Vol. 21 (2014), pp. 458-464.

[9] O.I. Abdelkarim and M.A. ElGawady: Concrete-Filled-Large Deformable FRP Tubular Columns under Axial Compressive Loading, Fiber, Vol. 3 (2015), pp. 432-449.

[10] A.L. Krishan, M.A. Astafyeva and R.R. Sabirov: Engineering Analysis and Design of Concrete Filled Steel Tube Columns: monograph, Saarbrucken, Deutschland: Palmarium Academic Publishing (2016), 261 p.

[11] A.L. Krishan, E.A. Troshkina and M.A. Astafeva: Strength of Short Concrete Filled Steel Tube Columns with Spiral Reinforcement, IOP Conference Series: Materials Science and Engineering, Vol. 262 (2017), No. 012048. http://dx.doi.org/10.1088/1757-899X/262/1/012048.

[12] M.R. Hamidian, M.Z. Jumaat, U.J. Alengaram, N.H. Ramil Sulong and P. Shafig: Pitch spasing effect on the axial compressive behavior of spirally reinforced concrete-filled steel tube (SRCFT), Journal Thin-Walled Structures, Vol. 100 (2016), pp. 212-223.

[13] J.B. Mander, M.J.N. Priestley and R. Park: Theoretical stress-strain model for confined concrete, Journal of Structural Engineering, ASCE, 114(8) (1988), pp. 1804-1826.

IOP Conf. Series: Materials Science and Engineering **652** (2019) 012038 doi:10.1088/1757-899X/652/1/012038

Compressed Concrete Filled Steel Tube Elements with Confinement Reinforcement of Concrete Core

Elvira P. Chernyshova[1,a], Vladislav E. Chernyshov[2,b] and Rustam R. Sabirov[1,c]

[1]Nosov Magnitogorsk State Technical University, 38, Lenin St., Magnitogorsk, 455000, Russia
[2]Saint-Petersburg Mining University, 21 line, 2, Vasilievsky Island, 199106, Saint-Petersburg, Russia
[a]ch-elvira@bk.ru, [b]ch_vlad99@mail.ru, [c]sabirov_rustam66@mail.ru

Abstract. The purpose of this paper is to study the main features of power resistance of compressed concrete filled steel tube elements (CFSTE) with spiral reinforcement. The spiral reinforcement of a concrete core makes it possible to increase the load-bearing capacity of such structures significantly. Based on the results of performed experiments, the coefficient of indirect reinforcement efficiency among CFSTE with spiral reinforcement made 1.48 to 1.62, depending on initial concrete class. This coefficient was in the range from 1.23 to 1.33 in similar samples without spiral reinforcement. The experiments showed that a high efficiency of these structures is maintained when high-strength concrete is used in them.

1. Introduction

In recent years, concrete filled steel tube elements have been widely used in the construction of buildings and structures. Most often they are used as vertical bearing structures of high-rise buildings [1-3], multi-storey buildings with an enlarged grid of columns [4, 5] and bridge structures [6-8]. This is facilitated by their numerous positive qualities [2, 9, 10], among which we can especially note the reduction of material consumption and the cost of the constructed facilities during the provision of the necessary bearing capacity, safety and survivability.

Due to the fact that CFSTE is mainly used as heavily loaded columns or supports, it is desirable to look for new ways to increase their bearing capacity. The use of high-strength concrete is promoted to the solution of this problem. This path should be considered a promising one, but it has its limitations.

Another option to increase the load-bearing capacity of compressed CFSTE is the development of a more rational design solution. To do this, its concrete core is proposed to be supplied with a reinforcing cage. The main working armature of a frame is spiral one, wound along longitudinal rods. Longitudinal reinforcement is also working. It can have both normal and high strength. The experiments recently carried out in Malaysia and China [11, 12] has proved a high efficiency of such designs.

The purpose of this work is to study the features of compressed CFSTE power resistance with a spiral reinforcement.

2. Experimental study technique

In order to achieve this goal, the experimental studies of centrally compressed CFSTE strength of circular cross-section with a spiral reinforcement have been performed. In order to demonstrate their effectiveness, reinforced concrete samples with spiral reinforcement and CFSTE without a concrete core reinforcement were studied simultaneously. 8 series of laboratory samples with different parameters were manufactured and tested in total. The basis of each series made 3 twin samples. The basic geometric and structural parameters of the samples are presented in Table 1. The length of all the samples made 1000 mm. They were made of two classes of concrete – C40 and C80. R2.40 and R2.80 series, as well as TR1.40, TR1.80 and TR2.40 had a confinement reinforcement in the form of a spiral. The parameters of the used frames with a spiral armature are given in Table 2.

The tests of CFSTE samples were carried out at the age of 28 days on a 500-ton hydraulic press 2PG-500 with a short-time compressive load. The transfer of the load to the samples was carried out over the entire cross section.

Table 1. Geometric and Structural Parameters of Laboratory Samples

Series	Section diameter, [mm]	Steel case		Spiral presence	Concrete prism strength, [MPa]
		Section dimensions, [mm]	Yield strength, [MPa]		
R2.40	209	-	-	+	41.1
R2.80	209	-	-	+	82.9
T1.40	112	112×1.2	300	-	43.8
T1.80	112	112×1.2	300	-	83.4
T2.40	219	219×5.0	285	-	42.0
TR1.40	112	112×1.2	300	+	43.8
TR1.80	112	112×1.2	300	+	83.4
TR2.40	219	219×5.0	285	+	41.7

Table 2. Parameters of Spiral and Longitudinal Frame Reinforcement

Series	Spiral reinforcement				Longitudinal reinforcement		
	Spiral diameter, [mm]	Diameter of rods d_s, [mm]	Coil step s, [mm]	Armature class	Rod diameter d_a, [mm]	Number of rods	Armature class
TR1.40	90	3	20	B500	3	4	B500
TR1.80	90	3	20	B500	3	4	B500
R2.40	185	5	30	B500	6	4	A500
R2.80	185	5	30	B500	6	4	A500
TR2.40	185	5	30	B500	6	4	A500

3. Experimental study results

The main test results, containing the data on the carrying capacity and deformability of the studied laboratory specimens averaged within each series, are presented in Table 3. The experimental values of the bearing capacity $N_u^{\exp}$ and relative longitudinal strains of the samples when the maximum load $\varepsilon_{cc1}^{\exp}$ are compared with their theoretical values – N_u^{th} and ε_{cc1}^{th}. The formulas for the calculation of these quantities are given in [13, 14].

The table also shows the total percentage of indirect reinforcement μ_c of samples, which is calculated by the following formula

$$\mu_c = \mu_p + \mu_{sc}, \tag{1}$$

where μ_p and μ_{sc} are the reinforcement percentage of an outer shell and spiral reinforcement, which are calculated by the following formulas

$$\mu_p = \left(\frac{A_p}{A}\right) \cdot 100\%; \tag{2}$$

Table 3. Main Results of Performed Studies

Series	μ_c, [%]	Destructive loading, [kN]		$\frac{N_u^{exp}}{N_u^{th}}$	m_c	Deformations $\times 10^5$		$\frac{\varepsilon_{cc1}^{exp}}{\varepsilon_{cc1}^{th}}$
		N_u^{exp}	N_u^{th}			ε_{cc1}^{exp}	ε_{cc1}^{th}	
R2.40	0.7	2167	1975	1.10	1.11	690	757	0.91
R2.80	0.7	3233	3368	0.96	1.10	670	567	1.18
T1.40	4.2	650	563	1.15	1.31	520	595	0.87
T1.80	4.2	933	866	1.08	1.23	510	459	1.11
T2.40	8.9	3217	3125	1.03	1.33	1150	897	1.28
TR1.40	5.0	700	733	0.95	1.60	1010	1191	0.85
TR1.80	5.0	1200	1103	1.08	1.48	930	808	1.15
TR2.40	9.6	4200	3914	1.07	1.62	1520	1779	0.85

$$\mu_{sc} = \left(\frac{2A_{sc}}{d_{ef}s}\right) \cdot 100\%. \tag{3}$$

In formulas (2) and (3) A, A_p and A_{sc} are the cross-sectional areas of a concrete core, a steel shell and a spiral reinforcement; d_{eff} is a spiral diameter; s is a spiral coil step.

In order to quantify the influence of a confinement reinforcement effect the coefficient m_c was calculated. It is equal to the ratio N_u^{th} to the sum of the maximum forces in a concrete core and a steel shell under the assumption of their work on uniaxial compression. For the samples of T1.40, T1.80, TR1.40 and TR1.80 series this calculation was made taking into account their flexibility [13].

The data in Table 3 show that the samples of all series have a confinement reinforcement effect. Its value depends on the presence of spiral reinforcement, a steel pipe, as well as on the percentage of confinement reinforcement and the strength of initial concrete. The lowest coefficient m_c are among reinforced concrete samples with a spiral reinforcement (R2.40 and R2.80 series), which is explained by a small percentage of reinforcement. The concrete filled steel tube samples without a spiral (T1.40, T1.80 and T2.40 series), have a higher percentage of reinforcement. Here, the efficiency of a confinement reinforcement is much higher. For the samples of TR1.40, TR1.80 and TR2.40 series, which have both a steel shell and a spiral reinforcement, the coefficient m_c is the largest one as compared to all others and it is reached 1.6. The average strength of the samples of the series TR2.40 is determined from the experiment results, which was 1.3 times higher in comparison with the samples of T2.40 series and 1.9 times higher in comparison with the samples of R2.40 series.

Thus, it has been experimentally proven that an additional reinforcement CFSTE with a spiral reinforcement can significantly increase their bearing capacity and efficiency. With the increase of concrete strength, the effect of a clip, although becomes somewhat weaker, nevertheless it remains significant.

The nature of studied sample destruction was mainly dependent on the presence of an outer steel shell. The destruction of samples of R2.40 and R2.80 series began with the detachment of concrete protective layer (Figure 1, a). With further load increase, their concrete core was crushed. The samples of all series with a steel shell, had a similar character of destruction. By the moment of the ultimate load reaching, they had the folds on a steel shell surface due to the loss of adhesion with the crumbling concrete along the perimeter of its cut surface (Fig. 1, b and 1, c). And right up to the stage of destruction the joint work of a concrete core, a longitudinal and a confinement reinforcement (if any), as well as a steel shell, were not violated. The process of destruction was flexible.

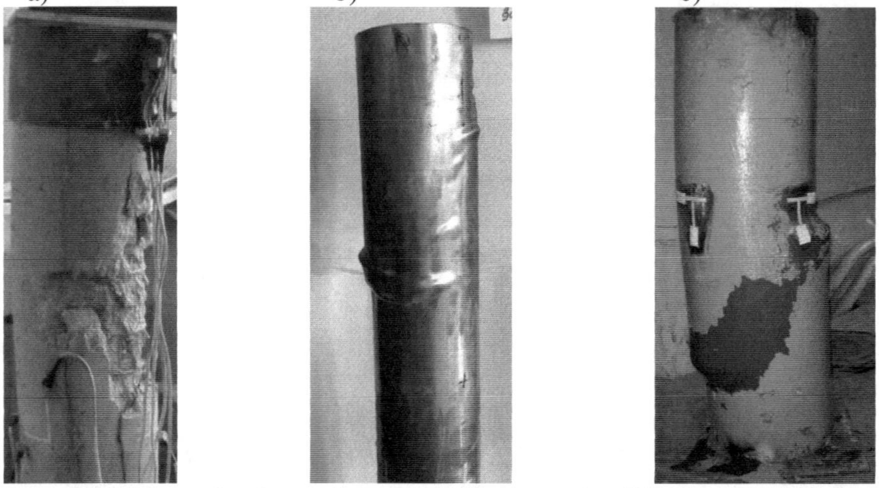

Figure 1. The nature of the samples destruction of the following series: R2.40 (a), T1.80 (b) and TR2.40 (c)

4. Discussion

The performed studies showed that the spiral reinforcement of a concrete core of CFSTE can significantly increase their bearing capacity and efficiency. When an outer steel shell is not operated (for example, during a fire), the confinement reinforcement effect can still be quite noticeable. Its magnitude is largely determined by the percentage of spiral reinforcement and is regulated by the structural parameters of a spiral armature.

The performed studies clearly demonstrated another important advantage of the columns with a steel shell and a spiral reinforcement of a concrete core. A large deformability of the prototypes with a high strength implies that a much larger amount of energy is required to destroy them in comparison with traditional reinforced concrete structures. This CFSTE property allows you to increase significantly the survivability of the building frames during earthquakes. This is very important to consider when you build buildings and structures in seismically active areas.

It should also be noted that the samples with a confinement reinforcement have an opportunity for an effective use of high-strength longitudinal reinforcement. For reinforced concrete structures in which concrete operates under uniaxial compression, the design resistance of the reinforcement is limited to 500 MPa. This value is conditioned by the deformability of the compressed concrete at the top of the deformation diagram (ε_{c1} = 0.2÷0.25%). In the studies carried out, the samples of T1.40 and T1.80 series showed the least deformability ($\varepsilon_{cc1} \approx 0.5\%$). But even with such longitudinal deformations of a concrete core, it is possible to achieve 100% use of compressed reinforcement of A800 class and above. Taking into account a very high deformability of CFSTE with a spiral reinforcement, it can be concluded that there is almost no restriction for them in the use of high-strength longitudinal reinforcement. Here there is a good opportunity and economic expediency to reduce the expense of steel, and thus to achieve the reduction of design cost.

5. Conclusions

The obtained experimental and calculated data confirm the efficiency of spiral reinforcement use for CFSTE concrete core. Due to the spiral reinforcement, there is a real possibility of fire resistance significant increase for such elements without the use of flame retardant coatings. A high efficiency of these structures is maintained when high-strength concrete is used in them. The next step in improving the CFSTE design may be changing the cross-section shape of the samples and spiral reinforcements, replacing the metal shell and spiral reinforcements with a fiberglass shell, reinforcing bars and spiral reinforcements. The relative tests are scheduled in the foreseeable future.

6. Acknowledgments

This article was prepared from the results of the implementation of a scientific project within the state task of the Ministry of Education and Science of the Russian Federation No. 7.3379.2017/4.6.

7. References

[1] L.H. Han and Y.F. An: Performance of concrete-encased CFST stub columns under axial compression, Journal of Constructional Steel Research, Vol. 93 (2014), pp. 62-76.

[2] A.L. Krishan, M.A. Krishan and R.R. Sabirov: Prospects for the application of concrete-filled steel tube columns at construction sites in Russia, Bulletin of Nosov Magnitogorsk State Technical University, No. 1(45) (2014), pp. 137-140.

[3] Cai Shaohuai: The latest experience of using concrete-filled steel tube in the PRC, Concrete and Reinforced Concrete, No. 3 (2001), pp. 20-24.

[4] L-H. Han, W. Li and R. Bjorhovde: Developments and advanced applications of concrete filled steel tubular (CFST) structures: Members, Journal of Constructional Steel Research, Vol. 100 (2014), pp. 211-228.

[5] R. Jayasooriya, D.P. Thambiratnam and N.J. Perera: Blast response and safety evaluation of a composite column for use as key element in structural systems, Engineering Structures, Vol. 61 (2014), pp. 31-43.

[6] B.C. Chen: New development of long span CFST arch bridges in China, Proc. Chinese-Croatian Joint Colloquium “Long arch bridges”, Brijuni Islands (2008), pp. 357-367.

[7] T.M. Mu, B.K. Fan, X.F. Zheng, Y.H. Zheng and B.Z. Xie: Wuxia Yangtze River Bridge in Wushan, China, Proc. Fifth International Conference on Arch Bridge, Madeira, Portugal (2007), pp. 911-918.

[8] Y. L. Yang and B.C. Chen: Rigid-frame tied through arch bridge with concrete filled steel tubular ribs, Proc. Fifth International Conference on Arch Bridge, Madeira, Portugal (2007), pp. 863-868.

[9] B. Uy, Z. Tao and L.H. Han: Behaviour of short and slender concrete-filled stainless steel tubular columns, Journal of Constructional Steel Research, Vol. 67, Iss. 3 (2011), pp. 360-378.

[10] Y.F. Yang: Behaviour of recycled aggregate concrete filled steel tubular columns under long-term sustained loads, Advances in Structural Engineering, Vol. 14, Iss. 2 (2011), pp. 189-206.

[11] M. Lai and J. Ho: Confinement effect of ring-confined concrete-filled-steel-tube columns under uni-axial load, Engineering Structures, Vol. 67 (2014), pp. 123-141.

[12] M.R. Hamidian, M.Z. Jumaat, U.J. Alengaram, N.H. Ramil Sulong and P. Shafig: Pitch spasing effect on the axial compressive behavior of spirally reinforced concrete-filled steel tube (SRCFT), Journal Thin-Walled Structures, Vol. 100 (2016), pp. 212-223.

[13] A.L. Krishan, M.A. Astafyeva and R.R. Sabirov: Engineering Analysis and Design of Concrete Filled Steel Tube Columns: monograph, Saarbrucken, Deutschland: Palmarium Academic Publishing (2016), 261 p.

[14] A.L. Krishan, M.A. Astafeva and E.P. Chernyshova: Strength Calculation of Short Concrete-filled Steel Tube Columns, International Journal of Concrete Structures and Materials, Vol. 12, Iss. 1 (2018), No. 84. https://doi.org/10.1186/s40069-018-0322-z.

IOP Conf. Series: Materials Science and Engineering **652** (2019) 012039 doi:10.1088/1757-899X/652/1/012039

Synthesis of hydrophilic silica nanoparticle and its application as self-cleaning coating for solar cell panels

S Boonsong[1], N Sowasod[1*], T Muangnapoh[2] , J Wongchaipanich[1], W Jarunsrirat[1] and W Tanthapanichakoon[3]

[1]Division of Chemical Process Engineering Technology, Faculty of Engineering and Technology, King Mongkut's University of Technology North Bangkok (Rayong Campus), Rayong 21120, Thailand
[2]National Nanotechnology Center (NANOTEC), National Science and Technology Development Agency, Pathum Thani 12120, Thailand
[3]Academy of Science, Royal Society of Thailand 10300, Thailand

Abstract. In past decades, nanotechnology has attracted much attention among researchers and industries due to its high potential of adding significant values to the original materials. Thin film coating with nanomaterial can provide new properties and enhance performances of pristine surfaces for certain applications. In this research, we investigated the development of superhydrophilic self-cleaning coatings for solar panels using colloidal silica nanoparticles. The coatings consist of colloidal silica nanoparticles with the average size of 30 nm, sodium dihydrogen phosphate and boric acid under various formulations. The coatings can be applied by simple painting or fine spraying. The light transmittance of the coated samples was characterized by UV–visible spectrophotometer. The water contact angle (WCA) was measured by a contact angle instrument. It was observed that the optimum weight ratio of silica nanoparticles : sodium dihydrogen phosphate : boric acid was 22:1:1 with contact angle of 1.83° and transmittance of 65%.

1. Introduction

In past decades, nanotechnology has attracted much attention among researchers and industries due to its high potential of adding significant values to the original materials. Thin film coating with nanomaterial can provide new properties and enhance performances of pristine surfaces for certain applications. In this research, we investigated the development of superhydrophilic self-cleaning coatings of the so-called mirror (material made from glass) with colloidal silica nanoparticles [1]. The mirror can be melted and easily formed into various shapes. When cooled down, it has a solid state that is amorphous and transparent, thereby making it popular for interior decoration that increases the brightness inside a building. Today it is also widely used in automotive, residential, and general industrial applications. However, one common problem associated with its use is that the glass is dirty from dust and water stains caused by rain, storm, air pollutants, PM2.5 from industrial plants, vehicles with combustion engines, and so on dirt gets out. Therefore, it is necessary to hire people to clean up the dirt. The mirrors installed on the exterior of high-rise buildings are much more difficult to clean.

Self-cleaning coatings may be classified into two major categories: hydrophobic and hydrophilic [2]. In this research, we aimed to develop superhydrophilic self-cleaning coatings for mirrors in general and solar cell panels in particular, [3] which consist of colloidal silica nanoparticles with an average size of 30 nm, sodium dihydrogen phosphate and boric acid in various formulations for testing and selection. The coatings can be applied by spraying. To save cost of material, the ideal coating should have a minimum thickness while displaying superhydrophilicity with strong surface adhesion and durability [4, 5]. The thickness of the applied coating and the microstructure of the nanoparticles in the coating were observed and characterized with a Scanning Electron Microscope (SEM). Its degree of hydrophilicity was represented by the water contact angle (WCA) of a static droplet on the coating measured with a contact angle meter. The transmittance spectra of the coated and uncoated samples were evaluated using a UV–vis spectrophotometer and assessment of the adhesion strength of the coating film, by the standard tape test method. In this investigation, microscope glass slides, which have excellent transparency, were used as substrate for the coating. The main objective was to develop an optimum spray-coating material formulation that would display excellent superhydrophilicity with strong adhesion, acceptable film thickness and reasonable transmittance. The durability and long-term characteristics of the optimized coating will be investigated and reported in the near future.

2. Materials and methods

2.1 Materials

Colloidal silica nanoparticles, 3%wt in isopropyl alcohol (IPA), were synthesized at NANOTEC using the Stöber process. Colloidal silica nanoparticles, 3%wt in water, were purchased from Nissan Chemical, Japan. Boric acid was purchased from ACROS, USA. Anhydrous sodium dihydrogen phosphate was purchased from QRëC, New Zealand.

2.2 Measurements

Characterization of microstructure, film thickness, water contact angle and adhesion strength

The static contact angle between a water droplet and the film surface was measured with a Contact Angle Meter (Model TC/TPC150, Dataphysics Instruments GmbH, Germany). At least 3 independent angle measurements were conducted for each thin film sample. The average contact angle and its standard deviation were reported. The microstructure of the thin film of hydrophilic nanoparticles was observed through scanning electron microscopy (SEM) (Model SU8230, Hitachi, Japan). The transmittance spectra of the coated and uncoated samples were evaluated using a UV– visible spectrophotometer and assessment of the adhesion strength of the coating film, by the standard tape test method

3. Results and discussion

Preparation of superhydrophilic coating solution

Table 1 lists the composition of all 18 formulations investigated. Taking the case of formulation no. 1 as example, 1 g of anhydrous sodium dihydrogen phosphate as additive agent and 1 g of boric acid for pH adjustment were put in a stainless-steel heating container, mixed with 10 ml of deionized water, and then heated at 80 ^{o}C under stirring [6]. After the heated mixture achieved complete dissolution, it became transparent and colorless. Then its weight was measured and the evaporated water was compensated, so as to gain the desired additive compound. Next 16 g of IPA containing 3%wt of silica nanoparticles was added to the additive compound and sufficiently stirred at room temperature to gain an inorganic aqueous coating solution having a final pH of 5. The other formulations in Table 1 were prepared in a similar manner.

Table 1 Composition of all 18 formulations

Formulation (no.)	3%wt Silica nanoparticles in IPA (g)	3%wt Silica nanoparticles in water (g)	Boric acid (g)	Sodium dihydrogen phosphate (g)	pH
1	16	-	1	1	5
2	18	-	1	1	5
3	20	-	1	1	5
4	22	-	1	1	5
5	24	-	1	1	5
6	26	-	1	1	5
7	28	-	1	1	5
8	30	-	1	1	5
9	32		1	1	5
10	-	16	1	1	7
11	-	18	1	1	7
12	-	20	1	1	7
13	-	22	1	1	7
14	-	24	1	1	7
15	-	26	1	1	7
16	-	28	1	1	7
17	-	30	1	1	7
18	-	32	1	1	7

Thin film coating on substrate

Microscope glass slides were used as substrate for thin film coating by spraying. Then newly coated microscope glass slides were dried at 80 °C for 30 min.

4.Water contact angle on the thin film coatings

To evaluate the hydrophilicity of the sprayed coating, static water droplet contact angle was measured for each of the 18 formulations and the measurement results are summarized in Table 2. To enhance the reader's understanding, photos of a droplet lying on a blank slide (without coating) and the slide coated with formulation no. 4 are shown in Figure 1. [7]

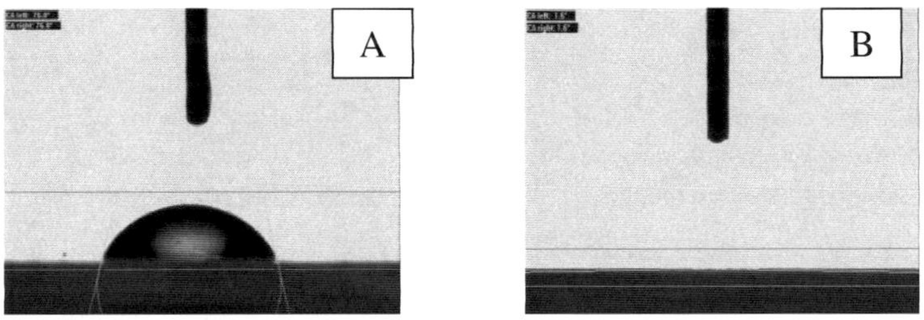

Figure 1. Contact angle measurement: (A) Blank slide, (B) Slide coated with formulation No. 4

Table 2 Average contact angles and standard deviations

No.	Average contact angle (°)
0 (blank)	70.87 ±13.43
1	6.73 ±2.71

2	6.17 ±0.40
3	5.87 ±0.70
4	1.83 ±0.21
5	3.10 ±1.65
6	8.32 ±2.76
7	9.30 ±1.35
8	8.70 ±2.63
9	8.37 ±2.91
10	31.03 ±8.45
11	35.03 ±5.97
12	29.40 ±2.71
13	21.57 ±3.88
14	28.03 ±3.66
15	16.50 ±4.39
16	15.87 ±0.35
17	13.97 ±2.32
18	18.50 ±1.13

As seen from Table 2, the WCA on the blank slide was widest at 70.86 °. Among formulations no. 1-9 which used IPA containing 3%wt silica nanoparticles, it was found that formulations no.3, 4 and 5 have the three smallest contact angles of 5.86, 1.38 and 3.10 degrees, respectively [4]. Similarly, among formulations 10-18 which used water containing 3%wt silica nanoparticles, formulations no.15, 16 and 17 have the three smallest contact angles of 16.50, 15.87 and 13.97 degrees, respectively. Therefore, it was decided to characterize formulations no. 3, 4, 5, 15, 16 and 17 with the other above-mentioned analytical tools.

5. Light transmittance through the thin film coating [8, 9]

As shown in Table 3, the transmittance spectra of the uncoated and 6 coated samples were evaluated using the UV–vis spectrophotometer [10].

Table 3 Transmittance spectra of the uncoated and 6 coated samples.

No.	Transmission values (%)
0 (blank)	100
3	35
4	65
5	60
15	75
16	75
17	85

The blank case (uncoated slide) had transmission value of 100%. The 3 selected IPA-based formulations no. 3, 4 and 5 were found to have lower transmittance than their 3 selected water-based counterparts. A similar trend was reported by Liu et al. [11]. More specifically, formulation no. 3 exhibited the lowest transmittance of 35% whereas formulations no.4 and 5 showed transmittance of 65% and 60%, respectively. In contrast, formulations no. 15, 16 and 17 showed transmittance values of 75%, 75% and 85%, respectively.

6. Conclusion

Based on the above experimental results, it may be concluded that formula no. 4 is the optimum formulation of the coating because it displayed excellent superhydrophilicity (the smallest contact angle of 1.83°) [7] with strong adhesion, acceptable film thickness 300 nm and reasonable transmittance of 65%.

7. Acknowledgements

This research was supported by King Mongkut's University of Technology North Bangkok (KMUTNB-62-DRIVE-09). The authors are grateful to KMUTNB (Rayong Campus) and the National Nanotechnology Center (NANOTEC) for the use of their lab facilities.

8. References

[1] X. Lu, Z. Wang, X. Yang, X. Xu, L. Zhang, N. Zhao and J. Xu: Surface and Coatings Technology. Vol. 206 (2011), p. 1490–1494

[2] X.Yang, L. Zhua, Y.Chenb, B. Baoc, J. Xuc and W. Zhou: Appl. Surf. Sci. Vol. 349 (2015), p. 916-923

[3] A. Syafiq, B. Vengadaesvaran, A. K. Pandey and N. Abd. Rahim: Journal of Nanomaterials. (2018), p. 1-10

[4] J. Son, S. Kundu, L. K Verma: Solar Energy Materials and Solar Cells. Vol. 98 (2012), p. 46-51

[5] J. Chen, L. Zhang, Z. Zeng, G. Wang, G. Liu, W. Zhao, T. Ren and Q. Xue. Colloids and Surfaces A: Physicochemical and Engineering Aspects.Vol. 509 (2016), p. 149-157

[6] Katsumi Kishimoto. (2012,August). United State Of America. US 8,241,416 B2

[7] Z. Hong, H. Yang, W. Yuanhao and Y. Hongxing: Energy Procedia. Vol. 105 (2017), p. 1077-1083

[8] C.S. Thompson, R. A. Fleming, M. Zou: Solar Energy Materials and Solar Cells. Vol. 115 (2013), p 108- 113

[9] L. Ye, Y. Zhang, C. Song, Y. Li, and B. Jiang: Materials. Letters, Vol. 188(2017), p. 316-318

[10] R. Prado, G. Beobide, A. Marcaide, J. Goikoetxea and A. Aranzabe: Solar Energy Materials and Solar Cells. Vol. 94 (2010), p. 1081–1088

[11] Y. Liu, W. Liu, D. Feng, Z. Wei, T. Guo, G. Wang, Z. Song : J Molecular Structure. Vol. 1177 (2019), p. 26-32

The Potential of Enzyme in Reducing the Maturation Time and the Use of Ammonia in Concentrated Natural Rubber Latex

Jiraporn Wongkhat [a], **Cattaleeya Pattamaprom** [b*]

Center of Excellence on Natural Rubber and Biopolymers Technology, Department of Chemical Engineering, Faculty of Engineering, Thammasat University, Pathumthani, 12120, Thailand
[a]jirawong163@gmail.com, [b*]corresponding author: cattalee@engr.tu.ac.th

Abstract. One of the most important parameters for indicating colloidal stability of concentrated natural rubber latex (CNRL) is the mechanical stability time (MST). The MST of CNRL after centrifugation is always low and slowly increases with storage time due to gradual hydrolysis of phospholipid on the surface of rubber particles into fatty acid soap. In this work, enzymatic approach was investigated to increase the colloidal stability of CNRL by using different grades and amounts of enzymes. The enzymes were tested with 2 grades of CNRL, which are the high-ammonia latex (HA latex) and low-ammonia latex (LA latex). We found that the MST of LA latex treated with 0.004% and 0.04% of Lipex reach the acceptable value (at least 600 sec.) within 14 and 10 days, respectively. In case of HA latex, the MST reached the acceptable value faster than that of LA latex (within 10 and 3 days at amount 0.004% and 0.04% of Lipex, respectively) while, the MST of commercial laurate-treated latex only increased slightly and could not pass the acceptable value after 31 days. In addition, the use of enzyme instead of laurate soap could reduce the use of ammonia in CNRL production.

1. Introduction

Fresh natural rubber latex (FNRL) tapped from rubber trees contain about 35% dry rubber content, which is unsuitable for producing rubber products. To increase the rubber content in latex, FNRL is usually converted into concentrated natural rubber latex (CNRL), which contains about 60% dry rubber content. The most popular commercial technique for producing CNRL is by centrifugation. There are two common grades of commercial latex, one is the commercial high-ammonia latex (commercial HA latex) and the other is the commercial low-ammonia latex with the presence of TMTD and zinc oxide (commercial LATZ latex). To preserve and stabilize the commercial HA latex, 0.65 wt% ammonia and 0.02-0.04 wt% ammonium laurate soap are added, while 0.25 wt% ammonia, 0.05 wt% ammonium laurate soap, 0.13 wt% TMTD and 0.13 wt% ZnO are added in commercial LATZ latex.

To ensure the latex stability of CNRL, the MST of above 600 seconds is generally specified in the standard Thai latex for sellable CNRL. However, the initial stability of CNRL freshly produced from the centrifugation process is low because the rubber particles are closer leading to an increasing attractive force and lower latex stability. Naturally, the MST of CNRL slowly increases with storage time due to gradual hydrolysis of phospholipid on rubber particles into fatty acid soap [1, 2, 3, 4]. Therefore, maturation of CNRL is necessary to allow the increase of MST. The practical method to expedite the increase of MST for commercial CNRL is by direct addition of negative ions such as laurate soap [5, 6] or other surfactants [7, 8] into CNRL. However, the increase of MST from this method leads to over-stabilization of the latex at long time, which would cause coagulation difficulty in the product forming processes [9]. To accelerate the MST of CNRL without over-stabilizing problem, a preliminary work in our group found that the enzyme lipase was more effective in increasing the MST of HA latex [10]. Moreover, the MST of HA latex reached a stable value after 10 days of storage time without causing an over-stabilized problem at long time.

Therefore, lipase addition is a potential method to reduce maturation time in latex production. In this work, we expand the research scope by exploring the potential of different grades and contents of lipase in 2 different grades of CNRL including the high-ammonia and the low-ammonia grades, and compared the results to commercial HA latex and LATZ latex. The potential of lipase usage in reducing the use of ammonia and other preservatives is also discussed.

2. Experimental

2.1 Materials

In this study, the high-ammonia latex without stabilizer was denoted as HA latex and low-ammonia grade without stabilizer and secondary preservatives (TMTD/ZnO) was denoted as LA latex. The freshly produced HA latex and LA latex were purchased from Thai Eastern Rubber Co., Ltd. The commercial HA latex and LATZ latex were also provided from the same company. The lipase enzymes obtained from Novozyme Co., Ltd. and Reach Biotechnology Co., Ltd were denoted as Lipex and CNLP, respectively.

2.2 Preparation of CNRL treated with lipase

To study the effect of lipase, the CNRL samples were treated with 2 grades of lipase (Lipex and CNLP) at the concentration of 0.004% and 0.04% by volume of CNRL. The treated samples were kept at room temperature and measured for the MST values, total solid content (TSC), dry rubber content (DRC), alkalinity (%NH_3) and pH. The untreated CNRL and the commercial CNRL treated with ammonium laurate soap were used as reference.

2.3 Determination of Properties

The MST was determined by using Mechanical Stability Test Machine (Klaxon MST test apparatus) operating at 14,000 rpm following the ISO 35:2004 standard. The TSC was determined from the weight of dried latex at 70 ^{o}C for 16 hr according to the ISO124:2014 and the DRC was determined by acid coagulation before drying following the ISO 126:2005. The alkalinity as ammonia (%NH_3) was determined by titration with 0.1 M HCl according to ISO 125:2011.

3. Results and discussion

3.1 Effect of lipase grades on the MST of CNRL

In this part, 2 grades of CNRL were used in this study including the high ammonia latex (HA latex) and the low ammonia latex without TMTD and zinc oxide preservatives (LA latex). The MST of CNRL samples treated with 2 grades of lipase were compared with the untreated latex samples. The MST values were measured after 41 days of maturation and the results are shown in Fig. 1. As can be seen, the MST of CNRL samples treated with Lipex was higher than that treated with CNLP in both types of CNRL. This indicated that Lipex was more effective than CNLP in accelerating the hydrolysis of phospholipids in CNRL. Therefore, Lipex was further used for accelerating the MST of CNRL in this research. Other properties of the CNRL samples in Fig 1 were reported in Table 1. It was found that, the values of TSC, DRC, %NH_3 and pH were not significantly affected by lipase treatment.

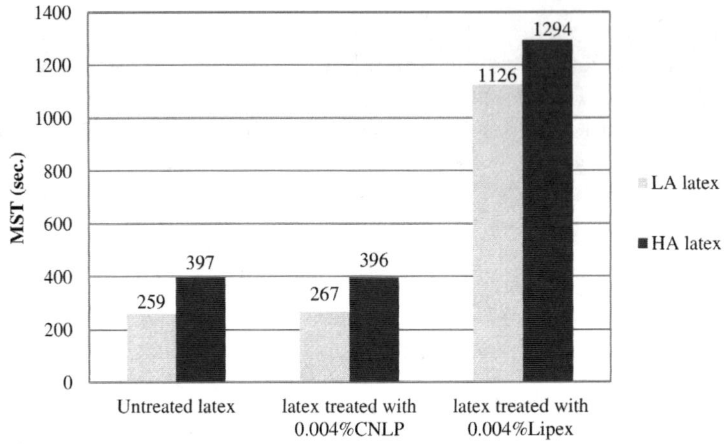

This implied that lipase addition did not cause any adverse effect on the CNRL.

Fig. 1 The MST (at 41 days) of untreated latex, CNRL treated with 0.004%CNLP and 0.004%Lipex

Table 1 Other properties of CNRL samples at 40 days

Sample	Property			
	TSC (%)	DRC (%)	$\%NH_3$	pH
Untreated LA latex	61.02	60.40	0.23	9.28
LA latex treated with 0.004%CNLP	61.68	60.42	0.25	9.33
LA latex treated with 0.004%Lipex	61.93	60.37	0.25	9.26
Untreated HA latex	61.64	59.75	0.55	9.73
HA latex treated with 0.004%CNLP	61.86	60.04	0.54	9.77
HA latex treated with 0.004%Lipex	61.07	59.97	0.54	9.67

3.2 Effect of lipase concentration on MST of CNRL

In this part, Lipex was added to CNRL (LA and HA latex) in the amount of 0.004% and 0.04% by volume of CNRL and the MST values of CNRL samples were followed for 31 days. The MST values of the Lipex-treated samples were compared with the commercial latex samples maturated for the same time as shown in Table 2. It was found that the MST of Lipex-treated CNRL increased more rapidly than the commercial samples in both LA latex and HA latex. The MST of LA latex treated with 0.004% and 0.04% of Lipex reach the acceptable value (at least 600 sec.) within 14 and 10 days, respectively. In case of HA latex, the result showed the same trend but the MST reached the acceptable value faster (within 10 and 3 days at amount 0.004% and 0.04% of Lipex, respectively). On the other hand, the MST of commercial laurate-treated latex only increased slightly and could not pass the acceptable value after 31 days.

Table 2 The MST (for 31 days) of commercial latex compared with latex treated with 0.004% and 0.04% Lipex

Maturation time (days)	**MST (sec.)**					
	LA latex			**HA latex**		
	commercial LATZ latex	**Treated with 0.004% Lipex**	**Treated with 0.04% Lipex**	**commercial HA latex**	**Treated with 0.004% Lipex**	**Treated with 0.04% Lipex**
1	-	214	315	-	270	396
3	-	364	426	-	498	**663**
6	85	447	545	86	555	774
10	91.5	544	**628**	134.5	**694**	949
14	-	**627**	591	-	1002	1023
17	119	620	841	164	975	1036
31	137	675	851	195	983	1179

From these results, it indicated that, lipase was more effective than the commercial laurate soap in improving the latex stability. Besides, the higher concentration of Lipex (0.04%) was more effective than at 0.004% as the higher lipase content could accelerate faster hydrolysis of phospholipids.

3.3 Effect of ammonia content on MST of CNRL

The effects of ammonia content on MST of Lipex-treated latex were compared with that of commercial latex samples in Table 2. As can be seen, HA latex (higher ammonia content) possessed higher MST than the LA latex in all cases, indicating that the ammonia content strongly affected the increase in MST of the Lipex-treated samples. This indicated that the hydrolysis of lipids was faster under more alkali condition. Nevertheless, all Lipex-treated latex possessed much shorter maturation time than the commercial laurate-treated ones.

4. Summary

In this research, lipase treatment was an effective method in accelerating the MST of both low- and high-ammonia CNRL without affecting other properties of CNRL. The results showed that, the MSTs of lipase-treated CNRL samples were significantly higher than that of the commercial laurate-treated HA and LATZ latex indicating that lipase could effectively hydrolyze lipids and phospholipids on rubber particles. We found that the MST of LA latex treated with 0.004% and 0.04% of Lipex reach the acceptable value (at least 600 sec.) within 14 and 10 days, respectively. In case of HA latex, the MST reached the acceptable value faster than that of LA latex (within 10 and 3 days at amount 0.004% and 0.04% of Lipex, respectively). This indicated that the hydrolysis of lipids was faster under more alkali condition.

5. Acknowledgement:

The authors are grateful to Thammasat University Research Fund for the financial support and also grateful to Novozyme Co., Ltd. for the support of lipase enzyme for this research.

6. References

[1] S .F .Chen and C .S .Ng: *Rubber Chemistry and Technology,* vol .57 (1984), p. 243

[2] L. Karunanayake, and G. P. Perera: *Journal of applied polymer science,* vol. 99(6) (2006), p. 3120-3124

[3] T .D .Pendle and A .D .T .Gorton: *Rubber Chem Technol,* vol. 57(5) (1978), p .986-1005

[4] J .Sansatsadeekul, J .Sakdapipanich and P .Rojruthai: *Journal of Bioscience and Bioengineering,* vol. 111(6) (2011), p .628-634

[5] B .Nithiuthai, P .Nithiuthai, P .Pongphai, S .Sumpaongern, N .Noreewong and C .Wuttichan: Research Report entitled "Acceleration of Natural Rubber, "Prince of Songkla University, Pattani, Thailand, (1995)

[6] K. M. D. Silva and S. Walpalage: *Rubber Research,*vol. 12 (2009), p. 59-70

[7] S. Manroshan, M.Y. AMIR-HASHIM and K. BOOTEN: *Rubber Research.,* vol. 10 (2007), p. 156-165

[8] Y .Tanaka, Y .Hioki, M .Hayashi, N .Ichikawa and T .Sakaki: *United States Patent,* U.S. Patent 5,610,212 (1997)

[9] D. M. Hill: *The Science and Technology of Latex Dipping* (Smithers Information Limited, United Kingdom 2018), p. 14

[10] D. Pimalai, Y. Wongngam and C. Pattamaprom: *Defect and Diffusion Forum,* vol. 382 (2018), p. 26-30

IOP Conf. Series: Materials Science and Engineering **652** (2019) 012041 doi:10.1088/1757-899X/652/1/012041

Identification of limiting deep drawing ratio using the energy criterion of fracture

Kyaw Zayar Soe[1], Feoktistov S I[2]

[1,2]Komsomolsk-on-Amur State University, 27 Lenin Ave., Komsomolsk-on-Amur, 681013, Russia

Email: [1]kyawzayarsoe52@gmail.com, [2]serg_feo@mail.ru

Abstract. The paper describes a method which has been developed for obtaining the limiting deep drawing ratio of titanium alloys and determines the moment of failure of the blank. This method is based on not only the use of Forming Limit Diagram (FLD) in predicting the failure of the blank but also the using method of variable parameters of elasticity in determining the stress-strain state in the deep drawing process.

1. Introduction

One of the most important tasks in the development of the technological process of sheet metal forming is deep drawing. It is the prediction of the moment of failure of the workpiece, that is, the determination of the limiting drawing ratio or the power drawing [1]. The modern development of computing technology is used for engineering calculations in the design of sheet metal stamping processes, and high accuracy to calculate the parameters of the stress-strain state of the blank. This process obtains to predict the failure of the workpiece by using the Forming Limit Diagram (FLD) [2,3,4,5,6,7,8]. This also allows FLD-diagrams to be widely used in determining the limiting capabilities of sheet metal forming processes.

In this paper, we offer an analytical method for constructing FLD diagrams by using strain energy criterion [9].

According to determine the stress-strain state, the method for variable parameters of elasticity accounts not only the thickness change in the deformation process but also the hardening of the material, which for titanium alloys is well described by the power function.

2. Model to the fracture of sheet blanks based on an energy criterion

In the construction of a model for the fracture of without defect isotropic sheet blanks, the energy criterion is widely used as the criterion of limiting deformation in the theoretical prediction. To determine the forming of sheet metal, Swift proposed a diffuse neck criterion [9], which described as diffuse neck starts when the load reaches a maximum along both principal directions. According to the transforming of the expressions obtained results in work [10], we can define an equation that allows us to construct the FLD diagram using the Swift criterion:

$$4(e_1-n)(e_2+2e_1)^3-3(e_1-2n)(e_1+2e_2)(e_2+2e_1)^2-$$
$$-3(e_1+2n)(e_1+2e_2)^2(e_2+2e_1)+2(2e_1+n)(e_1+2e_2)^3=0, \quad (1)$$

where e_1, e_2 are the major and minor logarithmic strains acting in the plane of the sheet blank; n is the exponent of the power approximation by the hardening curve of the third kind.

$$\sigma_s = A e_i^n .$$

Assuming the experimental and theoretical works, after the occurrence of a diffuse necking, the plastic deformation of the sample continues, and local necking can later appear, which differs from the diffuse not only in size but also in that its appearance and development take place under conditions of plane strain with intensive thinning sample thickness [11].

Hill [12] proposed that the criterion of forming limit should be the moment of local necking at which the increment of the total effort is zero. It is assumed that the local necking will be formed at an angle ψ to the direction of the major principal stress. The orientation angle of the neck with respect of loading axes can be expressed as follows:

$$\psi = tan^{-1}\left(\sqrt{-\alpha}\right) \tag{2}$$

Where $\alpha = e_2 / e_1$ is the ratio of main strains arising under conditions of simple (monotonic) loading. Due to the square root in equation (2), this equation has a physical meaning only with a negative value of e_2.

According to the transforming of the expressions obtained in work [12], we define an equation that allows us to construct the FLD diagram using the Hill criterion:

$$e_1 + e_2 - n = 0. \tag{3}$$

In practice, a combination of two criteria is used to construct FLD diagrams for energy criteria. The Hill criterion is applied for a negative value of e_2 by the equation (3) and the Swift criterion for a positive value of e_2 by the equation (1).

3. Identification of stress-strain state in the deep drawing process

When solving the problem of stress-strain state in the deep drawing process, we can use the method for variable parameters of elasticity [13, 14].

According to the physical relations of the deformation theory of plasticity and the compatibility equation for true strains, we can be written in Euler coordinates [15] and obtain the normal integral equation in stresses, which determines the stress state on the flange with an axisymmetric hood.

$$\sigma_\varphi = \frac{1}{2}\sigma_\rho - \frac{2}{3}E_{sec}\ln\left\{\frac{1}{\rho\sqrt{\rho}}\int_R^\rho \frac{3}{2}\sqrt{\rho}\, exp\left(-\frac{3}{4E_{sec}}\sigma_\rho\right)d\rho + \right.$$
$$\left. + \frac{R\sqrt{R}}{\rho\sqrt{\rho}} exp\left[-\frac{3}{2E_{sec\,R}}\left(\sigma_{\varphi R} - \frac{1}{2}\sigma_{\rho R}\right)\right]\right\} \tag{4}$$

Where, the index R indicates that the parameter value is considered at $\rho = R$, which corresponds to the outer edge of the flange in the deformation process.

Therefore, the drawing process is recommended to be solved in stresses. The solution is carried out by the method of successive approximations using a recurrent scheme by equation (4):

$$\sigma_\rho^{(i+1)} = \left(\int_R^\rho \frac{\left(\sigma_\varphi^{(i)} - \sigma_\rho^{(i)}\right)S(\rho)^{(i)}}{\rho} d\rho + S(R)^{(i)}\sigma_{\rho R}\right) / S(\rho)^{(i)} .$$

Where, the quantities with the sign of (i) and $(i+1)$ denote respectively their values in the ith and $(i+1)^{th}$ approximations.

The calculation results do not depend on the option of the initial approximation, but these results take approximately the values of the stresses obtained by the elastic solution.

$$\sigma_{\rho}^{(0)} = -\ln(R / R_0) \times \frac{E}{2}\left(\frac{R^2}{\rho^2} - 1\right);$$

$$\sigma_{\varphi}^{(0)} = \ln(R / R_0) \times \frac{E}{2}\left(\frac{R^2}{\rho^2} + 1\right).$$

By the conditions of deep drawing, we can express friction on the flange as follows [16, 17]:

$$by\ \rho = R, \sigma_{\rho} = \sigma_{\rho R} = \frac{\mu_{mp1}\left[R_0^2 - (r_0 + r_m)^2\right]q}{R \cdot S(R)};$$

$$by\ \rho = R, \sigma_{\varphi} = \sigma_{\varphi R} = \ln(R / R_0).$$

Where μ_{mp1} is the friction coefficient on the flange; q is the pressure.

After identification of the stresses, we can determine the strain by using the physical relationship of the deformation theory of plasticity.

According to the method for variable parameters of elasticity, after each approximation we can specify the value of the secant modulus as follows:

$$E_{sec}^{(i+1)} = \frac{A\left(e_i^{(i)}\right)^n}{e_i^{(i)}},$$

Where, $e_i^{(i)} = \frac{2}{\sqrt{3}}\sqrt{\left(e_{\rho}^{(i)} - e_{\varphi}^{(i)}\right)^2 + \left(e_{\varphi}^{(i)} - e_z^{(i)}\right)^2 + \left(e_z^{(i)} - e_{\rho}^{(i)}\right)^2}.$

To control the convergence of the process, the values of the secant modules are compared as follows:

$$E_{sec}^{(i+1)} - E_{sec}^{(i)} \le \Delta.$$

After reaching the given accuracy, the value of the stresses at the edge of the die can be taking into account the bending, friction, and flatness according to the following formulas [15,17]:

a) by $r_0 < \rho \le (r_0 + r_m)$

$$\sigma_{\rho}' = \left(\sigma_{\rho} + \sigma_{S1}\frac{S_1}{4r_m + 2S_1}\right)\exp\left(\mu_{mp2}\alpha\right),$$

b) by $\rho = r_0, \sigma_{\rho r_0} = \sigma_0$ and $\alpha = \frac{\pi}{2}$ are the radial stresses (tension in the wall Cup) that reach the highest value,

$$\sigma_{\rho\max} = \left(\sigma + \sigma_{S1}\frac{S_1}{4r_m + 2S_1}\right)\exp\left(\mu_{mp2}\frac{\pi}{2}\right) + \sigma_0\frac{S_2}{4r_m + 2S_2}$$

The tangential stresses in the area of the die edge change from some current value to zero, that is the workpiece leaves from the edge of the die, the stress state is close to linear tension. If we assume that the change occurs linearly depending on the angle, we obtain $r_0 \le \rho \le (r_0 + r_m)$.

$$\sigma_{\varphi}^{'} = \sigma_{\varphi}\left[1-\frac{2}{\pi}\arcsin\left(\frac{r_0+r_m-\rho}{r_m}\right)\right].$$

Where, σ_{φ} is tangential stress by using equation (4).

4. The identification for technological capabilities in the deep drawing process

The curve of strain hardening of a titanium alloy is approximated by a power function, in the form of the J. H. Hollomon's equation [18]:

$\sigma_s = \sigma_{sw}\cdot(e_i / e_w)^{e_w}$,

where, $\sigma_{sw} = \sigma_B \cdot \exp(e_w)$, and e_w is true (logarithmic) strain at the moment of the necking and σ_B is ultimate strength of material.

The technological process of deep drawing is set by the following parameters: r_0 is the radius of the die, R_0 is the radius of the workpiece, r_m is the edge radius of the die, q is the specific pressure on the flange, μ_{mp1} is the coefficient of friction on the flange and μ_{mp2} is the coefficient of friction on the edge of the die.

Initially, the radius of the workpiece R_0 is set that the deep drawing process is guaranteed to pass. For example, $R_0 = 1{,}25 r_0$. The paper represents the most complete view of the deep drawing process, which gives the position of the outer edge *R*. Therefore, another step is given to change the position of the outer edge *R*. According to the position of the outer edge that is taken by the method for variable parameters of elasticity, we obtain the stress-strain state on the flange of the blank.

Then, we can assume that $e_\rho = e_1$ and $e_\varphi = e_2$ which determine the position of the points characterizing the deformed state on the flange in the coordinates e_1, e_2. If all these points lie below Forming Limit Curve (FLC) by equations (1) and (3), then the process continues changing the position of the outer edge *R*. The calculation in accordance with the method of successive loading [19] is repeated until *R* becomes equal, i.e. the deep drawing process will end.

In a case any stage of loading one of the points characterizing the strain state on the flange in the coordinates e_1, e_2 will be higher than the FLC diagram, then the calculation is stopped and determined the limiting drawing ratio $m = r_0/R_0$ or the power drawing ratio $k = R_0/r_0$.

The results of calculations for the material BT1-1 are presented in the figure. 1. Where, curves 1, 2 and 3 describe the deformed state of the blank at the moment preceding of fracture. The solid thick line corresponds to the FLC constructed by using the equations (1) and (3).

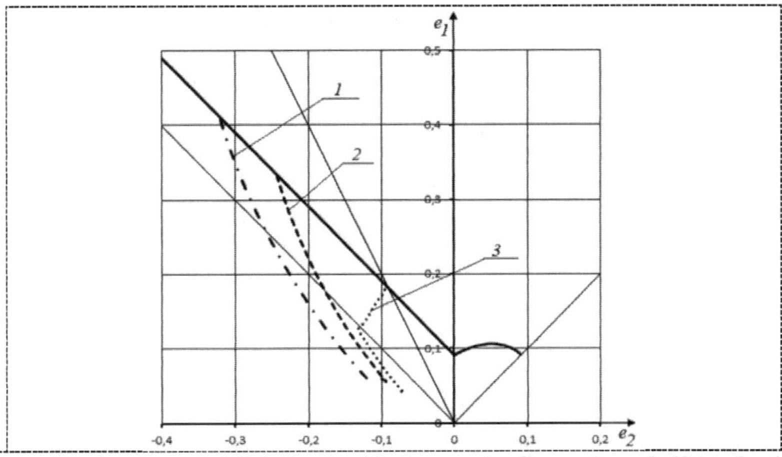

1 – without friction (m= 0,459), 2 – by friction on the flange (m= 0,504), 3 – by friction on the flange and bending, friction and flattening on the edge of the die (m= 0,574)

Figure 1. Identification of deep drawing ratio for material BT1-1 using the FLD-diagram Hill-Swift (energy criterion).

5. The conclusions

The results of this paper represent the methods of identification for the technological capabilities of titanium alloys in the deep drawing process, which allows predicting the friction of the blank by the Forming Limit Diagram (FLD).

References

[1] Romanovsky V.P 1977 *Handbook of cold stamping:* fifth ed,. M., Mechanical engineering, p.423 [in Russian]

[2] Dorel Banabic 2010 Sheet Metal Forming Processes. *Constitutive Modeling and Numerical Simulation.* London-New York: Springer Heidelberg Dordrecht. p.301

[3] Ma G.S, Wan M and Wu X.D 2008 Theoretical prediction of FLDs for Al-Li alloy at elevated temperature based on M-K model, *J. Chinese Journal of Nonferrous Metals*. **18** 980-984.

[4] Yang M and Dong X.H 2008 The Forming Limit Prediction Using a Rate-Dependent Crystal Plasticity Model, *J. Journal of ShangHai JiaoTong University*. **42** 720-723.

[5] Du P.M., Lang L.H and Liu B.S 2011 Theoretical prediction and parameter influence of FLDs based on M-K model, *J. Journal of Plasticity Engineering*. **18** 84-89.

[6] Chen G.N. and Hu S.G 1994 Damage, Instability and Construction of Forming Limit Curve in Sheet Metal Stretch Forming , *J. Journal of Plasticity Engineering*. **1** 31-46.

[7] Kami A., Dariani B.M 2015 Numerical determination of the forming limit curves of anisotropic sheet metals using GTN damage model, *J. Journal of Materials Processing Technology*. **216** 472-483.

[8] Ghosh A. K 1977 The influence of strain hardening and strain-rate sensitivity on sheet metal forming [J]. *Journal of Engineering Materials and Technology*, **99** 264–274.

[9] Swift H.W 1952 Plastic instability under plane stress, *Journal Mech Phys Soids*. **1** 1-18.

[10] Stoughton T.B., Zhu X 2004 Review of theoretical models of the strain-based FLD and their relevance to the stress-based FLD, *International Journal Plasticity*. **20** 1463-1486.

[11] Chumadin A.S 2014 *Theory and calculations of sheet-stamping processes (for engineers)*, second ed, M Exservice VIP, p. 216 [in Russian]

[12] Hill R. 1952 On discontinuous plastic states with special reference to localized necking in thin sheets, *Journal Mech Phys Solids*. **1** 19-30

[13] Malinin N.N 1975 Applied theory of plasticity and creep. *Textbook for university students*. second ed, "Mechanical Engineering", Moscow, p.400 [in Russian]

[14] Birger I.A 1961 *Round plates and shells of rotation*. – M. Oborongiz, p.368

[15] S.I Feoktistov 2001 *Automation of the design of technological processes and accessories of the blank-forming production of the aviation industry.* – Vladivostok: Dalnauka. p.183 [in Russian]
[16] Storozhev M.V. and Popov E.A 1977 *Theory of metal working with pressure, textbook for high schools*, forth ed, M., Mechanical Engineering. [in Russian]
[17] Popov E.A 1977 *Fundamentals of sheet punching theory,* second ed., Mechanical Engineering, Moscow p.278 [in Russian]
[18]] Hollomon J. H. 1945 "Tensile deformation," *Transaction of American Institute of Mechanical Engineering*, **162** 268–277
[19] Obraztsov I. F., Samples F. I., Bulychev A. A and Vasiliev V. V. 1986 *Structural mechanics of aircraft,* Mechanical engineering, Moscow, p.536 [in Russian]

IOP Conf. Series: Materials Science and Engineering **652** (2019) 012042 doi:10.1088/1757-899X/652/1/012042

A straightforward moment method to estimate the load and resistance factors

X Y Zhang
School of Civil Engineering, Central South University, 22 Shaoshannan Road, Changsha 410075, China

Email: zhangxuanyi@csu.edu.cn

Abstract. The load and resistance factor design is widely applied in engineering, where the load and resistance factors are introduced to account for the uncertainties in practice. Accurate estimation of the load and resistance factors is important for an appropriate design, which is generally conducted by the First Order Reliability Method (FORM). However, since the design point must be determined and derivative-based iterations have to be used, FORM is not practical for engineers. Furthermore, distributions of the load and resistance are required in FORM, which are usually unknown in practice. To overcome these deficiencies, moment methods are proposed, which estimate the load and resistance factors based on their moments. Although, the existing moment methods significantly reduce the iteration numbers, iterations are still necessary when higher accuracy is required. Therefore, a straightforward moment method is proposed in the present paper to estimate the load and resistance factors without iteration. The procedure of the proposed method is summarized in a flowchart with explicit expressions given. The accuracy of the proposed method is examined by comparison study among existing methods. It is shown that the proposed method is both accurate and efficient in estimating the load and resistance factors with no iteration required.

1. Introduction

In designing a structure, the insurance of the safety is the most important task that should be accomplished. To achieve this, the Load and Resistance Factor Design (LRFD) [1, 2] format is proposed and has been widely applied in practical engineering, where the safety of a structure is assured by making the nominal design resistances reduced by the resistance factors no less than the nominal design loads amplified by the load factors. The load and resistance factors are introduced to account for the uncertainties inherent in the determination of the nominal strength and the load effects. Accurate predetermination of the load and resistance factors based on specified reliability-based requirement is important for an appropriate design.

Generally, the load and resistance factors can be obtained by using the first order reliability method (FORM) [3], in which the "design point" must be determined, derivative-based iterations have to be used, and the problem of multiple design points has to be dealt with [4, 5]. The complexity of FORM would prevent the practicing engineers in general to perform it in engineering designs. AIJ [2] recommendation has provided a simple method based on the proposal of Mori [6], in which all the random variables are assumed to have known probability density function (PDF) and required to be transferred into lognormal random variables. However, in reality, the PDFs of some of the basic random variables are often unknown due to the lack of statistical data. ASCE [7] standard proposes simple equations to determine the load and resistance factors, where the sensitivity coefficients in the formulation uses approximate values. To obtain suitable load and resistance factors including random variables with unknown PDFs, Lu et al [8] proposed a method based on the moment method, where the load and resistance factors can be obtained including random variables with unknown PDFs. This moment method expands the application of LRFD into areas where the PDFs of random variables are unknown. However, it still needs iteration when high accuracy is required.

The objective of the present paper is to propose a straightforward moment method to determine the load and resistance factors with no iteration required. This paper is organized as follows. Firstly, the existing methods are reviewed. Then the proposed method is deduced with explicit formulas presented and procedure summarized. The accuracy of the proposed method is investigated with numerical examples. Finally, the findings of the present paper are concluded.

2. Review of existing methods

The LRFD format is expressed as follows

$$\phi R_n \geq \sum \gamma_i S_{ni} \tag{1}$$

where ϕ is the resistance factor; γ_i is the partial load factor to be applied to load effect S_i; R_n and S_{ni} are the nominal value of the resistance R and the load effect S_i, respectively.

The appropriate expressions of ϕ and γ_i are determined for reliability-based requirement:

$$P_f \leq P_{fT},\ \beta \leq \beta_T \tag{2}$$

where P_{fT} and β_T are the target probability of failure and target reliability index, respectively; P_f and β are the probability of failure and reliability index corresponding to the performance function:

$$G(\mathbf{X}) = R - \sum S_i \tag{3}$$

where R and S_i are the random variables representing the uncertainty in the resistance and load effects.

2.1. *FORM for LRFD*

By performing FORM, the LRFD format can be expressed as

$$R^* \geq \sum S_i^* \tag{4}$$

where R^* and S_i^* are the values of resistance R and load Si at the design point of FORM, respectively. And the load and resistance factors can be obtained as [1]

$$\varphi = R^* / R_n,\ \gamma_i = S_i^* / S_n \tag{5}$$

Since R^* and S_i^* are obtained using derivative-based iterations, explicit expressions of R^* and S_i^* are not available. Some simplifications have been proposed to avoid iterative computations [6].

2.2. *Existing moment method*

By using the second moment method, the reliability index can be obtained as

$$\beta = \beta_{2M} = \mu_G / \sigma_G,\ \mu_G = \mu_R - \sum \mu_{Si},\ \ \sigma_G^2 = (\mu_R V_R)^2 + \sum \sigma_{Si}^2 \tag{6}$$

where β_{2M} is the second moment reliability; μ_G and σ_G are the mean and standard deviation of $G(\mathbf{X})$, respectively; μ_R and μ_{Si} are the mean value of the R and S_i, respectively; V_R and σ_{Si} are the coefficient of variation (COV) of R and standard deviation of S_i, respectively. To make a balance between the economical and the reliability-based requirements, the second moment index β_{2M} is set to be equal to the targeted reliability index β_T:

$$\beta_{2M} = \beta_T \tag{7}$$

Substituting Eq. , $\mu_G = \mu_R - \sum \mu_{Si}$, $\sigma_G^2 = (\mu_R V_R)^2 + \sum \sigma_{Si}^2$ (6(6) into Eq. (7), Eq. (7) is rearranged as follows

$$\mu_R(1 - \alpha_{R\_M} V_R \beta_T) = \sum \mu_{Si}(1 + \alpha_{Si\_M} V_{Si} \beta_T) \tag{8}$$

where α_{R_M} and α_{Si_M} are the direction cosines of R and S_i, respectively, and are given as

$$\alpha_{R\_M} = \mu_R V_R / \sigma_G ,\quad \alpha_{Si\_M} = \sigma_{Si} / \sigma_G \tag{9}$$

Comparing Eq. (8) with Eq. (1), the resistance and partial load factors can be expressed as

$$\varphi = (1 - \alpha_{R\_M} V_R \beta_T)\mu_R / R_n ,\quad \gamma_i = (1 + \alpha_{Si\_M} V_{Si} \beta_T)\mu_{Si} / S_n \tag{10}$$

If R and S_i are mutually independent normal random variables, β_{2M} is the exact reliability index, and thus the load and resistance factors can be obtained exactly by using Eq. (10). However, in practical engineering, the R and S_i are usually non-normal variables and thus higher order reliability index are proposed. To take the advantage of the simple forms of Eq. (10), β_T is transformed into equivalent second-moment target reliability index β_{2T}, which is in the same space as β_{2M}, by using the higher order moment method. And then the reliability-based requirement given in Eq. (7) is transformed into

$$\beta_{2M} = \beta_{2T} \tag{11}$$

Table 1. The expressions of β_{2T} and μ_{R0} of existing methods

Name	Expression of β_{2T}	Expression of μ_{R0}
Zhao 2000 [9]	$3\{1-\exp[\alpha_{3G}(-\beta_T - \alpha_{3G}/6)/3]\}/\alpha_{3G}$	$\sum\mu_{Si} + \sqrt{\beta_T^{3.5}\sum\sigma_{si}^2}$
Zhao 2001 [10]	$\beta_T - \alpha_{3G}\left(\beta_T^2 - 1\right)/6$	$\sum\mu_{Si} + \sqrt{\beta_T^{3.5}\sum\sigma_{si}^2}$
Wang [11]	$3.8[1-\exp(-\alpha_{3G}\beta_T/3.8)]/\alpha_{3G}$	$\sum\mu_{Si} + \sqrt{\beta_T^{5}\sum\sigma_{si}^2}$
Zhao 2007 [12]	$k_2\beta_T^3 - l_1\beta_T^2 + k_1\beta_T + l_1$ $l_1 = \dfrac{\alpha_{3G}}{6(1+6l_2)}, l_2 = \dfrac{1}{36}\left(\sqrt{6\alpha_{4G} - 8\alpha_{3G}^2 - 14} - 2\right)$ $k_1 = \dfrac{1-3l_2}{1+l_1^2-l_2^2}$, $k_2 = \dfrac{l_2}{1+l_1^2+12l_2^2}$	$\sum\mu_{Si} + \sqrt{\beta_T^{3.3}\sum\sigma_{si}^2}$

The expressions of β_{2T} of different methods are listed in Table 1, where α_{3G} and α_{4G} are the skewness and kurtosis of the performance $G(\mathbf{X})$, respectively, and expressed as

$$\alpha_{3G} = [\alpha_{3R}(\mu_R V_R)^3 - \sum\alpha_{3Si}\sigma_{Si}^3] / \sigma_G^3 \tag{12a}$$

$$\alpha_{4G} = [\alpha_{4R}(\mu_R V_R)^4 + 6\sigma_R^2\sum_{i=1}^{n}\sigma_{Si}^2 + \sum_{i=1}^{n}\alpha_{4Si}\sigma_{Si}^4 + 6\sum_{i=1}^{n-1}\sum_{j>i}^{n}\sigma_{Si}^2\sigma_{Sj}^2] / \sigma_G^4 \tag{12b}$$

where α_{3R} and α_{3Si} are the skewness of the resistance R and load S_i, respectively; α_{4R} and α_{4Si} are the kurtosis of the resistance R and load S_i, respectively. With α_{3G} and α_{4G} given, β_{2T} can be easily determined with the aid of Table 1, and then the resistance factor ϕ and partial load factor γ_i are obtained by solving Eq. (11) and are expressed as

$$\varphi = (1 - \alpha_{R\_M} V_R \beta_{2T})\mu_R / R_n ,\quad \gamma_i = (1 + \alpha_{Si\_M} V_{Si} \beta_{2T})\mu_{Si} / S_n \tag{13}$$

Eq. (13) is in the same form as Eq. (10), while β_{2T} is applied in Eq. (13) instead of β_T in Eq. (10). When $\alpha_{3G} = 0$ and $\alpha_{4G} = 3$, $\beta_{2T} = \beta_T$ and Eq. (13) reduce to Eq. (10). Thus, Eq. (13) is used to represent the expressions applied to estimate load and resistance factors by moment methods.

According to Eqs. (9)-(13) and Table 1, the values of ϕ and γ_i depend on the value of μ_R, which is unknown in the design process. Therefore, Eq. (13) has to be evaluated iteratively, and the initial mean resistance, μ_{R0}, is listed in Table1. The accurate determination of μ_{R0} is necessary for an efficient method. However, according to Table 1, μ_{R0} are defined by experienced formula in all the existing moment methods, which are not precise enough and may cause the iterations. Thus a new formula is proposed in this paper to avoid the iterations required in the existing moment methods.

3. Proposed model

To simplify the procedure of estimating the load and resistance factors, a suitable formula to predetermine the μ_R is proposed. Once μ_R is determined, the load and resistance factors can be directly obtained with the aid of Eq. (9), (13), and thus the iterations are avoided.

If R and S_i (i=1, .., n) are independent normal random variables, the exact value of μ_R can be obtained by solving Eqs. (6) and (7), and is expressed as

$$\mu_R = \frac{1+\sqrt{1-\omega_R\omega_S}}{\omega_R}\sum\mu_{Si},\ \omega_R = 1-\beta_T^2V_R^2,\quad \omega_S = 1-\beta_T^2V_S^2 \tag{14}$$

where $V_S = \sum\sigma_{Si}^2/\sum\mu_{Si}$ is the COV of S, considering all S_i (i=1, .., n) as a whole load effect S.

In general, the resistance R and partial loads S_i (i=1, .., n) are not normally distributed, and then the exact value of μ_R is obtained by solving Eqs. (6) and (11) with the aid of Table 1 and given as

$$\mu_R = \frac{1+\sqrt{1-\omega_R'\omega_S'}}{\omega_R}\sum\mu_{Si} \tag{15a}$$

$$\omega_R' = 1-\beta_{2Tcheck}^2V_R^2,\quad \omega_S = 1-\beta_{2Tcheck}^2V_S^2 \tag{15b}$$

where $\beta_{2Tcheck}$ is the a checking value of β_{2T}, obtained at the checking value of μ_R with the aid of Table 1.To predetermine the value of β_{2T}, a checking value of μ_R, named as μ_{Rcheck}, is given as

$$\mu_{Rcheck} = 2\left(V_R^2+1\right)\sum\mu_{Si} \tag{16}$$

The procedure of the proposed method is depicted in Figure 1, where the procedure circled by bold dotted line is the first step of the proposed method to predetermine μ_R. As illustrated in Figure 1, with μ_R predetermined, the iterations required in the existing moment methods are avoided in the proposed method. And since the step to predetermine μ_R includes only explicit expressions, the proposed method is easy to conduct in practical engineering.

4. Application in structural reliability

4.1. Example 1

The first example considers the following performance function [7]

$$G(\mathrm{X}) = R-(D+L+S) \tag{17}$$

where R, D, L and S are the resistance, dead load, live load and snow load, respectively. The distribution and the first four moments of the random variables are listed in Table 2.

Table 2. Information of the random variables in Eq. (17).

Variables	Distribution	μ_i/D_n	V_i	σ_i	α_{3i}	α_{4i}	μ_R/R_n or μ_{Si}/S_{ni}	S_{ni}/D_n
R	Lognormal	—	0.09	—	0.271	3.131	1.06	—
D	Normal	1	0.25	0.250	0	3	1.0	1
L	Gamma	0.175	0.59	0.103	1.180	5.089	0.35	0.5
S	Gumbel	0.6874	0.21	0.144	1.140	5.4	0.982	0.7

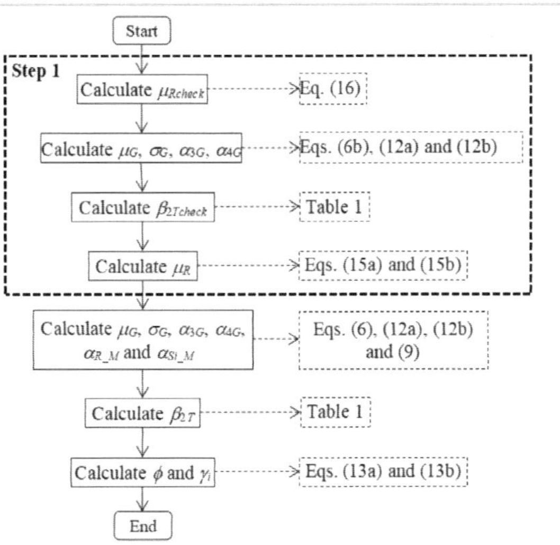

Figure 1. Procedure of the proposed method.

According to Figure 1, four steps are conducted to obtain the resistance factor ϕ and partial load factor γ_i. Different formulas in Table 1 are used, and the results are listed in Table 3. As can be observed from Table 3, the biggest relative differences among $\beta_{2T\text{check}}$ and β_{2T} obtained by different methods are relatively small. This indicates there is no big difference in the accuracy of these expressions of $\beta_{2Tcheck}$ and β_{2T} obtained by different formulas listed in Table 1. To make the proposed method more practical, Zhao 2001's formula is applied in the following example.

Table 3. Results of the proposed method.

	Zhao 2000	Zhao 2001	Wang	Zhao 2007
$\beta_{2Tcheck}$	2.9202	2.9190	2.9292	2.9667
β_{2T}	2.9810	2.9810	2.9827	3.0299
μ_R	3.0665	3.0659	3.0710	3.0898
ϕ	0.8697	0.8697	0.8694	0.8657
γ_D	1.4519	1.4519	1.4518	1.4577
γ_L	0.5038	0.5038	0.5038	0.5058
γ_S	1.1967	1.1967	1.1967	1.1995

4.2. *Example 2*

The second example considers the following performance function:

$$G(\mathrm{X}) = R - (D + L + S + W) \tag{18}$$

where W is the wind load. The statistical information of the random variables is listed in Table 4.

Table 4. Information of the random variables in Eq. (18).

Variables	Distribution	μ_i/D_n	V_i	σ_i	α_{3i}	α_{4i}	μ_R/R_n or μ_{Si}/S_{ni}
R	Lognormal	—	0.15	—	0.453	3.368	1.10
D	Normal	1	0.10	0.1	0	3	1
L	Lognormal	0.5	0.40	0.2	1.264	5.969	0.45
S	Gumbel	2.0	0.25	0.5	1.140	5.4	0.47
W	Gumbel	2.0	0.20	0.4	1.140	5.4	0.60

Since the distributions of the random variables are known, the FORM can be performed and the results are used herein as the benchmark. By using the moments of the random variables, the existing and proposed methods can be performed. The changes in the mean value of R (μ_R), the iteration numbers required and the resistance factor ϕ with the increase of β_T by using different methods are depicted in Figures 2 (a-c), respectively, which reveal that:

(i) As can be seen from Figure 2 (a), μ_R obtained by the moment methods are slightly larger than that obtained by using FORM, this error may come from the approximation of β_{2T} in the moment methods. When β_{2T} is relative small, the results obtained by the proposed method and the existing moment methods are almost the same. When the β_{2T} becomes larger, the difference between the proposed method and the existing method becomes larger, while the μ_R obtained by the proposed method becomes closer with that obtained by the FORM, which indicates the proposed method is suitable to conduct LRFD for β_T changes from 1 to 3.

(ii) As can be seen from Figure 2 (b), the derivative-based iteration numbers of FORM for LRFD varied from 9 to 22 as β_T changed from 1.0 to 3.0. The iteration numbers required in the existing method reduce significantly compared with those in FORM. However, one may have to conduct iteration steps to satisfy the tolerable error ($\varepsilon = 0.03$), except for some particular cases. There is no need of iteration in the proposed method.

(iii) As can be seen from Figures 2 (c), ϕ obtained by the existing methods and those by the proposed method coincide with each other, with the only exception of the resistance factor ϕ obtained by using Zhao 2007's formula. However, the difference in the resistance factor ϕ obtained by using the proposed and Zhao 2007's methods is not large. These again indicate the efficiency of the proposed method. The resistance factor ϕ obtained by the moment methods and FORM are different, this can be explained by that different combinations of load and resistances factors can result in the same design results. In design practice, if the resistance factor determined by a certain method is adopted, it is important that the corresponding load factors should be used.

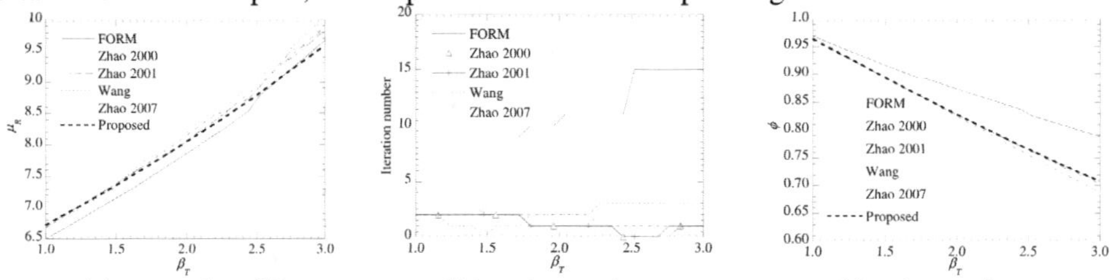

(a) mean value of R (b) iteration numbers (c) resistance factor

Figure 2. Comparison of the LRFD by using different method.

5. Conclusion

A straightforward method to estimate the load and resistance factors for reliability-based structural design without iterations is proposed. The procedure, summarized in flowchart, shows that the proposed method avoids iterations, which are needed by the existing methods. Thus the proposed method is much easier to apply. Comparison studies among the proposed and the exiting methods are conducted. The results of the proposed method are similar with those of the FORM which indicates the accuracy of the proposed method.

Acknowledgement

The study is partially supported by the National Natural Science Foundation of China (Grant nos.: 51820105014, 51738001) and the Fundamental Research Funds for the Central Universities of Central South University (Grant No.: 2018zzts030).

IOP Conf. Series: Materials Science and Engineering **652** (2019) 012042 doi:10.1088/1757-899X/652/1/012042

References

[1] Ang A, Tang W 1984 *Probability Concepts in Engineering Planning and Design* (New York: Wiley)
[2] AIJ 2002 *Recommendations for Limit State Design of Buildings* (Tokyo: Architectural Institute of Japan)
[3] Nowak A, Collins K 2000 *Reliability of Structures* (Singapore: MCGraw-Hill International Editions)
[4] Der-Kiureghian A, Dakessian T 1998 *Struct. Saf.* **20**(1) 37-49
[5] Barrabco-Cicilia F, Castro-Prates de Lima E, Sudati-Sagrilo L 2009 *Ingenieria Investigación y Tecnologia,* **10**(2), 87-97
[6] Mori Y 2002 *J. Struct. Constr. Eng.* **559** 39-46
[7] ASCE. *Minimum design loads for buildings and other structures*. ASCE Standard 7-10. ASCE, Reston, Va, 2010.
[8] Lu Z, Zhao Y, Ang A 2010 *Struct. Eng. Mech.* **36**(1) 19-36
[9] Zhao Y.G. et al. 2000 *J. Struct. Constr.* **530**, 21-8
[10] Zhao Y.G. et al. 2001 *J. Struct. Constr.* AIJ, **548**, 21-6
[11] Wang J, Lu Z, Zhang X, Saito T. Zhao Y 2017 *J. Asian Architect. Build. Eng.* **16**(1), 171-8
[12] Zhao Y, Lu Z, 2007 *J. Struct. Eng.* **133**(7), 916-4

Property assessment of concretes with graphene oxide mixed cement

Arash Radman[1,2] and Nazanin Joorabchi[1]

[1]University of Guilan, Faculty of Engineering, Khalije Fars Street, Gilan, Iran

Abstract. In the present study a number of properties of concrete samples are evaluated, in which, graphene oxide nanomaterial is mixed with ordinary Portland cement. Workability, compressive strength, flexural (tensile) strength, modulus of elasticity and chemical resistance to acidic environments of concrete samples, in which graphene oxide with weight ratios of 0.025% and 0.05% are added to ordinary Portland cement, are assessed and compared with plain concrete. This study shows that adding graphene oxide improves the compressive and flexural strength of concrete samples, while decreases their workability. It is found that the addition of graphene oxide slightly increases the modulus of elasticity of concrete samples. In 5 % hydrochloric and sulphuric acid solutions, samples with graphene oxide exhibit less mass loss at the age of 84 days, which suggests better resistance to acidic environments and durability for these types of concretes.

1. Introduction

Advances in construction technology put new demands on the properties of concrete. Therefore, in addition to main ingredients of concrete (i.e. cement, water and aggregates), admixtures are frequently used in mix designs, to improve and tailor concrete properties toward desired functions. These materials can provide concretes with properties that cannot be achieved by ordinary Portland cement (OPC) [1].

Nanotechnology is the science of tailoring materials' structures at atomic scale so that the desired properties at macro scale to be achieved [2, 3]. Together with advances in nanotechnology, nanomaterials have been introduced into the cement and concrete technologies with the aim at improving their properties. Effects of nanomaterials, in forms of nanoparticles, nanotubes, and nanofibres have been extensively examined in the concrete technology. The research on the application of nanotechnology in the cement industry has been reflected in several review papers [1, 4, 5].

Graphene oxide (GO) is a type of nanomaterial which is known for its extraordinary electrical, mechanical and thermal properties and has been studied for applications in cement industry [4, 6]. GO consists of a hexagonal carbon network with oxygen-containing functional groups. Single layer GO is usually manufactured using Hummers modified method [7]. The particle size of GO is smaller by four orders of magnitude than OPC. In contrary to the other forms of carbon nanomaterials, GO is hydrophilic and is highly dispersible in water [8, 9]. GO, with two-dimensional atomic structure, provides a large surface with high quality connection with calcium silicate hydrate (C–S–H) phases of OPC and in this respect has some superiority to other forms of carbon-based nanomaterials such as

[2] Corresponding Author: a.radman@guilan.ac.ir

carbon nanotube (CNT) [10, 11]. Presence of oxygen functional groups in GO reduces its elastic modulus and tensile strength in comparison with pure graphene [12]. Even though, the elastic modulus of GO ranges between 23 to 42 GPa and its tensile strength is around 130 MPa, which are still much higher than that of the cement matrix [13, 14].

One of the critical aspects of using any admixture in concrete is the control of hydration rate. It is known that nanomaterials generally decrease the setting time and increase the hydration rate of OPC [15, 16]. Oxygen functional groups of GO are the locations that have high reactivity with water and produce hardened phases at higher rates [11]. In addition to accelerated hydration, GO has a greater inclination towards adsorption of water onto its surface. One of the negative impacts of this characteristic is the reduction of workability of cement paste. Mini-slump tests indicate reduction in workability from 34% to 50% in different studies [6, 10, 17]; therefore, using superplasticizers are inevitable in most cases [6, 10, 8].

The increase in early age strength of cement pastes may imply that a long-term durability is achievable. Scanning electron microscope (SEM) and mercury intrusion porosimetry (MIP) experimental data indicate that the application of GO decreases the capillary porosity (that rages between 10nm to 10µm) while increases 1 to 10 nm pore volume in standard cement pastes [10, 11]. This is attributed to the higher hydration rate of GO [6, 11]. However, these opposite impacts make the effects of GO in concrete durability an open question. Moreover, oxygen functional groups of GO usually reduce the pH of the environment [19]; thereby, the alkaline phases are to some extent neutralized. Reduction of pH makes the concrete more vulnerable to environmental conditions.

The feasibility of application of GO as a nanoscale reinforcement to OPC has been studied by researchers [1, 10]. So far, majority of these studies have been conducted on mortars or pastes, in which GO has been added to OPC as a binder. The results of such studies on concrete samples may be quantitatively different. In this study, 0.025% and 0.05% (by weight of cement) GO is incorporated with OPC, and the mechanical properties of the resulted concretes in terms of workability, compressive strength, flexural strength (modulus of rupture) and modulus of elasticity have been measured and compared with plain concrete. The test procedures are based on the known standards of ASTM. In order to evaluate the chemical resistance of these types of concretes to acidic environments, samples with 0.05% of GO are tested in 5 % HCl and 5 % H_2SO_4 solutions and their mass reductions are compared with plain concrete samples.

2. Methodology

2.1. Materials Used

Single layer graphene oxide was used in this study. It was produced base on modified Hummer's method and provided in a slurry form. The elemental analysis of the used GO is presented in Table 1. The particle sizes of applied GO were ranging from 2.7 to 6.8 µm with dispersion pH of 3.5. The monolayer content was more than 95%. The atomic force microscopy (AFM) shows that the atomic thickness of the used GO sheets lied between 0.8 to 1.1 nm.

The cement used for this study was Portland cement type I, as per ASTM C150. Crushed granite aggregates were used in this study. The maximum size of aggregates were 13mm (Grain size 1) and 19 mm (grain size 2). In this study, it was necessary to use Polycarboxylate ether based superplasticizer, in some tests.

Table 1. Elemental analysis of applied GO.

Element	Atomic concentration
Carbon	60%
Nitrogen	1%
Sulphur	1%
Oxygen	38%

2.2. Mixing Procedure
In this study, 0.025% and 0.05% (by weight of cement) GO were incorporated with OPC concrete. Table 2 shows different mix proportions that were used in this study.
One of the key parameters in application of nanomaterials in concretes and mortars is the dispersion of these types of materials in fresh concrete mixes. In spite of the fact that GO is highly dispersible in water [8, 9], its dispersion in the fresh concrete is a challenge. It has been shown that using high speed shear mixer can help GO to remain dispersed in the fresh mix [17]. Although the applied GO was in slurry form, in this study a hand mixer with a rotation speed of 2000 rpm has been used for remixing the product with cement and water for 5 minutes, before adding to the rest of concrete ingredients.

Table 2. Mix proportions.

Mix No.	Water kg/m^3	Cement kg/m^3	GO g/m^3	Fine Aggregate kg/m^3	Course aggregate (Grain size 1) kg/m^3	Course aggregate (Grain size 2) kg/m^3	Superplasticizer kg/m^3 polycarboxylate ether solid content
OPC1	168	365	-	810	-	990	-
OPC2	168	365	-	810	990	-	-
GOPC1	168	365	182.5	810	-	990	-
GOPC2	168	365	182.5	810	-	990	0.65
GOPC3	168	365	91.25	810	-	990	0.325
GOPC4	168	365	182.5	810	990	-	0.65

2.3. Experiment Method
2.3.1. Compressive Strength. For determining the compressive strength of specimens, the standard test method of ASTM C 39-04a has been adopted. Tests were conducted on 150mm diameter by 300mm tall concrete cylinders. Specimens were cured under water at 20 ± 2°C and tested at the ages of 7, 14, 28, 56 and 84 days. Specimens from OPC1, GOPC1 and GOPC2 mixes were used for compressive strength tests (see Table 2).
2.3.2. Flexural Strength Test. A third-point loading test method was used for assessment of flexural strength (tensile strength). Specimens with dimensions of 550mm×150mm×150mm were used in this type of test. Span length of test apparatus was 500-mm. Other specifications of the test were in accordance with ASTM C 78-02.
2.3.3. Slump Test. Slump tests were in accordance with ASTM C143 test method.
2.3.4. Modulus of Elasticity. Chord modulus of elasticity was measured based on ASTM C469-02.
2.3.5. Chemical Resistance Test. Tests on chemical resistance were conducted in conformity with the ASTM C 267-01 ''Standard Test Method for Chemical Resistance of Mortars, Grouts, and Monolithic Surfacings and Polymer Concretes''. Specimens were cured under water for 28 days, at 20 ± 2 °C, and then they were immersed for 84 days in 5 % HCl and 5 % H_2SO_4 solutions. At the ages of 7, 14, 28, 56, and 84 days, the specimens were washed with water, cleaned, and dried in the air and then their weights were measured. Weight changes of specimens were calculated using equation (1):

$$Weight\ loss, \% = \left(\frac{W-C}{C}\right) \times 100 \quad (1)$$

in which, C is the conditioned weight of the specimen (gram), and W is the weight of specimen after immersion (gram).

3. Results and Discussion
3.1. Fresh Concrete Workability
Figure 1 compares slump test results of samples containing GO (GOPC1 and GOPC2 with 0.05% GO) and plain concrete (OPC1) at 15minutes and 120 minutes after mixing ingredients. As it can be seen from the figure, 15minutes after mixing, samples containing GO without superplasticizer (GOPC1) have considerable lower slump (5mm) in comparison with plain concrete (60mm); therefore, for the

rest of the tests polycarboxylate ether based superplasticizer has been applied (Table 2) to the mixes containing GO (GOPC2). The amount of superplasticizer has been determined experimentally to have identical slump with that of plain concrete at 15 minutes after mixing.
The results show that adding graphene oxide to the fresh concrete causes significant loss of workability. The reduction of workability, as a result of introducing graphene oxide to cement paste, has been reported in literature (for example in [6, 10]). This is generally attributed to the large specific surface of graphene oxide that adsorbs more free water onto its surface. The lack of free water increases the frictional resistance between mix ingredients.

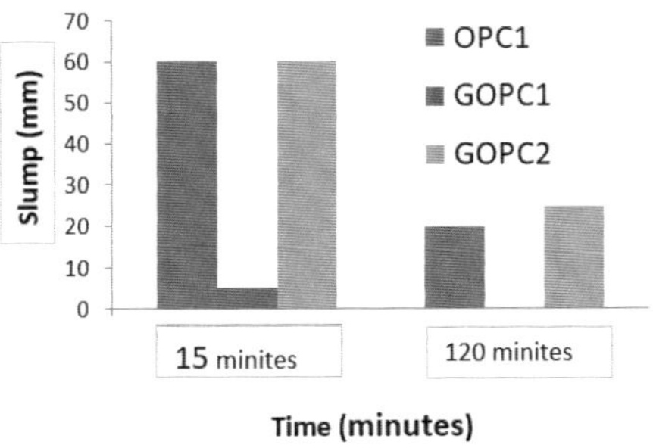

Figure 1. Slump loss versus time in plain concrete (OPC1), concrete containing 0.05% (of the cement weight) graphene oxide without superplasticizer (GOPC1) and concrete containing 0.05% graphene oxide with superplasticizer (GOPC2).

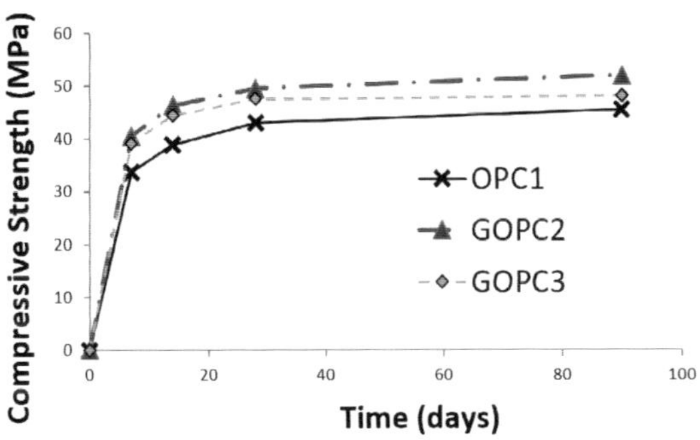

Figure 2. Compressive strength development of plain concrete (OPC1), concrete containing 0.05% (of the cement weight) graphene oxide (GOPC2) and concrete containing 0.025% graphene oxide (GOPC3).

3.2. Compressive Strength

Figure 2 shows the strength development of wet cured samples of plain concrete (OPC1), samples with 0.05% GO (GOPC2) and 0.025% GO (GOPC3) with time. Comparison between samples containing 0.05% GO with plain concrete shows that the addition of graphene oxide increases the compressive strength of samples to about 18% at the age of 14 days and about 15% at 28 days. The long-term strength development at the age of 90 days shows an increase in compressive strength of about 14% in samples with 0.05% graphene oxide. Moreover, the test results of concrete samples with 0.025% graphene oxide (by weight of cement) suggest that the increase of compressive strength is almost linear when GO content is within the range of 0% to 0.05%.
In reference [10] the compressive strength of OPC paste samples, mixed with graphene oxide, are compared with those of plain OPC paste samples; it is reported that the samples containing 0.03% (by weight of OPC) graphene oxide exhibit 46% improvement in compressive strength. It is generally known that the mechanical properties of cement pastes and concrete samples might be substantially different, partly because there are less cementitious materials in unit volume of concrete samples.
Moreover, as it was mentioned before, oxygen functional groups on the surface of GO are the locations that produce hardened connection with the cement matrix. It is believed that this mechanism enhances the mechanical properties, such as compressive and tensile strength [11, 15].

3.3. Modulus of Elasticity

Comparison between chord moduli of elasticity between plain concretes (OPC1) and concretes containing 0.05% graphene oxide (GOPC2) is given in Table 3. An approximately 7.1% increase in modulus of elasticity is observed in the concrete samples containing 0.05% GO. The samples were wet cured for 28 days before testing.

Modulus of elasticity increases with increase in the compressive strength of concrete, but there is no agreement on the form of the relationship [20]. It is also known that the modulus of elasticity is affected by the quality of the bond between cementitious matrix and aggregate phases of concrete. When the bond is stronger [20], as is in the case of concretes containing GO, the modulus of elasticity is usually higher.

3.4. Flexural Strength

The results of third-point loading tests on 28-day flexural strength of concrete samples with 0.05% GO (GOPC2) and plain concrete are shown in Table 3. Test results show that the flexural strength of the samples with 0.05% GO is approximately 9.4% higher than that of plain concrete.

In reference [10] it is reported that the flexural strength of the paste samples containing 0.05% GO is approximately 50% higher than that of the plain cement paste samples. The improvement in tensile strength could be a result of refinement of pore structure in samples containing GO [10]. In addition, it is generally recognized that GO is able to suppress propagation of nanoscale cracks in the cementitious matrixes [3], which is another possible reason for the increase in flexural strength of these types of concretes.

Table 3. Comparison between flexural strength and elastic modulus of plain concrete (OPC1) and concrete containing 0.05% graphene oxide (GOPC2) at 28 Days.

	OPC1	GOPC2
Flexural strength (MPa)	4.87	5.33
Elastic modulus (GPa)	39.19	41.97

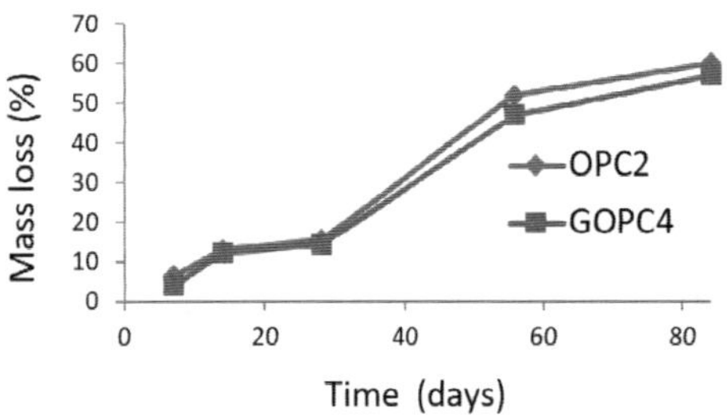

Figure 3. Mass loss resulted from immersion in 5% HCl.

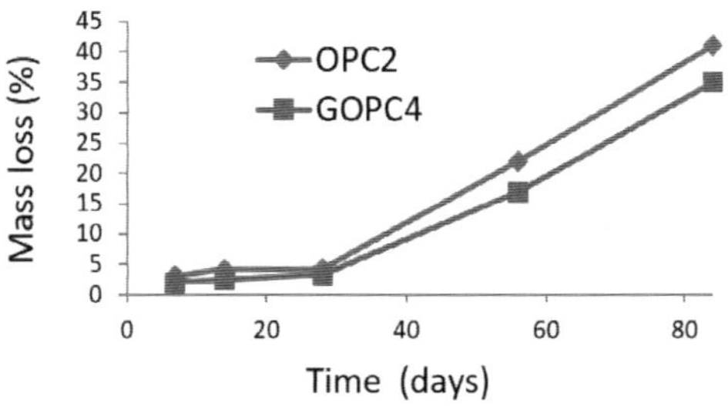

Figure 4. Mass loss resulted from immersion in 5% H_2SO_4.

3.5. Chemical Resistance

Test results on weight loss of specimens that were immersed in 5 % HCl and 5 % H_2SO_4 aqueous solutions, are shown in Figure 3 and Figure 4 respectively. All specimens' masses are decreased with time. At the age of 84 days, for specimens immersed in HCl the mass losses are 60 % for the samples made with OPC and 57% for the samples containing 0.05% of graphene oxide. Likewise, for specimens immersed in H_2SO_4 the mass losses are 41 % for the samples made with OPC and 35% for the samples containing 0.05% of graphene oxide at the age of 84 days.

An acidic attack to concrete involves chemical reactions between the acid and other components of the concrete. Acid attack dissolves the binder from the concrete surface. The acid permeation is highly dependent on the amount and structure of pore and cracks inside the concrete [21]. Previous studies show that inclusion of GO refines the capillary pores of cement paste [6, 10] thereby reducing the influence of acidic environment.

4. Conclusions

The results of this investigation indicate:
-Addition of graphene oxide significantly reduces the workability of concrete,
-Concrete samples containing 0.05% GO exhibit 18% , 15% and 14% increase in compressive strength at 14, 28 and 90 days of age with respect to plain concrete,
-An approximately 7.1% increase in modulus of elasticity is observed in the samples with 0.05% GO,
-Concrete samples containing 0.05% GO show 9.4% higher flexural strength at the age of 28 days with respect to plain concrete,
-Incorporation of GO slightly increases the durability of concrete in HCl and H_2SO_4 acid solutions.

References

[1] Chuah T J S Pan Z Sanjayan J G Wang C M and Duan W 2014 *Const. and Build. Mat.* **73** pp 113-124
[2] Grove J Vanikar S and Crawford G 2010 *J. of the Trans. Res. Board* **2141** pp 47-51
[3] Singh V Joung D Zhai L Das S Khondaker S I and Seal S 2011 *Prog.in Mat. Sci.* **56** no 8 pp 1178-1271
[4] Sanchez F Sobolev K 2010 *Cons. and Build. Mat.* **24** Issue 11 pp 2060-2071
[5] Raki L Beaudoin J Alizadeh R Marker J and Sato T 2010 *Materials* **3** no 2 pp 918-942
[6] Pan Z Duan L Li D and Collins F 2013 *WO Application Patent* WO2013096990A1
[7] Li D Müller M B Gilje S Kaner R B Wallace G G 2008 *Nat Nanotechnol* **3** no 2 pp 101-105
[8] Qiu L Yang X Gou X Yang W Ma Z F Wallace G Li D 2010 *Chemistry* **16** pp 10653-10658
[9] Konstantin S Miguel F G 2014 Prc. *Am. Cer. Soc.* https://doi.org/10.1002/9780470588260.ch16
[10] Gong K Pan Z Korayem A H Qiu L Li D Collins F Wang C M and Duan 2015 *J. of Mat. in Civil Eng.* **27**
[11] Lv S Ma Y Qiu C Sun T Liu J and Zhou Q 2013 *Con. and Build. Mat.* vol 49 pp 121-127
[12] Garg A Sinnott S B 1998 *Che. Phy. Let.* **295** pp 273-278
[13] Zhu Y Murali S Cai W Li X Suk J W Potts J R and Ruoff R S 2010 *Adv. Mat.* **22** pp 3906-3924
[14] Dikin D A Stankovich S Zimney E J Piner R D Dommett G H Evmenenko G Nguyen S T Ruoff R S 2007 *Nature* **448** no 7152 pp 457-60
[15] Vera-Agullo J Chozas-Ligero V Portillo-Rico D García-Casas M J Gutiérrez-Martínez Mieres-Royo J M and Grávalos Moreno J 2009 *Nanotechnology in Construction* Springer Berlin Heidelberg
[16] Makar J Margeson J and Luh J 2005 *Proc. 3rd Int. Con. on Const. Mat* Vancouver pp 1-10
[17] Gong K Tan T Dowman M Qiu L Li D Collins F G and Duan W 2012 *Proc. 6th int. comp. conf. ACUN-6* Monash University Melbourne Australia
[18] Swamy R 1994 In: ICE Proceedings ICE Virtual Library
[19] Li D Müller M B Gilje S Kaner R B and Wallace G G *Nat. Nanotec.* **3** pp 101–105
[20] Neville A M 2011 *Properties of Concrete* Fifth Edition Essex England
[21] Newman J Choo B S 2003 *Advanced Concrete Technology* Great Britain

IOP Conf. Series: Materials Science and Engineering **652** (2019) 012044 doi:10.1088/1757-899X/652/1/012044

Effect of curing temperature and particle size distribution on unconfined compressive strength of raw and treated fly ash-lime modified phosphogypsum waste

T Mashifana[1,2], F Okonta[2] and F Ntuli[1]

[1]University of Johannesburg, Department of Chemical Engineering Technology, P.O. Box 17011, Doornfontein 2088, South Africa
[2]University of Johannesburg, Dep. of Civil Engineering Sciences, P.O. Box 524, Auckland Park 2000, South Africa
Email: tmashifana@uj.ac.za; fnokonta@uj.ac.za; fntuli@uj.ac.za

Abstract. The Voluminous quantity of phosphogypsum waste is generated by phosphoric acid manufacturing industry and end up being landfilled, resulting into environmental pollution. This study was done to determine applications for raw and treated waste phosphogypsum in building and construction, to determine the best conditions to produce the final product. Two readily available wastes namely phosphogypsum and fly ash were used and blended with hydrated lime. Best strengths was obtained from a mixture of phosphogypsum-lime-fly ash specimens were at elevated temperatures. The strengths obtained at the temperature of 80°C were above the 3.5 MPa permissible strength for building masonry with unconfined compressive strength up to 4.8 MPa. At lower temperature of 40°C the optimum strength obtained was at the minimal phosphogypsum content of 30%. The optimum strength was achieved at the PG content of 50% for raw PG and 60% for treated phosphogypsum. The strengths obtained for the treated phosphogypsum were lower than the minimum permissible load bearing strength due to the non-uniform particles distribution and the presence of impurities and forces of adhesion between citric acid and phosphogypsum. Increasing the curing temperature also reduced the radionuclides initially present in the raw PG without any prior treatment.

1. Introduction

Utilization of solid waste is extremely significant to the sustainable development of the economy and society due to the depletion of raw resources. Phosphogypsum (PG) and fly ash (FA) are the wastes produced in large quantities in South Africa. The quantities of this waste occupy notably large land and discharged to the environment without any prior treatment, and result in bringing on heavy air, soil and water pollution [1]. PG stockpiles with a quantity of 2 million tonnes are located at the now non-operational plant of AECL in Chloorkop and 5 million tonnes with an increase by 240 000 tonnes per annum at the Potchefstroom plant [2]. PG is a by-product of the manufacture of phosphoric acid by a wet chemical process and normally have a $CaSO_4.H_2O$ content higher than 95%. Studies have been conducted to make PG into binders and wall bricks or partitions [3-6].

Around 6.5 billion m^3 of road base courses are constructed per annum in China. Although a few applications of fly ash, Portland-fly ash cement, clay—fly ash bricks, sand—lime bricks, etc., have been developed enormous quantity of this waste is still unutilized [6]. These volumes of unutilized industrial wastes can be reused in combination of phosphogypsum, fly ash and lime, for the building and construction industry [6]. Kumar investigated the use of gypsum-fly ash-lime material to make bricks and hollow blocks of adequate strength, an economical alternative to burnt clay hollow bricks, burnt clay bricks and concrete hollow blocks [6]. Mashifana et al. (2018), studied the geotechnical properties and application of lime modified phosphogypsum [7]. The results obtained showed that, the stabilization of mild acid treated phosphogypsum with lime-fly ash significantly improved the unconfined compressive strengths, liquid limit and plastic limit of the material [7]. In this study a new type of material from the mix design of PG-Lime-FA which yielded the strengths above the minimum permissible strength for building brick were produced and the best conditions to produce these material were investigated.

2. Experimental details

Phosphogypsum was collected from Phalaborwa mine in South Africa. To determine the optimum mix design that yielded the highest unconfined compressive strength (UCS) and the optimum curing temperature, material designated Raw 1 was produced by blending 30% phosphogypsum (PG), 50% fly ash (FA) and 20% hydrated Lime (L). Mix design designated Raw 2 was the mixture of 40% PG, 40% FA and 20% L. The third mix, Raw 3, contained 50% PG, 30% FA and 20% L. The last mix prepared, designated Raw 4 was made up of 60% PG, 20% FA and 20% L. Only waste PG and FA were varied in the mix designs and hydrated lime as a commercial product was kept constant at a proportion of 20%. The test specimens were prepared by blending the raw phosphogypsum, raw fly ash, lime powder and water at optimum water content. Standard proctor compaction tests were conducted on the produced mix designs, to determine the maximum dry density (MDD) and optimum moisture content (OMC) for the unconfined compressive tests.

The brick specimen was cast in a 100mm x 100mm mould and cured at different temperatures of 40 0C and 80°C, for 4 days. The unconfined compressive strength for the different mix designs was determined. The paste produced was characterised using X-ray diffraction (XRD), Energy dispersive X-Ray spectroscopy (EDS), X-ray fluorescence (XRF) and Scanning electron microscope (SEM) to determine the chemical composition and the morphology of the final produced material. PG was treated with citric acid, oxalic acid, sodium bicarbonate and sodium bicarbonate and the best reagent that reduced the radionuclides in the material was found to be citric acid. Citric acid was used as a leaching reagent to treat the material at room temperature, leaching for 24 hours. The slurry was then filtered, dried in the oven for 24 hours at 80°C and then analysed for chemical composition and morphology by XRD, XRD and SEM. The final product obtained was blended with FA and lime with the same proportion as the raw material and followed the same procedure.

3. Results and discussion

3.1. Elemental composition of raw material

The elemental compositions of the materials utilized to prepare the mix designs for raw and treated phosphogypsum, fly-ash and lime are presented in Table 1. The raw phosphogypsum was laden mainly with sulphur trioxide and calcium oxide and consisted of radionuclides. After the chemical treatment majority of the contaminants that hinder some of the processes in civil engineering were reduced, thus radionuclides, phosphorus and fluorides. Radionuclides are nor desirable in any building material as they are radioactive and can contaminate the environment and affect people health. The hydrated lime utilized was dominated with calcium oxide and fly ash consisted of constituent such as silica, calcium oxide, ferrous iron and titanium oxide.

3.2. The effect of curing temperature on the UCS

The UCS obtained for raw PG at different temperatures is presented in Figure 1(a). The strength of the composites Raw PG30, Raw PG 40, Raw PG 50 and Raw PG 60 increased with the increase in curing temperature. The highest strength is evident in composite with the concentration of 20% L, 50% PG and 30% FA at the temperature of 80°C. At lower temperatures UCS was the highest at PG 30. At

80°C, the UCS almost doubled and is the highest at PG 50. Although there was a slight decrease in the UCS from 80°C – 100°C, the UCS at 100°C is higher than the lowest temperature investigated and the highest UCS is also observed at PG 50. For raw PG the optimum curing temperature that yielded the highest UCS of 4.98 MPa is 80°C for the composite containing 50% PG. The optimum strength obtained at PG 50 is more than the minimum permissible burnt masonry clay according to The South African standard (SANS 227, 2007) which requires a minimum of 3.5 MPa [8]. Therefore the material can be used in a load bearing wall. Increasing the content of PG resulted in increased UCS, a trend clearly visible when curing at 80°C. For lower temperatures, the UCS decreased with increasing PG content.

The UCS results obtained for treated PG, the specimen cast at the respective MDD and OMC and cured at lower and elevated temperatures of 40°C, 80°C and 100°C are shown in Figure 1(b). The highest strength is observed at the minimal PG content of 30% for all the curing temperatures. The PG content of 50 % yielded the lowest strength and a clear observation of a decrease in UCS with the increasing PG content is evident, the opposite of what was obtained for raw PG. Similar observation occurred with the elevated temperature, the UCS obtained at 80°C and 100°C is higher than the UCS at 40°C.

The highest strength obtained is 1.5 MPa at 80 °C. Therefore, curing temperature has a direct effect on the unconfined compressive strength of PG-Lime-Fly Ash composite for both treated and untreated PG.

The unconfined compressive strengths for the treated PG specimen is lower that the strength for the raw PG, this can be attributed to the particles disintegration with each other in the presence of citric acid, due to morphological change. Ionization carboxylic groups giving rise to the highest anionic charge density exist in citric acid, resulting in maximizing citric adsorption on the surface of phosphogypsum; this was proven by the increase in the surface area of the treated PG.

The two oxygen ions in carboxylic group of citric acid are exactly the same at 4.0 and the forces of adhesion in gypsum crystals are higher in magnitude [9]. Lanzón and García-Ruiz (2012), found that the treatment of gypsum with citric acid significantly reduces the compressive strength of gypsum [10]. The measurement of the material hardness also revealed a decreased with citric acid treated gypsum. Researchers reported that gypsum crystal habit determines the physical properties and compactly of the material. In study when citric was used to treat PG, the material showed a reduction in UCS and became less rigid due to the lower degree of interlocking [11-14]. These results are similar to those reported by [10].

Lime combined with fly ash reacted with silica and alumina pozzolans and formed strong cementitious matrix that characterizes a lime-stabilized layer. To study the mechanism that has occurred when raw PG and treated PG were stabilised with lime and fly ash and subjected to elevated temperatures and to understand what influenced and contributed to higher strength for the raw PG composite and lower strength for treated PG composites, a thorough study was conducted on the mineralogy of the composites to know the hydration products formed and on the particle size distribution. The obtained composites after strength determination were analysed and the results are presented in the following sections.

Only comparison between the lower temperatures of 40°C and only 80°C, were investigated further as the optimum elevated curing temperature that yielded the highest UCS is 80°C and no further tests on curing temperature of 100°C were conducted. The standardized documents reviewed by Delgado and Guerrero (2007) indicate that compressive strength standards range between 1.3 and 2.1 MPa for use in non-load bearing walls [15].

In 2011, Deboucha and Hashim's study suggest that in practice typical compressive strength in up to 1 storey construction range from 1 – 4MPa [16]. Some of the national standards prescribing minimum compressive strength of solid clay and concrete masonry units, where the minimums are given as 8.6, 5 and 3.5 MPa respectively are reported [17-19]. The raw PG composites attained strengths more than the minimum permissible by SANS 1215 [19]. For chemically stabilised pavement material following

IOP Conf. Series: Materials Science and Engineering **652** (2019) 012044 doi:10.1088/1757-899X/652/1/012044

97% modified AASHTO, the required minimum and maximum unconfined compressive strengths for C1-C4 material are as follows; 6-12 MPa, 3-6 MPa, 1.5-3 MPa and 0.75-1.5 MPa [20].

Table 1. Elemental analysis of the raw material (wt%).

Component (%)	Raw PG	L	Treated PG	FA
F	1.0633			
Al_2O_3	0.2276	0.279	0.091	28
SiO_2	1.3702	0.503	0.857	47.9
P_2O_5	1.2839	0.0288	0.719	0.725
SO_3	51.0129	0.191	54.4	0.606
CaO	43.6526	73.4	42.75	5.13
TiO_2				2.47
Fe_2O_3	0.1214	0.225	0.048	4.83
Total				
Radionuclides	0.6198		0.405	

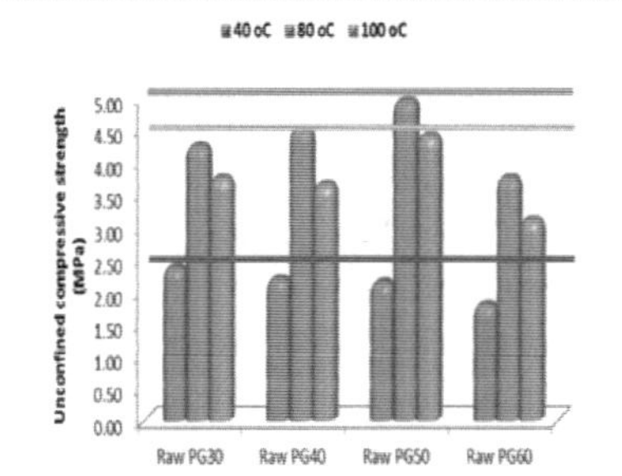

Figure 1(a). Figure 1(a). UCS of mix designs at lower and elevated temperatures for raw PG.

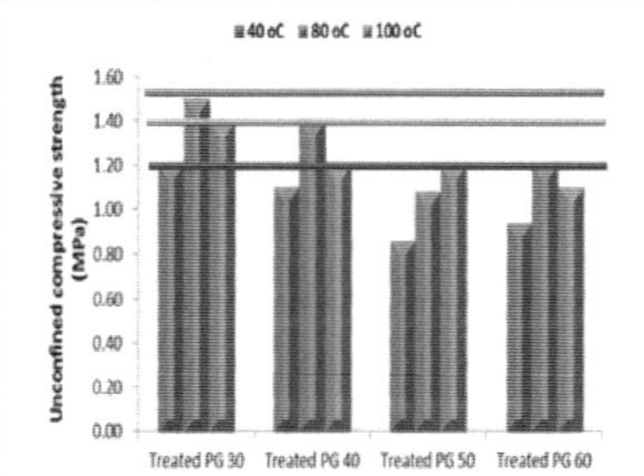

Figure 1(b). Figure 1(b). UCS of mix designs at lower and elevated temperatures for raw PG.

3.3. Mineralogy analysis of the composite (raw PG) at elevated temperatures

The gypsum content is increasing with the increase in PG content for composites cured at 80 oC, Figure 1(a). A significant increase of gypsum is observed in the PG composites cured at elevated temperature, ranging from 81% to 94%. At raw PG 50, the constituents obtained with their respective contribution are those of gypsum, $CaSO_4.0.5H_2O$ (93.69%), calcium aluminium sulphate, Ca6A12$(SO_4)_3(OH)_{12}\cdot 26H_2O$ (4.64%), hypothetical silica, SiO_2 (0.41%), calcium iron oxide, $CaFeO_3$ (0.09%) and hedenbergite calcium, $CaFeSi_2O_6$ (1.17%). The predominant constituents in the composites produced at lower and elevated temperatures are that of gypsum and silica. The highest unconfined compressive strength was achieved in a composite containing 20% lime, 50 % PG and 30% FA, when the material was cured at 80 0C. The results also indicate that with increasing temperature to 80 0C, gypsum ($CaSO_4 \bullet 2H_2O$) initially present in the material, was transformed to $CaSO_4 \bullet 0.5\ H_2O$, indicating a decrease in water absorption, due to the voids and pores in the matrix being filled by the new constituents that were formed. There were new hydration products formed at PG 50 namely; calcium aluminium sulphate $Ca_6Al_2\ (SO_4)_3(OH)\ _{12}\cdot 26H_2O$, siliminanite, Al_2SiO_5, calcium iron oxide, $CaFeO_3$, hedenbergite calcium, $CaFeSi_2O_6$. A hydrated calcium sulphate has a distinct powder

diffraction pattern, with a different structure and content. Curing consumed the hydrated lime by the pozzolanic reactions, as the result calcium silicates, aluminates, and alumina silicates the durable constructional material compounds were precipitated during the curing process. The results obtained agree with the study conducted by [21]. The study conducted by Plowman and Caprera (1984) reveals that lime addition activates the alumina and silica phases and the final material contained silimanite and hypothetical silica with higher proportion than the raw material [22]. The hydration product in a poriferous PG-lime-FA specimen result into particles integration with each other. 50% (1:1.5) lime to fly ash content to 50% PG content yielded the highest strength. Thus indicating that at PG 30 and PG 40, fly ash is in axcess and lime is not sufficient for complete pozzolanic reaction to take place.
When the fly ash -lime ration to phosphogypsum is reduced from 70:30 to 50:50, maximum strength is achieved, showing that the optimum pozzolanic reaction between lime-fly ash which contributes to strength development takes place at fly ash content of 30%. Below 30% fly ash content, lime content to fly ash might be in excess for the pozzolanic reaction to occur for the stabilization of phosphogypsum, hence the drop in strength from PG 50 to PG 60. Reddy and Gourav (2011) reported similar result when they investigated the strength of lime–fly ash, compacted using different curing techniques and gypsum additive [23]. There was a decrease in the composite strength attributed to lime: fly ash ratio. The highest peak on the raw PG composite cured at elevated temperature ranging from 1180 cps to 2710 cps at 2θ is associated to gypsum.
The XRD analysis of the treated PG 50 at elevated temperature of 80 oC is shown in Figure 2(b). The identified constituents are Gypsum (G), $CaSO_4.2H_2O$ (32.35%), calcium aluminium sulphate, $Ca_6Al_2(SO_4)_3(OH)_{12}\cdot 26H_2O$ (33.18), wollastonite (W), $CaSiO_3$ (17.93%), hedenbergite (H), CaFeSi2O6, (1.48%) and keatite (K), SiO_2 (15.01%). The gypsum content in the treated PG composite is less pure than PG in the raw composite and less pure than the PG before curing. The new hydration products formed are, calcium aluminium sulphate, wollastonite and hedenbergite. The XRD result agrees with the XRF as the constituents in the treated PG mineralogy are associated with silicate, iron and calcium, the chemicals also present in high weight % as reported using XRF. There was a reduction in the gypsum content as compared to the treated PG prior curing, a reduction of 65% showing that during the curing process, gypsum was consumed. Due to the addition of fly ash, an increase of 87% silicate oxide is observed, a compound which has silicate as the predominant compound. No radionuclides were detected in the treated PG composite. In 2010, Huang and Lin on their investigation on phosphogypsum–steel slag–granulated blast-furnace slag–limestone cement, reported an increase in peaks of ettriggite and a decrease in that of PG due to the consumption of PG during hydration and formation of more ettringite [24].

3.4. pH and relative density of the raw and treated PG

The composites the density for treated PG composites was higher than that of the raw PG composites. Therefore, the specific gravity did not play a major role in contributing to the high strength obtained for the raw PG composites and the lower strengths for the treated PG material. Relative density is the most representative measure of the degree of compactness of soil [25]. Thus the low relative density accounted directly for the low MDD, porous inter granular packing and low mobilized strength (UCS – bond stiffness of the binder and granular interlocking friction of fabric). Both the raw and treated PG composites represents an alkaline materials. Although the raw PG and treated PG materials initially had an acidic pH of 4.17 and 2.31 respectively, blending the material with fly ash and lime and the development of composites, changed the pH to alkaline. This is due to the presence of hydroxide ion from lime used to stabilize PG which contributed to the alkalinity of the material. Lime stabilization causes a significant improvement in soil texture and structure by reducing plasticity and by providing pozzolanic strength gain [26]. Maintenance of high pH above 10 indicates the ability for the pozzolanic reaction to occur and continue over a long term promoting further strength gain [26].

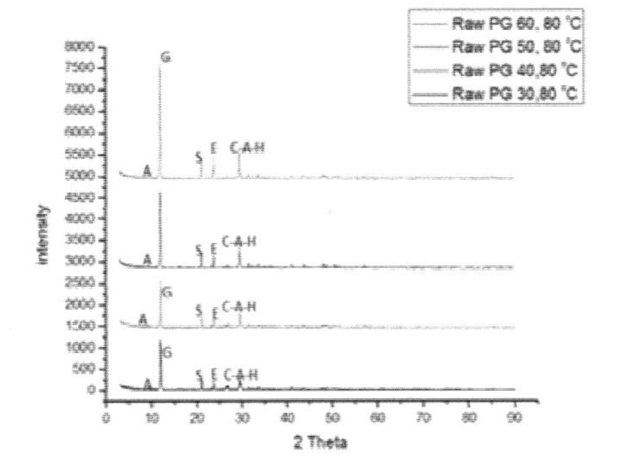

Figure 2(a). XRD analyses of raw PG composites at elevated temperature (80 oC) , (G-gypsum, A-siliminate, E-enttringite, C-A-H-calcium aluminium sulphate, S-hypothetical silica).

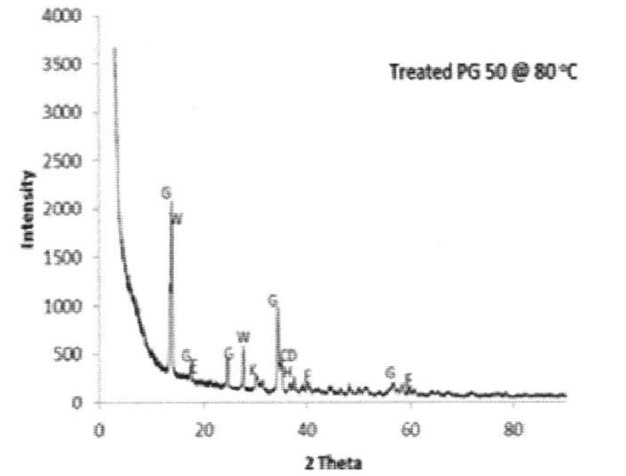

Figure 2(b). XRD analyses of treated PG 50 composite at elevated temperature (80 ^{o}C).

3.5. Particle Particle size distribution (PSD)

There was a significant modal size shift in all the composites from the raw PG to treated PG. A bimodal size shift to the right was evident with all the composites, indicating a change in the proportion of the larger sized particles of the PSD, thus enlargement of particles due to conglomeration. Nelson (1983) presents a study indicating a strong dependence of strength on grain size and found out that a smaller grain size means greater strength; this is due to the grain contact models [27]. Well graded soils are generally denser than poorly graded or uniformly graded soils, and mobilized greater confined (biaxial or triaxial) and unconfined strength. This also provides explanation on the UCS results obtained.

3.6. SEM image analysis

A smooth surface is observed for the raw PG composites and unlike the treated PG surface which looks bristly. This might be a result of the new hydration products formed and covering the surface of the PG. Fairly large and stronger particles are clearly visible in the raw PG and the particles in treated PG seem to be more like flakes. The overall SEM analysis of raw and treated PG composites supports the particle size distribution analysis results obtained and proved that number of particles, surface area and volume of particles had a significant role in the higher unconfined compressive strengths (figure 3).

Figure 3(a). SEM micrograph of (a) raw PG 50 composite.

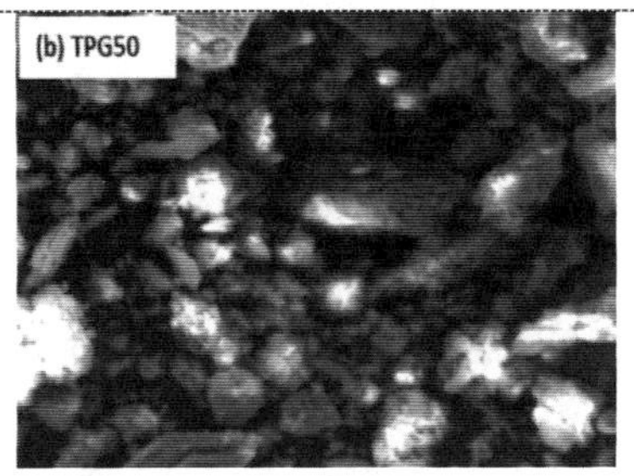

Figure 3(b). SEM micrograph of treated PG 50 composite

4. Conclusion
Stabilizing the raw PG with fly ash and lime and curing at elevated temperature resulted in the formation of new hydration products. The raw PG composite with the high percentage of hydration products, yielded the highest strength. Particle size, number of particles and the surface area also played a significant role on the UCS obtained. The TPG shows that at higher temperature (80°C) UCS is improved and the final strength obtained are lower but within a range of 0.3 – 2 MPa and applicable to be used as backfill bearing material. From the moments of PSD for both the raw and treated PG composite, it is evident that particle size and particle surface area played a significant role on the unconfined compressive strengths obtained. The developed raw PG composites meet the minimum strengths requirements to be classified as a C2-C3 material and the treated PG composite falls under C4 material. The C2/C3 class material can be used as subbase material and C4 as subgrade material [28]. The obtained results also show that the geotechnical properties of the stabilized phosphogypsum, both the raw and treated by lime and fly ash, were significantly improved and in respect to strength the materials are suitable for base/sub-base materials in road construction.

5. Acknowledgement
The authors would like to thank the University of Johannesburg and National Research Foundation of South Africa for the financial support.

6. References
[1] Shen W, Zhou M and Zhao Q 2007 *Constr. Build. Mater.* 21
[2] Green P and Geiber S 2000 *APSA. 99 2*
[3] Rusch AK, Guo T and Seals RG 2000 *J. Hazard. Mater.* **93** 2
[4] Singh M and Garg M 1999 *Cement. Concrete. Res.* **29**. 2
[5] Kumar S 2003 *Build. Environ,* **38** 2
[6] Kumar S 2000 *Constr. Build. Mater.,* **16** 8
[7] Mashifana TP, Okonta FN and Ntuli F 2018 *Mater. Sci-Medzg,* **24** 3
[8] SANS 227 2007 *SABS*
[9] Finot E, Lesniewka E, Goudonnet J and Mutin JC 2000 *Appl Surf Sci.* 161
[10] Lanzón M and García-Ruiz PA 2012 *Constr. Build. Mater.* 28 **1**
[11] Yasar E and Erdogan Y 2004 *Int. J. Rock Mech. Min. Sci.* **41** 5
[12] Pereira A, Palha F, de Brito JD and Silvestre JD 2011 *Constr. Build. Mater.* **25** 4
[13] Lee HK, Lee KM, Kim YH, Yim H and Bae DB 2004 *Cement. Concrete. Res.* **34** 4
[14] Dalui SK, Roychowdhury M and Phani KK 1996 *J. Mater. Sci.* **31** 5
[15] Delgado MCJ and Guerrero IC 2007 *Constr. Build. Mater.***21** 2
[16] Deboucha S and Hashim R 2011 *Sci. Res. Essays.* **6** 3
[17] ASTM C62 2008 *ASTM Int.*
[18] BS EN 771-3 2003 *BSI.*
[19] SANS, 1215 2008 *SABS.*
[20] Green P 2009 *TREMTI.*
[21] Sinapullaiah PV and Baig Moghal AL 2011 *J. Mater. Civil. Eng.*
[22] Plowman C and Cabrera JG 1984 *Cement. Concrete. Res.,***14** 2
[23] Reddy BV and Gourav K 2011 *Mater. Struct.* **44** 10
[24] Huang Y and Lin Z 2010 *Constr. Build. Mater.* **24** 7
[25] Townsend FC 1972 *AEWES.*
[26] Little DN 1999 *Nat. Lime. Assoc.*
[27] Nelson R, 2001 *GPP.*
[28] TRH4 1996 *DOT SANRAL.*

IOP Conf. Series: Materials Science and Engineering **652** (2019) 012045 doi:10.1088/1757-899X/652/1/012045

Stabilization of gold mine tailings: The effect of hydrated lime on the unconfined compressive strength

T Mashifana, N Sithole and E Mkhonto

University of Johannesburg, Department of Chemical Engineering Technology, P.O. Box 17011, Doornfontein 2088, South Africa
*Corresponding author, Email: tmashifana@uj.ac.za

Abstract. Stabilization of gold mine tailings for application in the building and construction industry, provides a significant solution to the shortage of building and construction materials for the country's growing population. The use of gold mine tailings in the production of building blocks / bricks, is examined in this study. The tailings used in the study were characterized by having an optimum moisture content (OMC) of 14.92% and a maximum dry density (MDD) of 1 757 kg/m3. The specimens were cured at a temperature of 80oC and hydrated lime was added as a stabilizing additive to enhance the chemical and geotechnical properties of the material. The use of lime as a stabilizing additive was successful in increasing the unconfined compressive strength (UCS) of the material. UCS was improved from 0.04MPa to a maximum strength of 3.02MPa. As far as the UCS is concerned, the optimum tailing: lime was obtained at a ratio of 70:30 and the produced material is suitable as masonry.

1. Introduction

In large volumes, materials considered as wastes or by-products are generated by numerous industrial activities and most are disposed into a sensitive environment [1]. South Africa is a mineral rich country, by far the most gold that has been mined in South Africa (98%) has come from the Witwatersrand goldfields [2]. The gold mines in this area are situated around an ancient sea, where the age of uraninite grains provide valuable evidence in the old placer-hydrothermal controversy regarding the origin of gold and uranium mineralization [3]. The extensive exploitation of the gold resources led to numerous mine tailings scattered around the Witwatersrand Basin [2]. The tailings dams, particularly in Johannesburg and its neighbouring cities, are the main source of airborne particulate matter pollution, dust blown into the surrounding communities can potentially have adverse effects on human health [4]. Mine waste dump sites become a permanent toxicological problem in the ecosystem and are deleterious to human health [5]. The long history of gold mining of led to an increase in the generation of gold tailings. Historically, South Africa's economy was built chiefly on the mining sector. This resulted into the need for building dams for tailing disposal, which in turn occupy large areas that can better be used for human settlement. Another major problem associated with gold tailings, is the adverse effects on the natural balance of the environment and ecological system. The mine leachates from gold mine wastes impoundments mostly contain high concentrations of sulphates and metals which negatively affect the quality of ground water [6]. In 1998, Rösner and Van Schalkwyk identified 270 tailings dams related to gold mining and discovered that these tailings occupy a total area of about 180km2 in South Africa [7].

Over the years, studies were conducted by many researchers with the aim of finding applications for mine tailings in the construction and building sector. Any study focusing on utilizing the gold mine tailings for building and construction purposes is a potential solution to combat the building of dams for tailing disposal as well as mitigating environmental pollution. The production of building and construction materials from gold mine tailings offers solutions to utilise the resources that are readily available to help solve the environmental challenges, add valued to the waste, supplement the supply of building material and build low-cost housing from wastes. According to statistics South Africa, South Africa's population increased from 40.6 million in 1996 to 51.7 in 2011 and 55.6 million in 2016 [8]. The growth in population has resulted in an increasing demand for housing, which places severe pressure on natural resources used for construction materials, due to depletion of raw materials. There is therefore an urgent need to find innovative alternatives to supplement these natural resources. Gold mine tailings can be used in brick making and cement production to benefit the country regarding low cost housing. The utilization of gold mine tailings can be used to produce conventional bricks which are manufactured using fired clay in high temperature kilns and ordinary Portland cement (OPC) [9]. The most commonly used material for the production of bricks, clay, is in most cases made by artisanal, small- scale manufacturers utilising sand and gravel. These quarrying operations are energy intensive and have a negative impact on the landscape, as they generate large volumes of waste [9]. The potential for re-use of gold mine tailings as secondary construction materials and in phytoremediation was investigated by Mapinduzi et al., (2016), the major tailings oxides are: silicon dioxide (SiO_2), which forms more than 50% w/w, aluminium oxide (Al_2O_3) (9.61 to 14.60); iron oxide (Fe_2O_3) (10.4 to 17.47); and sulphur trioxide (SO_3) (11.40 to 12.10). The pH varies from 7.2 to 7.5, typical for soils with large amounts of calcium and magnesium [10]. Preethi et al., 2017, conducted a study on gold ore tailings as a partial replacement for fine aggregates in concrete [11]. Compressive and flexural strength increased marginally from 5% to 15% replacement. There was a slight decrease in the corresponding compressive and flexural strengths at 20% replacement. Good correlation was observed between compressive strength and flexural strength. It was observed that the addition of gold ore tailings that would replace the fine material at a particular proportion, demonstrated an enhancing effect on mechanical properties of concrete. An investigation of cemented backfill technology with ultra-fine tailings in a gold mine was comprehensively conducted. The analysis results of chemical components of tailings showed that the content of SiO_2 is relatively low, i.e., 33.08%, but the total content of calcium oxide (CaO), magnesium oxide (MgO) and Al_2O_3 is relatively high i.e., 36.5%. After settlement of 4–6 hours, the tailing slurry with the initial concentration of 40% achieved the maximum settling concentration of 54.692%, and the corresponding maximum settling unit weight was $1.497 g/cm^3$.

The two major gold mines in the Paleoarchean Barberton Greenstone Belt (BGB) of Southern Africa are Fairview and Sheba mines [12]. Gold at these mines is associated with quartz–carbonate ± rutile veins and occurs both as invisible gold finely dispersed in sulphides (primarily pyrite and arsenopyrite), and as visible electrum grains hosted in pyrite [12]. The gold distribution in the BGB is structurally controlled [13], and recent work has shown that the main phase of gold mineralisation in the BGB was controlled by pre-existing faults and fractures, and postdates the main structural and metamorphic episodes [14-16]. Altigati et al., (2016) identified sulphides (mainly pyrite), either as inclusions (10-30μm) or as sub-microscopic gold and large gold grains ($\geq$ 100μm) within the silicates (mostly quartz) as the two main types of mineralization in BGB [17]. This study was carried out to investigate whether the gold mine tailing from the Barberton mine can be applied and effectively used for building and construction applications.

2. Experimental details

The gold mine tailings were collected from Pan African Resources Gold Mine situated in Mpumalanga. The material was dried and then the initial work was conducted to characterize the gold mine tailings by conducting the baseline studies and chemical characterization. Sieve analysis was undertaken to study the particle size distribution of the material. The X-ray fluorescence, X-ray diffraction and scanning electron microscope were used to study the elemental composition, mineralogy and morphology of the material, respectively.

A standard compaction test, following ASTM D698 [18], was carried out on the material to obtain the optimum moisture content (OMC) and maximum dry density (MDD) of the tailings. The brick specimens were cast in a 100mm x 100mm x 100mm mold and cured at different temperatures at room temperature for 7 days and an elevated temperature of 80oC for 4 days. The unconfined compressive strength (UCS) was determined before and after stabilisation by using the UCS machine to measure the strength with a loading rate of 15kN/min. Lime was then used as a stabilising additive and added at ratios of 90:10, 80:20 and 70:30 to gold tailings. The UCS of the stabilised gold tailings was measured thereafter. To ensure consistent results, three composites were cast for each test and results presented are an average of the three casts.

3. Results and discussion

3.1. Relative density and pH of gold tailings

The relative density of the gold tailings was found to be 2.8064 g/cm^3. The pH of the material was found to be 8.94, showing that the gold tailings are not acidic, but moderately alkaline. The gold tailings proved to have a relative high density. The relative densities of the soils fall between 2.5 for clays, 2.65 for silica beach sand and above 2.65 for red/brown sand. When compacted, these materials could be up to 1 600 kg/m^3 and the material of such densities could mobilise UCS greater than 6 MPa in most cases with lime and fly ash.

3.2. Elemental analysis of un-stabilised gold tailings

The elemental composition (major oxides) results of the gold mine tailings determined by X-Ray fluorescence (XRF, Rigaku ZSX Primus II) are presented in Table 1. From the results in Table 1, it is evident that the results of the gold mine tailings exhibit results similar to other researchers, and are composed mainly of SiO2. Using the classification system employed by Vassilev and Vasseleva, 2007, the major elements (>1%) are: silica (SiO2); iron oxide (Fe2O3); alumina (Al2O3); calcium oxide (CaO); magnesium oxide (MgO); potassium (K2O), and arsenic trioxide (As2O3). The relative concentration of calcium is a result of liming, which is added to the acid slime before disposal onto the tailings dam [7]. In Table 1, the results indicate that the elemental composition of the cyanide treated gold tailings bear close similarities with the composition of the conventional materials used for commercial specimen making where the major oxides are silica, aluminium and calcium oxide - as the tested gold mine tailings confirm [19]. The material does not contain uranium, this is advantageous as it is undesirable in specimen- making materials. Some previous studies showed the presence of uranium oxide, but as a trace element. Although it can be found in traces, its presence is worth noting as uranium is a highly radioactive element and therefore can present health and safety implications [2].

Table 1. XRF analysis of gold tailings before stabilisation, mass %.

Oxide Content	SiO_2	Al_2O_3	Fe_2O_3	CaO	As_2O_3	MgO	K_2O
Mass (%)	49.63	12.22	13.48	8.31	1.03	5.01	3.62

Table 2. XRD analysis of gold tailings before stabilization

Minerals	Quartz, SiO_2	Anorthite, $CaAl_2Si_2O_8$	Magnetite, Fe_3O_4
Raw gold tailings	35.93	63.39	0.68

Table 3. XRF analysis of gold tailings cured at elevated temperature.

Oxide content	**Mass (%)**	**Oxide content**	**Mass (%)**	**Oxide content**	**Mass (%)**
F	0.109	Cl	0.020	Fe_2O_3	14.80
Na_2O	0.399	K_2O	4.004	CO_2O_3	0.010
MgO	4.647	CaO	9.756	NiO	0.097
Al_2O_3	11.84	TiO_2	0.802	CuO	0.021
SiO_2	45.75	V_2O_5	0.042	ZnO	0.024
P_2O_5	0.010	Cr_2O_3	0.409	As_2O_3	1.209
SO_3	5.414	MnO	0.329	Rb_2O	0.021
SrO	0.023	ZrO_2	0.033	Nb_2O_5	0.005
BaO	0.125	PbO	0.009		

Table 4. XRF analysis of gold tailings cured at lower temperature.

Oxide content	Mass(%)	Oxide content	Mass (%)	Oxide content	Mass(%)
F	0.129	Cl	0.019	Fe_2O_3	14.04
Na_2O	0.394	K_2O	3.920	CO_2O_3	0.013
MgO	4.929	CaO	9.540	NiO	0.097
Al_2O_3	13.05	TiO_2	0.767	CuO	0.020
SiO_2	44.78	V_2O_5	0.037	ZnO	0.023
P_2O_5	0.087	Cr_2O_3	0.337	As_2O_3	1.390
SO_3	5.974	MnO	0.309	Rb_2O	0.018
SrO	0.018	Y_2O_3	0.022	ZrO_2	0.028
BaO	0.043	PbO	0.008		

Table 5. Proportions of gold mine tailings and lime.

Tailings		Lime	
%	Mass (g)	%	Mass (g)
90	1 581.3	10	175.7
80	1 405.6	20	351.4
70	1 229.9	30	527.1

3.3. X-Ray diffraction (XRD) analysis of gold tailings

The mineralogical analysis by means of XRD shows that the gold tailings are composed of three phases; mainly quartz, anorthite and magnetite. The material is comprised of quartz (SiO2) and quartz bearing constituent, as the primary major mineral (See Table 2). This is due to quartz being highly resistant to chemical weathering. The chemical composition of the gold mine tailings depends on the mineralogy of the ore deposits, the characteristics of processing fluids used from the processing plant [20-21]. The XRD analysis did not show the presence of any radionuclides, such as uranium, indicating that the gold mine tailings are eco-friendly for building and construction applications.

3.4. SEM and EDS of gold tailings

Figure 1 and Figure 2 show the gold mine tailings SEM micrograph and EDS. The SEM micrograph indicates the broad distribution of particles of finer particles. The gold mine tailings consist of irregular shaped flakes. Which represent how the material looks and it presents as flat, intermediate and granular particles [22]. SEM confirms that particles of the material have a porous and cracking

IOP Conf. Series: Materials Science and Engineering **652** (2019) 012045 doi:10.1088/1757-899X/652/1/012045

structure associated with numerous gold mine tailings properties. The EDS shows that the material contains constituents such as silica, calcium, iron, potassium, aluminium and magnesium, whose oxides and mineralogy were also detected in the elemental composition by XRF and XRD, respectively.

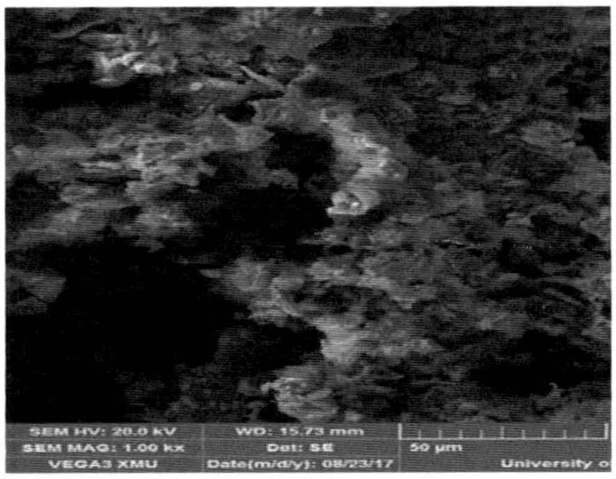

Figure 1. Gold mine tailings SEM image.

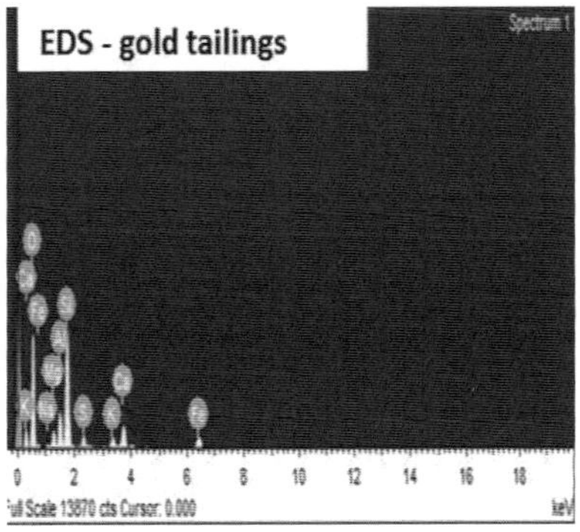

Figure 2. Gold mine tailings EDS.

3.5. Particle size distribution and standard proctor compaction test of gold tailings

Figure 1 and Figure 2 show the gold mine tailings SEM Mass loss during sieve analysis was calculated and the percent of mass loss was found to be 0.4 %. This was also reflected by the total cumulative percentage of 99.6%. The results are presented in percentage cumulative form, in which the total amount of sizes retained or passed by a single notional sieve is given for the range of sizes. The results indicate that most of the particles fell into 53μm to 300μm range. Cumulative percentage shows that 91% of the material passed the 212 μm screen aperture while about 20% passed the 75 μm screen. The particle size range used in standard commercial specimen making includes coarser sand particles as well as fine particles. The material used for the test was, in comparison, relatively finer. The compaction test was conducted to determine the maximum dry unit weight/maximum dry density (MDD) of compaction of gold mine tailings as well as the optimum moisture content (OMC). The MDD of the material is 1757 kg/m^3 and the OMC is 16%. This shows a significant alterations of the specimen resulted from the lime-stabilization process.

3.6. Unconfined compressive strength (UCS)

A total of 6 specimen were prepared and 3 specimens were cure for 7 days at room temperature and a further 3 were cured for 4 days in the oven at a curing temperature of 80°C. The tests were repeated thrice for consistency. When measuring the strength of a specimen, the specimen failed at once. The maximum load that was measured and recorded was 0.1kN. In order to calculate the strength, equation 2 was used and the strength from the maximum load was found to be 0.04MPa. The material is regarded as having a fine texture. Load =0.1kN; Area = 2 500 mm^2; 1N/mm^2 = 1MPa.

$$q_u = \frac{Load}{Corrected\ Area} \quad (1)$$

3.7. XRF of gold tailings after curing at low and elevated temperature

Table 3 shows the elemental composition of the material after compaction. The specimen was cured for four days at elevated temperatures. Using classification used by Vassilev and Vasseleva (2007), the major elements found (>1%) were: silica (SiO_2),Iron oxide (FeO); alumina (Al_2O_3); calcium oxide (CaO); magnesium oxide (MgO); sulphur trioxide (SO_3); potash (K_2O) and arsenic trioxide (As_2O_3) [22]. The minor elements range is (0.1% – 1%) and the trace elements range (<0.1%). The relative proportion of silica, which was predominant in the tailings, decreased. The results after compaction still indicate that the chemical composition of the cyanide treated gold tailings bear close similarities with the composition of the conventional materials used for commercial specimen making, with major oxides similar to those of the raw material [19]. Table 4 shows the elemental composition of the material after compaction of the specimen that was cured for seven days at room temperature. The major elements are still similar to the raw sample with minor differences.

3.8. PSD of gold tailings after curing at low and elevated temperature
The volume size distribution of raw gold mine tailing before compaction of the specimen that was cured for 7 days at room temperature and for 4 days in the oven at a temperature of 80°C was studied. The data was obtained using a Malvern analyser. A significant modal size shift in the volume distribution was evident, when comparing the raw and cured specimen. The raw sample and the sample of specimen that was cured for 7 days at room temperature showed a unimodal volume distribution around 500 μm whereas the sample of the specimen that was cured in the oven showed a unimodal volume distribution around 200μm.

3.9. XRF analysis of lime stabilised gold tailings at different lime ratios and PSD of raw and stabilised tailings
From Figure 3, the elemental composition of the lime stabilised gold tailings is presented. The chemical composition of the sample showed a drastic decrease in the relative proportion of silica from 49.6 mass% of the raw sample to 35.6wt% of the stabilised sample, due to the significant increment in the relative proportion of calcium oxide, which increased from 8.31wt% to 24.8wt% at of the ratio 90:10. For the 80:20 ratio, there was a decrease in the relative proportion of silica 49.6 mass% of the raw sample to 27.7 wt%, while calcium oxide increased from 8.31 mass% of raw sample to 37.3%. This clearly shows that the addition of lime as a blending material has an effect on the chemical composition of the material. In the 70:30% ratio, the relative proportion of silica was reduced from 49.6 mass% of the raw sample to 30.4% and calcium oxide increased from 8.31 mass% of raw sample to 32.8%. Figure 4 shows the volume size distribution of raw gold mine tailing before compaction, a sample of 90%tailings:10%lime specimen, sample of 80% tailings: 20%lime specimen and a sample of 70% tailings: 30%lime specimen. The specimens were cured for 7 days in the oven at a temperature of 80°C. The raw sample showed a unimodal volume distribution around 500μm whereas the samples of ratios of tailings: lime specimens showed a bimodal volume distribution around 300μm.

3.10. Unconfined compressive strength of lime stabilised gold tailings
Figure 5 shows the strengths of the stabilized sample with different lime ratios. The strength of the raw material increased drastically after stabilization with lime. Lime has an exceptional content for the maximum possible strength gain, which mainly depends on the type of gold mine tailings and mineralogy [23].The UCS obtained for the raw material was 0.04MPa. According to the South African standard, the minimum strength required for load bearing is 3.5MPa [24]. The ratio of 70% tailings to 30% lime had the highest strength of 3.02MPa which is closer to 3.5MPa. The difference between the two strengths was calculated to be (3.5 – 3.024 = 0.476MPa). The strength of the raw material was found to be 0.04MPa which clearly showed that the material was weak without the addition of a stabilizer. Lime reacted with silica and alumina pozzolans and formed a strong cementitious matrix that characterizes a lime-stabilized layer. In respect to the UCS obtained, there is also potential for the material to be used in road construction as pavement material. The strength requirements for using lime stabilised layers as structural layers in pavement systems vary considerably from agency to agency [25]. Lime stabilization causes a significant improvement in the tailings UCS.
In a study conducted by Plowman and Caprera (1984), the addition of lime in a mix design activates the alumina and silica phases; these are the constituents that play a role in improving the strength of the material [26]. The high lime content required to stabilize the gold tailings and yield the highest strengths in Figure 5, shows that lime was consumed in the curing process when the content of gold tailings was reduced from 90% to 70%. There was therefore a higher demand for lime to further

improve the unconfined comprehensive strength when tailings content was reduced in the mix design [27]. This also indicate that in the high content gold tailings mix design lime reacted with the constituents in the tailings and was significantly consumed, hence the greater demand. The standardized documents reviewed by Delgado and Guerrero (2007) indicate that compressive strength standards range between 1.3MPa and 2.1MPa for use in non-load bearing walls [28]. Unconfined compressive strength within a range of 0.3MPa – 2MPa is indicated when used as backfill bearing material; therefore 10% lime modification can be utilized in that area. The results obtained yielded UCS of 1.7 MPa at 90% gold tailings and lower lime content of 10% investigated, showing that the material can be used in bulk for applications that require low strengths material such as backfilling (Figure 5). At 80% gold tailings to 20% lime, the UCS obtained is 2.9 MPa. Although 30% lime mobilized higher strength of 3.02 MPa, this option however result into utilization of lower content of the gold tailings (70%) and it may not be cost effective. Therefore for bulk utilisation of gold tailings from Barberton the lower content of lime is therefore highly recommended.

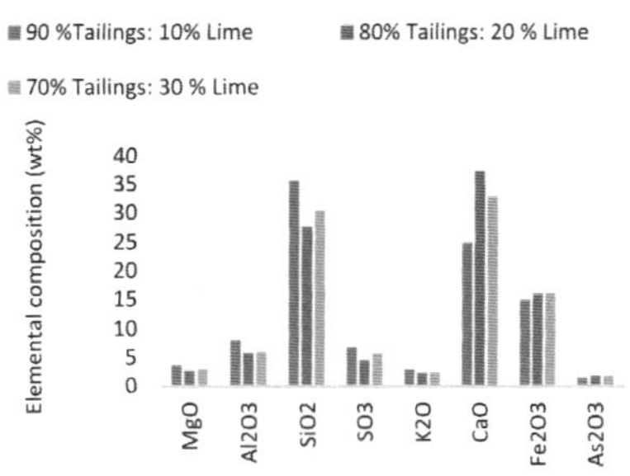

Figure 3. XRF analysis of stabilised gold tailings.

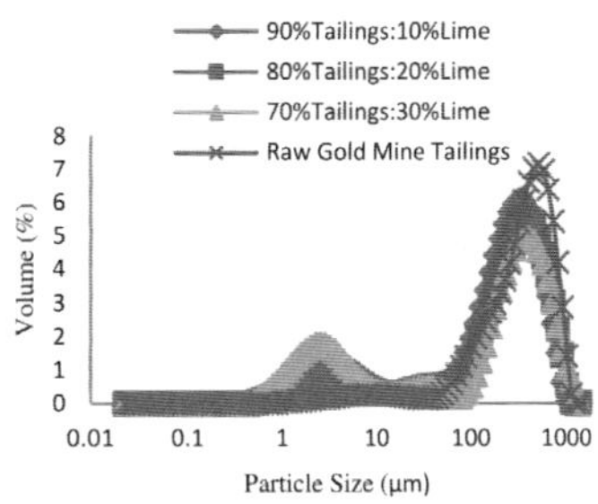

Figure 4. PSD of raw and stabilized gold tailings

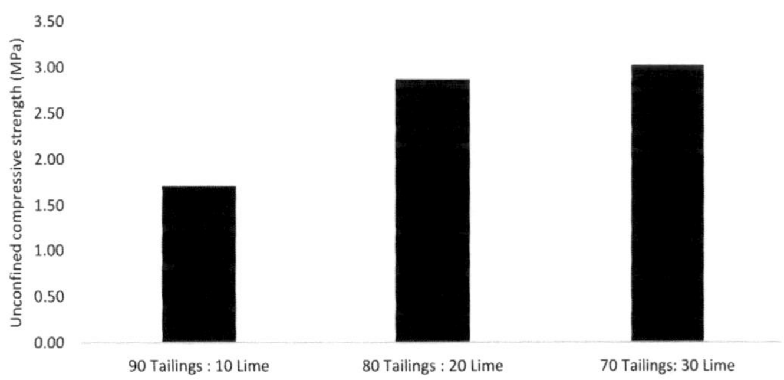

Figure 5. UCS of raw and stabilized gold tailings.

4. Conclusion

Vast quantities gold tailings are discarded in the natural environment resulting into environmental pollution and covering a considerable area. This study was conducted to investigate the potential of stabilising Barberton gold mine tailings and produced a material that can be used for building and construction. The material structure and composition was studied by XRD, XRF and SEM. Utilising lime as a stabilising additive showed that with the incremental addition of lime concentrations, the proportion of calcium oxide also increased drastically, as expected. There was a significant enhancement of unconfined compressive strength when gold tailings were stabilised with lime. The strength of the material improved from 0.04MPa to 1.7MPa with 10% lime, 2.87MPa with 20% lime to the highest strength of 3.02MPa with 30% lime. The higher the lime content used, the higher the strength. In respect to the strengths obtained after stabilisation using 10%-30% lime, the stabilised tailings are indicated for use as masonry, as a backfill bearing material and for construction of pavements. Large volumes of gold tailings can therefore be stabilised by improving their strength by using a lime content as low as 10% in the process. This option does not only mobilise the strength of the gold tailings, but it utilised higher volume of tailings and is cost effective as it uses low lime content.

Acknowledgement
The authors would like to thank the University of Johannesburg and National Research Foundation of South Africa for the financial support.

References

[1] Mashifana TP, Okonta FN and Ntuli F, 2018 *Mater. Sci-Medzg* **24** 3
[2] Malatse M and Ndlovu S 2015 *J. South. Afr. Inst. Min. Metall* **115** 4
[3] Pretorius D A 2012
[4] Nkosi V, 2016
[5] Netshiongolwe K E 2018
[6] Grover BPC and Johnson RH, Tutu H 2016 *Water SA* **42** 1
[7] Rösner T and Van Schalkwyk 2000 *Eng. Geol. Environ* 592
[8] Christopher AJ, A.J 2017 *Afric. J. Stud.*, **38** 1
[9] Zhang L 2013 *Constr. Build. Mater.* 47
[10] Mapinduzi RP, Bujulu PM, Mwegoha WJ 2016 *Int. J. Environ. Sci.* **7** 1
[11] Preethi AV, Rajendra S, Navneeth S and Kumar P. 2016. *IJLTEMAS* 5
[12] Agangi A, Hofmann A and Przybyłowicz W 2014 *Ore. Geol. Rev.* 56
[13] Dirks PHGM, E.G. Charlesworth EG and Munyai,MR 2009 *S.Afr. J. Geol.* 112
[14] Dziggel A, Poujol M, Otto A, Kisters AF, Trieloff M, Schwarz WH and Meyer FM 2010 *Precambrian. Res.* 179
[15] Munyai MR, Dirks PHGM and Charlesworth, EG 2011 *J. Geol.* **114** 2
[16] Dirks PH, Charlesworth EG, Munyai MR and Wormald R 2013 *Precambrian.Res* 226
[17] Altigani MAH, Merkle RKW and Dixon RD 2016 *Ore. Geol. Rev.*, 75
[18] ASTM, D 698 2012
[19] Ahmari S and Zhang L 2012 *Constr. Build. Mater* 29
[20] Zhao H, Xia B, Qin J and Zhang J 2012 *J.Environ. Sci*, 24
[21] AECOM. 2014
[22] Vassilev SV and Vasseleva CG 2007 Fuel, 86
[23] El-Mahllawy MS and Kandeel AM 2014 *HBRC J.* **10** 1
[24] South African National Standards 2007 *SANS 227*
[25] Little DN 1999 *Nat. Lime. Ass*
[26] Plowman C, Cabrera JG 1984 *Cement. Concrete. Res. Research* **14** 2
[27] Sivapullaiah PV and Moghal AAB 2010 *J. Mater. Civil Eng.* **23** 2
[28] Delgado MCJ and Guerrero I.C 2017 *Constr. Build. Mater.* **21**, 2

Silicon and aluminium leaching kinetics from acidic gold mine tailings

T Falayi[1,], T Mashifana[21] and F Ntuli[2]

[1]Civil Engineering Sciences, University of Johannesburg, South Africa
[2]Chemical Engineering Technology, University of Johannesburg, South Africa

Abstract. Si and Al were leached from gold mine tailings (GMT) in order to determine if the GMT could be used as precursors for the synthesis of geopolymers. Possible synthesis of geopolymers from GMT would help reduce their potential pollution effects through heavy metal immobilisations. The effect of type of alkali (NaOH/KOH), alkali concentration, temperature, milling and calcination of GMT and solid liquid ratio (S/L) on the leaching of Si and Al was investigated. The optimum conditions were 10 M KOH, S/L of 0.5% m/v and a leaching temperature of 95°C. The leaching obeyed the shrinking core model with the surface chemical reaction being the controlling step. The leaching of Si was linked to the Al leaching since the Si/Al ratio of the leachate was around 2 irrespective of leaching conditions. KOH yielded more Si and Al. as compared to NaOH, with 54.7 and 26.7 Si and Al leached respectively from as received GMT. Calcination at 900°C resulted in 18% and 22 % increase in Si and Al yield respectively for KOH based leaching. The successful leaching of Si and Al from GMT using KOH provides opportunities for the geopolymerisation of GMT thereby allowing the minimisation of their pollution effects.

1. Introduction

Geopolymerisation is a process where materials containing Al and Si undergo dissolution and polycondensation of Al and Si species under alkaline conditions to form three dimensional structure which is amorphous or semi crystalline [1]. Geopolymers have been shown to shown to have high strength, low density, thermal and high chemical stability [2]. There is an estimated 600 000 tonnes of gold mine tailings (GMT) dotted around the Gauteng province of South Africa exposing people to pollution effects of these tailings dams [3]. It has been postulated that any alumina-silicate material can undergo geopolymerisation and as such GMT can also be used a precursors for the synthesis of geopolymers. GMT has been shown to contain silica and alumina [4, 5]. With so much GMT it is imperative to turn the waste into useful civil engineering materials through geopolymerisation.

The first step of the geopolymerisation process is thought to be the alkaline dissolution of Si and Al [6]. The leachability and mechanism of leaching of alumina and silica from the mine tailings is therefore an important parameter to be studied as to determine the aluminosiliate material can be used as a precursor for synthesis of geopolymers. A survey of literature does not show the kinetic mechanism and thermodynamics of Si and Al leaching from gold mine tailings. Most geopolymerisation processes look at the use of NaOH only as the alkali [7, 8] and this study seeks to establish the effectiveness of KOH as an alkaline activator.

[1] Corresponding author: tmashifana@uj.ac.za

The reactivity of a material to synthesise geopolymers is dependent on its capacity to easily dissolve in alkaline media [9]. This paper seeks to establish the reactivity of acidic GMT in geopolymer systems. This paper seeks to establish the kinetics and thermodynamics of aluminosilicate leaching from acidic gold mine tailiings. The paper seeks also to establish the effect of type of alkali (KOH/NaOH) on the leaching process. The understanding if these parameters may lead to increased utilisation of GMT as precursors for the synthesis of geopolymers. The effect of solid loading, leaching temperature, alkali concentration, GMT particle size and GMT calcination on the leaching of alumina and silica were also investigated.

2. Materials and Methods

2.1. Materials

The GMT was obtained from a GMT dam in the West Rand of Gauteng province of South Africa. KOH and NaOH were supplied by Rochelle chemicals. Table 1 shows the XRF analysis of the GMT respectively. The pH and SG of the GMT were found to be 3.84 and 2.99 respectively.

Table 1. XRF analysis of GMT

Component	Result
MgO	5.26
Al_2O_3	6.98
SiO_2	74.5
SO_3	3.05
K_2O	1.26
Fe_2O_3	7.03

The major oxides were found to be SiO_2, MgO, Al_2O_3, K_2O and Fe_2O_3 which were about 95% of the material.

2.2. Equipment

A thermostatic shaker was used for alkaline leaching of silica and alumina from the acidic GMT.. X-ray fluorescence (Rigaku ZSX primus II) was used for the determination of metal oxide content of the GMT. Mineralogical structure was determined using an X-ray diffraction technique (Ultima IV Rigaku XRD). Leached Si and Al in the leachate were measured using an Atomic Absorption Spectrometer (AAS) using a nitrous oxide-acetylene flame. FTIR (Fourier transform infrared spectroscopy) Thermoscientific Nicket IS10 was used to characterize the GMT before and after leaching.

2.3. Methods

Leaching experiments were conducted in a thermostatic shaker at a stirring speed of 250 rpm. Silicon and aluminium leaching was optimised parametrically by varying concentration of alkali (KOH/NaOH) from 2 to 10 M, solid liquid loading (S/L) from 2 to 10%, leaching time from 30 to 300 mins, leaching temperature from 50 to 95°C (95°C being the maximum safe temperature of the thermostatic shaker).Optimisation was done with varying one parameter at a time whilst the rest were kept constant. At the end of the appropriate leaching time the slurry was filtered and the filtrate was acidified with HCl before analysing for Si and Al using the AAS.

2.4. Statistics

The results reported are an average of three readings. The error bars in graphs are at 95 confidence interval of the mean.

3. Results and Discussion

3.1. *Effect of solid loading and alkali type*

Figure 1 shows the variation in yield of silicon and aluminium with solid loading and type of alkali

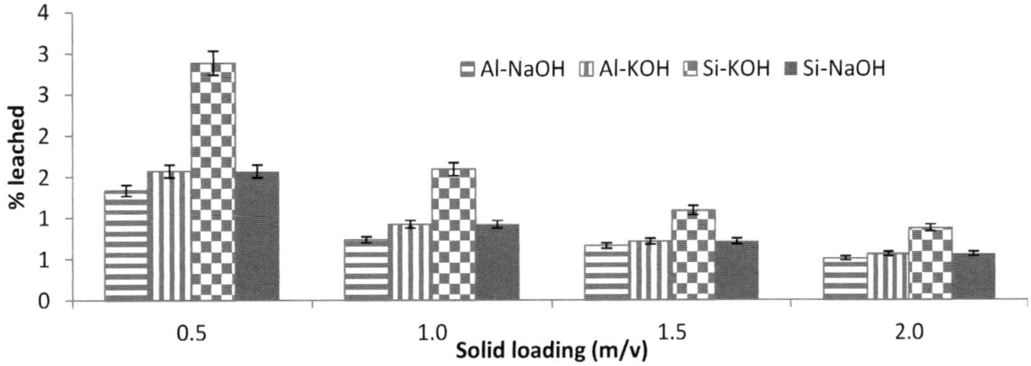

Figure 1: Si and Al yield (Leaching temperature 40°C, time 300 mins, agitation speed 250 rpm).

Leaching with KOH yielded more Si and Al as compared to NaOH based leaching. This was because the leaching with KOH is more alkaline as compared to NaOH and these results were in agreement with some similar experiments but on different materials [10]. More Si was leached than Al and this may be due to that there is more silica than alumina in the GMT (Table 1) although there have been suggestion that the intrinsic dissolution of Si is higher than Al [10]. There was a decrease in Si and Al yield with an increase in solid loading (S/L) and this may be due to reagent starvation at high S/L [11] leading to a decrease mass transfer efficiency. Though geopolymerisation reactions occur at high S/L (25-50%), the low S/L of 0.5% can still be used as an indication of the reactivity of a possible geopolymeric material.

3.2. *Effect of alkali concentration*

Figure 2 shows the the variation of Si and Al yield with alkali concentration. There was an increase in Si and Al yield with increase in alkali concentration from 5 to 10 M. The reaction for the dissolution of Si and Al can be written as in equation 1

$$Al - O - Si + 7(OH) \rightarrow Al(OH)_4^- + {}^-OSi(OH)_3 \qquad (1)$$

From equation 1 the increase in alkali concentration favours higher OH^- ions which then favour the forward reaction leading to more Si and Al dissolution. However at 15 M alkali concentration there was drop in Si and Al yield (though insignificant as the concentrations at 10 and 15 M were within the error limits). This was due to that at 15 M the slurries became very viscous and leading to an impediment in the mass transfer process.

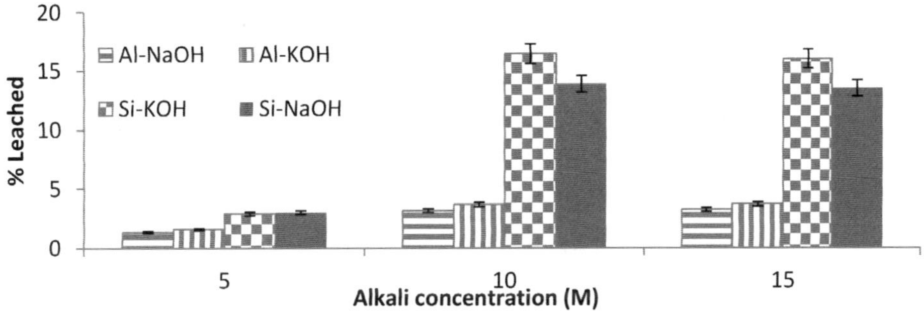

Figure 2. Variation in Si and Al yield with alkali concentration (Leaching temperature 40°C, time 300 mins, agitation speed 250 rpm, S/L 0.5% m/v).

IOP Conf. Series: Materials Science and Engineering **652** (2019) 012046 doi:10.1088/1757-899X/652/1/012046

3.3. Effect of leaching temperature

Figure 3 shows the variation of Si and Al yield with temperature.

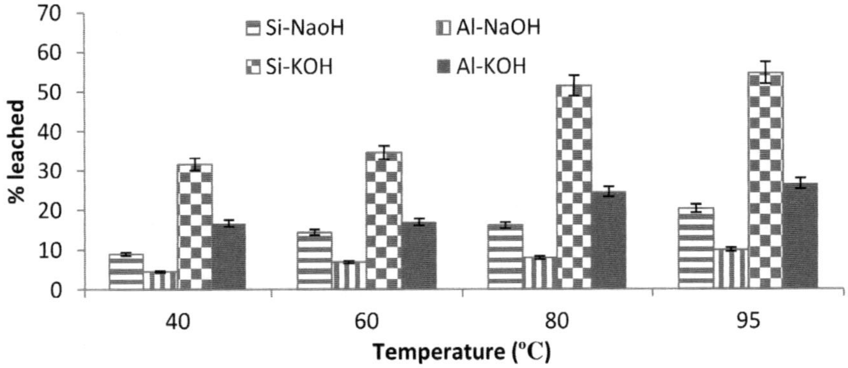

Figure 3. Variation of yield with temperature (Time 300 mins, agitation speed 250 rpm, S/L 0.5% m/v, alkali concentration 10 M).

There was an increase in Si and Al yield with an increase in temperature. The increase in temperature provided thermal energy to overcome activation energy for the dissolution process. One interesting phenomenon was that though the Si/Al ratio of GMT was 9.4, the leachates of both KOH and NaOH had a Si/Al ratio of about 2. This then shows that the dissolution of Si depends on the dissolution of Al.

3.4. Effect of milling

Figure 4 shows the variation in yield of Si and Al with milling time.

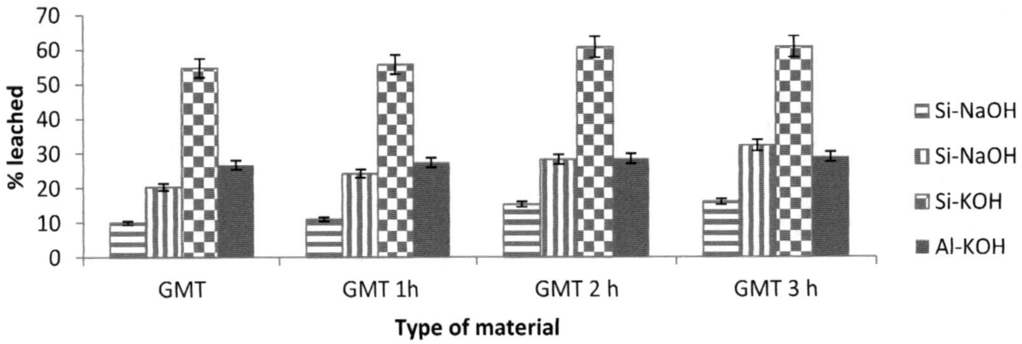

Figure 4. Variation in yield with milling time (Time 300 mins, agitation speed 250 rpm, S/L 0.5% m/v, alkali concentration 10 M, leaching temperature 95°C)

Milling resulted in at least a 10% increase for Si yield for KOH based leaching using a GMT sample milled for 2 h. Milling resulted in a decrease in particle size.

3.5. Effect of GMT calcination

Figure 5 show the variation in Si and Al yield with calcination at 750 and 900°C for 1 h respectively

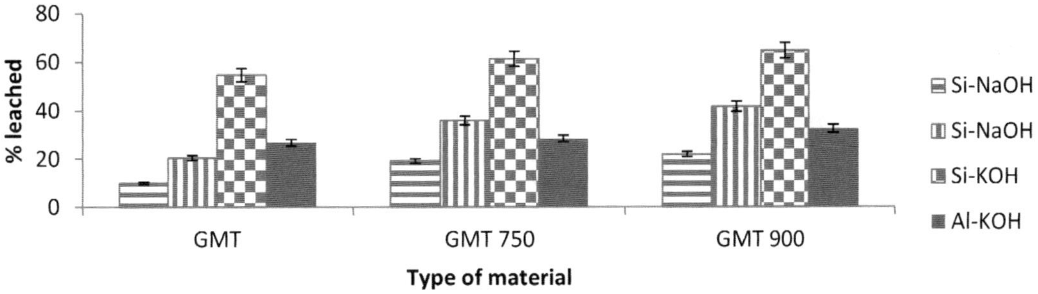

Figure 5. Variation in yield with milling time (Time 300 mins, agitation speed 250 rpm, S/L 0.5% m/v, alkali concentration 10 M, leaching temperature 95°C

For the KOH based leaching there was 18 and 22% increase in Si and Al yield respectively with the use of GMT calcined at 900°C as compared to as received GMT. The reason for this is shown in Figure 6.

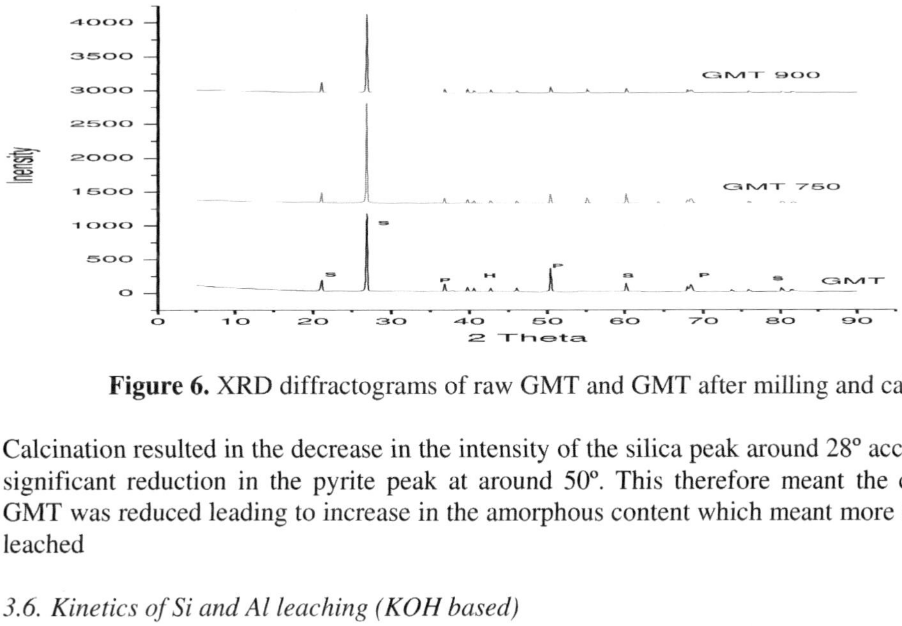

Figure 6. XRD diffractograms of raw GMT and GMT after milling and calcination

Calcination resulted in the decrease in the intensity of the silica peak around 28° accompanied with the significant reduction in the pyrite peak at around 50°. This therefore meant the crystallinity of the GMT was reduced leading to increase in the amorphous content which meant more Si and Al could be leached

3.6. Kinetics of Si and Al leaching (KOH based)

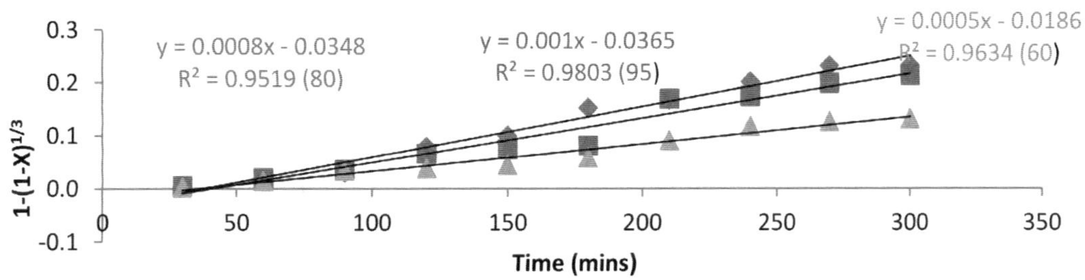

Figure 7. Chemical surface reaction plots for Si (Time 300 mins, agitation speed 250 rpm, S/L 0.5% m/v, KOH concentration 10 M)

The kinetics of Si and Al leaching were tested using KOH since it gave the highest yield. Figures 7 and 8 show the leaching kinetic plots based on surface chemical reaction of the shrinking core model.

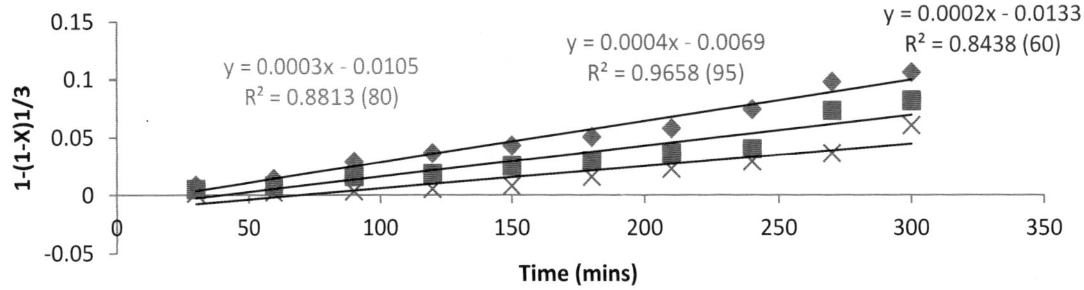

Figure 8. Chemical surface reaction plots for Al (Time 300 mins, agitation speed 250 rpm, S/L 0.5% m/v, KOH concentration 10 M)

The leaching of Si and Al was controlled by the surface chemical reactions between the alkali and Si and Al [12]. The equation for the plot is represented as

$$1-(1-X)^{1/3} = k_r t \qquad (2)$$

Where X is the fraction leached, k_r is the reactant constant, t is time. Equation 2 was chosen to represent the leaching of Si and Al using KOH because the equation representing phenomenon when solvent diffusion through liquid film is the slowest step had correlation coefficients below 0.5 and were not plotted.

4. Conclusion

Si and Al can successfully be leached from GMT using KOH and hence GMT can be used as a precursor for the synthesis of geopolymers. KOH leaching is more effective than NaOH leaching. The optimum leaching conditions are a time of 300 mins, agitation speed of 250 rpm, S/L of 0.5% m/v, KOH concentration 10 M and 95°C. The leaching of Si and Al obeys the shrinking core model where the surface reaction is the controlling step. Calcination increases the amorphous content resulting in the increase of Si and Al yield. Milling of GMT reduces particle size and increases surface area resulting in the increase in Si and Al yield. This research therefore provides insight into the possible solidification of gold mine tailings using alkaline activation.

Acknowledgements

The authors would like to thank Rebotile Nare for as siting with laboratory work. The authors also thank the University of Johannesburg for providing the Global Excellence and Stature research fellowship to fund the author.

References

[1] F. Pacheco-Torgal, J. Castro-Gomes S. Jalal, Const. *Buil. Maters,* **23**, 20 (2009).
[2] A.P.D Panias I Giannopoulou, *Proceedings of European Metallurgical Conference. Dusseldorf,* Germany June 11-14, 2007
[3] F. Winde, Water Sa, 36, 239 (2010).
[4] J. Kiventera, L. Golek, J. Yliniemi, V. Ferreira, J. Deja, M. Illikainen, Int.J.Min.Pro. **149**, 104 (2016).
[5] B.N. Yalala, *Characterization, bioavailability and health risk assessment of mercury in dust impacted by gold mining,* PhD thesis , University of Witwatersrand (2015)
[6] E. Ch. Panagiotopoulou, K. Kerraki G. Kakali, J Mater Sci, **42**, 2967 (2007)
[7] L. Zheng, W. Wang Y. Shi, *Chemosphere,* **79**, 665 (2010)
[8] X. Jiao, Y. Zhang T. Chen, *Const. Buil. Mater,* **38**, 43 (2013)
[9] J.N.Y Djobo, A. Elimbi, H.K. Tchakoute, S. Kumar, *Envir. Sci Poll. Res,* 2016.

[10] H. Xu J.S.J. Van Deventer. *Int. J. Miner. Process.* **59**, 247 (2000).
[11] E. Ntumba Malenga, A.F. Mulaba-Bafubiandi and W. Nheta. *Hydro.* **155**, 69 (2015).
[12] F. Habashi, Principles of Extractive Metallurgy. Science Publishers, Quebec, 1980

Durability of desilicated fly ash lime composites

T Falayi[1], N.T Sithole[2] and F Okonta[1]

[1]Civil Engineering Sciences, University of Johannesburg, South Africa
[2]Chemical Engineering Technology, University of Johannesburg, South Africa
Email: tfalayi@gmail.com; nastassias@uj.ac.za; fnokonta@uj.ac.za

Abstract. Remediation of residual fly ash from coal fired electric power plants is beneficial to the environment. The chemical, physical, geotechnical and radiological properties of desilicated fly ash (DFA) were investigated. The DFA: lime ratio of 70:30 composite was found to have an average UCS of 8.8 MPa showing 19.5% water absorption after a 24 h soak with a corresponding 23.5% reduction in UCS. The composite was found to meet the minimum UCS requirements after 8 wet and dry cycles but with failure at 10 cycles. The composite was found not to be resistant to acid attack at pH 2 and hence could not be used in the vicinity where landfill leachate may come into contact with the composite. The composite had relative stability against acid rain and normal rain. The composite met the minimum requirement of ASTM C34-13, and thus provide an opportunity for high volume utilization of the residual ash for construction of load bearing walls.

1. Introduction

Fly ash (FA) is a residue from the combustion of coal for electricity generation. FA is a well- known pollutant as long term exposure to FA through inhalation causes fibrosis of lungs, silicosis and pneumonitis bronchitis [1, 2].The increase in the need for silica has led to the use of FA as a source of silica [3, 4]. The leaching of silica from fly ash is a well-studied topic [3, 4]. Silica leaching is achieved using either KOH or NaOH. The leached silica can then be used to produce zeolites, coatings and mesoporous silica adsorbents [5-7]. Not much attention has been given to the residue of silica leaching from FA. If the silica leaching residue, desilicated fly ash (DFA), is not utilised then it becomes a secondary pollutant whose effect on the environment are unknown.

DFA has been shown to be alkaline with a pH of 12.28 [8]. The alkaline pH is due to basic oxides (especially K2O) it contains. DFA is characterised by low silica to alumina ratio of 0.51 due to removal of silica during leaching. One research has shown that DFA mixed with lime can be used to stabilise expansive soils [8]. This study therefore seeks to explore the durability of the DFA: lime composite with regards to its potential application in masonry. The most competent DFA: lime composite was found to be the one with a ratio of DFA: lime of 70:30 wt respectively [8].

This composite was then subjected to wet unconfined compressive strength (UCS) tests, alternate wet and dry cycles tests and exposure to different acidic media. The criterion for determination of durability was UCS using ASTM standards. The research also focused on the use of microstructure to explain physical properties of the composite.

2. Materials and Methods

2.1 Materials

FA was supplied by Camden power station in South Africa. DFA was obtained through leaching of silica from FA.

2.2 Equipment

Mineralogical analysis was done using a Rigaku Ultima IV difractometer. Unconfined compressive strength (UCS) was tested on all samples using a Cyber plus evolution compression machine at 0.25 MPa/s [9]. The reported UCS was an average of 3 samples

2.3 Methods

The cured DFA: lime (70:30) composite was subjected to a variety of tests to determine durability. ASTM D559- (ASTM, 2015) [10] was used to determine the variation of UCS with alternate wet and dry cycles.The cured composites were also immersed for a period of 28 days in various acidic conditions. A pH of 2, 4.88 and 6 were chosen. The solution were synthesised by mixing 1 M citric acid and 1 M sodium citrate until the required pH was attained. pH 2 represented the landfill leachate pH, pH 4.88 represented acidic rain and pH 6 represented normal rain [11]. Water absorption of the cured composites was determined using ASTM C373 (ASTM 2014) [12].

2.4 Background

Previous studies by the authors [8] have shown that a DFA: lime (70:30) cured at 80°C for 96 h produced the most competent composite with a UCS of 8.8 MPa. The composite had an optimum moisture content of 30%. An increase in lime content to 40% was accompanied with a 24.4% drop in UCS. The environmental footprint of the composite was assessed using Toxicity Characteristic leaching Procedure (TCLP) and the results are shown in Table 1.

Table 1. TCLP analysis if the DFA: lime composite

Element	DFA (ppm)	DFA: Lime composite (ppm)	Statutory Limits (ppm)
Cr	0.116	0.028	5
Mn	0.388	0.015	10
Fe	5.98	1.25	5
Ni	0.215	0.022	0.4
Cu	0.315	0.035	2
Zn	0.125	0.001	4
Ba	10.3	0.821	100
Pb	0.286	0.0141	5

The composite was shown not to be toxic as all analysed elements were within the TCLP limits. Of interest too was that the curing of the composite was accompanied with the reduction in leachability of metals. This showed that the DFA: lime composite was capable of immobilising metals in its structure. The XRD analysis of the DFA: lime composite is shown in Figure 1.

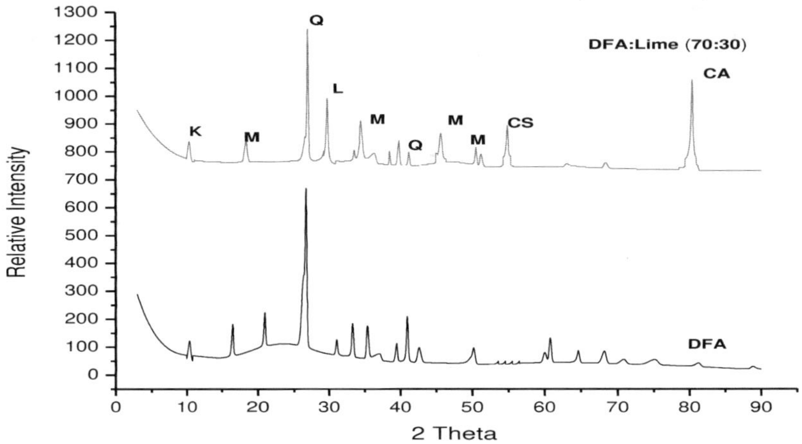

Figure 1. XRD analysis of DFA and DFA :lime composite (cured at 80°C for 4 P: Philipsite K, M: Mullite, Q: Quartz, L: CaO, CS: Calcium silicate hydrate, CA: tri-calcium aluminate hydrate

The obtaining hydration products of the DFA: lime composite were calcium silicate hydrate and tri-calcium aluminate hydrate, which were responsible for the strength gain at the end of the curing period.

3. Results and Discussion

3.1 Durability using wet compressive strength

Table 2 shows the variation in wet compressive strength of the composite with soaking over a period of 24 h. The composite showed a low water absorption rate of 17%. This is within the acceptable limits ATMM C34-13 where the maximum water absorption should be less than 25%. The low water absorption rates can be attributed to increase in crystalline structure of the composite as compared to DFA as shown in the XRD analysis (Figure 1).

Table 2. Wet compressive strength of the DFA: lime composite

Composite (DFA:Lime)	70:30
Mass of cast (g)	1504.83
Mass of cast after 24 h soak (g)	1765.25
UCS before soaking (MPa)	8.6
UCS after soaking (MPa)	6.5
% water absorption	17.3
% reduction in UCS	24.4

3.2 Durability using wet and dry cycles

Figure 2 shows the change in UCS with wet and dry cycles

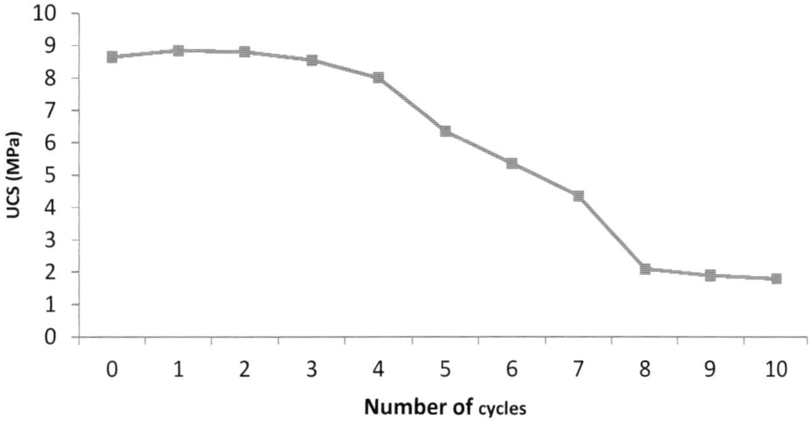

Figure 2. Variation in UCS with number of wet and dry cycles for the 70:30 DFA: lime composite.

The DFA: lime composite was resistant to alternate wet and dry cycles of up to 8 cycles. The initial increase in UCS from cycle 1 to cycle 3 may be due to availability of moisture during soaking which then allows hydration to take place during the drying part of a cycle. The composite fails after 8 cycles as the UCS drops to below 3.5 MPa. This is the minimum UCS allowed by the South African Building council for non-facing plastered brick (SANS 227:2007). The decrease in UCS is due to an increase in average porosity of the composite as is shown in Table 3. Figure 3 also shows the microstructural change that happen during the wet and dry cycles. It is important to note that there is a reduction in the intensity of the lime (CaO) peak after 8 cycles. Lime is weak filler and is therefore dissolved by water during the cycles due to water ingress into the composite. During the dry phase of the cycle, voids are left where lime unoccupied space leading to reduction of UCS.

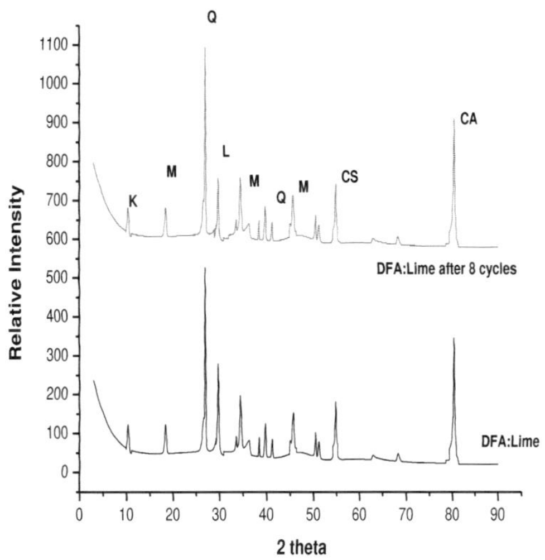

Figure 3. Variation in UCS with soaking days and pH for the DFA: lime composite (70:30)

Table 3. Variation in porosity with number of wet and dry cycles

Cycle	Average porosity
0	0.25
1	0.24
2	0.27
3	0.27
10	0.29

3.3 Durability using various acidic environments

Figure 4 shows the variation of UCS with exposure to different acidic environments.

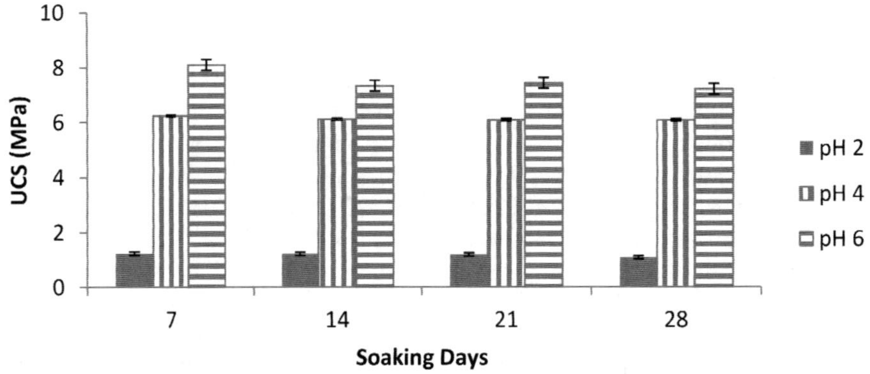

Figure 4. Variation in UCS with soaking days and pH for the DFA: lime composite (70:30).

There was an 87% reduction in UCS with a 28 day exposure to simulated landfill conditions. This therefore meant that the composite is not suitable for building near landfill sites where there might be a risk of exposure to landfill leachate runoff. The 28 day exposure to simulated acid rain resulted in a 30% reduction in UCS. Simulated normal rain showed a UCS reduction of 18%. The UCS reduction was also accompanied with weight loss of the composite. Figure 5 shows the microstructural changes after 28 days of immersion in different acidic conditions.

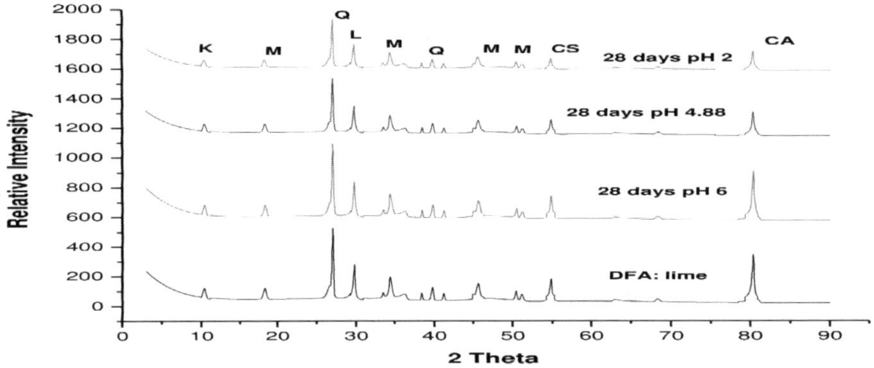

Figure 5. Change in microstructure of the DFA: lime composite with 28 days of immersion in various acidic solutions

There was a reduction in the intensity of all peaks with immersion in acidic media. There was significant reduction of the tri-calcium aluminate hydrate and calcium silicate hydrate for the composite immersed in pH 2 solution. This would explain the 87% reduction in UCS. The reduction for the respective peaks in the composite immersed in pH 6 solution was not significant and hence the less than 20% reduction in UCS.

4. Conclusion

The following conclusion can be derived from this work: Calcium Silicate hydrates and tricalcium aluminate hydrates are responsible for UCS growth in DFA lime composites. The presence of lime as a weak filler in the composite is responsible for the loss of UCS during 24 h soak and wet and dry cycles. The average porosity increase due to dissolution of lime also leads to UCS decrease. The consumption of calcium silicate and tricalcium aluminate hydrate during acid soak is responsible for the reduction of UCS on exposure to acidic environments. The DFA lime composite can be used in areas where acidic rain is experienced and not where landfill leachates can be experienced. This study provides opportunities for use of waste in the construction industry thereby reducing environmental impact of the waste. The DFA lime composite met the minimum requirements of ASTM C34-13[13].

References

[1] C.J. Schilling, A.G. Davison, S. Durham, S. T.A.G. Newman, Brit. Med. J. 292, 1561 (1986)
[2] K. Cho, Y.J. Cho, D.L. Shrivastara, S. Kapre, *Chest*, 106, 309 (1994)
[3] O. Font, N. Moreno, S. Díez, A. Querol, A. López-Soler, P. Coca, F. García Pena, *J.Haz. Mater*, 166, 94 (2009).
[4] N. Moreno, X. Querol, J.M. Andrés, K. Stanton, M. Towler, H. Nugteren, M. Janssen-Jurkovicová, R. Jones, R., Fuel, 84, 351(2005)
[5] K.S. Hui, C.Y.H. Chao *J. of Haz. Mater.*, B137, 1135, (2006)
[6] T. Liou, H. Lin, JIEC, 18, 1428 (2012)
[7] H.W. Nugteren, *Secondary industrial minerals from coal fly ash and aluminium anodising waste solutions.* PhD thesis, Delft University of Technology (2011)
[8] T. Falayi, F.N. Okonta, F. Ntuli, *Mater. and Struct.* 49, 4881 (2016)
[9] ASTM D5102 *Standard Test Methods for Unconfined Compressive Strength of Compacted Soil-Lime Mixtures* (2009)
[10] ASTM D559 *Standard test methods for wetting and drying compacted soil cement mixtures* (2015)
[11] K. Phoungthong, Y. Xia, H. Zhang L. Shao, P. He, Front. Environ. Sci. Eng. 10, 399 (2016).
[12] ASTM. C373 *Standard Test Method for Water Absorption, Bulk Density, Apparent Porosity,* and Apparent Specific Gravity of Fired Whiteware Products, Ceramic Tiles, and Glass Tiles. (2014).
[13] ASTM. C34 Standard Specification for Structural Clay load Bearing Wall Tile. (2013).

Reuse of pulverized fired clay brick wastes as cement substitute in mortar for sustainable construction

Oluwarotimi Olofinnade[1,*], Anthony Ede[1], Julius Ndambuki[2], David Omole[1] Precious Francis[1], Cynthia Chigere[1] and Kehinde Oyeyemi[3]

[1]Department of Civil Engineering, Covenant University, Km 10, Idiroko Rd., Ota, Nigeria
[2]Department of Civil Engineering, Tshwane University of Technology, Pretoria, South Africa
[3]Department of Physics, Covenant University, Km 10, Idiroko Rd., Ota, Nigeria
[*]Corresponding author: rotimi.olofinnade@covenantuniversity.edu.ng (OM Olofinnade)

Abstract. Recently, there has been an increase on the use of waste materials for development of new materials for construction. This is due to the need for sustainable development which emphasizes on waste reduction, reduction of greenhouse gases (GHG), greener clean environment and conservation of raw resources. This study examines the appropriateness of clay brick waste in powder form as a suitable material for partial substitute of cement in mortar. The clay bricks were obtained as generated waste materials from a brick producing factory in Lagos metropolis and then subjected to grinding. Tests which include; specific gravity, particle size distribution, x-ray fluorescence and scanning electron microscopy were carried out on the pulverized clay brick material (CBP). Further, it is used to partially replace cement in mortar at replacement percentage of 0, 5, 10, 20, 25, 30, 40 and 50% using a mix ratio of 1: 2.75 at 0.45 water-cement ratio. Compressive strength and strength activity index tests were carried out on the cube specimens after days of curing. The obtained results clearly indicate that the pulverized clay brick is an amorphous material and contains similar chemical oxide compounds of a pozzolana. Furthermore, it was observed that the compressive strength of the mortar was further improve at an optimum value of 20% cement replacement with clay brick powder compared to the control mortar. Therefore it implies that generated clay brick wastes can be pulverized and used as pozzolanic admixture in mortar production instead of discriminate dumping in landfill.

1. Introduction

In recent years, increasing concerns on the needs to protect the environment necessitate the creation of the concept of sustainable construction in the construction industry. According to [1], the practice of sustainable construction implies responsible management of the built environment considering efficiency of material resources and ecology. Moreso, the origin of sustainable construction can be traced to the concept of sustainable development which emphasized on the needs for the present generation to meet their own needs without compromising the ability of future generations to meet theirs [2]. In addition, the amount of generated wastes and by-products from households and industries are increasing at an alarming rate due to increase population and urbanization. Although wastes generation are unavoidable aspects of human action, however, it is important that urgent steps be taking to reduce the amount of generated wastes and also finding a viable alternatives of recycling and reusing produced wastes. Furthermore, it was emphasized by [3] that recycling and reusing of waste is very necessary from the view point of protecting and preserving the environment. In addition, study by [4] stated that the 3R (reduce, reuse and recycle) initiatives is now a globally recognized tool that can

contribute to the reduction of greenhouse gases (GHG) emission. The construction industry, is a major consumer of natural resources and also producing large amount of wastes. A study by Cachim [5] indicated that about half of the wastes generated by the developed nations have their sourced in construction and demolition wastes (C&DW). The C&D wastes mainly consist of materials such as; wood, concrete, clay brick, mortar, glass, plastic, aluminium and steel that are commonly used in civil construction. However, at same time, there is also an increasing concern with the consumption rate of the natural resources in the construction industry and the impact on the environment. Concrete is one of the main construction material in the construction industry and will remain so in the future [6, 7]. According to Naik [8], concrete is versatile and economical, and has today become a major construction material, however, it has impacts on the environment [8]. For instance, cement, is one of the vital component material of concrete and mortar, however, its production process generates a great deal of dust and carbon dioxide (CO_2) [9, 10]. A study by [6] and [11] pointed out that in year 2000, about 1.25 billion tons of CO_2 was released to the atmosphere only from production of cement resulting in serious environmental pollution. This implies that finding a way of limiting the portion of cement in mortar or concrete can extenuate environmental pollution. In addition, [5, 20] emphasized that the impacts of using concrete and mortar on the environment can be minimize by reusing some of the generated wastes as alternative materials to partially replace cement in the concrete or mortar to ensure resource preservation and reduction of GHG emissions.
There are some C&D wastes such as clay brick that possess the required chemical compositions for a binder substitute. In addition, several studies have reported that using mineral admixture such as; fly ash and slag which are sourced as wastes to partially substitute cement in concrete is a good way to turn waste into wealth, as well as achieving a sustainable environment. Moreover, because of the composite nature of concrete, it provide a real means of managing wastes in good quantities [3, 12]. Clay brick materials are largely used in Nigeria, for building construction [9]. Consequently, large amount of wastes are generated by the brick factories and the construction industry (as demolition wastes) and often time these wastes are dumped in landfills. However, previous research studies have reported on the possibility of reusing crushed clay brick wastes as coarse aggregate, fine aggregate and mineral admixture in concrete [9], [13-18]. Moreover, studies by [9], [15], [18], [19] mentioned that clay brick wastes are potential pozzolana, which can be utilize as cement substitute to help reduces the impact of cement on the environment. Besides, studies by [24] stated that about 30-50% of the generated 5 billion tons of construction waste all over the world are made up of clay brick wastes. Moreover, in Nigeria, the ceramic and clay brick industries produced large amount of brick and ceramic products annually, and therefore generated huge amount of brick wastes which are mostly disposed in landfills, or dumping sites posing as environmental challenge [9], [18]. Studies by [15, 18] reported that comminuted fired clay brick wastes possess pozzolanic cementitious properties, and a more valuable alternative resources for construction instead of discriminate dumping in landfill. Therefore, this present study is aim to characterize and investigate the effect of using pulverized locally sourced clay brick wastes as partial substitute for cement on the properties of mortar for sustainable construction.

2. Methodology

The cement used in this study for the mortar production is the ordinary Type I Portland cement (OPC) of grade 42.5 conforming to the specification requirements of [21]. The clay brick materials was obtained locally as generated wastes from a brick producing factory located within Lagos metropolis, Nigeria. The clay brick materials were crushed and ground to the required fineness, and then sieved through a BS mesh of 75 micron opening so as to achieve a uniform particle sizes. Physical, chemical and microstructural characterisation were conducted on the cement and clay brick powder (CBP) samples. The sand used in the mortar mixes was sourced commercially. For the production of the reference mortar, sand (passing through 4.75 mm sieve but retained on 0.075 mm sieve) combined with ordinary Portland cement were used for the mixes. The particle size distributions and fineness modulus of the sand was determined by sieve analysis. The mix proportion of the mortar mixes was selected in compliance to ASTM C109 [22], using mix ratio of 0.45:1:2.57 (water: cement: sand) by mass for each test property. For each mix, the cement content consisted of OPC and blended mix of

OPC and CBP. The portion of the clay brick powder was used to partially replaced cement by the proportions: 0, 5, 10, 20, 25, 30, 40 and 50% (Table 1). The binders was thoroughly mixed together in order to ensure even blending and distribution of all constituents. Mortar cubes of dimension 100 x 100 x 100 mm were casted in mould, covered with plastic sheets and removed after 24 hours. Each specimen was appropriately labelled for identification and cured in water by immersion until the age of test. The cubes specimens were tested for compressive strength at ages of 3, 7 and 28 days for each percentage replacement of cement with CBP in the mortar [22]. The average strength of three specimens was determined for the various tests. The compressive strength was carried out using YES-2000 digital display compression machine.

Table 1. Batching of mortar mixes

Materials	0%	5%	10%	20%	25%	30%	40%	50%
Sand (kg)	11	11	11	11	11	11	11	11
Water (kg)	1.8	1.8	1.8	1.8	1.8	1.8	1.8	1.8
Cement (kg)	4	3.8	3.6	3.2	3.0	2.8	2.4	2.0
CBP (kg)	-	0.2	0.4	0.8	1.0	1.2	1.6	2.0

Table 2. Physical properties of mortar constituents

Properties	OPC	CBP	Sand
Fineness Modulus			2.69
Specific gravity	3.15	2.52	2.62
Water absorption		0.15	0.42

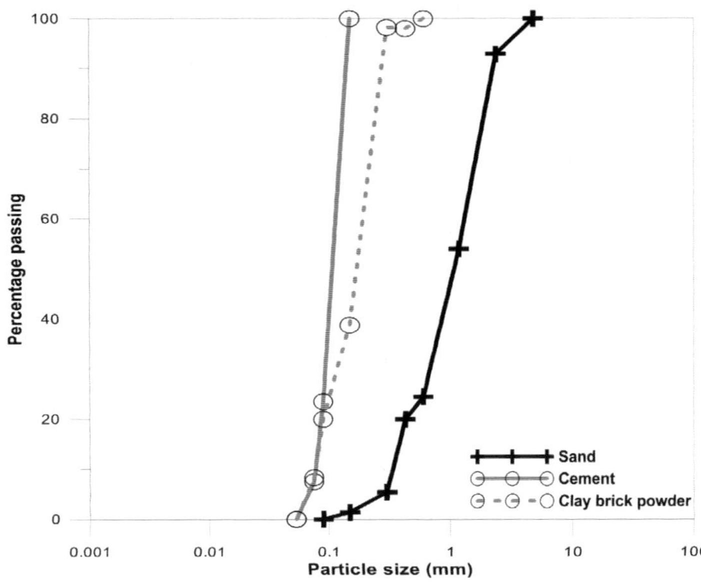

Figure 1. Particle size distribution for sand, cement and CBP

3. Results and Discussion

Figure 1 shows the particle size distribution of the sand, CBP and cement used in this study. The figure clearly indicate that the sand can be classified as a uniformly graded sand (SP). An indication that the sand material is good for producing mortar. Meanwhile, the particle size distribution for the

CBP was noticed to be slightly similar to that of cement, implying the promising property of the CBP to be blend with a binder material as mineral addictive. The physical properties of the CBP, cement and sand are presented in Table 2. Meanwhile the results of metallic oxides composition tests using the X-ray fluorescence (XRF) conducted on the cement and clay brick powder are depict in Table 3. The result clearly depicts the high silica and alumina content (60.64% and 14.23%) in the CBP material as required for a natural pozzolanic material. In addition, based on the recommendation of ASTM C618 [23], CBP can be classify as class N pozzolan since the addition of the percentage compositions of Fe_2O_3, SiO_2 and Al_2O_3 (79%) is above the minimum requirement of 70% for a class N pozzolan. The loss on ignition (LOI) value is less than 10%, while the SO_3 is less than 4%. It is expected that class N pozzolan will show more pozzolanic tendency compare to any other pozzolans.

Table 3. Chemical composition for cement and CBP materials

Compositions	Cement (%)	Clay brick powder (CBP) %
SiO_2	24.08	60.64
Al_2O_3	19.40	14.23
Fe_2O_3	6.28	4.93
CaO	74.25	0.27
MgO	3.96	1.72
K_2O	0.85	1.44
Na_2O	1.74	1.94
TiO_2	0.62	0.98
P_2O_5	1.21	0.90

Furthermore, the results of the scanning electron microscope (SEM) carried out on the CBP and cement materials are presented in Figure 2.

Figure 2. Particle morphology (a) Cement; (b) Clay brick powder (CBP)

The SEM micrograph clearly reveal the morphology of the CBP particles as rounded grains with rough geometry compare to the broad and round solid shapes of the cement particles. The SEM shows the existence of amorphous solid mass structures of the CBP materials. Moreover, the energy dispersive x-ray spectrometer (EDX) results also reveal silicon, aluminium, and oxygen as dominant elements in the CBP. This shows the suitability of CBP material to be adopted as pozzolanic addition in mortar for sustainable construction. The strength activity index (SAI) determined in accordance with ASTM C311 [25] on mortar specimens for 7 and 28 days clearly depicts that fired clay brick powder satisfied the requirements to be used as pozzolana in concrete or cement mortar for partial substitute for Portland cement.

3.1. Strength

The strength development of the mortar containing CBP is vital because it gives an indication to the quality of the mortar mix and subsequent usage as construction material. The results of the compressive strength for reference mortar and mortar containing different percentage dosages of CBP are shown in Figure 3. The obtained results indicate an increase in the compressive strength of the mortar at various mixes as the age of curing increases. However, it was observed that an increase in CBP content in the mortar mixes resulted in the reduction of the compressive strength from above 20% CBP. Further, a gradual increase in the strength was noticed at lower CBP addition of 5 and 10%. Meanwhile, the addition CBP at 20% replacement resulted in higher strength at 28 days curing compare to the other mixes. The dosage level of 20% is considered the optimum level for partial cement replacement with CBP.

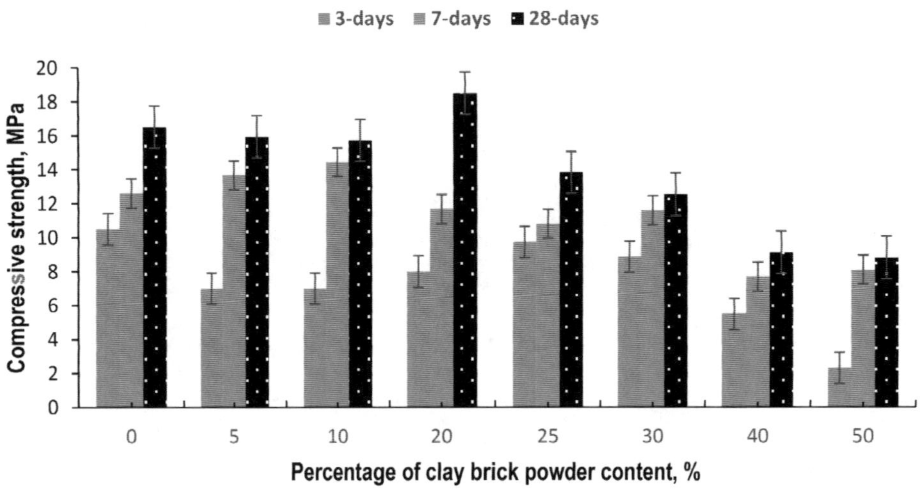

Figure 3. Strength development of control and mortar specimens containing CBP

4. Conclusion

This study examines the sustainable reuse of clay brick wastes pulverized into powder as substitute for cement in mortar production. The following conclusions can be drawn from this study:

i. The chemical compositions of clay brick powder suggest that the material can be classify as class N pozzolan material following the suggestions of ASTM C618 report on pozzolanic materials.

ii. The strength of mortar was significantly improve with the incorporation of clay brick powder (CBP) up to 20%. However, the highest strength was achieved at 20% CBP content after 28 days of curing, implying that the possible optimum CBP content in mortar can be up to 20%. This study clearly indicate that clay brick powder can be adopted as mortar component for innovative construction material in sustainable building construction.

Acknowledgment

The authors wishes to appreciate the Management of Covenant University for their financial support.

References

[1] Plessis CD, 2007 Constr. Manage. Econ. 2007 **25** 67–76.

IOP Conf. Series: Materials Science and Engineering **652** (2019) 012048 doi:10.1088/1757-899X/652/1/012048

[2] WCED, 1987 Our Common Future: Report of the World Commission in Environment and Development, University Press, Oxford
[3] Olofinnade OM, Ede AN, Ndambuki JM, 2017 *J Mater Env. Sci.* **8** 1139-1152.
[4] Sakai S, Yoshida H, Hirai Y, Asari M, Takigami H, Takahashi S, Tomoda K, Peeler MV, Schmid-Unterseh T, Douvan AR, Hathaway R, Hylander LD, Fischer C, Oh GJ, Jinhui L, Chi NK, 2011 *J Mater Cycles Waste Manag.* **13**, 86–102.
[5] Cachim PB, Portugal Sb07-Sustainable Construction, Materials and Practices: Challenge of the Industry for the New Millennium, 2007 **2**, 950-956.
[6] Toledo Filho RD, Goncalves JP, Americano BB, Fairbairn EMR, 2007 *Cem. Concr. res* **37** 1357 – 1366
[7] Liu Q, Tong T, Liu S, Yang D, Yu Q, 2014 *Constr. Build. Mater.* **73**, 754–763
[8] Naik TR, 2008 P*ractice Periodical on Structural Design and Construction* **13** 98-103.
[9] Olofinnade OM, Ede AN, Ndambuki JM, Bamigboye GO, 2016 *Mater Sci. Forum* **866** 63-67
[10] Kline J, Barcelo L, 2012 Cement Industry Technical Conference, *IEEE Press*, 1-14.
[11] Shakir AA, Naganathan S, Mustapha KN, 2014 *IJST Trans Civ Eng.* **38**, 51–60.
[12] Olofinnade OM, Ndambuki JM, Ede AN, Booth C, 2017 *Int. J Eng Res. Afr.* **31**, 77-93.
[13] Akhtaruzzaman AA, Hasnat A, 1983 *Concr Int* **5**, 58–63.
[14] Bektas F, 2007 *PhD dissertation, Iowa State University*
[15] Zong L, Fei ZY, Zhang SP, 2014 *J Clean Prod.* **70** 175–182
[16] Aliabdo AA, Abd-Elmoaty AM, Hassan HH, 2014 *Alexandria Eng J* **53**,151-168
[17] Naceri A, Hamina MC, 2009 *Waste manage.* **29**, 2378 – 2384
[18] Olofinnade OM, Ede AN, Booth CA, 2018 *Handbook of Env. Mat. Manage.* Springer
[19] Goncalves JP, Tavares LM, Toledo Filho RD, Fairbairn EMR, 2009 *Constr Build Mater.* **23**, 1971–1979
[20] Joshua O, Olusola KO, Ogunde AO, Ede AN, Olofinnade RM, Nduka DO, 2017 *Int J App Eng Res*, **12**, 13959-13565, (2017)
[21] NIS 444: Part 1, 2003 composition, specifications and conformity criteria for common cements, Abuja
[22] ASTM C109, 2005 Standard test method for compressive strength of hydraulic cement mortars (using 100 mm cube specimens) ASTM
[23] ASTM C618-12a, 2012 Standard test methods for coal fly ash and raw or calcined natural pozzolan for use in concrete, ASTM
[24] Lu J, Lu BZ, Cao SG, Li PX, Wang HT, Shen HY, 2012 *Fly Ash Compr Util.* **37** 37–42
[25] ASTM C311 / C311M-13, 2013 Standard Test Methods for Sampling and Testing Fly Ash or Natural Pozzolans for Use in Portland-Cement Concrete ASTM

IOP Conf. Series: Materials Science and Engineering **652** (2019) 012049 doi:10.1088/1757-899X/652/1/012049

Sustainable utilization of crushed waste glass as sand replacement for production of eco-friendly interlocking paving stones

Oluwarotimi Olofinnade[1,*], Anthony Ede[1], Julius Ndambuki[2], David Omole[1] Kayode Jolayemi[1], Kehinde Oyeyemi[3], Adekunle Ajao[4] and Anietie Ukoh[1]

[1]Department of Civil Engineering, Covenant University, Ota, Ogun State, Nigeria
[2]Department of Civil Engineering, Tshwane University of Technology, Pretoria, South Africa
[3]Department of Physics, Covenant University, Ota, Ogun State, Nigeria
[4]Department of Building Technology, Covenant University, Ota, Ogun State, Nigeria
[*]Corresponding author: rotimi.olofinnade@covenantuniversity.edu.ng (OM Olofinnade)

Abstract. In an attempt to ensure sustainability of construction materials, efforts were being made to reuse some non-biodegradable waste materials as substitute materials in mortar production. Sand is a major constituents in the production of mortar, however, using waste materials such as glass as partial or complete substitute for sand in mortar production will ensure the reduction on the needs to dredge for it, thereby preserving the very valuable natural sand resources within the environment. Furthermore, the use of crushed waste glass as a replacement for sand in the production paving stones will ensure that the paving stones are eco-friendly thereby eliminating the challenges of discriminate dumping of glass wastes in landfill. This research investigates the utilization of crushed waste glass as partial and complete replacement for natural sand in cement mortar for the production of interlocking paving stones (IPS). The waste glass was varied at various percentage proportions by weight to replace the fine aggregate in the cement mortar mixes. Laboratory tests which includes chemical composition, particle size distribution, specific gravity, and compressive strength were carried out on the crushed glass materials, sand and paving stone samples respectively. The obtained results show that the both the load and non-load bearing interlocking paving stones containing 20% crushed waste glass exhibited better performance in strength development than the control and other paving stone mixes containing various content of waste glass materials. This study demonstrates clearly that waste glass can be adopted to produce eco-friendly interlocking paving stones, and this will help both in the sustainable management of glass wastes and greening of the environment.

1. Introduction

Annually, over a million tons of glass wastes are being produced all over the world. Report by the International Energy Agency (IEA) [1] stated that as of 2005, the total amounts of waste glass generated was about 130 metric tons. Furthermore, from the report, European Union (EU), China and the United States of America (USA) were reported to be generating approximately 33 metric tons, 32 metric tons and 20 metric tons of the waste glass respectively [1, 2]. However, waste glass being a non-biodegradable waste material, constitute a major nuisance to effective landfill or dump sites management and operations [3]. It was reported by [2 – 6] that glass is an amorphous material, with large percentage of silica compounds used for producing different forms, types, and colours of glass products. Meanwhile, as generated waste materials, due to its nature occupy huge portions of spaces in landfill/dump sites, often triggering serious environmental pollutions [2]. Consequently, in view of rising scarcity of valuable spaces for locating landfill/dumping sites and the need to lower generated wastes, eradicate pollution and protect the environment, finding alternate solution to indiscriminate disposal of waste glass in dumping sites has become necessary. In addition, studies by Olofinnade et al. [3, 6], [7] mentioned that one of the best alternatives of solving the menace glass wastes on the environment is to reuse and recycle them as alternative construction material. In addition, recently there has being a rising demand for high performing sustainable materials in the construction industry, which constitutes a move towards eco-friendly and sustainable practices [8 – 10]. Around the world today, there is a general awareness that the construction industry depends so much on conventional materials such as granite, sand and cement, for concrete and mortar production. However, it was emphasized by researchers that continued exploitation of these resources raises huge environmental and economic costs especially in developing nations, and also contributing heavily to continued depletion of natural resources and greenhouse gases emission (GHG) [8-10]. However, studies by [9], [11 – 13] mentioned that the inception of sustainable development and renewable materials prompted mainly through regulations, landfill shortages, environmental concerns, limited availability of natural aggregates, innovations and cost reduction, the construction industry is now committed to the use of supplementary cementitious materials (SCM) from environmental wastes in concrete and mortar production. According to [13], [14 – 15], utilization of waste materials such as waste glass will helps achieving green concrete production, contributes to reduction of GHG emission and have been observed to be effective in both structural and non-structural engineering applications. Waste glass provides equal alternative options as a substitute material for traditional aggregate in concrete. Using crushed waste glass materials as aggregate and SCM in concrete and mortar benefits the environment and helps in reducing the needs for extracting raw resources. A study by Ismail et al. [16] mentioned that waste glass and sand have almost similar physical and chemical properties. In addition, several studies have been carried out on reusing waste glass as aggregate and cement replacement in cement mortar and concrete by examining the influence and optimum percentage content of waste glass that can be used as aggregate and cement substitute in the production of mortar and concrete [2 – 7], [10], [15 – 17]. Moreover, reusing wastes such as waste glass in concrete and mortar works for structural and non-structural engineering structures such as road pavements and buildings contributes to achieving sustainable construction [15, 18]. Therefore, this present study investigates the suitability of crushed waste glass (WGS) as sand replacement in cement mortar for the production of eco-friendly interlocking paving stone units that can be used for paving walkways and sidewalks for pedestrians and landscapes. The option of interlocking paving stones has over time proved effective for roadways in areas of high water table without losing its strength and durability.

2. Methodology

2.1. Cement

The cement used in this study was Ordinary Portland Cement (OPC) of grade 42.5 which was sourced commercially within the local market in Ota, Ogun state, Nigeria. The grain size distribution of the Portland cement is shown in Figure 1, while its physical properties is shown in Table 3. The chemical compositions of the cement was analysed using the X-ray fluorescence (XRF) is presented in Figure 3.

2.2. Sand

The natural sand (fine aggregate) material was sourced commercially. The grain size distribution analysed through sieve analysis is also depicted in Figure 1. Moreso, the physical attributes of the sand are shown in Table 3. The oxide compositions of the natural sand materials was analysed using the X-ray fluorescence (XRF) is presented in Figure 3.

2.3. Crushed waste glass aggregate (WGS)

The crushed waste glass employed as substitute for the natural sand (fine aggregate), was obtained as disposed mixed coloured glass wastes from dump sites and waste collection points within Sango-Ota, Ogun state, Nigeria. The glass wastes comprises mainly bottle and container glasses and flat glasses.

The collected waste glass materials were thoroughly washed with water to remove contaminants and impurities, then air-dried before crushing to required particle sizes using the milling machine. The particle size distribution of the milled glass sand material (WGS) is presented in Figure 2. Moreover, Figure 1 show samples of the crushed glass material and the natural sand. Table 3 depicts the physical properties of the waste glass materials, while the chemical compositions of the WGS materials are shown in Figure 3.

2.4. Water

Portable water obtained from the Structures and Materials laboratory in Covenant University was used for mixing the various constituents to produce the mortar for moulding the interlocking paving stone (IPS) samples.

2.5. Materials proportioning, Mixing and Testing

The cement, sand, waste glass sand (WGS) and water used in this research were all weighed using a digital weighing balance. The mortar constituents were batched by weight in compliance to ACI Standard [19]. Two (2) mix ratios (cement: fine aggregate: water) adopted for this study are in the range of 1:2.75 for load bearing IPS and 1:5 for non-load bearing IPS with corresponding water-cement ratios of 0.485 and 0.50 respectively. The mixing process was carried out manually to ensure good mixing of the mortar components. Crushed waste glass sand (WGS) was used to partially and completely replace the natural sand in percentage proportions of 0, 5, 10, 20, 30, 40, 50 and 100%. The required batching quantities of the various mortar constituents for the production of the interlocking paving stone (IPS) are presented in Table 1 and 2 for the load bearing paving stones and non-load bearing paving stones respectively. Before, pouring the mortar in the plastic moulds, the moulds were thoroughly greased with oil for easy demoulding of the IPS units after hardening. Each layers of the freshly prepared mortar was manually compacted twenty-five times with a 25 mm taping rod during the process of casting.

Table 1. Proportioning of mortar material for load bearing paving unit

Percentage substitute of sand with waste glass, %	Mortar constituents (kg)			
	Cement	Sand	Waste glass	water
0	25	68.75	-	12.13
5	25	65.31	3.44	12.13
10	25	61.88	6.88	12.13
20	25	51.56	17.19	12.13
30	25	48.13	20.63	12.13
40	25	41.25	27.50	12.13
50	25	34.38	34.38	12.13
100	25	-	68.75	12.13

Table 2. Proportioning of mortar material for non-load bearing paving unit

Percentage substitute of sand with waste glass, %	Mortar constituents (kg)			
	Cement	Sand	Waste glass	water
0	25	125	-	12.50
5	25	118.75	6.25	12.50
10	25	112.50	12.50	12.50
20	25	100.00	25.00	12.50
30	25	87.50	37.50	12.50
40	25	75.00	50.00	12.50
50	25	62.50	62.50	12.50
100	25	-	125	12.50

The IPS's were demoulded after 2 days to ensure some level of curing and strength gain before curing in water by total immersion. The IPS moulds used are of the H type having a surface cross-sectional area of 0.032 m^2 and thickness of 0.06 m. After the IPS were cured in water for 7 and 28 days, they were further openly dried and compressive strength test was carried out on the IPS samples. Three specimens were tested for each tests.

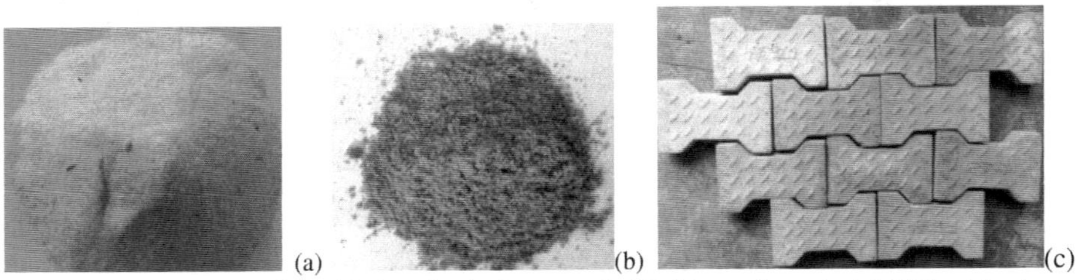

Figure 1. Specimen materials (a) Crushed glass sand; (b) Natural sand; (c) IPS with glass sand

3. Results and Discussion

Waste glass sand exhibited a very low water absorption of 0.40% and compare favourably well with the recorded water absorption of 0.42% for the natural sand as shown in Table 3. The particle size grading of the fine aggregates; that is for both natural sand and waste glass sand (WGS) are closely similar as presented in Figure 2. Similarly, the estimated specific gravity and fineness modulus for both materials are close (Table 3). The obtained results on the physical properties for the natural sand and glass sand (WGS) clearly shows that both materials have similar physical properties as also mentioned in the study of Ismail et al. [16].

Table 3. Physical properties of materials

Properties	Portland cement	Sand	Waste glass sand (WGS)
Fineness Modulus	0.60	2.69	2.99
Specific gravity	3.15	2.62	2.50
Water absorption		0.42	0.40

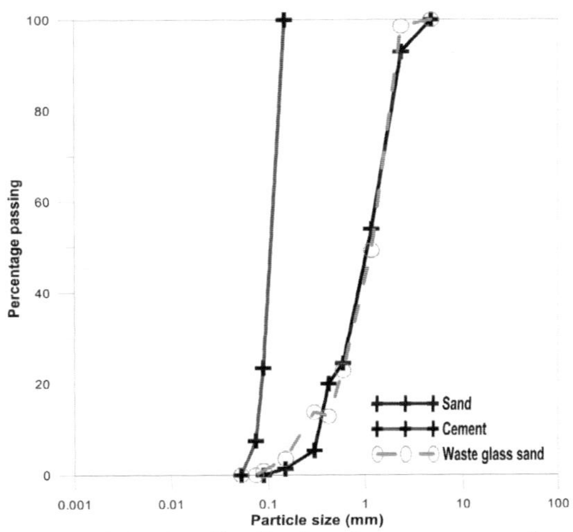

Figure 2. Particle size distribution for cement, sand and WGS

The oxide compositions of the Portland cement as displayed in Figure 3 shows the alkali compound of the Portland cement was 0.18%, using $Na_2Oeq = Na_2O + 0.658K_2O$, indicating a low-alkali cement content. Moreso, the chemical composition clearly shows the cement material having a high percentage of CaO content, while the crushed waste glass (WGS) particles indicate the WGS has a high percentage of silica (SiO_2) content as the natural sand, while the sodium compound content indicate the waste glass material to be soda-lime (Figure 3).

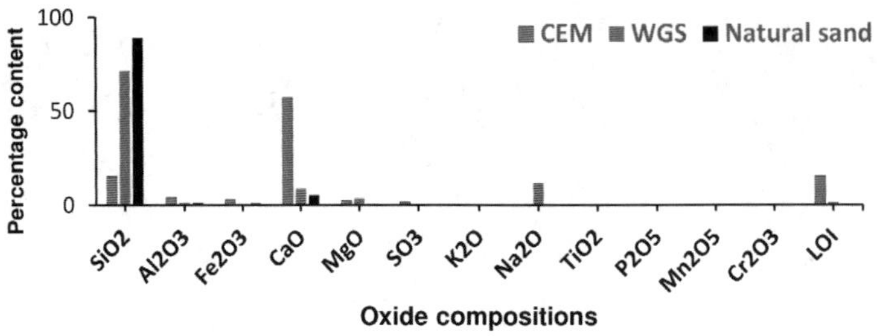

Figure 3. Chemical compositions for cement, sand and crushed glass materials

Figure 1(c) shows the picture of the interlocking paving stone (IPS) containing waste glass sand (WGS) as replacement for natural sand. The obtained results as displayed in Figures 4 and 5 indicated a declining trends in the compressive strength of the interlocking paving stone units as the percentages of glass content increases for both the load bearing paving stones and non-load paving stone units. This implies a loss in the strength of the paving units with increasing replacement dosages of glass content irrespective of the mix ratio. Moreover, a more significant increase in the compressive strength for the IPS was observed as the curing ages increases from 7 to 28 days curing age. This can be attributed to the pozzolanic effect of the glass at prolong curing. However, an appreciable increase in compressive strength was observed at 20% replacement level compare with the control paving stones as indicated in Figure 4 and 5 for both paving stones produced at different mix ratio.

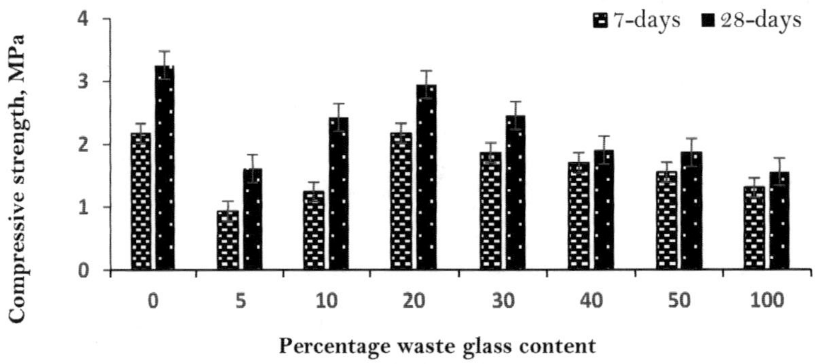

Figure 4. Compressive strength for load bearing interlocking stones with glass content

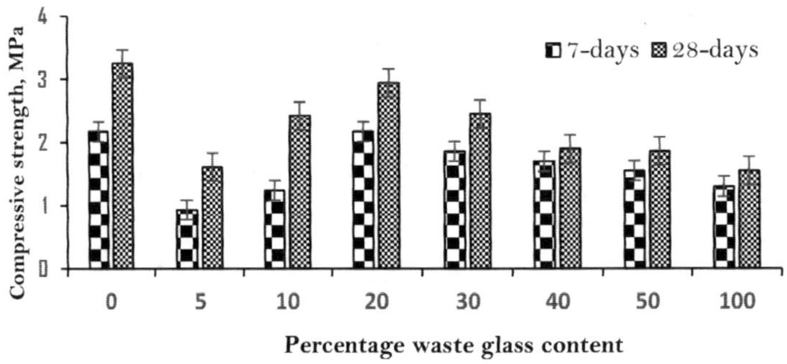

Figure 5. Compressive strength for non-load bearing interlocking stones with glass content

In addition, the obtained results at 20% replacement level agree with the standard requirements suggested by NBRRI [20] on compressive strength of interlocking paving stone units.

4. Conclusion

This study investigates the strength performance of interlocking paving stones produced with waste glass sand as substitute for natural sand. The following conclusions can be deduced from this study:

i. The physical properties and chemical compositions for the natural sand and glass sand (WGS) shows close similarity.
ii. The strength of the interlocking paving stones were significantly improved with the incorporation of waste glass sand at 20% level.
iii. The interlocking paving stones containing WGS meet the standard requirements specified for compressive strength of interlocking paving stone units that can be adopted for constructing pedestrian/non-traffic paving and light roadways route such as landscape, building environments and premises, public and car parks, low-traffic pedestrian sidewalks and stabilizing sloppy terrain. This study demonstrates clearly that waste glass can be adopted to produce eco-friendly interlocking paving stone, and this will help both in the sustainable management of glass wastes and greening of the environment. In addition, it is recommended that further studies be carried out on the long term performance and durability of IPS produced with cement mortar containing crushed glass particles.

Acknowledgments
The authors wish to appreciate the Management of the Covenant University for their financial support.

References

[1] IEA, Tracking Industrial Energy Efficiency and CO_2 Emissions 2007 OECD/IEA 1–321
[2] Rashad AM, 2014 *Constr. Build. Mater.* **72** 340–357)
[3] Olofinnade OM, Ede AN, Ndambuki JM, 2017a *J Mater Env Sci.* **8**, 1139-1152
[4] Ling Tung-Chai, Poon Chi-Sun, Wong Hau-Wing, 2013 *Resour Conserv Recycl.* **70** 25-31
[5] Siddique R, 2008 *Waste materials and by-products in concrete*, Springer, 147–175
[6] Olofinnade OM, Ndambuki JM, Ede AN, Olukanni DO, 2016 *Mater. Sci Forum* **866** 58-62
[7] Afshinnia K, Rangaraju PR, 2016 *Constr. Bui. Mat.* **117** 263–272
[8] D'Alessandro F, Asdrubali F, Baldinelli G, 2014 *Constr. Bui Mater.* **68** 277–284
[9] Sojobi AO, 2016 *Cogent Eng* **3** 1255168
[10] Olofinnade OM, Ede AN, Ndambuki JM, 2017b *Int J Eng Tech innov.* **7** 280 – 291
[11] Bamigboye GO, Ede AN, Raheem AA, Olofinnade OM, Okorie U, 2016 *Mater Sci Forum* **866** 73-77
[12] Joshua O, Olusola KO, Ogunde AO, Ede AN, Olofinnade RM, Nduka DO, 2017 *Int J Appl Eng Res* **12** 13959-13565
[13] Eisa A, 2014 *Int J Conc Struc. Mat* **8** 251–258
[14] Park S, 2012 *Int J of Conc Struc Mat.* **6** 231–237
[15] Olofinnade OM, Ndambuki JM, Ede AN, Booth C, 2017c *Int J Eng Res Afr* **31** 77 – 93
[16] Ismail ZZ, Al-Hashmi EA, 2009 *Waste Manag.* **29** 655-659
[17] Tan KH, Du H, 2013 *Cem. Concr. Compos.* **35**, 109–117
[18] Oyedepo OJ, Oluwajana SD, Akande SP, 2014 *Civil Env Res* **6** 35–42
[19] ACI 211.1-91, 2002 *ACI Committee 211 Report* ACI Farmington Hills MI
[20] NBRRI, 2006 *Nigerian Building and Road Research Institute Newsletter* **1** 15–17

Evaluation of performance of foam produced with different methodologies for use in foam concrete production

S S Sahu[1] and I S R Gandhi[1,*]

[1]Research Scholar, Department of Civil Engineering, Indian Institute of Technology, Guwahati, Assam, India
[1,*]Assistant Professor, Department of Civil Engineering, Indian Institute of Technology, Guwahati, Assam, India

Email: gindu@iitg.ac.in

Abstract. For the production of foam concrete which finds applications in many areas, generally as a function of its relatively lightweight, the use of stable and quality foam is the key requirement. However, the literature available on the influence of characteristics of foaming agent and foam on the properties of foam concrete are rather limited. Hence a more systematic research is needed in this direction. The present work focuses on the relative performance evaluation of foam (initial foam density, foam stability, viscosity of surfactant solution and bubble size) using a typical synthetic surfactant and additive using two different foam production methodologies namely stirrer and foam generator. Comparative studies on foam production methods indicated that foam produced based on compressed air method (foam generator) is of better quality with lesser liquid fraction and drainage compared to that of foam produced with stirrer. Also studies have proved that there is a strong relationship between viscosity of surfactant solution, initial foam density, foam stability and bubble microstructure.

1 Introduction

It is a well-known fact that pure liquids cannot foam and the presence of surfactant is essential for making and stabilising a foam [1,2]. One traditional and most familiar way of preparing foam is by using a dispersion technique, which consists of mechanical shaking of surfactant solution or whipping a surfactant solution mechanically with a high speed stirrer or fan [3]. Researchers [4,5] have proved that the mean bubble size of foam produced decreased with increase in rotational speed of stirrer. However, the most widely used method of foam production for use in concrete is compressed air method where the foam is generated by mixing compressed air and surfactant solution in high density restrictions [6]. Hence, in order to compare the influence of foam production methods on the behaviour of foam, two different foam production methods selected for the present study are mechanical method (high speed stirrer) and compressed air method (foam generator). The relative characteristics of foam and surfactant solution viz., initial foam density, bubble size and viscosity of surfactant solution produced with a synthetic surfactant and additive has been studied for two different foam production methods.

2 Experimental Program

2.1 Materials and Methods

This study investigates two different foam production methodologies using a synthetic surfactant namely Sodium lauryl Sulphate (SLS) and an additive namely Carboxy Methyl Cellulose (CMC). As discussed earlier, mixing speed is important in case of stirrer method (M1), hence for the present study maximum speed of 1136 rpm was adopted for M1. Similarly for generator method (M2) the pressure at which the foam is generated was varied from 353 kPa to 490 kPa.

For evaluating the relative characteristics of foam, a range of SLS concentration from 1% to 10% and CMC concentration from 0.1 % to 0.3% were adopted. The initial foam density was measured immediately after its generation while the stability of foam was assessed by free drainage test prescribed by Def-Standard 42-40 [7]. A drainage pan of 1612 ml nominal volume with a conical base rounded to accept externally a 12.7 mm bore by 25 mm long polymethyl methacrylate tube with a 1.6mm bore brass cock at its lower end was used. The pan was filled with foam and the volume of the solution drained was measured at various time intervals. The influence of addition of additive CMC on viscosity of surfactant solution was measured using Anton Par Rheometer with the speed of rotation varied from 150 to 200 rpm at constant temperature of 250C. Bubble microstructure was studied by placing a thin layer of foam on a glass plate (immediately after the generation of foam) under optical microscope and the images captured were analyzed with a computer with image analysis software (Image J) [8].

3 Results and Discussion

3.1 Initial foam density

The initial foam density of foam produced with stirrer (Method M1) and foam generator (Method M2) are shown in Figure 1. The initial foam density of foam produced using stirrer (Method M1) decreased significantly with increase in SLS concentration up to 7% beyond there was not much variation.

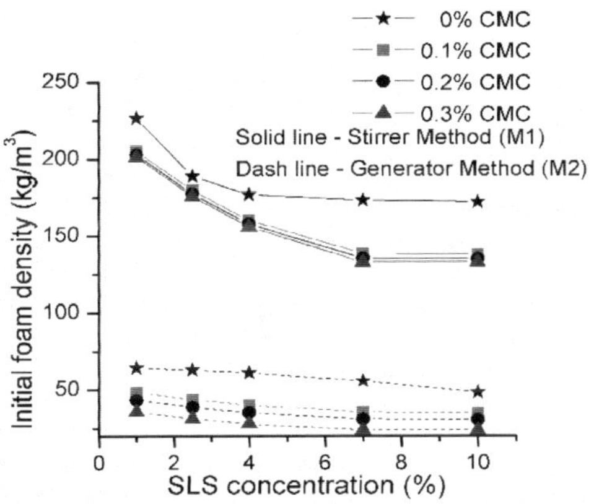

Figure 1. Variation of initial foam density with SLS concentration for M1 and M2 (490 kPa)

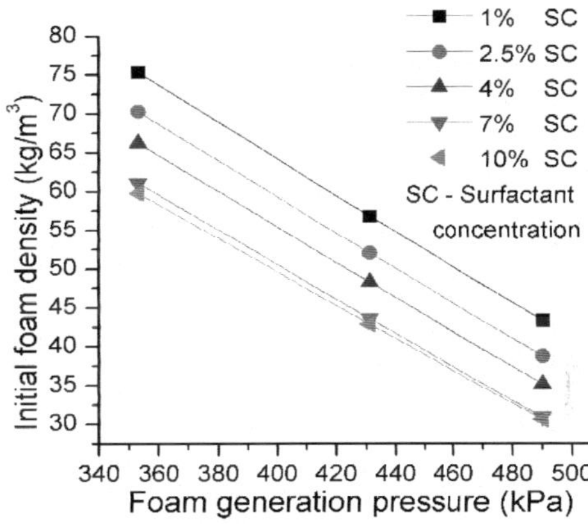

Figure 2. Variation of Initial foam density with foam generation pressure for M2 (CMC 0.2%)

The reason for this behaviour could be ascribed to the production of foam with relatively good quality (more amount of air and lesser liquid fraction) at higher SLS concentration. The variation in initial foam density with CMC concentration was very less compared to that with SLS concentration. The variation of initial foam density of foam with SLS concentration produced using foam generator (Method 2) also showed trend similar to that of foam produced using method M1.

Also it is noticed from Figure 1, that the effect of variation of CMC concentration on initial foam density is more significant unlike foam produced using method M1. This shows that the potential of CMC in improvement of quality of foam (reduction of liquid fraction) is utilized to a greater extent in foam produced using foam generator based on compressed air method.

For a given SLS concentration, an increase in foam generation pressure reduces the initial foam density (IFD) significantly for foam produced using method M2 (Figure 2). Also it is to be noted that the effect of foam generation pressure is very predominant when compared to other parameters for foam produced using method M2. Hence for the foam generator model used for the present study, to produce foam of good quality (lower liquid fraction) higher pressure is needed for sufficient mixing of air and liquid. Further, the experimental results indicated that the foam produced using generator is relatively drier foam and meets the ASTM C 796 requirement of initial foam density of 32 to 64 kg/m^3 for use in preformed foam production [9].

3.2 Foam Stability

Figure 3 shows the percentage solution drained for foam produced using method M1 and M2 at 5 minute after foam generation. Foam produced using stirrer being wet foam evidently shows higher drainage. Percentage solution drained of foam produced using both the methods reduces significantly with increase in SLS and CMC concentration. The reduction in percentage solution drained or improvement in stability of foam could be ascribed to the production of good quality of foam (with lesser liquid fraction) at higher levels of SLS and CMC concentration. In particular, the addition of CMC has resulted in enhancement of viscosity of surfactant solution and production of stable foam with smaller size bubble microstructure as explained in upcoming sections. Furthermore the reduction in solution drained due to addition of CMC is more predominant for foam produced using method M2 which could be again attributed to efficient role of CMC in foam produced based on compressed air method when compared to that of stirrer. Also the early rate of drainage in the first five minutes is very low for foam produced using method M2 compared to that of M1. In addition it is to be noted that the effect of variation of SLS and CMC concentration is more significant on foam drainage when compared to that of initial foam density. Figure 4 shows the variation of percentage solution drained with SLS concentration and foam generation pressure (FGP) at fixed level of 0.2% CMC concentration. Similar to earlier observations on the effect of FGP on IFD, the foam drainage reduces significantly with increase in FGP.

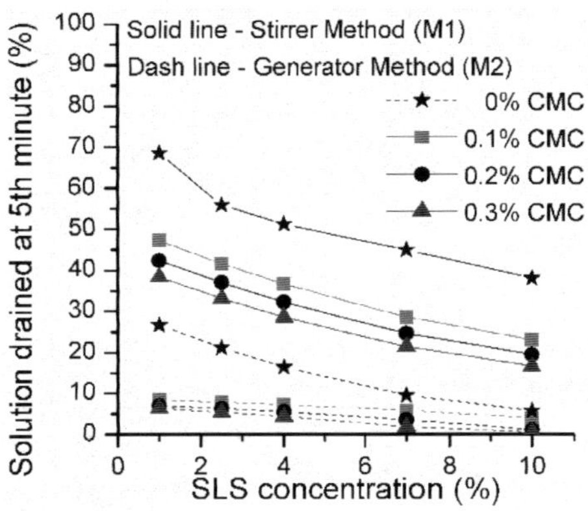

Figure 3. Variation of percentage solution drained with SLS concentration for M1 and M2 at 5th minute after foam generation (490 kPa)

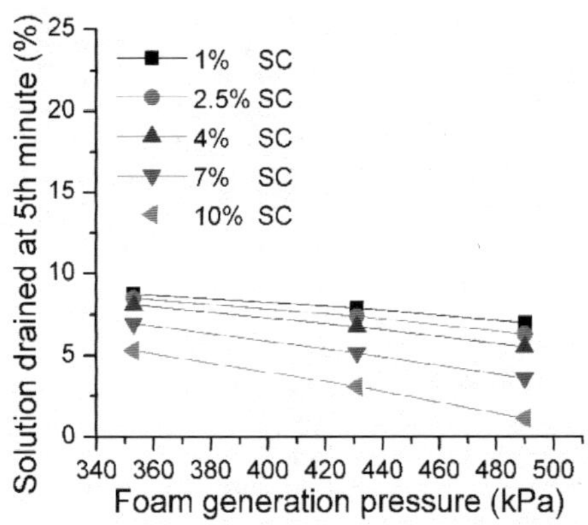

Figure 4. Variation of percentage solution drained with foam generation pressure for M2 (CMC 0.2%)

Hence again from stability point of view, it is preferable to choose higher SLS, CMC concentrations and higher FGP for the foam generator used. However it was observed that adoption of FGP beyond 490 kPa resulted in interrupted foam production affecting the foam output rate. Hence for the current foam generator model the recommended pressure was restricted to 490 kPa.

3.3 Viscosity of surfactant solution

The viscosity of surfactant solution is plotted in Figure 5. It is noticed from Figure 5 that the viscosity of surfactant solution at low levels of SLS and CMC concentration is very less and this got enhanced significantly with increase in SLS and CMC concentration. Furthermore, results indicated that the significant increase in viscosity of SLS surfactant solution with increase in SLS concentration was achieved only in presence of CMC. This could be attributed to the tight packing of surfactant molecules achieved due to the addition of CMC which thereby enhanced the viscosity of surfactant solution significantly [10]. Hence as the viscous surfactant solution can result in stable foam with smaller bubble microstructure as discussed in upcoming sections, it is preferable to adopt higher SLS and CMC concentration. However as the viscosity of surfactant solution can also affect the workability of foamed concrete mix, the optimization of CMC dosage must also consider the workability and stability of foamed concrete mix.

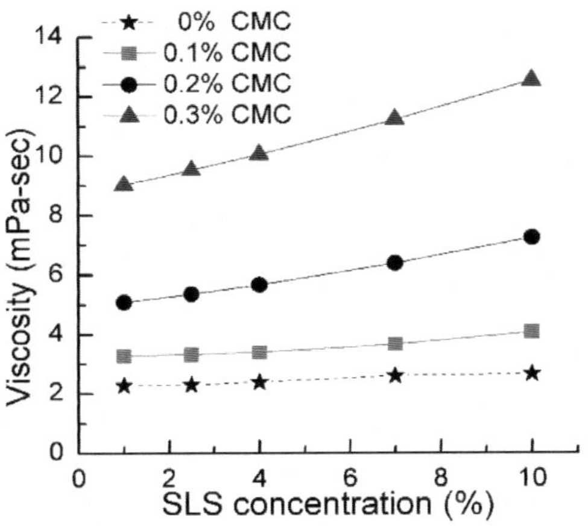

Figure 5. Variation of viscosity of surfactant solution with SLS concentration at various CMC

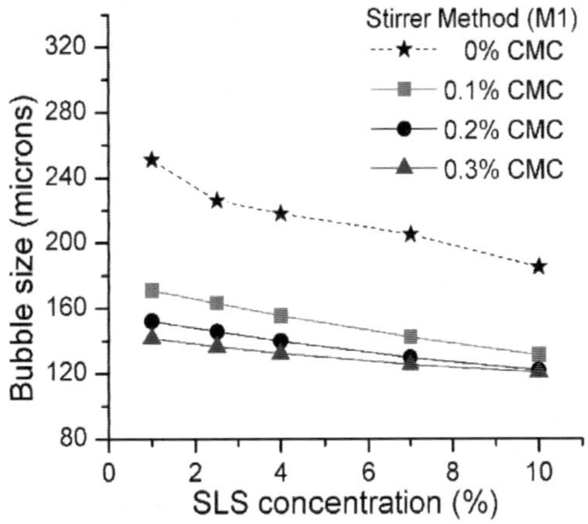

Figure 6 (a) Stirrer method M1

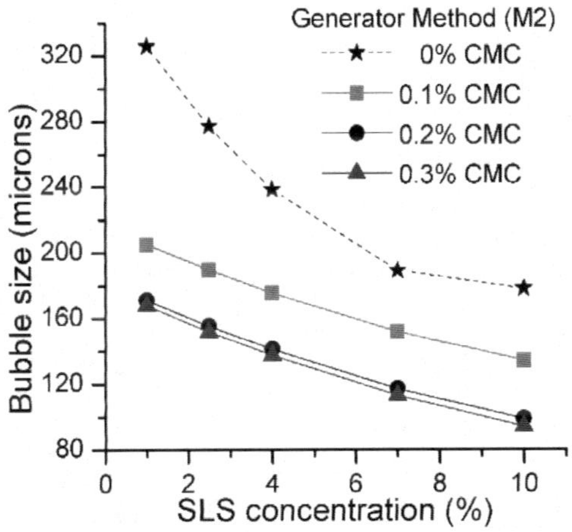

Figure 6 (b) Generator method M2

Figure 6. Variation of average bubble size with SLS concentration for M1 and M2 (490 kPa)

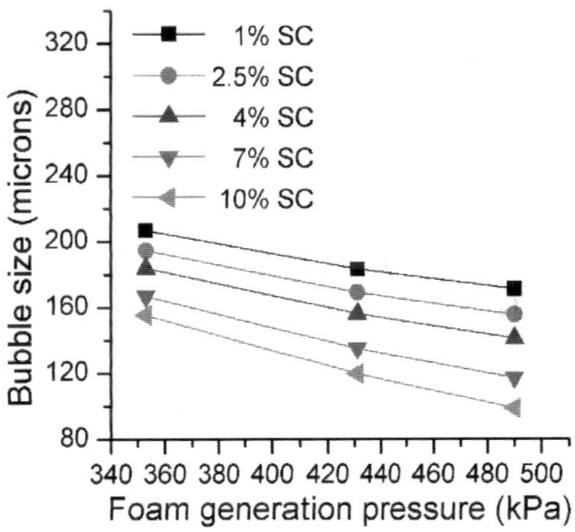

Figure 7. Variation of bubble size with foam generation pressure for M2 (CMC 0.2%)

3.4 Bubble size

Figure 6 (a, b) shows the variation in average bubble diameter of foam produced using methods M1 and M2. It is observed for both M1 and M2 that the adoption of higher SLS and CMC concentration results in foam with reduced bubble diameter. The reduction in bubble size could be attributed to the increase in viscosity of surfactant solution due to addition of CMC. Hence the above results are in agreement with the earlier observations by other researchers [11] that highly viscous surfactant solution can produce foam with smaller bubble microstructure [12, 13].

Figure 7 shows the variation of average bubble diameter with SLS concentration at fixed level of 0.2% CMC concentration. A percentage reduction of 17% in average bubble diameter was observed for variation in FGP from 353 to 490 kPa at higher levels of SLS concentration. The reduction in bubble sizes could be ascribed to improvement in viscosity of surfactant solution at higher FGP. Comparison of bubble sizes of foam showed that the stirrer foam which is basically a wet foam (with liquid fraction 20%) has finer bubble structure (Figure 6 (a) compared to dry foam produced using foam generator. Dispersion of more volume of air per unit volume of surfactant solution in dry foam could be the reason for coarser bubble structure in dry foam with lesser liquid fraction [14]. However at very high SLS and CMC concentrations foam produced by method M2 is found to have relatively smaller size bubbles compared to M1. Hence, in order to produce foam with smaller bubble size microstructure, higher surfactant concentration is needed for generator method M2 (Figure 6 (b).

3.5 Correlation between viscosity of surfactant solution, initial foam density and foam stability

A scatter plot showing the relation between viscosity of surfactant solution and initial foam density of foam produced with methods M1 and M2 are plotted in Figure 8. It was observed that an increase in surfactant concentration (at a constant CMC concentration) associated with some increase in viscosity resulted in significant reduction of the initial foam density up to SLS concentration of 7% after which there is not much variation for foam produced with both methods M1 and M2. This trend of reduction in initial foam density with increase in SLS concentration could be attributed to the significant improvement in quality of foam due to increase in surfactant concentration. On the other hand, the significant increase in viscosity of surfactant solution achieved due to increase in CMC concentration (at a constant SLS concentration) did not result in much variation in initial foam density (% reduction in IFD less than 4%) for method M1. However for method M2, significant increase of 208 % in viscosity achieved due to increase in CMC concentration from 0.1 to 0.3% at constant SLS concentration of 10% resulted in 32% reduction in initial foam density.

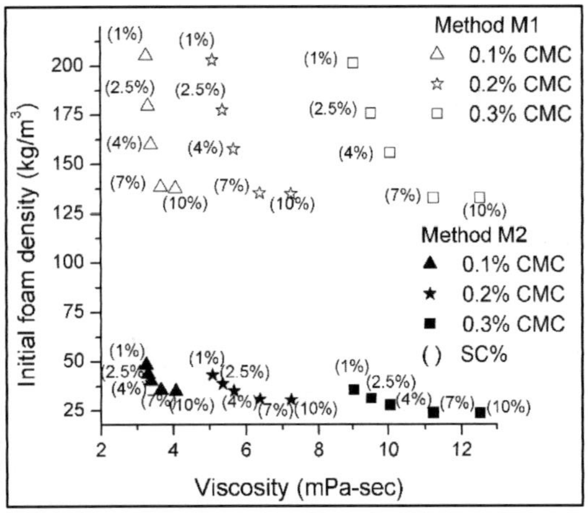

Figure 8. Variation of Initial foam density with viscosity of surfactant solution for M1 and M2

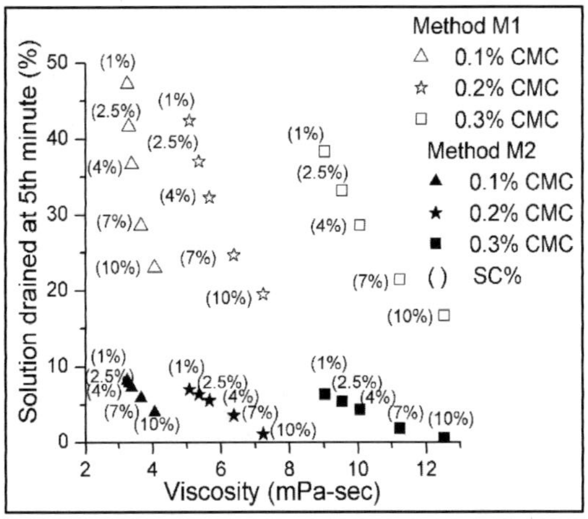

Figure 9. Variation of percentage solution drained with viscosity of surfactant solution for M1 and M2

Hence based on the above observations it can be concluded that the influence of increase in viscosity achieved due to addition of the CMC on improvement of quality of foam (reduction of liquid fraction) is realized to a greater extent in foam produced using foam generator based on compressed air method. The relation between viscosity and percentage solution drained at 5th minute after foam generation for foam produced with stirrer (M1) and generator (M2) are plotted in Figure 9. Similar to discussion on foam density, it was observed that an increase in surfactant concentration (at a constant CMC concentration) associated with some increase in viscosity resulted in significant reduction in percentage solution drained for both M1 and M2. Furthermore, the significant increase in viscosity of surfactant solution achieved due to increase in CMC concentration (for a constant SLS concentration) resulted in significant reduction in foam drainage. It could be observed apparently for foam produced with M2, that the percentage solution drained is reduced significantly to less than 1% particularly at higher concentration of CMC which confirms the earlier observation that role of CMC in foam quality improvement is more efficient in compressed air method of foam production. The reason for reduction in drainage due to increase in viscosity of surfactant solution could be ascribed to the cushioning effect of viscous lamellae around the bubble and reduced bubble size which improved foam stability.

4 Conclusions

Following conclusions are drawn based on the present study.

Based on comparative studies on foam production methods, it can be concluded that foam produced based on compressed air method (foam generator) is of better quality with lesser liquid fraction and drainage compared to that of foam produced with stirrer. Also the potential of CMC in improvement of quality of foam (reduction of liquid fraction) is utilized to a greater extent in foam produced using foam generator based on compressed air method. Studies on relative foam behaviour for two different foam production methods indicated that the initial density and drainage of foam produced using both the methods (M1 and M2) decreased significantly with increase in surfactant concentration resulting in relatively good quality of foam with lesser liquid fraction and more stability. For the compressed air method of foam production, the effect of FGP is predominant compared to that of other foam production parameters. Hence it is recommendable to adopt higher levels of SLS, CMC concentration and FGP to produce highly viscous surfactant solution which can result in stable foam with lesser liquid fraction and smaller bubble size microstructure.

References

[1] Birdi K S 2009 *Surface and Colloid Chemistry: Principles and Applications* (New York: CRC Press)
[2] Pugh R 1996 Foaming, foam films, antifoaming and defoaming *Adv. Colloid Interface Sci.* **64** 67–142
[3] Krzan M 2013 Rheology of the wet surfactant foams and biofoams-a review *Tech. Trans. Chem.* **1** 9–27
[4] Gido S, Hirt D, Montgomery S, Prud'homme R and Rebenfeld L 1989 Foam bubble size measured using image analysis before and after passage through a porous medium *J. Dispers. Sci. Technol.* **10** 785–793
[5] Hirt D, Prudhomme R and Rebenfeld L 1987 Characterization of foam cell size and foam quality using factorial design analyses. *J. Dispers. Sci. Technol.* **8** 55–73
[6] Gandhi I S and Ramamurthy K 2010 Relative assessment of density and stability of foam produced with four synthetic surfactants *Mater. Struct.* **43** 1317–1325
[7] Def-Standard 42-40 2002 *Foam Liquids, Fire Extinguishing (Concentrates, Foam, Fire Extinguishing)* Ministry of Defence, United Kingdom
[8] Image J https://imagej.nih.gov U.S. National Institutes of Health, Bethesda, MD, USA (visited on 3/6/2016)
[9] ASTM C 796 1997 *Standard Test Method for Foaming Agents for Use in Producing Cellular Concrete Using Preformed Foam.* ASTM International, United States
[10] Sharma M K, Sharma G D and Society F P 1991 *Particle technology and surface phenomena in minerals and petroleum* (London: Plenum Press)
[11] Kroezen A, Wassink J and Schipper C 1988 The flow properties of foam *J. Soc. Dyers Colour* **104** 393–400
[12] Ramamurthy K, Nambiar E K and Gandhi I S 2009 A classification of studies on properties of foam concrete *Cem. Concr. Compos.* **31** 388–396
[13] Sahu S S, Gandhi I S and Khwairakpam S 2018 State-of-the-art review on the characteristics of surfactants and foam from foam concrete perspective *J. Inst. Eng. (India) Series A* **99** 391-405
[14] Bonsu K, Shokri N and Grassia P 2016 Fundamental investigation of foam flow in a liquid-filled Hele-Shaw cell *J. Colloid Interface Sci.* **462** 288–296

New texture with black grey color of fired clay wall tiles produced from rice husk ash, brown glass cullet, and local plastic clay

W Ubolrat[1] * and P Duangkamol [2]

[1,2] Faculty of Engineering, Production Engineering Department, 1518 Pracharat 1 Rd., Bangsue, Bangkok, Thailand

Email: ubolrat.w@eng.kmutnb.ac.th; bmind2316@gmail.com

Abstract. Developing fired clay wall tiles by valorising waste materials and local plastic clay was objective of this study. Rice husk and glass cullet are waste materials which had been selected. They were mixed with local plastic clay of Angthong province in Thailand. All materials used were characterized their composition by X-ray Fluorescence technique. Specimens had been prepared by dry pressing and then fired at 950°C and 1050°C. Physical properties, bending strength and water absorption, were tested and compared with Thai Industrial Standard (TIS 2508-2555). In addition, X-ray Diffraction technique and Scanning Electron Microscope were used to analyse microstructure of selected specimens. The results of this experiment shows that utilizing of 60-70% brown glass cullet, 10-30% rice husk ash fired at 950° C is feasible to produce wall tiles for commercial. It provides high bending strength and low water absorption and can pass the TIS 2508-2555. Furthermore, it provides new texture with black grey colour of wall tiles.

1. Introduction

Currently, there are many community industries in Thailand. Thai government institution dedicated for SMEs. It had allocated THB 1.2 billion ($USD 37 million) to support the growth of SMEs and estimated that the fund will help SMEs to create innovation up to 100,000 products. This can increase value to the economy [1]. To comply with this policy, this paper aimed to improve competency of community industries in Angthong province of Thailand by developing new texture ceramic products. Waste materials i.e. rice husk ash from biomass power plant and brown glass cullet were selected. They were mixed with local plastic clay of Angthong province.

For saving environment, rice husk ash; waste material from the biomass power plant has been selected. Due to consuming rice husk as the fuel of boiler, the power plant generates abundant of rice husk ash. In addition to having negative effects on the environment, it leads a high cost burden to the manufacturers to dispose it. Many researches about rice husk ash (RHA) have been carried out. Effect of rice husk and rice husk ash to properties of bricks was determined by Sutas et al [2]. They summarized that bricks adding rice husk ash (RHA) have high compressive strength and density than that of adding rice husk (RH). Haslinawati et al [3] studied the effect of temperature on ceramic from rice husk ash. They founded that RHA had sintered at 1000oC contained high percentage of cristobalite. In addition, rice husk ash and fly ash were utilized to manufacture red bricks [4]. This research proposed that 15% RHA, and 15% fly ash mixed with plastic clay were appropriated to produce bricks which fired at 1050oC. Eliche-Quesada et al [5] also studied utilization of rice husk ash and wood ash for developing clay matrix brick. The proportion of 10 wt. % rice husk ash and 30 wt. % wood ash can fulfil the standard of clay masonry units.

Moreover, glass cullet also one of waste materials have been generated from production of glass industries and discarded glass products i.e. glass bottles. From the yearly report by the pollution control department in 2016 [6], 42% of waste glass in Thailand is not utilized, amounting to 1,014,653 tons. Owing to containing fluxing agent, glass cullet can be employed as a material for reducing the firing temperature [7]. Many researchers proposed that waste glass has the potential for enhancing mechanical properties of ceramic products i.e. wall and floor tiles [8], fired clay bricks [9, 10, and 11]. In this study, combination between rice husk ash, brown glass cullet, and plastic local clay (Angthong clay) were conducted. The expected outcome was to manufacture new texture with black grey colour of fired clay wall tiles.

2. Materials and Method

2.1 Materials

Waste materials used in this research were Rice Husk Ash (RHA) and Brown Glass Cullet (BGC) which mixed with Angthong clay (Ang Clay). All of them were characterized the chemical composition by X-ray Fluorescence (XRF) which are shown in table 1.

Table 1. Chemical analysis of materials used by XRF technique

	SiO_2	K_2O	CaO	MgO	Fe_2O_3	Al_2O_3	SO_3	Na_2O	TiO_2
RHA	92.67	1.96	1.18	0.78	0.54	0.56	0.34	0.07	0.03
BGC	72.1	0.2	10.6	2.4	0.1	1.6	0.15	12.9	0.05
Ang Clay	64.95	2.46	0.59	1.43	6.45	21.95	0.23	0.41	0.99

Fluxing materials i.e. K_2O, Na_2O, and CaO (See table 1) are most contained in BGC. This indicated that fluxing materials can promote the low sintering temperature of ceramic bodies. However, SiO2 is the majority oxide containing in all materials. Mixtures of materials used are conducted by exploiting the tri axial diagram. It can formulate 36 formulas and classified into 8 groups: A, B, C, D, E, F, G, and H (See table 2).

Table 2. Mixture compositions for 36 formulas (wt. %)

	A								B							C						D					E				F			G		H
	A1	A2	A3	A4	A5	A6	A7	A8	B9	B10	B11	B12	B13	B14	B15	C16	C17	C18	C19	C20	C21	D22	D23	D24	D25	D26	E27	E28	E29	E30	F31	F32	F33	G34	G35	H36
RHA	10	20	30	40	50	60	70	80	10	20	30	40	50	60	70	10	20	30	40	50	60	10	20	30	40	50	10	20	30	40	10	20	30	10	20	10
BGC	80	70	60	50	40	30	20	10	70	60	50	40	30	20	10	60	50	40	30	20	10	50	40	30	20	10	40	30	20	10	30	20	10	20	10	10
Ang Clay	10	10	10	10	10	10	10	10	20	20	20	20	20	20	20	30	30	30	30	30	30	40	40	40	40	40	50	50	50	50	60	60	60	70	70	80

Note that, formulas in highlight color could not be molded, due to containing RHA higher than 30 wt. %. This made the mixtures difficult to be formed. Therefore, there were 25 formulas remained to be investigated.

2.2 Method

2.2.1 Preparation the specimens. When drying until having constant weight and milling by a laboratory ball mill, all materials were sized by 297 micrometer sieve. And then, they had mixed together by dry mixing process for 30 minutes. Mixtures of formulations were sprayed with 10 wt. % of water and molded by uniaxial pressing technique in a 50x100 mm rectangular die at 100 bars and dried again at 150° C.

In order to form the hard body, all specimens were fired by 100° C/h at maximum temperature 950°C and soaked for 1 h.

2.2.2 Testing properties of specimens. All specimens were heated up by the electric laboratory kiln, then they were suitable to be tested properties. The properties of fired specimens to be investigated are expressed as follow.

- Three points bending was exploited to determine **bending strength** of specimens. **Water absorption test** was conducted by placing specimens in water until heated up and boiled for 2 h, and soaking them under water for 24 h. Specimen's water absorption was determining by calculating difference of dry and wet weight of them. **Linear firing shrinkage** of specimens was carried out by measuring specimens' width with Vernier calliper. Difference of their before and after firing width were calculated for determining firing shrinkage. **Weight loss** of fired specimens was determined by weighting specimens before and after firing, then calculating their different weight. Specimens' bulk density was investigated by Archimedes' technique.
- **Characterization microstructure of specimens** Scanning Electron Microscopy (SEM) series Hitachi SU3500 and X-ray Diffractometer (XRD) series D8 ADVANCE. (BRUKER) were employed to analyse microstructure of fired specimens. However, formulas obtaining the highest and lowest bending strength were selected to investigate their micro structure.

3. Results and discussion

3.1 Fluxing materials, bending strength, and linear firing shrinkage of tested samples

Na_2O, K_2O, and CaO are fluxing contained in each mixture formulas have been calculated. They are compared with bending strength and linear shrinkage of tested specimens. It is shown in table 3. Formulas No. A4-A8, and B12-B15 are unable to be tested. Due to having high content of RHA and low content of Ang clay, they cannot be molded. All 8 formula groups show the trend of increasing bending strength and linear shrinkage, when increasing content of fluxing materials (see table 3).

Table 3. Comparison of % fluxing material, bending strength, and % shrinkage of 36 formulas

Group	No.	% Fluxing	Bending St. (MPa)	%Shrinkage
A	A1	19.63	8.91	0.96
	A2	17.58	18.18	3.82
	A3	15.53	16.32	3.67
	A4	13.48	0.00	0.00
	A5	11.43	0.00	0.00
	A6	9.38	0.00	0.00
	A7	7.33	0.00	0.00
	A8	5.28	0.00	0.00
B	B9	17.60	17.81	1.89
	B10	15.55	18.07	3.71
	B11	13.51	14.52	2.57
	B12	11.46	0.00	0.00
	B13	9.41	0.00	0.00
	B14	7.36	0.00	0.00
	B15	5.31	0.00	0.00
C	C16	15.58	21.99	2.53
	C17	13.53	15.08	2.36
	C18	11.48	5.11	2.09
	C19	9.43	2.36	1.67
	C20	7.38	0.00	1.65
	C21	5.33	0.00	1.21
D	D22	13.56	15.18	2.16
	D23	11.51	7.36	2.09
	D24	9.46	2.42	1.53
	D25	7.41	1.69	1.45
	D26	5.36	0.00	1.08
E	E27	11.53	11.92	2.03
	E28	9.48	5.10	1.92
	E29	7.43	2.82	1.72
	E30	5.38	1.38	0.98
F	F31	9.51	10.79	1.96
	F32	7.46	4.55	1.90
	F33	5.41	2.49	1.42
G	G34	7.48	11.79	1.88
	G35	5.43	5.12	1.43
H	H36	5.46	10.43	1.19

Table 4. Comparison of % water absorption, % wt. loss, and density of 36 formulas

Group	No.	% Water Ab	% Wt. loss	Density (g/cm3)
A	A1	9.76	2.05	1.19
	A2	17.74	4.34	1.61
	A3	19.85	4.88	1.64
	A4	0.00	0.00	0.00
	A5	0.00	0.00	0.00
	A6	0.00	0.00	0.00
	A7	0.00	0.00	0.00
	A8	0.00	0.00	0.00
B	B9	8.82	2.84	1.64
	B10	17.30	4.55	1.82
	B11	25.84	6.16	1.91
	B12	0.00	0.00	0.00
	B13	0.00	0.00	0.00
	B14	0.00	0.00	0.00
	B15	0.00	0.00	0.00
C	C16	13.44	2.91	2.17
	C17	23.20	5.09	2.15
	C18	32.05	6.81	2.14
	C19	36.26	7.41	2.12
	C20	40.72	9.48	2.06
	C21	47.38	10.75	2.00
D	D22	16.17	3.78	2.24
	D23	24.43	7.52	2.23
	D24	31.05	7.57	2.15
	D25	34.04	8.09	2.13
	D26	41.11	9.63	2.07
E	E27	15.55	3.57	2.31
	E28	22.53	6.48	2.27
	E29	28.25	8.22	2.21
	E30	35.17	9.79	2.14
F	F31	15.51	5.18	2.35
	F32	21.85	6.99	2.30
	F33	27.79	7.92	2.23
G	G34	15.13	5.73	2.37
	G35	21.52	7.06	2.31
H	H36	16.52	6.22	2.42

This can summarize that effect of fluxing materials promote the low sintering temperature of ceramic bodies [7-11]. They further make ceramic tiles obtain higher bending strength. Porosity of tiles are decreased and lead to increase shrinkage of specimens.

Note that, although formula No. A1 and B9 have the highest fluxing material, but they have lower bending strength and % shrinkage than the other formulas of their group. This phenomenon can analyse that these two formulas have the sintering temperature less than 950° C. Therefore, instead of higher banding strength and shrinkage, it makes them tend to expand their body and having more porosity.

3.2 Water absorption, % weight loss, and bulk density of tested samples

As shown in table 4, % water absorption, % wt. loss, and density have a similar trend. Water absorption is increased when bending strength is low. However, there is exception in formula No. A1 and B9. Although, they obtain the lower bending strength but they still have low water absorption. This phenomenon is the effect of close pore in their bodies which make them having lower absorption. When analysing % wt. loss of formulas in each group, it shows that higher RHA content in the formula makes the samples have high weight loss. Considering bulk density, the formulas having high content of Ang. Clay provide the higher density. As such, formulas in group A, B have lower density than the other group that having higher Ang. Clay content.

3.3 Scanning Electron Microscopy (SEM) analyzing

In order to analyze microstructure of fired samples, the formulas having highest and lowest bending strength within a group had been selected to be characterized. Formula No. A1, A2, C16, C21 were selected to be investigated.

Microstructure of No. A1 has various pores (see figure1). Comparing between A1 and A2 (see figure 2), it found that A2 has the sinter body of glassy phase and has less pores than that of A1. This phenomenon has consistent with the value of bending strength which A1 has lower strength than that of A2.

Figure 1. Scanning Electron Microscope at x3000 of No. A1

Figure 2. Scanning Electron Microscope at x3000 of No. A2

Microstructure of No.C16 and C21 are shown in figure 3 and 4. These figures obviously represent that formula No, C16 has a sintered body filling with glassy phase. This leads to the highest strength of samples. On contrarily, formula No.C21 has zero bending strength (see figure.4), it shows the porous structure of sample body.

Figure 3. Scanning Electron Microscope at x3000 of No.C16

Figure 4. Scanning Electron Microscope at x3000 of No.C21

3.4 X-ray Diffractometer (XRD)

Crystal structure of fired samples were investigated by XRD. Likewise, SEM analysis, some formulas had been selected to be analysed. XRD analysis of formula No. A1, A2, C16, and C21 are expressed in table 5.

Table 5. Crystalline phase of selected formulas comparing with bending strength

Formula No.	% XRD crystalline phase				Bending St. (Mpa)
	Quartz	Cristobalite low	Nepheline	Wollastonite 2M	
A1	5.93	42.53	6.71	44.83	8.91
A2	3.68	64.11	4.74	27.46	18.18
C16	25.16	23.08	27.78	23.99	21.99
C21	43.3	42.1	6.17	8.42	0

Bending strength of these 4 formulas are also indicated in table 5. It can be notice that higher content of nepheline and wollastonite phase can promote higher bending strength i.e. formula No. A2 and C16. Exception for formula No. A1, although having high content of nepheline and wollastonite phase, it obtains low bending strength. This can be summarized that formula No. A1 has sintering point less than 950o C. To be fired at this temperature, this make it expand the shape and lead to have high porosity.

3.5 Comparison physical properties of optimal formulas with Thai Industrial Standard (TIS 2508-2555)

Table 6. Comparison of bending strength, water absorption with TIS 2508-2555, and composition of optimal formulas

		Bending St (MPa)	% Water absorption >10%	% Wt Composition		
TIS 2508-2555 (BIII)		15	Each sample must not exceed 20%	RHA	BGC	Ang Clay
Formula No.	A2	18.18	17.74	20	70	10
	A3	16.32	19.85	30	60	10
	B9	17.81	8.82	10	70	20
	B10	18.07	17.3	20	60	20
	C16	21.99	13.14	10	60	30
	D22	15.18	16.17	10	50	40

The optimal formulas i.e. No. A2, A3, B9, B10, and C16 were selected because they have high bending strength. They are compared with standard of TIS as shown in table 6. They all can achieve the bending strength and water absorption comparing with TIS 2508-2555. In addition, the composition of these formulas are expressed again in table 6.

4. Conclusion

The objective of this study was to valorise waste materials which were RHA and BGC. They all the wastes from industrial process. In addition, local clay in Angthong province was selected to be mixed with these waste materials for manufacture fired clay wall tiles. The results as described in the previous section can be concluded as follow.

- Effect of BGC can reduce sintering temperature and provide the high bending strength of ceramic bodies. Moreover, water absorption of samples were also decreased. It illustrated in SEM analysis which make samples contain high glassy phase (see figure 3).
- Utilizing of RHA provides black grey colour for wall tile ceramics. However, composition of RHA in samples should not higher than 30% (see table 6). If utilizing higher than 30%, it will lead the lower bending strength and higher water absorption (see table 3, 4). Figure 5 shows the fired ceramic specimens of formulas that achieve TIS 2508-2555. Obviously, formulas No. A2, A3, B9, and B10 express the black grey effect which attract the community company for commercial production.

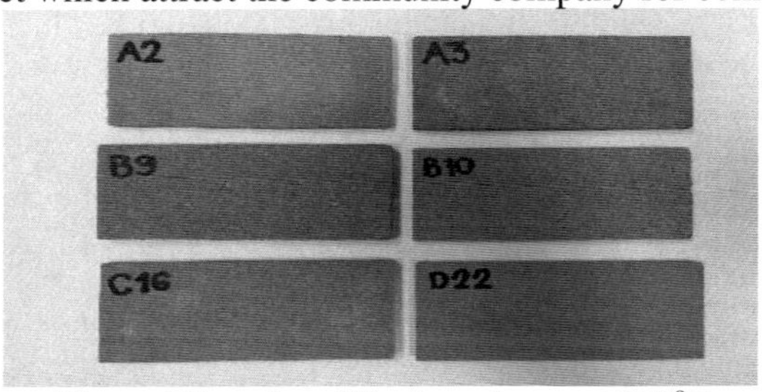

Figure 5 Samples of fired specimens at 950°C

- Combination of Ang Clay in formulation, it plays a role to facilitate the plasticity of samples that make them easily moulded. In this study, it can be utilized by composition up to 40%.
- Benefits of this study are enhancing the value of waste materials for developing ceramic tiles fired at low temperature. This action is one of efforts that can save social environment.
- The further study should to investigate the lower sintering temperature of formula No. A1. Because it showed the phenomenon of expanding shape that can be modified by reducing firing temperature.

Acknowledgements

This research was funded by the King Mongkut's University of Technology North Bangkok. Contract no. KMUTNB-61-DRIVE-027.

References

[1] Asia: Thailand to increase SMEs by over 3 million in 2018. Retrieved from https://www.rfigroup.com/rfi-group/news/asia-thailand-increase-smes-over-3-million-2018
[2] Sutas J, Mana A and Pitak L 2012 *Procedia. Eng.* **32** 1061-1067
[3] Haslinawati M M, Matori K A, Wahab Z A, Sidek H A A and Zainal A T 2009 *Int J Basic Appl Sci.* **9** 111–117
[4] Sultana M S, Hossain M I, Rahman M A and Khan M 2014 *J. Sci. Res.* **6(3)**
[5] Eliche-Quesada D, Felipe-Sesé M, López-Pérez J and Infantes-Molina A 2017 *Ceram. Int.* **4**

IOP Conf. Series: Materials Science and Engineering **652** (2019) 012051 doi:10.1088/1757-899X/652/1/012051

[6] Pollution Control Department, Ministry of Natural Resources and Environment, Thailand State of Pollution Report 2016. http://www.pcd.go.th. (accessed 3 November 2017).
[7] Braganca S R and Bergmann C P 2005 *Mat Res*, **8**
[8] Youssef N F, Abadir M F and Shater M A O 1998 *J Eur Ceram Soc*. **18**(12)
[9] Chidiac S E and Federico L M 2007 *CAN J CIVIL ENG*. **34**(11)
[10] Loryuenyong V, Panyachai T, Kaewsimork K and Siritai C 2009 *Waste Manage*, **29**
[11] Kazmi S M, Abbas S, Nehdi M L, Saleem M A, Munir M J and Mater J 2017 *Civ. Eng*, **29**(8)

IOP Conf. Series: Materials Science and Engineering **652** (2019) 012052 doi:10.1088/1757-899X/652/1/012052

Flow-Induced Vibration Response of the Chenderoh Dam Bottom Outlet Section Due to the Effects of Water Spilling

Mohamad Hazwan Mohd Ghazali[1], Ahmad Zhafran Ahmad Mazlan[1*], Muhammad Aqil Azman[1], Mohd Hafiz Zawawi[2] and Mohd Rashid Mohd Radzi[2]

[1] School of Mechanical Engineering, Universiti Sains Malaysia, Engineering Campus, Nibong Tebal, Penang, Malaysia
[2] Department of Civil Engineering, College of Engineering, Universiti Tenaga Nasional, Selangor
Corresponding author, Email: zhafran@usm.my

Abstract. Hydropower-type dam can generate an electricity to the surrounding areas during the release of water from the upstream to the downstream. However, in some cases, the flow-induced vibration during the water spilling might induced significant vibration effects to the dam structure. This is a common phenomenon that happened in any of dam structures during the operating condition. In this study, the flow-induced vibration response at the bottom outlet section of the real scale Malaysian Chenderoh Dam model is investigated using ANSYS software. The results of frequency domain response and operational deflection shapes (ODS) from the flow-induced vibration are compared with the natural frequencies and mode shapes of the bottom outlet section. From the study, the transient vibration responses induced from the flow of water occurred at the operating frequency of 23.14 Hz while the natural frequency of the bottom outlet section is located at 9 Hz of frequency. This indicates that for the normal case of water spilling at the bottom outlet section, there is no resonance phenomenon happened. However, this result is useful for the dam operation section in order to avoid any disaster of the dam structure in the future.

1. Introduction

The vibration induced from the flow of water to the dam structure can lead to catastrophic failure if its natural frequencies coincide with the shedding frequencies of the flow due to the resonance phenomenon [1, 2]. The force values induced by the water depends on the pressure and velocity of the flow as well as the material properties of the actual structure. Studies regarding the flow-induced vibration on various samples have been conducted over the years but only few involved the dam structures.

The studies on the radial gate that based on its opening were conducted by Salazar [3] and Lee [4]. Simulations for both studies were performed using ANSYS software. Based on the simulation results from Salazar, the flow does not directly induced vibration on the gate but rather on the spillway structure. Lee also found that maximum acceleration occurred at gate opening height of 0.08 m. Another study was conducted by Ilgar *et al.* [5] to investigate the dynamic behaviour of the gate for different shape of valves.

Fluid-structure interaction (FSI) analysis can be applied using both monolithic or partitioned methods. Partitoned method can be further broken down into one-way and two-way FSI [6]. Comparison study between monolithic and partitioned approaches have been conducted over the years and information regarding both methods can be found in literature [7]. For one-way and two-way FSI, Ezkurra *et al.* [8] and Benra *et al.* [9] have studied both methods and they found that two-way FSI gives more reliable results.

Generally, there are two conditions of water flow phenomenon in Chenderoh dam; water spilling and surging. Water spilling occurs when all the gates are fully opened, and maximum amount of water is released from upstream to downstream area. For the water surging, all the gates are suddenly closed during emergency. In this paper, the water spilling case is studied where the bottom outlet section is modelled exactly as the original drawing of the Chenderoh dam. The FEM model used the properties of the real concrete materials and two-way FSI simulation study is performed using ANSYS transient structural and fluent workbenches and the results are compared with the modal and harmonic analyses.

2. Methodology

2.1. Modelling of the bottom outlet section of the dam structure

The three-dimensional (3D) model of the bottom outlet section is constructed using SolidWorks software, as shown in Figure 1. Bottom outlet section is the second source of power generation for the Chenderoh dam and it has the underground penstock for the release of water from upstream to downstream area. The release of water is controlled by two butterfly gates, as shown in Figure 2.

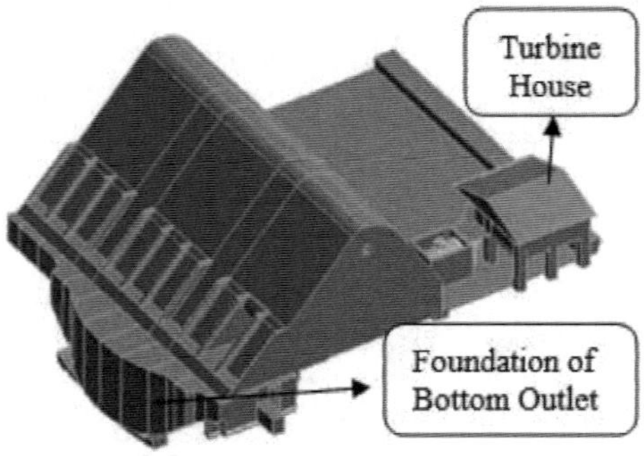

Figure 1. 3D model of the bottom outlet section of Chenderoh dam.

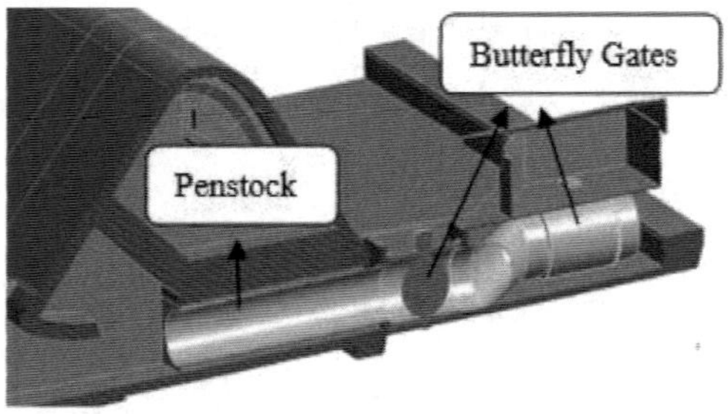

Figure 2. Locations of the butterfly gates of the bottom outlet section.

2.2. Boundary conditions of the structure and fluid of the dam

For the bottom outlet structure modelling in the ANSYS Transient, the boundary condition that been set for all the bottom faces are fixed support, as shown in Figure 3.

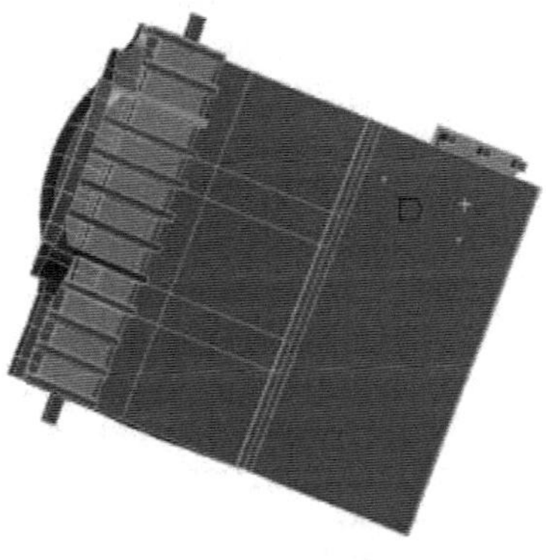

Figure 3. Boundary condition of the bottom outlet section.

For fluid section in ANSYS Fluent, the type of boundary conditions for inlet and outlet sections are velocity inlet and pressure outlet, respectively. The inlet velocity magnitude for water is set as 1 m/s. Figure 4 and 5 show the inlet and outlet areas of the bottom outlet section respectively.

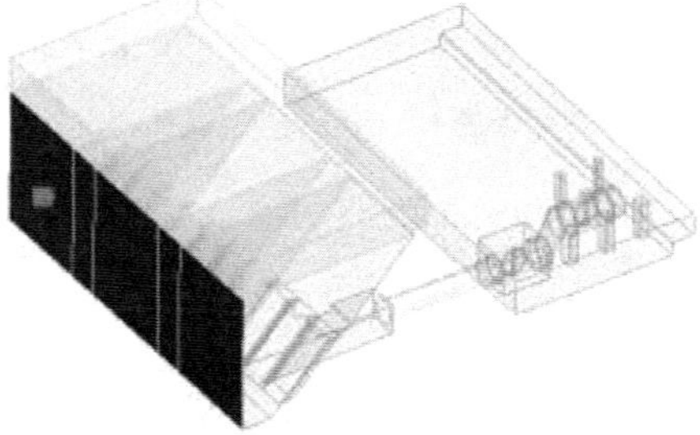

Figure 4. Inlet area for the fluid domain of bottom outlet section.

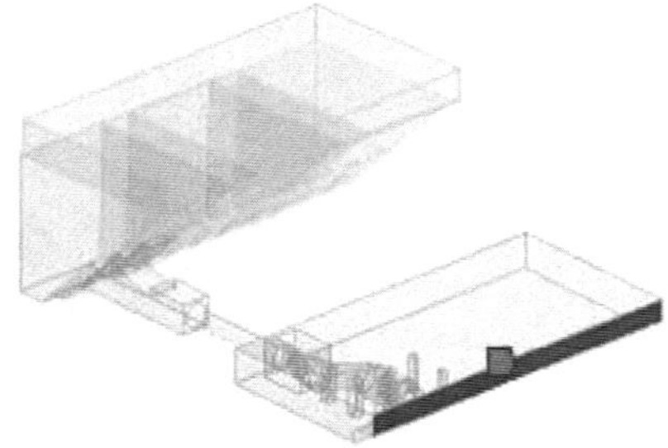

Figure 5. Outlet area for the fluid domain of bottom outlet section.

2.3. Two-way fluid-structure interaction (FSI) of the bottom outlet section

Two-way FSI of the bottom outlet section is made up of two systems analyses (Transient Structural and Fluent) and one integrated system (System Coupling) located in the ANSYS software. These systems are linked together as shown in Figure 6. Compared to one-way method, the displacement of the structure is also transferred to the fluid solver in two-way FSI [10]. Although it has higher computational cost than one-way FSI, it can produce more accurate and stable results [11].

For the Transient Structural system, the type of mesh used is tetrahedral mesh. In the analysis of bottom outlet section, the step end time is set to 5 seconds and the step control is defined by substeps. After setting-up the boundary condition as shown in Figure 3, the standard earth gravity acceleration of 9.8 m/s^2 is applied at $-y$ direction. In the Fluent system, an automatic mesh is applied in order to reduce the simulation costs in terms of time and number of elements.

A 2nd order upwind is selected for the solution method of turbulent kinetic energy and dissipation rate. 0.01 second is set as a time step size with maximum iteration of 20 times. In the System Coupling (integrated system), the end time is set similar to Transient Structural system which is 5 seconds and the time step size used is 0.01 second, similar to Fluent system.

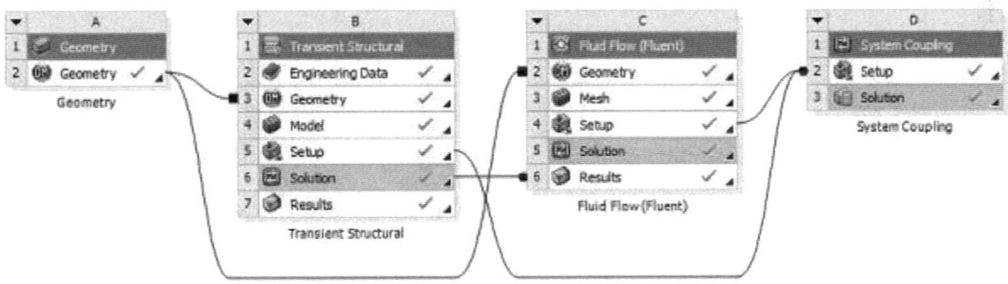

Figure 6. Linked setup between systems in two-way FSI.

3. Results and discussions

3.1. ODS results of the bottom outlet section

Figure 7 shows the ODS result of the Chenderoh dam bottom outlet section. From the figure, the deflection is mainly occurred at the upper region of the structure and there is no significant effect at the penstock and turbine house locations caused by the water flown.

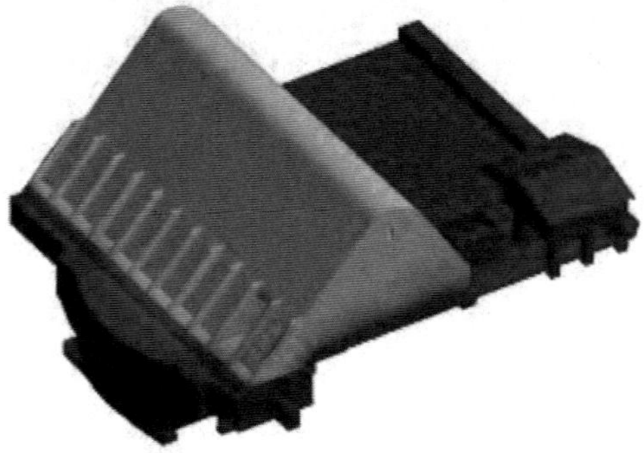

Figure 7. ODS of the bottom outlet section.

It can also be observed in Figure 8 and 9, the ODS results of the bottom outlet section in terms of time and frequency domains. From the frequency domain graph, the highest overall deflection occurs at the operating frequency of 23.14 Hz with maximum displacement value of 5.36×10^{-1} m.

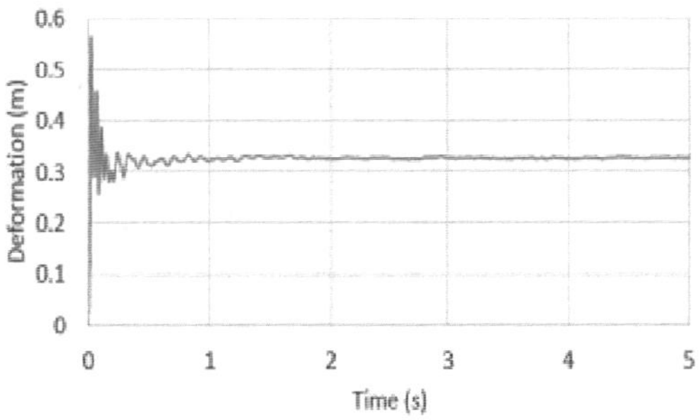

Figure 8. Time-domain graph of the deformation of the bottom outlet section.

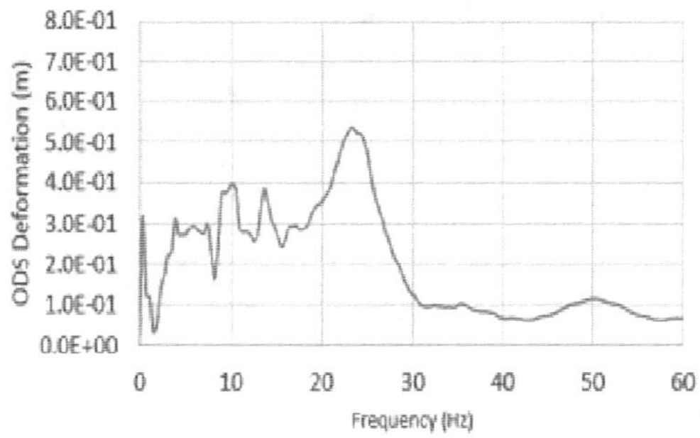

Figure 9. Frequency-domain graph of the deformation of the bottom outlet section.

3.2. Comparison between ODS and modal analysis (MA) of the bottom outlet section

Table 1 shows the ODS result from the Transient Analysis (FSI) and mode shape result from the Modal Analysis (MA) respectively. It can be observed that the ODS and the mode shape exhibit almost similar result where the deformation occurred at the upper region of the bottom outlet structure.

Table 1. Comparison between ODS and mode shape for bottom outlet section.

ODS	Mode Shape
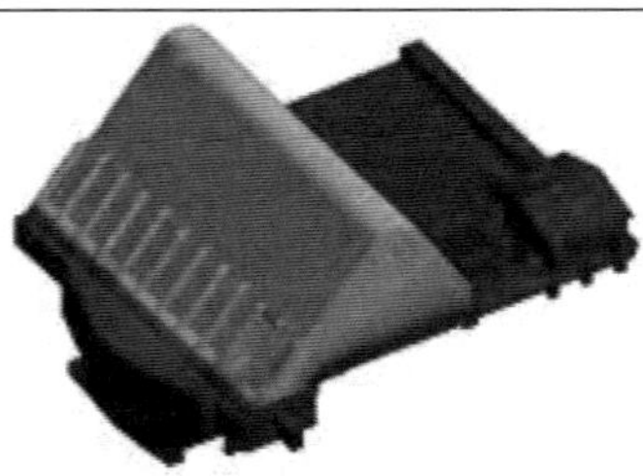	

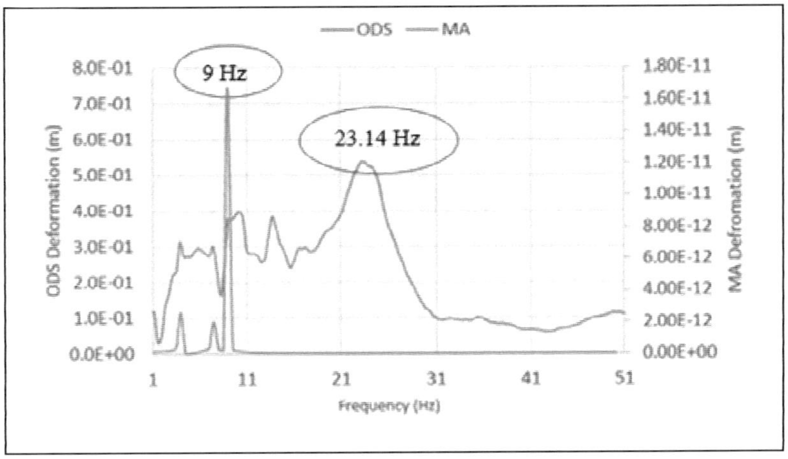

Figure 10. Relationship between ODS and MA deformations at the bottom outlet section in frequency domain.

Figure 10 shows the frequency response from the ODS and MA results. From the figure, the ODS deformation occurred at the operating frequency of 23.14 Hz and MA deformation occurs at natural frequency of 9 Hz. This means that both are in the different frequency peaks. Based on the frequency response function (FRF) of the harmonic response analysis, the highest deflection occurs in z-direction with amplitude value of 1.68×10^{-11} m. Due to the fact that these two frequencies do not coincided, resonance phenomenon did not happen. Thus, there is no significant failure can be expected at the bottom outlet section of Chenderoh dam for the normal water spilling condition.

4. Conclusion

In this study, the effect of flow-induced vibration on the Chenderoh dam bottom outlet section was successfully analysed by two-way FSI and MA using ANSYS. Based on the ODS result, the highest overall deflection occurs at the operating frequency of 23.14 Hz whereas MA deformation occurs at 9 Hz. Since both deformations occur at different frequencies, it can be concluded that there is no critical failure to the bottom outlet section because of the absence of resonance phenomenon. Similar analysis on the sections that involve the turbines or gates are still a challenge due to the complex dynamic mesh. For the future research, the effect of tunnel surging (gates are fully closed) on the bottom outlet section can be analysed.

Acknowledgments

The authors would like to acknowledge Universiti Sains Malaysia and Uniten R&D Sdn. Bhd. for providing the facilities and financial assistance under accounts (304/PMEKANIK/60313052, 203/PMEKANIK/6071370 and U-841).

References

[1] Soljic I, Dejanovic I and Matijasevic L 2009 *Chem. Biochem. Eng. Q.* **10** 287-94
[2] Castro A and Botero F 2017 *Ingenieria y universidad* **21** 155-76
[3] Sanchez D and Salazar J E 2010 *Proc. ASME 10th Biennial Conf. Eng. Systems Design Analysis (Istanbul)* 1-5
[4] Lee O S, Seong H and Kang J W 2018 *Eng. App. Comp. Fluid Mech.* **12** 567-83
[5] Javanshir I and Javanshir N 2015 *J. Vibroengineering* **17** 478-86
[6] Ahamed M F, Atique S, Munshi M A and Koiranen T 2017 *Am. J. Eng. Res.* **6** 86-9
[7] Michler C, Hulshoff S J, Van Brummelen E H and De Borst R 2004 *Comput. Fluid* **33** 839-48
[8] Benra F K, Dohmen H J, Pei J, Schuster S and Wan B 2011 *J. Appl. Math.* B **2011** 16

[9] Ezkurra M, Esnaola J S, Martinez-Agirre M, Etxeberria U, Lertxundi U, Colomo L, Begiristain M and Zurutuza I 2018 *Int. J. Mech. Mater. Eng* **12** 4
[10] Chen Y B, Wang Z K and Tsai G C 2015 *Int. J. Eng. Tech. Innov.* **5** 33-44
[11] Vaassen J M, De Vincenzo P, Hirsch C and Benoit L 2011 7th *Eur. Symp. Aerothermodynamics (Brugge)* **692** 128

A case study on the plan for soil reinforcement by TCS in south korea

Seokhyun Seo, Wondong Kang, Jungsang Lee, Euiyoup Jung

325 Hwangsaeul Ro, Bundang Gu, Seongnam Si, Gyeonggi Do, Denver Korea E&C, South Korea

Email: terzaghi7@naver.com; action1288@naver.com; ljungs5815@naver.com; jenpcom@naver.com

Abstract. Compaction grouting was developed in the 1950s as a remedial measure for the correction of building settlement, and used almost exclusively for that purpose for many years. Over the past 25 years, however, compaction grouting technology has evolved to treat a wide range of subsurface conditions for new and remedial construction. Compaction grouting is used foundation rehabilitation, ground improvement, railroad subgrade stabilization, seismic/liquefaction mitigation, sinkhole remediation, tunneling stabilization, void filling. It also has merit that makes possible to construct structures having narrow working spaces. But the effects of ground improvement depending on the type of soil must be studied in order to adopt in various soil(granular soil and cohesive soil). In this paper, field and laboratory tests were performed to understand effects of ground improvement for soft ground and seismic mitigation. To know the characteristics of bearing capacity in adjacent ground, plate loading test for behind the improving ground. Strength characteristics of TCS were indicated by uniaxial compression test. Ground improvement effect by TCS together strength increment effect by TCS itself was proved that TCS can be used for soil improvement method using bearing capacity increment.

1. Introduction

The TCS(Total Compaction Grouting System) method is a low-mobility mortar with 2 inch(50 mm) or less slump value, whose grouting material is composed of fine soil to secure plasticity and coarse soil to increase internal friction force, whose basic material is soil cement. The grouting material is an improvement method to compactify the ground by reducing pores between soil particles through consolidation of soil via applying a radial pressure to the ground in a solid type, not permeating into the pores in the surrounding ground([1]). This is a completely different non-discharge replacement grouting method from the discharge replacement in the venation concretion of cement, the permeation concretion of liquid and the high-pressure injection.

This exerts an excellent effect for securing a structural stability by reinforcing the soft ground and the sand, stone and gravel stratum. Therefore, this is widely used for the reinforcement of land/marine soft ground, the seismic and liquefaction measure of sand ground, the reinforcement of lighters wharf and quay wall structure, the seismic reinforcement of shore, the side-movement blocking of piers and bank bodies, the reinforcement and restoration of differential settlement, the reinforcement of pores, and the refilling of sand, stones and gravels([2-4]).

The TCS method can freely form an improvement range and concrete shape with almost no damage to the surrounding areas at a low noise and low vibration. In addition, the equipment is miniature to

enable construction at a narrow space(town, residential area, apartment, etc.) to have a high applicability and to be eco-friendly without slime.
This paper aims to take a look at the ground improvement characteristics appearing via an application of the TCS method, and to grasp the settlement suppression effect and bearing capacity improvement according to the reinforcement method during ground improvement, the countermeasure method for the section with liquefaction and the ground reinforcement effect through a quality verification after reinforcement through construction cases

2. Mechanism

The TCS method is a method to increase ground density by compressing and reinforcing the surrounding ground via forming a cylindrical homogeneous concrete through compaction of low-mobility mortar grouting materials in the ground. To emphasize this low mobility of grouting materials, the Michael Byle argued that it is necessary to change "Compaction grouting" to "Low Mobility Grouting" in 1997([5]). The TCS methods are divided into a mortar compaction grouting method and a mortar non-compaction grouting method. The mortar compaction grouting method is a method having both concepts of soft ground plus difference reinforcement and seismic reinforcement through the grouting to improve soft ground and the grouting to fill the pores with sand and stones at a grouting rate of 0.06m^3/min or less([6]). The mortar non-compaction grouting method performs grouting such as filling of large cavities and filling of large differences at a grouting rate of 0.06 m^3/min or more regardless of dissipation of pore pressure.
The mortar compaction grouting method is classified into constant-pressure grouting and constant-rate grouting are shown in Fig.1. The constant-pressure grouting keeps the bearing capacity of the ground constant to form a cylindrical non-reinforced concrete column in the ground, increase the density of the surrounding ground according to the compressed reinforcement of the soil around it to four directions, and form a complex ground, securing a stability against bearing capacity and settlement. The constant-rate grouting forms a constant-root pile to get an effect of consolidating the surrounding soft ground by establishing a pile while forming an improvement via compression of soft ground through discharging a proper amount of grouting material in the ground considering the soil proper of the site and an effect of improving the ground by making the consolidated surrounding soft ground strong

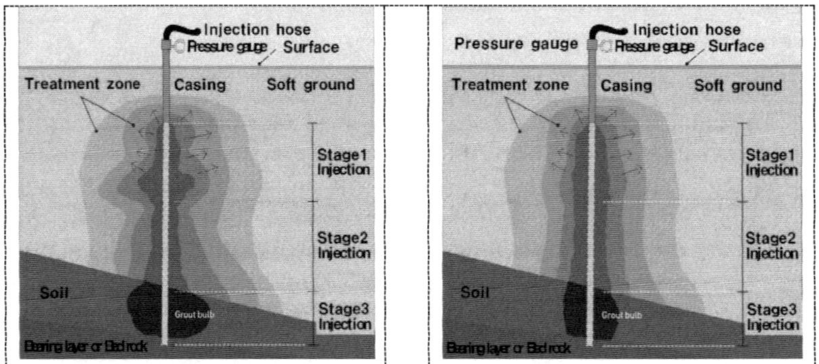

(a) (b)

Figure 1. Injection method of TCS (a)constant-pressure grouting (b)constant-rate grouting

3. Construction cases of the TCS method

3.1. Seismic ground reinforcement

The seismic performance reinforcement methods can be largely divided into the method of increasing stiffness of the ground by executing grouting, the method of integrating the structure and the foundation ground by constructing a micro-pile, the method of reducing the back soil pressure by installing a new wall structure and the method of reducing the back soil pressure by replacing the back of quay walls using a light material.

In application of methods, we selected and applied the TCS method that can minimize interference in operation of the wharf, is excellent in securing stability of structures, has abundant cases of similar construction, and is excellent in environmental efficiency and economic efficiency, as the optimal reinforcement method.

Because eccentric bearing capacity was insufficient as a result of seismic analysis on the foundation sand and stones in the quay wall, compaction grouting was applied to the sand and stone stratum and it was reinforced with 2 rows of Ø1,800 mm, center to center 1.6 m pier-type wall to secure stability of bearing capacity are shown in Fig.2.

After completion of construction, confirmation boring was performed to check the diffusion range of the improved object and conduct a compressive strength test are shown in Fig.3. The confirmation boring positions were 1 bore at the central part and 2 pores at the surrounding part, and the survey result showed a core recovery rate of 70.8 % and a compressive strength of 11.44 MPa, which met the design criterion of 4.8 MPa are shown in Fig4.

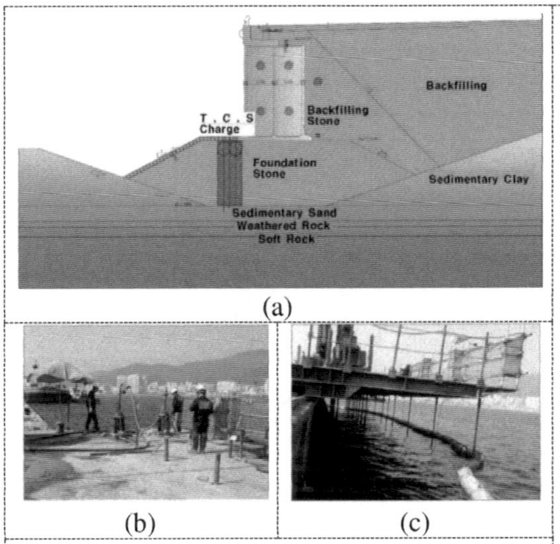

Figure 2. Cross-sectional diagram and site situation (a)reinforced cross-sectioni diagram (b)grouting work (c)installation of contamination-preventivenets

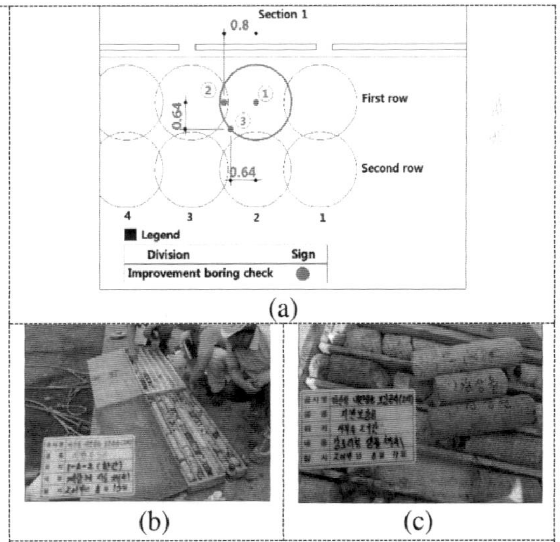

Figure 3. Construction position diagram and sampling (a)construction position diagram (b)improved object core confirmation (c)sampling for uniaxial compression test

Quality inspection report

Sample Name	T.C.S(Core)	Issue number	Test No. 2898
Collection place	Masan Harbor	Date received	08/11/2014
Name of the project	Masan Harbor seismic reinforcement work		
ordering organization	Masan Regional Maritime harbor Authority		
Constructor	Young Jin Construction		
Client	Denver Korea E&C		

- result -

No	Kind	Days	Sample Size(mm)			OutCome (MPa)	Test Methods
			Diameter	×	Height		
1	Compressive strength	28	50.8	×	101	9.17	KS F 2328
			50.4	×	101	10.48	
			50.3	×	101	14.66	
			Average			11.44	

Figure 4. Result of uniaxial compression test

To confirm the result of seismic reinforcement, boreholes were drilled and gamma-gamma(density verification) was conducted before and after the execution of TCS methods at the ○○1 port and the ○○2 port. Count rates were measured and density values were calculated by depth, which were compared and analyzed. The analysis result showed that density increased at the section to which the TCS method was applied. The results of density measurement by depth are shown in Fig. 5.

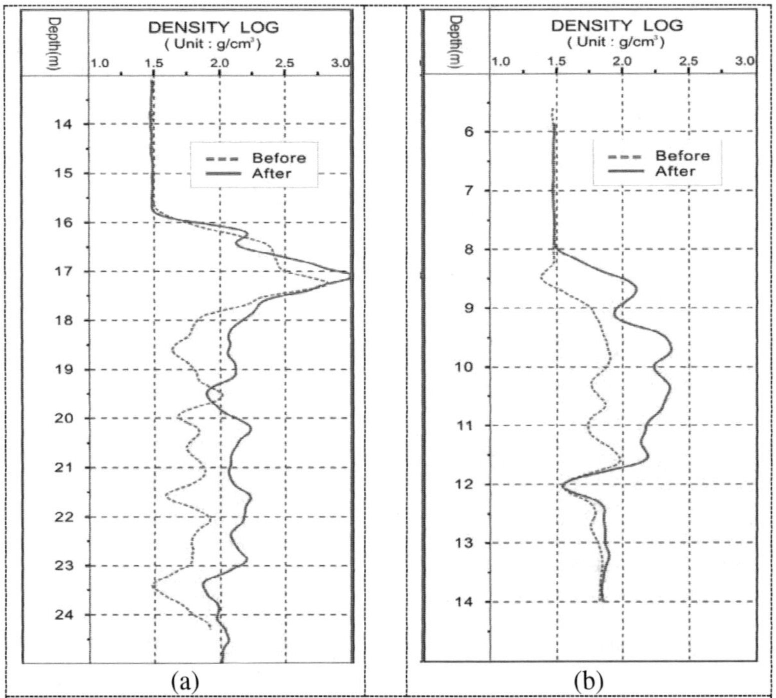

Figure 5. Change of densities by depth (a) ○○1port (b) ○○2port

For the ○○1 port, the densities of the buried layer into which grouting was conducted were 1.771 before grouting and 2.094 after grouting, an about 18.2 % increase. For the ○○2 port, the densities of the buried layer were 1.827 to 2.171, an about 18.8 % increase. This is because the grouting materials filled the vacant spaces and crack gaps between the sands and stones of the buried layer to increase the unit weight.

3.2. Soft ground reinforcement case

The shore ground of the dredging soil reclamation site at the Tando port is a very soft marine clay stratum with a N value of 2 and a depth of 16.5 m, which is very deep to need an application of suitable ground improvement method for safe construction of shore. To protect the soft ground and prevent sinking of shore, the TCS method was applied to the cross-section of the breakwater to secure stability.

For this site, both land work and marine work were conducted simultaneously, and boring equipment was installed on the work stand for smooth marine work to conduct grouting work on a barge on which a barge plant was installed.

A total of 4,675 boreholes were drilled in center to center 2.1 × 2.1 m up to a depth of N>20 from the soft ground reinforcement position, and the replacement rate of TCS(Ø1,000 mm) is about 17.8 % of the object ground volume are shown in Fig.6. When grouting materials were mixed during construction at this site, specimens were always made for tests by us and national authorized institutions for quality control, and the result of strength tests met the design strength enough.

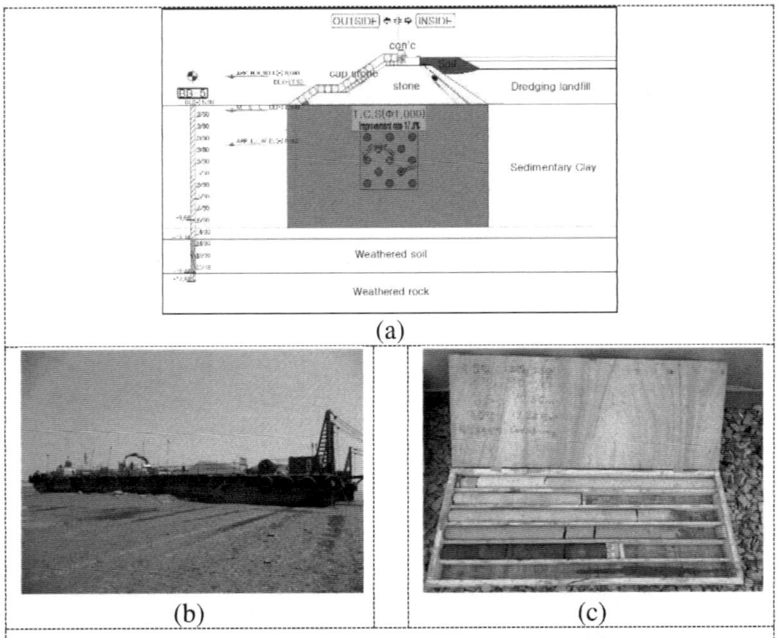

Figure 6. Cross-sectional diagram and site situation (a)reinforced cross-sectional diagram (b)grouting work (c)confirmation boring

3.3. Structure reinforcement and restoration

For the MY-TV parking tower at Deungchon-dong, differential settlement happened due to digging work during construction of a neighboring building. Before reinforcement, the structure of the parking tower was inclined 13 cm, and its depth of 3 m was composed of mud to cause settlement continuously. The construction site of the neighboring building was balanced in a LW method, so its ground was improved to some degree.

The ground was composed of buried layers, accumulated layers, residual soil layers and weathered rock layers from the top, and the buried layer was composed of fine soil or sand containing gravels and silts, and the ground was very loose. To restore the inclined building, the A row was grouted up to 2~2.5 m depth in a top-down type and the A, B, C, D rows were grouted up to 6.0~8.0 m depth in a bottom-up type are shown in Fig.7.

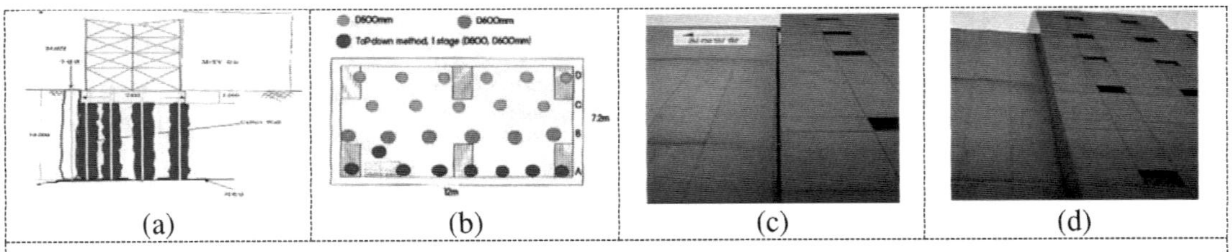

Figure 7. Reinforced floor plan and building situation (a)reinforced cross-section diagram (b)reinforced plane (c)before reinforcement (d)after reinforcement

For the settlement having happened at a house in Gangwon-do, reinforcement work was conducted using the TCS method to suppress the additional settlement expected in the future, restore the vulnerable part to a possible range, and solve stability in beauty and uncomfort in use.

For this house, maximum 30.0 cm settlement happened in the diagonal direction on the road side due to heavy rains because of storms. By applying both the top-down method and the bottom-up method of this method, the building was successfully restored to recover its original structural functions are shown in Fig.8.

Table 1. Change of heights before and after construction.

Measured point	Before construction(m)	After construction(m)	Increase/ decrease(cm)
No.1	+11.995	+12.024	▲2.9
No.2	+11.930	+11.989	▲5.9
No.3	+11.870	+11.965	▲9.5
No.4	+11.999	+12.062	▲6.3
No.5	+12.140	+12.162	▲2.2
No.6	+12.204	+12.215	▲1.1
No.7	+12.259	+12.258	▲0.1
No.8	+12.137	+12.150	▲1.3

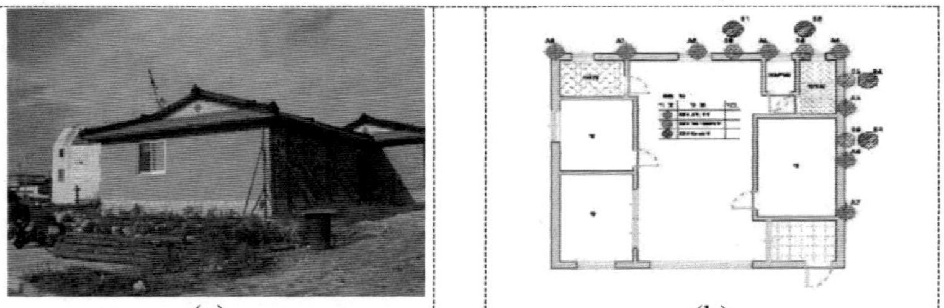

(a) (b)

Figure 8. Field situation and reinforced positions (a)reinforced structure view (b)reinforced positions

Because a result of safety diagnosis for the ○○ station building showed that its major members were cracked by friction forces in the piles according to the consolidated settlement of the surrounding ground, the building was reinforced using the TCS method to prevent cracks. Grouting was conducted for a total of 174 boreholes to a depth of 50 m beside the existing piles with respect to the mat of the building foundation, in a bottom-up type up to GL-2.0 m and in a top-down type up to the remaining 2 m section are shown in Fig.9.

As a result of conducting a static loading test to check the construction quality of the formed piles and obtain engineering data such as allowable bearing capacity and settlement quantity, the minimum allowable bearing capacity was 43.4 tons/ea or higher, which is higher than the design strength of 42.0 tons/ea without displacement of the existing structure are shown in Fig.10.

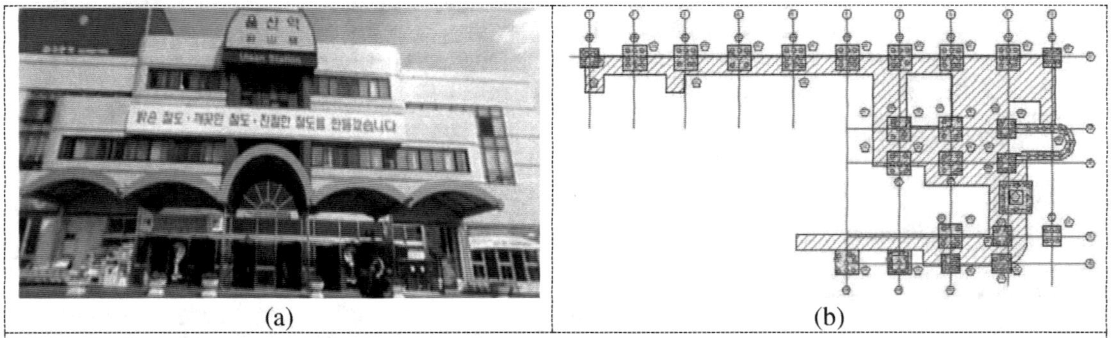

(a) (b)

Figure 9.○○Station building view and reinforced positions (a) ○○station building view (b)reinforced positions

IOP Conf. Series: Materials Science and Engineering **652** (2019) 012053 doi:10.1088/1757-899X/652/1/012053

(a) (b)

Figure 10. Loading test and core sampling

At the ○○apartment complex in Gangreung city, ground settlement happened(10-69 cm). The structure- installed area caused almost no settlement due to its pile foundation, but the road, above-ground parking lot and flower garden using a soft ground treatment method by sand trains caused consolidated settlement in general. This site obtained a composite ground improvement effect of piles plus ground reinforcement by grouting 1,988 boreholes of 800 mm diameter and 354 boreholes of 600 mm diameter to improve the soft ground, and as a result of grasping the soft ground improvement effect due to a change in physical and mechanical property, the generated settlement quantity was reduced to an allowable settlement quantity of 25 mm or less. In addition, the structure and road having lost its existing function were repaired and improved to a pleasant residential environment.

For the mechanical properties of ground improvement after construction, the result of standard penetration test showed a generally increasing(1-6) trend of N value after improvement, and the result of piezo-cone penetration test showed an increase in ground properties such as cone resistance, over-compaction ratio and undrained strength compared to before improvement are shown in Fig.11. The result of site vane test showed the shear strength by region increased by about 43~79 %. For the physical properties according to the room soil test, the natural water content ratio decreased 25%, the plasticity index decreased 7 %, the initial spacing ratio decreased 0.07, the moisture density increased 0.05 ton/m³, and the cohesion increased 0.06 ton/m2 compared to before improvement to confirm that the strength characteristics of the original were improved. The leading compaction load increased 0.24 and the compression index decreased a little.

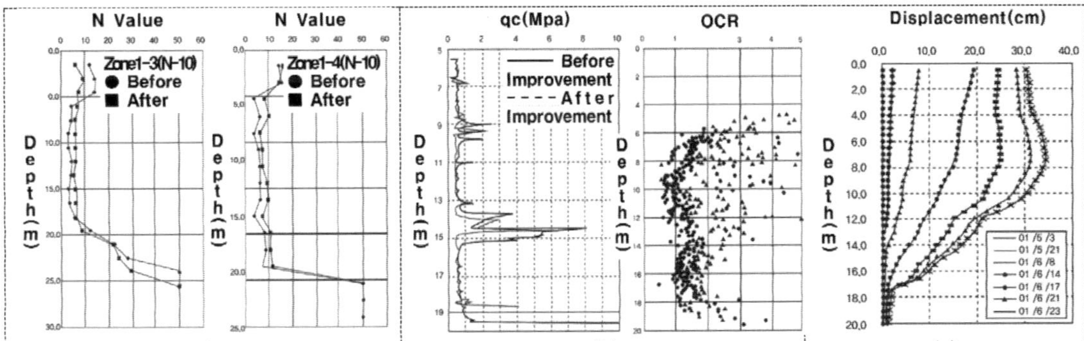

(a) (b) (c)

Figure 11. Test result (a)N value change (b)piezo-cone penetration test (c)horizontal displacement change in the ground

4. Conclusion

This study grasped the ground improvement effect via a case study for applications of the TCS method, and derived the following conclusions.

1. The grouting quality could be confirmed through a site loading test, a standard penetration test and a room uniaxial compressive strength test after construction of the TCS method as a measure of seismic reinforcement and soft ground reinforcement.
2. It could be confirmed that applying the TCS method to a soft ground showed a structure bearing capacity effect and a settlement suppression effect as well as a strength improvement effect of the ground around the improved object before and after application of the method.
3. It can be seen that this method has a good applicability to the soft ground reinforcement, liquefaction measure, structure restoration and reinforcement method through various construction cases.

Acknowledgments
This research was supported by the Korea Agency for Infrastructure Technology Advancement under the Ministry of Land, Infrastructure and Transport of the Korean government(Project Number: 19SCIP-B130947-05-000000).

References

[1] ASCE/GI, 2010. "Compaction grouting consensus guide", ASCE Standard ASCE/G-I 53-10, ASCE, Reston, VA.
[2] Berry, Richard M. and Farrar, Robert R. 1997. "Grouting Around Trenchless Repairs - The Forgotten Technology", Proceedings of the Pipeline Division Conference, ASCE, page 342.
[3] Baez, J. I. and Henry, J. F., 1983. "Reduction of liquefaction potential by compaction grouting at Pinolpolis West Dam, S. C.", ASCE, Geotechinical special publication No. 35; Proceedings Geotechnical Practice in Dam Rehabilitation, pp. 493-506.
[4] Henry, James F., 1989. "Ground modification techniques applied to sinkhole remediation", Proceedings of the Third Multidisciplinary Conference on Sinkholes, p327.
[5] Byle, M. J., 1997. "Limited mobility displacement grouting; When 'compaction grout' is not 'compaction grout'", Grouting: Compaction, Remediation, Testing,ASCE Geotechinical special publication No. 66, pp. 32-43.
[6] Bandimere, Samson W., 1997. "Compaction grouting state of the practice - 1997", Grouting: Compaction, Remediation, Testing, ASCE Geotechinical special publication No. 66, p. 18.

Process Modelling and Economic Evaluation for NanoLignin Production

Udari Prasadini Perera[1], Mei Ling Foo[1], Khang Wei Tan[2, 3] and Irene Mei Leng Chew[1, 4]

[1]School of Engineering, Monash University Malaysia, Jalan Lagoon Selatan, 47500 Bandar Sunway, Selangor, Malaysia
[2]School of Energy and Chemical Engineering, Xiamen University Malaysia, Selangor Darul Ehsan 43900, Malaysia
[3]College of Chemistry and Chemical Engineering, Xiamen University, Xiamen 361005, China

[4]E-mail: irene.chew@monash.edu

Abstract. Lignin nanoparticles (nanolignin, NL) are sustainable, green material which can be derived from industrial and agricultural waste. NL has enhanced properties compared to standard lignin due to a larger surface to volume ratio. Hence, it's suitable for a wide range of applications including drug delivery, stabilising agent, substitute for silver nanoparticles and reinforcement material. This study analyses the commercial feasibility of NL production from high pressure homogenization. The production process was simulated based on the basic laboratory results using commercial simulation software, SuperPro Designer. Manufacturing cost and minimum selling price for NL was estimated to be 5.1 USD/kg and 6.1 USD/kg for a continuous plant with 15t/day production rate. The sensitivity of three factors; Lignin concentration to the homogenizer, production rate and lignin price on process economics was analysed and results shown lignin concentration has a severe impact on manufacturing cost and selling price. The production cost reduces to 2.46 USD/kg when the lignin concentration increases to 6%.

1. Introduction

Lignin is the second most abundant polymer from biomass and the primary source for aromatic polymers. The presence of different organic functional groups such as phenolic, aliphatic hydroxyl, methoxy and terminat aldehyde make lignin a potential component to synthesize environmentally friendly polymer based materials [1]. However, lignin is still hugely underutilised, and most of the industrial lignin use as a low cost fuel for energy generation. A small fraction of Lignin has been used as carbon black substitute, rubber reinforcer, phenolic resin component and pre-curser for carbon fiber [2]. Lignin is a complex heterogeneous polymer whose structure and characteristics highly depend on the source and the pulping method. Also, poor reactivity, hydrophobicity and difficult processability (due to high branched and hydrogen bonded three-dimensional structure) cause challenges in developing lignin applications [3].

Lignin in nanoscale can reduce some of the problems related to lignin processing. Nanolignin (NL) offer different or improved properties compared to parent material [4] due to the larger surface to volume ratio. NL polymer composites have high mechanical performance [5], enhanced thermal

stability [6] and higher anti-oxidant activity [7]. Water-dispersed NL can be used to stabilise Pickering emulsions, carry silver ions in antimicrobial applications and drug delivery applications [8].
Many approaches have been reported to synthesize NL which include chemical methods (solvent shifting, pH shifting, carbon dioxide antisolvent and cross-linking/polymerisation), mechanical methods (homogenization and ultrasonication) or other methods such as ice-segregation, template base synthesis, aerosol processing and electrospinning [9]. Among the above methods, the most investigated methods are solvent and pH shifting. But both approaches require a massive amount of solvents and involve downstream processing which may not be cost effective in the industrial scale. In nanoparticle synthesis processes, many mechanical or physical methods are less energy intensive as compared to chemical pathway [10]. Hence, for this study mechanical homogenization was selected to produce NL. Currently, the NL is in lab production and a techno-economic analysis has only been done for aerosol lignin [11] and colloidal lignin manufacturing process [12].
Process simulation is an essential tool for the chemical industry to forecast the economic feasibility of new process model at the initial stage of process design. This study presents the preliminary economic analysis for NL production via mechanical homogenization. The process is simulated in SuperPro Designer v9.5 based on the laboratory data. The simulation data consists of estimated capital and operational costs and minimum selling price, which suggests the potential implementation of industrial scale NL plant.

2. Laboratory scale NL synthesis

The Alkali lignin, obtained from palm oil empty fruit bunch black liquor was dissolved in water using magnetic stirring. The solution was sent to high-pressure homogenizer to obtain NL. Later the NL solution was frozen at -40° C for 24 h and dried using a spray dryer to get solid NL particles. The particle size distribution of NL was examined by Malvern Nano-ZS Zetasizer (Malvern Instruments Ltd, UK) using dynamic light scattering (DLS) method. The Z-average was 283 nm, and polydispersity index (Pdi) was 0.390.

3. Process simulation

The production of NL in laboratory scale was conducted in batch mode. To improve the capacity and cost efficiency, the NL production was simulated in a continuous manner assuming that the plant would have operated 330 days in a year. The NL production rate was set at 625 kg/h (4950 t/y). The value was based on 10% lignin production from a general pulp mill [13]. Also, the lignin loss during the process was assumed to be negligible. Figure 1 shows the base case process flowsheet simulated in SuperPro Designer. First, lignin (625kg/h) was diluted in a blending tank with water.(1420 kg/h fresh water and recycled water 29194 kg/h). The concentration of lignin after the dilution was 2 wt. % (625kg/h lignin in 30614kg/h water). Then the dissolved lignin was pumped to high-pressure homogenizer. During the homogenization process, lignin size would be reduced to nanoscale. The NL solution was sent to spray dryer. The moisture content of the dried NL was less than 5wt. %. The solid particles in the vapour were removed by air filters. Then the vapour stream was subsequently cooled and air was emitted through a vent. The recovered water from the vapour stream (S-5) was reused for lignin dissolution.
Major costs considered for NL process were given as; Lignin 300USD/t [11], water 0.57USD/m^3 and electricity 0.07USD/kWh (based on Malaysian water and electricity tariff),chilled water 0.4 USD/t and steam 12 USD/t (based on SuperPro 2019 values).
The details of the equipment sizing, number of units and purchase cost of 2019 were obtained from SuperPro designer and summarised in table 1. Figure 2 shows the percentage of contribution of each equipment to the total equipment purchase cost. The homogenizer unit is the most expensive equipment, and for the base case it contributes to 51% of the total equipment purchasing cost. Also, 13 units of homogenizers are required to maintain the process. This cost can be reduced by varying the homogenization condition such as lignin concentration and production rate, which will be studied in detail in the subsequent sensitivity analysis.

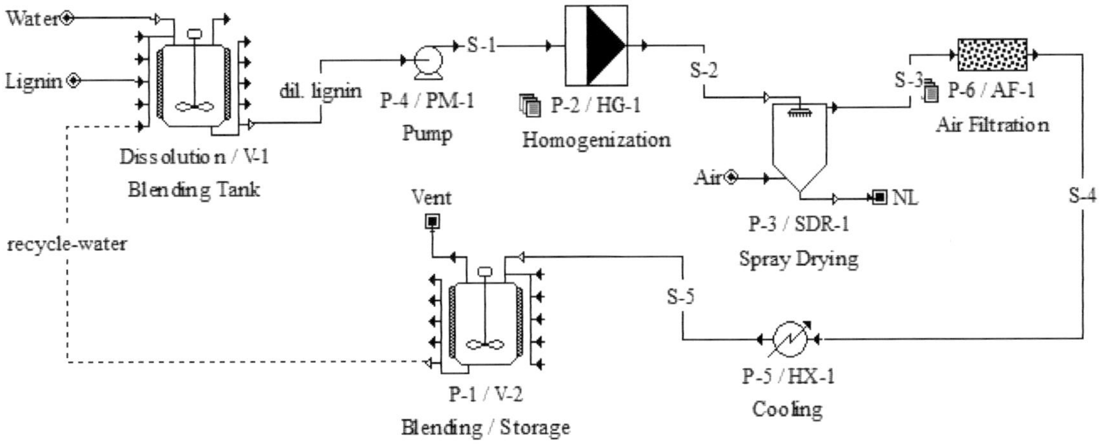

Figure 1. Simulation flowsheet for industrial scale NL production process from homogenization.

Table 1. Capacity and Purchase cost of equipment.

Equipment	Blending Tank (V-1)	Pump (PM-1)	Homogenizer (HG-1)	Spray Dryer (SDR-1)	Air Filter (AF-1)	Heat Exchanger (HX-1)	Storage Tank
Units	1	1	13	1	7	1	1
Capacity	80,000 L	1.23 W	20,000 L/h	305,751 L	14,163,872 L/h	100 m^2	80,000
Unit Price ($ x 1000)	311	22	121	378	60	104	371

4. Process economics

The economic evaluation was performed using the inbuilt function of SuperPro Designer v9.5. The minimum selling price (MSP) of NL was based on the desired payback period (5 years). Production cost for the base case was 5.1USD/kg, and MSP was 6.1 USD /kg. Internal rate of return (IRR) needed to be more than 10% after the taxation to the project to be profitable. NL production plant with 4950t/y capacity claimed a capital investment of investment $ 27.6 Million and operating cost of $ 26.5 Million. Calculated annual revenue of the project was $ 31.7 Million. The IRR and the ROI values for the project calculated as 12.7% and 19.6%, respectively. The project could be considered profitable since the IRR was more than 10%, and ROI showed considerable value.

5. Sensitivity Analysis

Lignin concentration affects the capital cost (equipment sizing) and energy requirement significantly. Unit production cost for different lignin concentrations was calculated (figure 3). When the lignin concentration was low (less than 2 wt. %), larger capacity equipment needs to produce the same amount of NL. Hence the unit production cost increased significantly due to increment in capital investment and operational cost. On the other hand, when the lignin concentration was high, the unit cost of NL decreased as the capacity of equipment needed are lowered. For 6 wt. % lignin concentration, unit production cost and MSP are 2.46 USD/kg and 2.96 USD/kg. For this scenario, the production cost decreased by 48% compared to the base case (2 wt. %). Therefore, the lignin concentration at the homogenizer seems to be a very sensitive parameter for project economics. Thus, optimization on lignin concentration should be considered to ensure the process remains cost effective.

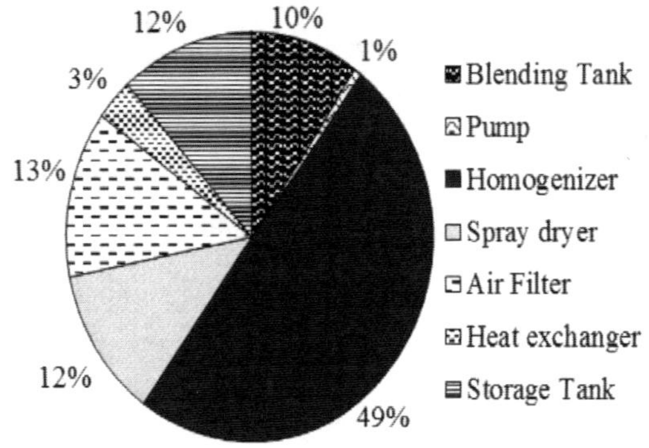

Figure 2. Percentage contribution of equipment to total equipment cost.

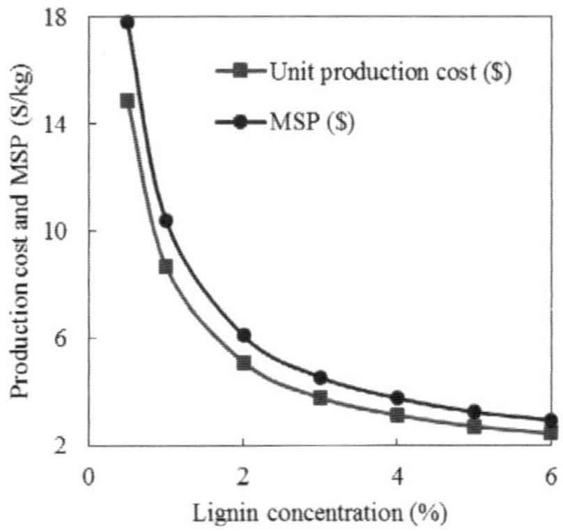

Figure 3. Influence of lignin concentration on production cost and MSP of NL.

The production rate can influence the unit production cost of the product. For the base case simulation, the NL plant is assumed to integrate with pulp mill and consume approximately 10% of generated lignin to produce NL. Since NL falls under speciality material, it can be delivered as an independent product/precursor for another product in less volume. Also, the production capacity can be higher than the base scale, depending on the market demand. Hence different production rates vs unit production cost and MSP was simulated and presented in figure 4. When the production rate was lowered to 1 t/day, unit production cost increased by 3-fold. When the production rate heightened to 30 t/day, unit production cost decreased by 8% as compared to the base scale (15 t/day).

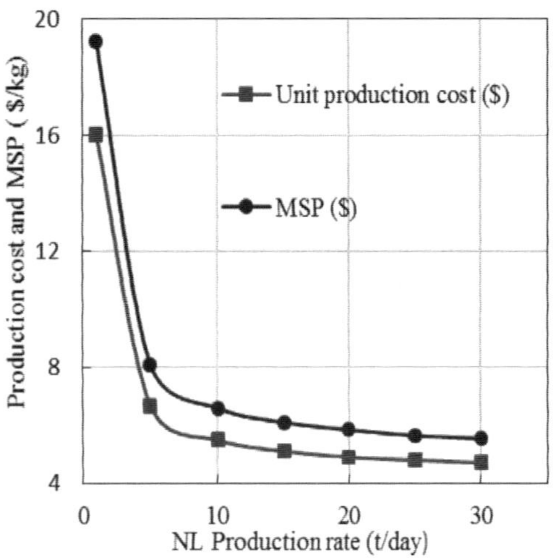

Figure 4. Influence of NL production rate on production cost and minimum selling price (MSP) of NL

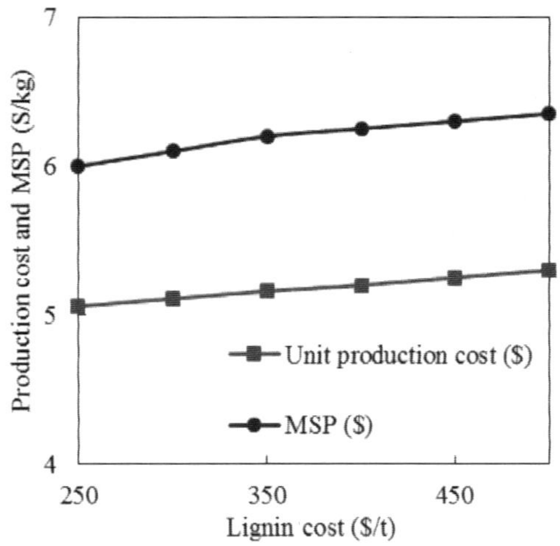

Figure 5. Impact of lignin price on production cost and minimum selling price (MSP) of NL

Lignin was the primary input material of the process. Lignin market prices can vary depending on seasonality and purity. For the base scale, kraft lignin was selected (due to the abundance in the market). In figure 5, the lignin price is varied from 250 USD/t to 500 USD/t. Unit production cost reduces 1% from the base case (300 USD/t) when the lignin price is 250 USD/t. If the lignin price rise to 500 USD/t, there would be a 4% unit production cost improvement.

6. Conclusion

A commercial feasibility study was conducted for industrial scale NL process using SuperPro Designer. For a NL plant with production capacity 15 t/day, the selling price of NL with 5 years payback period is $6.1/kg. The IRR and ROI values for the project is found to be 12.7% and 19.6% respectively, which indicate the feasibility of the process. The lignin concentration in the homogenizer has been recognised as the most sensitive parameters for project economics. The selling price is at $2.96/kg when the lignin concentration is increased to 6 wt. %. Even though the production rate and lignin market price may influence the manufacturing cost and selling price, they are not as prominent as the lignin concentration. According to the production and selling prices in this study, NL has the potential to use in high-value low volume applications such as a substitute for silver nanoparticles and stabilising agents for Pickering emulsions. However, to substitute lignin using NL for the high volume applications, i.e. Formaldehyde resins, the NL manufacturing cost requires to be lowered. To reduce the cost research on using low-grade lignin as feedstock for high volume low-value applications, optimising homogeniser conditions (pressure and number of passes) according to NL applications and investigating the effect of high lignin concentration at the homogeniser on the process and the final NL properties are suggested.

References

[1] Rahman O U, Shi S, Ding J, Wang D, Ahmad S and Yu H 2018 *New J. Chem.* **42** 3415-25.
[2] Kannangara M S 2015 *Development and Integration of Acid Precipitation Based Lignin Biorefineries in Kraft Pulping Mills* (Montréal: Université De Montréal)
[3] Gonzalez M N G, Levi M, Turri S and Griffini G 2017 *J. Appl. Polym. Sci.*, **134** 1–10
[4] Zhao W, Simmons B, Singh S, Ragauskas A and Cheng G 2016 *Green Chemistry* **18** 5693–00
[5] Tian D, Hu J, Bao J, Chandra R P, Saddler J N and Lu C 2017 *Biotechnol. Biofuels* **10** 1–11
[6] Nair S S, Sharma S, Pu Y, Sun Q, Pan S, Zhu J Y and Ragauskas A J 2014 *ChemSusChem* **7** 3513–20
[7] Yearla S R and Padmasree K 2016 *J. Exp. Nanosci.* **11** 289–02
[8] Figueiredo P, Lintinen K, Kiriazis A, Hynninen V, Liu Z, Bauleth-Ramos T, Rahikkala A, Correia A, Kohout T, Sarmento B and Yli-Kauhaluoma J 2017 *Biomaterials* **121**, 97–108
[9] Beisl S, Miltner A and Friedl A 2017 *Int. J. Mol. Sci.* **18** 1244
[10] Stark W J, Stoessel P R, Wohlleben W, and Hafner A 2015 *Chem. Soc. Rev.* **44** 5793–5805
[11] Abbati De Assis C, Greca L G, Ago M, Balakshin M Y, Jameel H, Gonzalez R and Rojas O J 2018 *ACS Sustain. Chem. Eng.* **6** 11853–68
[12] Bangalore Ashok R P, Oinas P, Lintinen K, Sarwar G, Kostiainen M A, and Österberg M 2018 *Green Chem.* **20** 4911–19
[13] Culbertson Jr C G 2017 *Commercialization of sustainable bio-refinery projects for the pulp & paper industry* (North Carolina: North Carolina State University)

Oil Well Cement for high temperature-A review

Pankaj Munjal, Kong Kian Hau*, Abhinav Prabhakar, and Cheng Chuen Hon Arthur

Civil and Environmental Engineering, National University of Singapore (NUS), Singapore
Corresponding author, Email: ceekkh@nus.edu.sg

Abstract. Oil well cement (OWC) are widely used in oil industry both in exploration and in production of oil and gas. The purpose of OWC is to seal the annular space between metal casing and walls of a borehole. OWC has other application such as grouting and repair of offshore structure, and in cementing of other wells likes as water, waste disposal, geothermal, injection or storage. There are numerous problems such as high temperature and high pressure in the downhole condition of oil well. OWC slurry undergoes a reduction in strength (strength retrogression) with time when exposed to high temperature and high pressure. This paper reviews the OWC and its classifications, hydration process at ambient and higher temperature. Moreover, this study provides a comprehensive insight on anti-strength retrogression additives, which can potentially be use with OWC at high temperature and pressure.

1. Introduction

Oil well cementing is the process of placing a cement slurry in the annulus space between the well casing and the geological formation surrounding to the well bore [1]. The purpose of cementing is (i) to provide the zonal isolation between different geological layers, (ii) to support the casing, (iii) to protect the casing from corrosion. The exposure conditions of oil well cement (OWC) is different from the conventional construction industry. Sometimes OWC slurries are pumped to the depth of 5000 metres where the temperature and pressure can exceed of 150 °C and 100 MPa, respectively [2]. The static bottom hole temperature increases by 0.8 °C for every 30 m of the well depth in Gulf coast region [3].

The temperature can be as high as 400 °C [4] depending up on the depth and location of geothermal well. Consequently, the OWC can face additional challenges in contrast to conventional construction. Further, OWC must be able to withstand with weak and porous formation, corrosive fluid etc. at this much high temperature and high pressure (HTHP). Several additives have been explored to enhance the physical and chemical properties of OWC slurry for different purposes.

The purpose of additives are [5] (i) to alter the setting time of cement slurry under HTHP conditions of the well, (ii) to disperse cement particles, (iii) to control the fluid loss from the cement slurry, (iv) to control the shrinkage of cement after its set and hardens (v) to improve the interfacial bonding between casing and geological layer of the well, (vi) to control influx and migration of developed fluids into the cement column during setting.

Over the last decades, various types of additives are used to design the cement slurry for successful fill the space between casing and geological formation. In this paper, behaviour of cement paste at ambient and high temperature is briefly described. Moreover, an insight into the different additives explored by various researchers for oil well cementing at high temperatures and high pressures is provided.

2. Oil Well Cement (OWC)

Oil well cement is a Portland cement developed for the usage in oil and gas industry. The different types of Portland cement are widely used in the oil & gas industry to meet specific physical and chemical property set up by American Petroleum Institute (API).

2.1 Classification of OWC

Since 1924, the American Petroleum Institute (API) is establishing and maintaining standards for the worldwide oil and natural gas industry. Generally, API standards are review and revised by at least every five years. The latest API standard (10 A, 24th edition) specified the six classes (i.e., Class A, B, C, D, G and H) of oil well cement with three grades ordinary (O), moderate sulphate resistant (MSR), and high sulphate resistant (HSR) [6]. The details of available API classes of cement are present in Table 1. This latest API specification (24th edition) is identical to International Organization for Standardization (ISO), ISO 10426-1:2009. The cement of class E, class F, and class J are deleted in this latest specification. Class E and F contain the chemical retarders for controlling the thickening time under the defined conditions [7]. In reality, the identical condition as per API specification is rarely achieved and these cement is incompatible with the additional additives. Further, Class J cement is based upon a β-calcium silicate-silica composition, which is not a true Portland cement [7]. The hydration process of class J cement was very slow compare to other classes of cements.

Table 1. API classification of cement [6].

Cement Class	Description
A	This class of cement intended to use up to a depth of 1830 m from surface when special properties are not required and is available only in O grade. This cement is similar to ASTM C150, Type I cement.
B	This class of cement intended to use up to a depth of 1830 m from surface when conditions require moderate or high sulphate resistance and is available in both MSR and HSR grades. This cement is similar to ASTM C150, type II.
C	This class of cement intended for use up to a depth of 1830 m from surface when conditions require high, early strength and is available in O, MSR and HSR grades. This cement is similar to ASTM C150, type III.
D	This class of cement is obtained by grinding clinker, consisting essentially of hydraulic calcium silicates, usually containing one or more forms of calcium sulfate as an interground additive. There are some additives may be used during the manufacturing of class D cement [6]. This class of cement intended for use up to a depth of 1830 to 3050 m from surface when conditions require high temperature and pressures. It is available in MSR and HSR grades.
G	This class of cement is obtained by grinding clinker, consisting essentially of hydraulic calcium silicates, usually containing one or more forms of calcium sulfate as an interground additive. There are no additives other than calcium sulfate or water, or both, shall be interground or blended with the clinker during manufacture of class G well cement [6]. This class of cement intended for use up to a depth of 2440 m from surface as manufactures. It is available in MSR and HSR grades. It can be used a wide range of well depth with additional additives and retarders.
H	This class of cement is obtained by grinding clinker, consisting essentially of hydraulic calcium silicates, usually containing one or more forms of calcium sulfate as an interground additive. There are no additives other than calcium sulfate or water, or both, shall be interground or blended with the clinker during manufacture of class H well cement [6]. The composition of Class G and H cement are similar and the difference is the surface area. Class H is coarser than the class G cement. This class of cement intended for use up to a depth of 2440 m from surface. It is available in MSR and HSR grades.

2.2 Hydration of OWC

The four mains well know compounds present in the Portland cements are depicted in Table 2. The major constituent of Portland cement is silicate phases (C_3S and C_2S) which is more than 80 % of total materials [7]. These compounds are anhydrous. When these compounds exposed to water then they are decomposed and forming hydrated compounds. The reactions of cement compounds with water and their products can be represent in the symbolic equations as [7]:

$$2C_3S + 6H \rightarrow C_3S_2H_3 + 3Ca(OH)_2 \quad (1)$$

$$2C_2S + 4H \rightarrow C_3S_2H_3 + Ca(OH)_2 \quad (2)$$

$$C_3A + 3C\bar{S}H_2 + 26H \rightarrow C_3A.3C\bar{S}.32H \quad (3)$$

$$C_3A + 3C\bar{S}.32H + 2C_3A + 4H \rightarrow 3C_3A.C\bar{S}.12H \quad (4)$$

where, the standard cement chemistry abbreviations are used, where CaO is denoted as C; Al_2O_3 as A; SiO_2 as S, SO_3 as $\bar{S}$; Fe_2O_3 as F; and H_2O as H.

The product $C_3S_2H_3$ represent calcium silicate hydrate and commonly called "C-S-H gel". The fully hydrated cement constitutes approximately 65 % of CSH gel at ambient conditions [7]. This CSH gel contemplate as the principal binder of hardened cement, which is responsible for stability, and strength of the cement. Typically, the hydration process of Portland cement has separated into five stages as shown in Figure 1. (i) Preinduction period: This is the first stage after cement comes in contact with water and its duration is only a few minutes. This stage shows the wetting of powder and initiation of hydration, (ii) Induction period: The hydration activity has commenced in this stage and CSH gel formation initiated. The duration of this stage is a few hours at ambient temperature, (iii) Acceleration period: This stage represents the setting period of cement. During this stage, the hydration process is accelerated and formed CSH gel deposits into the available water-filled space, (iv) Deceleration period: The hydration rate of cement is decrease in this period. The duration of this stage is several hours, (v) Diffusion period: The hydration process is continuing at a slow rate and strength increases in this stage. The duration of this stage is indefinite at ambient temperature.

Table 2. Composition of Portland Cement [7].

Compound	Formula	Abbreviated Formula
Tricalcium silicate	$3CaO.SiO_2$	C_3S
Dicalcium silicate	$2CaO.SiO_2$	C_2S
Tricalcium aluminate	$3CaO.Al_2O_3$	C_3A
Tetracalcium aluminoferrite	$4CaO.Al_2O_3.Fe_2O_3$	C_4AF

2.3 Hydration at high temperature

The hydration process is effectively influenced with the increased temperature. The morphology and stability of the CSH gel are strongly dependent upon the temperature. The hydration products remain the same up to the temperature of 40 °C. As the temperature increases from 40 °C to 110 °C, certain changes may occur in the morphology of CSH gel. When the temperature increases beyond 110 °C, the metamorphosis of CSH gel will be introduced and that result in increases in permeability and decreases in the compressive strength of cement paste. This phenomenon called as "strength retrogression". At higher temperature, the CSH gel often convert to α-C_2SH (alpha dicalcium silicate hydrate) which is highly crystalline, [7]. Consequently, the permeability of cement will increase which cause severe problem for the oil well condition.

2.4 Prevention of strength retrogression

The solution for prevention of strength retrogression is reducing the lime-to-silica ratio (C/S ratio) in the CSH gel. Generally, the C/S ratio in the CSH gel is around 1.65 [2]. Many studies have shown that strength retrogression can be prevented by addition of silica-based materials [8–9]. The conversion of α-C_2SH at higher temperature can be prevented by addition of 35 to 40 % silica contents (i.e., silica fume and silica flour) by weight of cement (BWOC), reducing the C/S ratio to about 1.0 [9]. In the presence of silica, tobermorite mineral ($C_5S_6H_5$) is formed which preserved the high strength and low permeability at temperature of 110 °C [9–10]. As the temperature increase above 150 °C, this tobermorite mineral is transform to xonolite (C_6S_6H) with up to an 11.5 % decrease in absolute volume [11]. Sometimes, tobermorite can persist up to temperature of 250 °C because of aluminium contents [11]. As the temperature further increase above 250 °C, the phase truscottite ($C_7S_{12}H_3$) originates and is stable up to the temperature of approximately 350°C with silica quartz or 390 °C with silicic acid

[4]. When the temperature approaches 400°C, the stability of both truscottite and xonotlite get disrupt and they would dehydrate which results in the disintegration of the set cement [7]. These nature of phase changes (Figure 2) at high temperature is dependent upon the mineralogical composition of the cement, its C/S ratio (mole of lime per mole of silica; CaO/SiO_2), the amount of fine particles (silica fume), and the temperature and pressure levels [12].

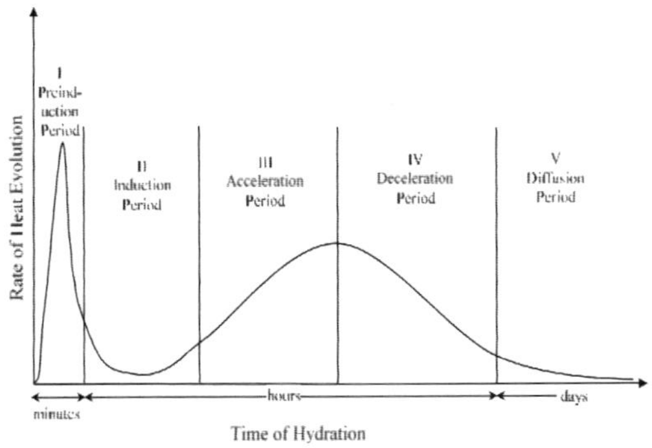

Figure 1. Hydration Process of Portland Cement [7].

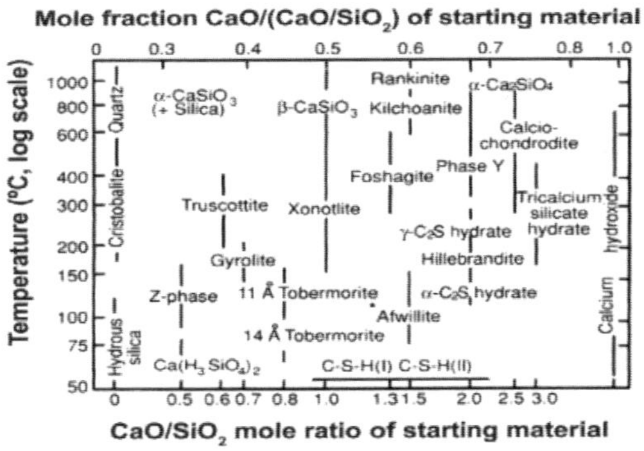

Figure 2. Portland cement phases with temperature versus the C/S ratio. The vertical lines shows the presence of a different phase obtained at a certain temperature condition [13].

3. Additives for anti-strength retrogresssion

The additive used as anti-strength retrogression in OWC slurry is high silica content, which reduce the C/S ratio. The high silica content can be obtained from several sources such as silica flour, silica fume, Nano-silica, rice-husk ash, sugarcane biomass etc. An insight on various source of silica content used as anti-strength retrogression additives is presented in the following sub-sections.

3.1 Silica Flour

Silica flour simply made by grinding pure silica sand to a very fine powder. The silica flour constitutes SiO_2 approximately 99 % in the chemical composition [13]. It is widely used in oil well cementing industry as anti-retrogression agent. Many researchers [13–14] have studied by use of silica flour as anti-retrogression agent at high temperature. Costa et al. [13] investigated the compressive strength behaviour of class G cement slurry with addition of silica flour (0, 30, 35 & 40 % BWOC) at temperature of 38 °C and 300 °C after 7 and 28 days. Authors [13] observed that the silica flour reacted at high temperature of 300 °C, forming the xonotlite phase and improving the mechanical behaviour, especially in the slurries with 35 % and 40 % BWOC.

3.2 Silica Fume

A very fine (grey coloured) non-crystalline silica produced in electric arc furnace as a by-product of the production of elemental silicon metal or ferrosilicon alloys. Silica fume particles are extremely small, with more than 95 % of the particles being less than 1 μm (100 times finer than cement particle) with high specific surface area 20 m^2/g [14]. Some other names of silica fume are condensed silica fume and micro silica [15]. Silica fume particles have a high water demand and absorb excess water in cement slurry. The silica fume is very effective anti-strength retrogression additive for OWC slurry and reacts rapidly with CH to form a more compact form of CSH [14].

3.3 Nano Silica

Nowadays, Nano Silica (NS) is widely used in construction and oil well industries. NS is extremely fine silica particle with nearly about 1000 times smaller than the average cement particles [16]. The addition of NS in cement slurry, react with calcium hydroxide (CH) and increases the amount of CSH gel, leading to a compact microstructure and improves the mechanical properties [17]. Many researchers [16–17] have investigated the NS as anti-strength retrogression additive for OWC slurry for different temperature conditions and stated that the NS has a high pozzolanic activity, which enhanced the fire resistance of cement specimens [18].

3.4 Sugarcane biomass

Many countries worldwide are producing sugarcane for many purposes. The sugarcane biomass is the residues after the sugarcane crushed and used for generally cogeneration. The composition of such residues mainly comprises of silica either amorphous or crystalline, which implies a potential application in concrete, mortars and cement slurries [19]. Anjos et al. [20] investigated the effect of the addition of sugarcane biomass waste on the compressive strength of oil well cement slurries subjected to temperature of 280 °C. Authors [20] concluded that the sugarcane biomass waste can be used as a substitute for silica flour in oil well cements subjected to high temperatures.

3.5 Ceramic Waste Powder

Ceramic waste powder (CWP) is the waste product produced during the process of polishing ceramic tiles, floor tiles and sanitary ware etc. CWP constitutes approximately 80 % silicon dioxide and aluminium oxide which makes it useful for concrete technology [21]. There is limited information reported for use of ceramic waste powder for the application of OWC at high temperature. Souza et al. [8] studied the effect of ceramic waste for application in OWC slurry at high temperature and concluded that the use of ceramic waste isolated is not adequate to prevent the strength retrogression of OWC slurry at higher temperature. However, it can be use as partial replacement of silica flour in certain proportion.

3.6 Ground Granulated Blast-furnace Slag

Ground granulated blast furnace slag (GGBFS) is the slag generated from iron producing blast furnaces, which rapidly chilled in water and then pulverised into a powder. GGBFS is widely used in many cement applications where it provides enhanced durability, resistance to sulphate attack, high resistance to chloride penetration [22]. Researchers [9, 8 & 23] investigated the effect of GGBFS in OWC slurry at high temperature. Alp and Akin [23] stated that it is possible to prepare GGBFS blended OWC slurries with higher compressive strength than conventional silica blended slurries at high temperature.

3.7 Rice Husk Ash

Rice Husk Ash (RHA) is an agricultural by-product of burning rice husk in the boiler under controlled temperature. The process produces about 25 % ash containing 85 to 90 % amorphous silica plus about 5 % alumina, which makes it highly pozzolanic [24]. Vidal et al. [25] studied the RHA as anti-strength retrogression additive for OWC slurry subjected to 300 °C temperature. Vidal et al. [25] concluded that the RHA is effective on inhibiting the cement compressive strength retrogression and it could be a sustainable alternative to commercial silica flour.

3.8 Palm Oil Clinker

Palm Oil Clinker (POC) is a waste material produced upon burning of palm oil shells and Mesocarp fibres in a boiler of palm oil mill [30]. POC generally contains 60 to 75 % of silica (SiO_2) content. Karim et al. [26] investigated the effect of high temperature on the compressive strength and microstructure of cement slurry containing POC and concluded that the compressive strength is significantly improved.

4. Conclusions

A comprehensive review on anti-strength retrogression additives used in oil well cement at high temperature is performed. The paper shows the hydration process of Portland cement at ambient and higher temperature. The morphology and stability of the CSH gel are strongly dependent upon the temperature. At higher temperature, the CSH gel often convert to α-C_2SH which is major cause of strength retrogression. The conversion of α-C_2SH at higher temperature can be prevented by addition of 35 to 40 % silica contents by weight of cement (BWOC), reducing the C/S ratio to about 1.0. It is observed that there are many anti-strength retrogression additives available in worldwide, which enriched in high silica content. It concluded that the available waste material near to oil and gas industry, which are richest in silica content, could be use in OWC slurry for preventing strength retrogression. In the future work, authors are planning to investigate the effect of combination of these waste materials for preventing the anti-strength retrogression of OWC slurry at high temperature and high pressure.

Funding and acknowledgments

This work was supported by the Singapore Maritime Institute (ID: SMI-2016-OF-05). Grateful acknowledgement is made to Nautec for collaborative support with access to the company's OWC laboratory, equipment and materials. The authors would like to acknowledge key personnel from Nautec: Dr Kok-Seng Chia, and Mr Elo Yde, for their advisory role in this project.

References

[1] Shahriar A 2011 *Investigation on Rheology of Oil Well Cement Slurries Ph.D Thesis* (The University of Western Ontario Canada 1-52.
[2] Neville A 2011 *Properties of Concrete* (Enland: Pearson) 1-215.
[3] Mehata P and Monteiro P 2006 *Concrete: Microstructure, Properties, and Materials* (McGraw-Hill) 120-250.
[4] Luke K, Taylor H F and Kalousek G L 1981 Cem. Concr. Res. **11** 197–203.
[5] Cowan K M and Eoff L 1993 *SPE Int. Symp. Oilf. Chem.* 317–327.
[6] API Specs 10A 2010 *Specification for Cements and Materials for Well Cementing* (USA: API).
[7] Nelson E B 2006 *Well Cementing* (Elsevier) 25-235.
[8] Souza P, Soares R, Anjos M, Freitas J, Martinelli A, Melo D 2012 Brazilian J. Pet. Gas., **6** 105–113.
[9] Kuzielová E, Žemlička M, Másilko J, Palou M T 2018 J. Therm. Anal. Calorim. **133** 63–76.
[10] Hewlett P C 1998 *Lea" s Chemistry of Cement and Concrete* (Elsevier) 25-811.
[11] Chow S Y and Kalousek G L 1976 Soc. Pet. Eng. J. **16** 307–309.
[12] Hager I 2013 Bull. Polish Acad. Sci. Tech. Sci. **61** 1–10.
[13] Costa B L, Souza G, Freitas J C, Araujo R G and Santos P H 2017 J. Pet. Sci. Eng. **158** 626-633.
[14] Grabowski E and Gillott E 1989 Cem. Concr. Res. **19** 333-344.
[15] Campos W and Brandao E M 1999 15th Brazillian Congr. Mech. Eng. 1-10.
[16] El-Gamal S M A, Hashem F S and Amin M S 2017 Constr. Build. Mater. **146** 531-546.
[17] Heikal M, Ali A I, Ismail M N and Ibrahim S A 2014 Constr. Build. Mater. **70** 339-350.
[18] Heikal M, Al-Duaij O K, Ibrahim N S 2015 Constr. Build. Mater. **93** 1067–1077.
[19] Anjos M, Martinelli A E, Melo D M, Renovato T, Souza P D, Freitas J C 2013 J. Pet. Sci. Eng. **109** 291-297.
[20] Anjos M, Martinelli A E, Melo D M 2011 Mater. Sci. Eng. A. **529** 49-54.
[21] El-Dieb A S, Taha M R, Kanaan D, Aly S T 2017 Proc. Inst. Civ. Eng. - Constr. Mater. **171** 109-116.
[22] Wild S, Kinuthia J M, Robinson R B, Humphreys I 1996 Clay Miner. **31** 423-433.
[23] Alp B, and Akin S 2013 38th Work. Geotherm. Reserv. Eng. Stanford Univ. Stanford USA.
[24] Zareei S A, Ameri F, Dorostkar F, Ahmadi M 2017 Case Stud. Constr. Mater. **7** 73-81.
[25] Vidal A V, Araujo R G, Freitas J C 2018 J. Clean. Prod. **204** 292-297.
[26] Karim M R, Chowdhury F, Zabed H, Saidur M R 2018 Constr. Build. Mater. **183** 376-383.

IOP Conf. Series: Materials Science and Engineering **652** (2019) 012056 doi:10.1088/1757-899X/652/1/012056

Performance of Alum Sludge as partial replacement for cement adding superplasticizer

Gunalaan Vasudevan

Department of Construction Management and Real Estate, Faculty of Built Environment, Tunku Abdul Rahman University College Malaysia
Email: gunalaanv@tarc.edu.my

Abstract. The production of alum sludge from waste water treatment plant is increasing all over the world. Disposal of alum is becoming a serious environmental problem. Another alarming issue is attributed to the high usage of concrete in the construction industry where concrete is one of its main material. This will lead to increased cement production and emission of carbon dioxide emission. Hence, the research for replacement of cement by alum sludge is essential to reduce both the emissions of carbon dioxide and the disposal problem of alum sludge. An attempt has been made to investigate the use of sewage sludge ash as partial cement replacement in concrete. The alum sludge is incinerated at the temperature of 600 and a duration of 3hours.The incinerated alum sludge is sieved through size of 150. Four different percentages of 5%, 10%, 15% and 20% of alum sludge (AS) is used to replace the cement in the concrete. XRD and XRF tests were carried out to determine the similarity component present in the AS. Water absorption test and slump test has also been carried out to check the capacity of water absorption, durability and workability of the AS concrete. The compressive strength of AS concrete is also determined through rebound hammer and compression test. Result show that AS has the potential to replace cement as the major chemical component of AS is similar to that of cement.10% of AS concrete shows a lower absorption value than the control sample which shows that 10% SSA would have a better durability. However, an increase in the percentage of AS decrease the workability and compressive strength and yet, the 10% AS concrete possesses compressive strength higher than 25N/mm2 which can be applied in structural application. The findings suggested that the suitability of alum sludge concrete in different cement replacement ratio is applicable for different usage. The application of alum sludge in concrete will directly reduce the quantity of the cement used which also decreases the emission of carbon dioxide and solving the disposal problem of alum sludge.

1. Introduction

This project is about on the replacement of alum sludge as part of the cement in concrete .Alum sludge is a waste product produce of water treatment plant. Because of the urbanization and development of populace, the measure of alum sludge has expanded quickly finished the years and is required to increment further [1]. Higher measures of alum sludge may influence the earth. In ebb and flow inclines, the alum sludge is arrange through land filling, sea disposal, arrive application and farming use .Late research has discovered that flow disposal strategies acted ecological issues such like water, air and air contamination. Land disposal is the least expensive approach to arrange the alum sludge among the strategies for the disposal since it likewise help to treat the land for developing products. Sludge really has the greatest amount when contrasted with the other item created by the procedure of

alum treatment, and it is discovered that is contains substantial metals in its synthesis[2].Other than that, this report consist of the three capter.Chapter1,introduction.Chapter2, literature review and chapter 3 research methodology. After the three chapter, the experiment will show the result of the strength and workability of the concrete that mix with alum sludge
Cremation wound up noticeably one of the option for the disposal of alum sludge. One of the research at the Nanyang Technological Institute demonstrates that 550°C is a great condition for sludge cremation. The alum sludge was dewatered amid the channel procedure from different water treatment plants in Singapore and was auto warm amid burning at 550°C. The motivation behind burning alum sludge 2 at the temperature over 550°C is to expel any natural issue which is not appropriate in concrete generation.

2. Problem Statement

Expanding in populace will impact that more alum sludge will be produced. Thus, disposal of residential alum sludge turns into a huge issue because of time and space required and it is expanding the budgetary weight of wastewater treatment organizations. Disposal of local alum sludge without any appropriate way may cause various natural and welbeing hazard related issues. Along these lines, the disposal of alum sludge must be direct to keep some significant issues in our nation. From the exploration we anticipated that in the time of 2020 that the created of alum sludge is around 1.32 billion every year. From the data of Indah Water Konsortium Malaysia Expressed that Malaysia creates more than 4.5 million cubic meters of residential alum sludge every year since 2005 and the volume of the alum sludge produce is expanding step by step. Alum sludge is one of the biggest waste material in Malaysia. Alum Sludge additionally a neighbourhood ecological issue. The general sludge treatment had been forms before the sludge cake is transported to endorsement site for disposal .One of the real difficulties in the preparing of local alum sludge is much related with the administration of high dampness content .It is an exceptionally shaky natural issue in which set out to create awful gasses. The utilization of alum sludge as a non-customary development material had been examined and investigate because of the issues and confinements of disposal. The alum sludge is utilized as a filler and mixed cement material for concrete and it had been utilized as crude material to deliver development items, for example, blocks, lightweight total, tiles and clearing squares, and cementitious material.

3. Literature Review

3.1 Pozzoloric Materials

3.1.1 Alum Sludge

Alum sludge is a by-product of water treatment plant. Due to the urbanization and growth of population, the amount of alum sludge has increased rapidly over the years and is expected to increase further. Higher amounts of alum sludge may affect the environment. In current trends, the alum sledge is disposed through land filling, ocean disposal, land application and agriculture use. Recent research has found that current disposal methods posed environmental issues such as water, air pollution. Land disposal is the cheapest way to dispose the alum sludge among the methods of the disposal because it also helps to fertilize the land for growing crops. Sludge really has the greatest amounts when contrasted with the other item delivered by the procedure of alum treatment, and it is discovered that it contains substantial metals in its structure. Other than that, there are some insoluble issue likewise found in alum sledge, for example, natural issue, pathogens supplements, and metals. Alum sludge additionally contains solvent substances, for examples are minerals, salts and natural chemicals. Incineration became one of the alternative solutions for the disposal of Alum sludge[3]. One of the researches at the Nanyang Technological Institute shows that 550°C. The purpose of incinerating alum sledge was dewatered during the filter process from various water treatment plants in Singapore and was auto-thermal during combustion at 550°C. The purpose of incinerating alum sludge 2 at the temperature above 550°C is to remove any organic matter which is not suitable in concrete production.

3.1.2 Superplasticizer

SuperPlasticizer or also known as a high range water reduces. It is a chemical admixtures utilized where very much scattered molecule suspension is required. It utilized as dispersants to maintain a strategic distance from molecule isolation (Gravel, coarse and sands), and to enhance the stream qualities of suspensions. For example, in solid applications. When it includes into cement or mortar, it enables the lessening of the water to cement proportions however it does not influence the workability of the hardening of the fresh concrete. From research, we know that the strength of concrete increase when the water to cement ratio decreases. The addition of SuperPlasticizer in the truck during delivering the concrete is a fairly new development within the industry. Admixtures added in transit through automated slump management systems, such as Verify allows concrete producers to maintain slump until discharge without reducing concrete quality.[4]

3.1.3 Use of alum sludge as construction materials

Sludge that deliver from a water treatment plant may influence the groundwater quality and making domain issue. To tackle this sort of issue, the sludge can be utilized for concrete as fine total in the development territory that can decrease the negative impact on lands and can approach towards eco-accommodating development [5]. Research centre scale tests were done which cantered around the properties of alum sludge fiery debris as replacement materials in their capability to be utilized as to blend with cement. In the outcome demonstrate that STP sludge have a compound quality to the cement.

Howard examined experimentally reuse water treatment infrastructure university kuala lumpur research journal vol 2 no.1 (2014).26 sludge from water treatment plant had been use as fine aggregates. Utilizing 10% and 20 % of water treatment sludge as a blend to make a heap bearing concrete piece [6]. The outcome demonstrate that it can decrease the cost and half water treatment sludge proportion in the blend. This could expedite a considerable measure of favourable circumstances transfer it will be the most noteworthy esteem workable for years to come. [7]

4. Result and Discussion

4.1. Workability

Workability is defined as the easement of the freshly mixed concrete can be placed, compacted and finished. In this study, slump test was chosen as the workability test for the fresh concrete. Every different batch of concrete. Due to its simplicity and convenience, slump test was carried out before the moulding of the fresh concrete. The slump test result is shown in Table 1

Table 1. Workability in term of slump cone test

Alum Sludge	Slump Level, mm
0%	130
5%	115
10%	280
15%	164
20%	157

When we replace 5% cement with alum sludge, the value of slump result is 90mm but the slump is started to increase with replacement of 10% alum sludge, the result shows the slump level reach 120mm.The slump level reached the highest point which is 280mm when the replacement of alum sludge reached 15%.15% of replacement is the best workability when concrete mix with alum sludge

IOP Conf. Series: Materials Science and Engineering **652** (2019) 012056 doi:10.1088/1757-899X/652/1/012056

as replacement of cement. Unfortunately, the slump level start to decrease to 157mm at 20% replacement as table shows. Alum sludge which have the ability to absorb water compared to cement powder has slightly reduced the slump of the concrete mix. The decrease in workability can be explained by two factors, which is the irregular morpho0logy of alum sludge particle surface. The shape of alum sludge particles is not spherical which caused negative influence on workability. Alum sludge could slightly decrease the workability and strength of cement concrete due to high water demand of alum sludge.

4.2. Compressive Strength Test

Compressive strength test compressive strength of concrete is the most common performance measure used by the engineer in designing building and other structures. In this study, the test age of the concrete was 7 days, 14 days and 28 days 3 cubes were tested for each of the same concrete proportion in order to obtain the average data. The compressive strength result of various. Based on the result shown in Figure 1 the product which mix with alum sludge shown lower initial strength as compared with the control. The cement replacement of 10% alum sludge has shown the highest compressive strength of 36.41 Mpa at 28 day while the lowest compressive strength is the 20% replacement of cement, which is 15.48 Mpa. The replacement of alum sludge at 5% has increase the compressive strength up to 10% as compared with the control concrete. The compressive strength decrease as the percentage of cement replacement by alum sludge increases. The alum sludge concrete shows increase compressive strength with curing time and the compressive strength loss is proportional to the percentage increment of sludge.

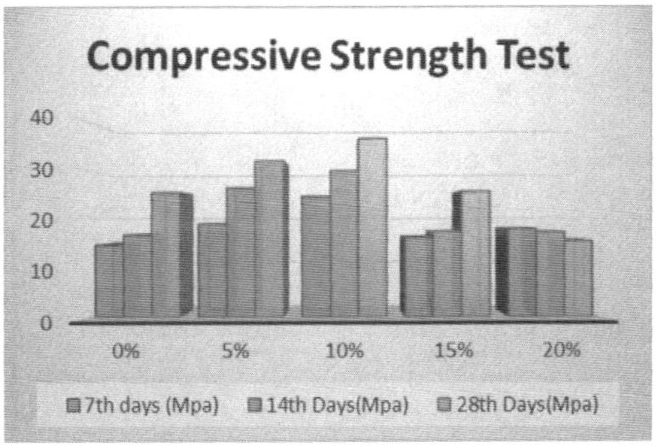

Figure 1. Compressive Strength at different Curing age

4.3. Ultrasonic pulse velocity (UPV)

The results UPV test of high performance concrete made from thermally activated alum sludge and pozzolanic materials in binary and ternary blended binders that were tested and cured at 7 days, 14 days and 28 days show ranges from 15us to 35us, with readings that increase with age and curing time. Also all the measured UPV values were greater than 34us. The results are comparable with previous works for concrete with pozzolanic materials with the replacement of 5%, 10%, 15% and 20%. Based on the line graph above, the best result of UPV test is at the 10% of replacement. Result indicate that the trend of the UPV result of binary and ternary blended binders in high performance concrete with alum sludge is similar to the trend of compressive strength. It is found that the trend in UPV values shows an increase with increasing compressive strength for all the mixture. In other words, an increase in UPV is accompanied by an increase in the compressive strength. Other researchers have results have obtained similar results too.

IOP Conf. Series: Materials Science and Engineering **652** (2019) 012056 doi:10.1088/1757-899X/652/1/012056

Table 2. Ultrasonic Pulse Velocity Test at different curing age

	7th days		14th days		28th days	
Percentage replacement	Direct	semi-direct	direct	semi-direct	direct	semi - direct
0%	25.23	19.87	28.2	21.57	29.27	21.11
5%	32.1	21.4	32.3	21.2	32.43	22.53
10%	32.9	22.9	33.4	23.8	34.7	24.32
15%	21.9	20.46	22.87	21.26	24.09	22.24
20%	25.34	19.63	26.77	19.86	28.08	20.15

4.4. Splitting Tensile Strength Test

The test result for the HPC across all mixtures are presented in Figure 2 that the tensile strength of behaved similarly to the compressive strength. The average tensile strength was within the permissible values, in accordance with design specification. The result revealed that concrete containing 6% AS has greater splitting tensile strength than the control concrete across all ages. Beyond this replacement level, the tensile strength decrease .The ratio of splitting tensile strength to compressive strength at 28 days was between 6.7 % and 7.2 %,which is lower than the result obtained from normal and medium strength concrete ,which range from 8% to about 10 %.This result indicates that as the compressive strength of concrete increase, the ratio of splitting tensile strength to compressive strength decreases, which is consistent with the result of order studies on high strength concrete.

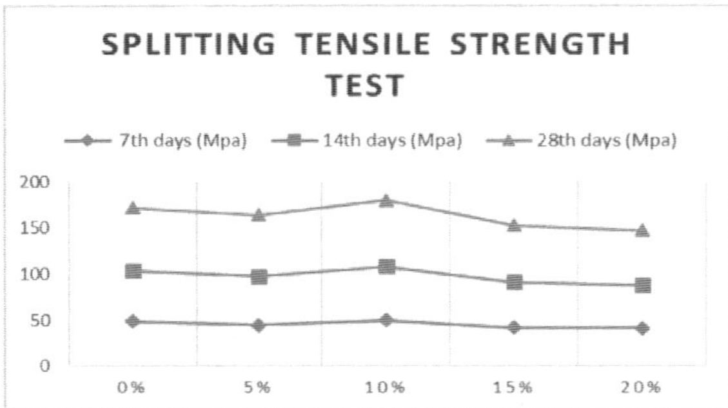

Figure 2. Splitting Tensile Strength Test Different Curing age

4.5. SEM Test

Microstructures of sewage sludge concrete. After SEM analysis of alum sludge concrete it can be observed the surface of the sludge concrete has dense enamel layer, SEM picture, the surface is almost not observed the existence of tiny openings hole. This surface characteristic determined the low water absorption of the self-made concrete. In the Figure 3, through the internal pore structure inside. The honeycomb pore structure wide speared in concrete which speared in concrete which determined high strength and light performance characteristics. Sludge concrete X-ray diffraction analysis shown in figure 3 & 4. From the enamel layer of alum sludge concrete wean found XRD pattern the main crystalline phases of the surface on self-made concrete is mullite.Millit is a very important product for sintering concrete, it is skeleton for concrete as strength microstructure formation. Cristobalite and mullite make the concrete have high strength and low water absorption, Sludge concrete performance. Using alum sludge can produce green high performance concrete. Internal pore structure of workability mineral admixtures make strength and workability of light weight aggregate concrete improve, but excessive addition could reduce concrete strength particularly in the early stage, so there is a reasonable dosage light strength concrete strength is increase with the decrease of water cement ratio, but for the limit of strength of light weight aggregate is not obvious when water cement ratio reduced to a certain extent increasing sand ratio can improve concrete strength , but have adverse impact on the workability of fresh concrete, Using the study results,optimizied the concrete mix proportion and prepare grade 40 light weigh aggregate concrete with density level 1700kg/m3 slump degree over 18cm,extended degree over 50cm

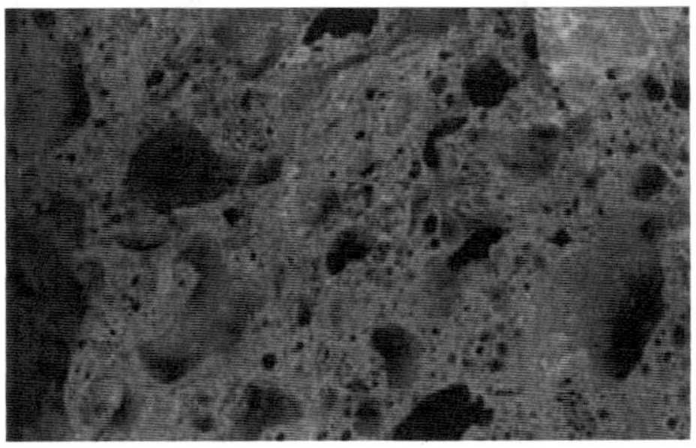

Figure 3. Internal Pore Structure of alum Sludge

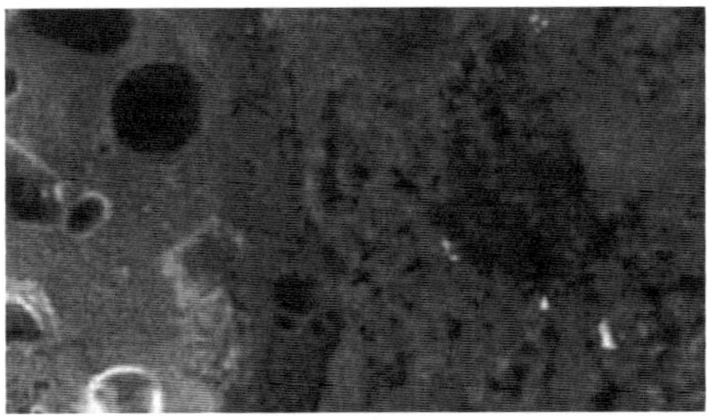

Figure 4. Surface Crystal Structure of alum Sludge

5. Conclusion and Recommendation

The overall objective of the work was to investigate the feasibility of incorporating alum sludge ass an additive material to concrete mix. This study included the preparation of concrete mixes containing alum sludge and the evaluation of sludge concrete properties in fresh and hardened states. The studying properties involved mix workability, compressive strength and density. According to experimental results, the usage of alum sludge in concrete mixes as an alternative of disposal for alum sludge is possible. The current type of sludge accumulated in dumping sites and the expected future type were used in making concrete mixes. The influence of both types on concrete properties was studied. In all cases there was an optimum quantity of sludge which can be used without introducing any change in mix preparation and acceptable properties were still be produced. The study showed that the dry sludge can be used in interlock brick production without changing the normal industrial process. Recommendation for further research The following recommendations are proposed for further research and study in order to from a complete picture of using sewage sludge in concrete mixes:

1) The durability of sludge concrete need comprehensive researches
2) Further research is needed to determine alum sludge effects on reinforcement if sludge concrete
3) The effect of drying temperature of alum sludge on concrete properties needs further research
4) Other studies are encouraged to obtain the effect of using different percent of sludge with concrete.

Reference

[1] Haider Mohammed owaid, 2013, Phyical and Mechanical properties of high performance concrete with alum sludge as partial cement replacement on 22 March 2017
[2] JamshidiMehdadi N,Jamshidi M,2011,Application of sewage dry sludge as fine aggregate in concrete
[3] J.h.Tay,K.Y.Show,S.Y.Hong,C.Y.Chien and D.J.Lee, »Potential reuse of wastewater sludge for innovative application in construction industry,Bulletin of the college of Engineering N.T.U,No.86October 2002.
[4] ASTM C192.Standard Practice for marking and curing Concerte Test Spécimens in the Laboratory.American Society for Testing and Materials : 2004
[5] Chih Cho Wu.(1997).Effets of polymer dosage on alum sludge dewatering characteristics and physical properties

[6] Mohammed Owaid, H., Hamid, R., Sheikh Abdullah, S. R., Tan Kofli, N., & Taha, M. R. (2013). Physical and mechanical properties of high performance concrete with alum sludge as partial cement replacement. Jurnal Teknologi (Sciences and Engineering)
[7] Taha,M.(2013).Physical and mechanical ; properties of high performance concrete with alum sludge as partial cement.Jurnal teknologi (sciences and Engineering,65(2),105-112

Interactive system to control the kinematic error of a five-axis milling machine

Supatpong Narmnark[1, a], Stanislav S. Makhanov[2,b]

[1,2]School of Information, Computer and Communication Technology, Sirindhorn International Institute of Technology, Thammasat University, Pathumthani 12121, Thailand
[a]s.narmnark1993@gmail.com, [b] Makhanov@siit.tu.ac.th

Abstract. The kinematic error (KE) of a five-axis milling machine is an important measure of the accuracy. The proposed software allows to interactively position the cutter contact (CC) points to reduce the KE. The system helps to optimize the tool path by a trial and error approach and provides a learning experience for the users of five-axis machines. The code works with the STL and OFF files as well as with parametric surfaces.The Interactive system (IS) was offered to unexperienced users as a game with a goal to reduce visible kinematic errors while keeping the number of available CC points. The users were able to achieve a decrease of of the maximum KE by 91.73% and of the average KE by 43.70%. The program has been verified with a post processor from HAAS VF-2TR machining center (Trunnion VMCs).

1. Introduction

Five-axis milling machines (Fig 1.) become popular and common in manufacturing, bio-medical, automobile and mold industries [1], [2]. Compared to three-axis milling machines, the two additional rotary axes provide higher accuracy and improve machinability of complex shaped parts [3].

The five-axis geometric errors can be attributed to three main types. The first type is related to the kinematic of the machine and includes the systematic errors and errors attributed to the initial set-up. The machines with three linear axes have a total of 21 linear independent geometric error components whereas the five-axis milling machine has 42 components (twice of the three-axis machine!)[4].The second source is the imperfection of the movements the machine components and its fixtures. Finally, an important source of the geometric errors is an inaccurate tool positioning which leads to a curvature interference and gouging (see, for instance, survey [5]). In many cases, this type of error is unavoidable but can be considerably reduced.

Some other, less prominent but still important, sources of errors are: machine operating conditions such as the material removal rate (feed rate), the depth of the cut, wet or dry cutting, clamping conditions, the tool wear and other tool imperfections. In this paper, we offer a software to analyze and reduce the KE of the five-axis machine with the understanding that the above-mentioned sources of errors exist. Under certain conditions, some of the errors could actually exceed the KEs. However, our numerical experiments and the actual machining show that these errors are important and could lead to significant inaccuracies (even collusions) in the five-axis mode [6]. At the moment there exists a variety of tools to generate a tool path and simulate the five-axis machines such as MasterCam [7], NX11[8] and SolidWorks[9]. However, up to now an IS that allows to interactively change the tool

IOP Conf. Series: Materials Science and Engineering **652** (2019) 012057 doi:10.1088/1757-899X/652/1/012057

path has not been developed. In the future the system can be integrated with a knowledge database which combines the experience of the machinist, operator and G-code programmers.

Figure 1. Five-axis milling machine (HAAS VF-2TR)

2. Methodology

Our five-axis machining operation includes the following steps.

1. Input data (Fig. 2a): part surface (STL, OFF, etc.), tool type, tool size, required machining accuracy δ_{max}.

2. Flatten the surface using a MATLAB library [10] and detect the corresponding parametric domain (Fig. 2b).

3. Establish a minimum bounding rectangle (MBR) and map it onto $\Delta = [0,1]\times[0,1]$. Note that at this point the surface has been parameterized i.e. $S = S(u,v) \equiv (x(u,v), y(u,v), z(u,v))$, where $(u,v) \in \Delta$ are the parametric coordinates.

4. Create a tool path on Δ. In this paper we show the system applied to correct a zigzag tool path [11]. However, it is applicable to a spiral and a radial tool path as well.

5. Use the barycentric interpolation to find the tool path (CC points) on the STL part surface.

6. Find the normals (Fig. 2c) and the curvature at the CC points.

7. Detect the curvature interference. Find an appropriate inclination of the tool for the flat end cutter. In case of a ball nose tool replace it by a tool with a smaller radius.

8. Create an interference free tool path and check the scallop height h [12] between the tracks. If the system shows a violation of the scallop constraint $h < \delta_{max}$, insert interactively new paths between the corresponding tracks.

9. Use a specific postprocessor (in our case HAAS-2TR) evaluate the KE. The system shows the violations of the KE in the 2D parametric domain and the actual tool trajectories on the 3D part surface.

10. The user interactively changes the positions and the number of the CC points along the tool path in order to reduce the KE

11. Generate a G-code and feed it into a solid modelling simulation software such as Vericut or NX - 11in order to eliminate global collisions.

12. The solid model is verified visually and compared with the design surface.

14. The G code is verified on a real machine using a soft material.

15. The resulting part is measured against the design surface.

16. The real industrial part is produced and verified.

The proposed IS to control the KE is used by steps 8-10, whereas steps 1-7 can be interpreted as preparation of the data and steps 11-14 as post processing steps.

The software works with STL or OFF surfaces. Alternatively, the user inputs parametric equations or polynomial approximations such as the Bezier patch or the Gregory patch. The STL/OFF surface is flattened and assigns an initial equal spaced CC points on a zigzag toolpath (Fig. 2d). The user manipulates the positions of the CC points interactively with in order to achieve a better accuracy.

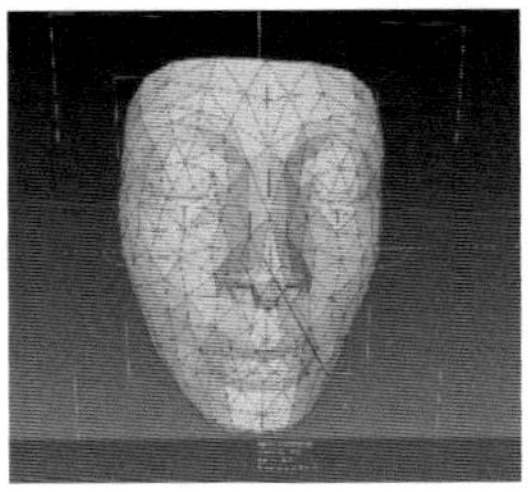

(a) Original STL surface

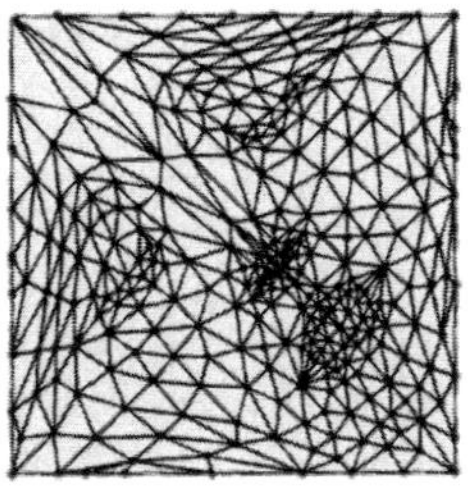

(b) Flattened surface and the boundarv

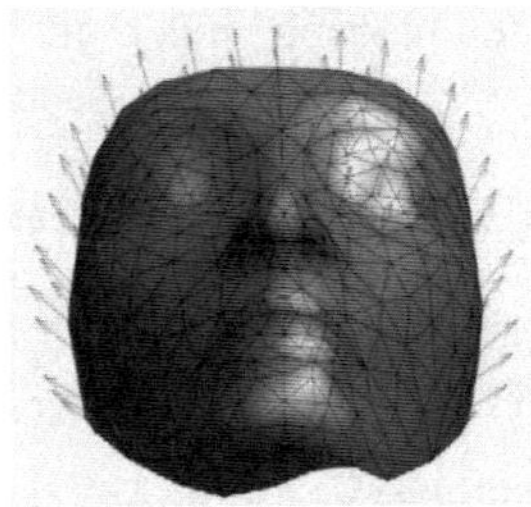

(c) The normals at CC points

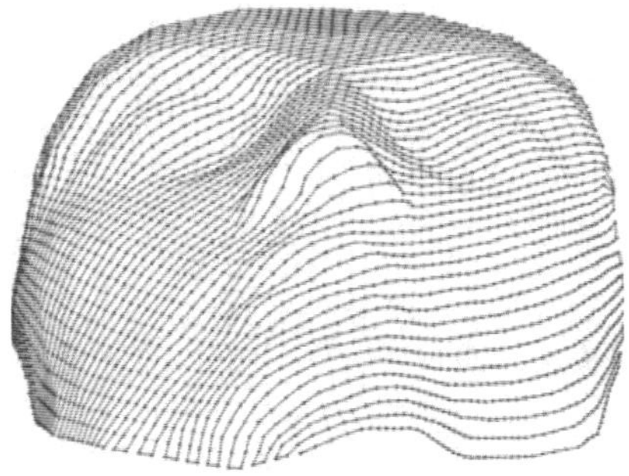

(d) Tool path on the STL surface

Figure 2a-e. Tool path generation from triangular meshes

3. Interactive System to control the KE

The software is characterized by a minimalistic design and simplicity. The users who are not familiar with the intricate theory of five axis machining should be able to intuitively generate a reasonable tool path, verify it visually and perform a solid modeling. Our simple menu and the user screen are depicted in Fig. 3.

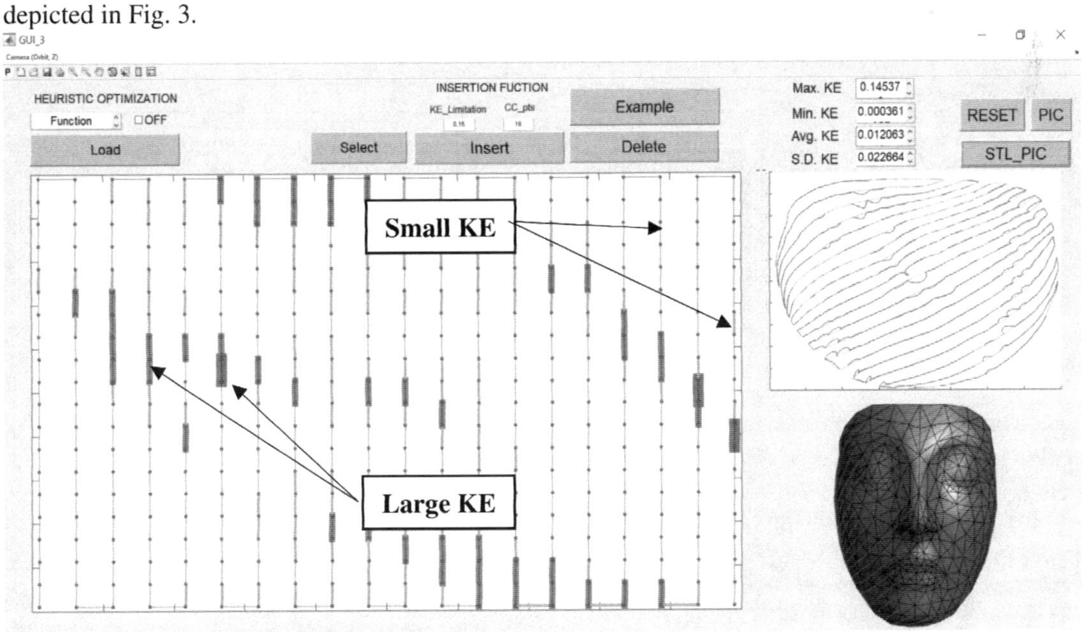

Figure 3.The interface of interactive systems software for .STL

The IS provides a heuristic adaptation and automatic point insertion proposed in [13],[14]. The heuristic adaptation includes four functions as follows

1. Move : the user moves the CC points in the parametric space (u,v) along the u –axis (see Fig. 3) in order to re-distribute the points and reduce the KE without increasing their number.
2. Add: the user adds CC points to the toolpath. The function reduces the KE but also increases the operation time in the case of high speed machining.
3. Delete : the user deletes CC points in order to reduce the machining time.
4. Select: the user selects the CC points in order to perform operations 1-3 on a set of points.

The system evaluates the KE as follows: let $W^D(s_p, s_{p+1}, t) \in S(u,v)$ be a space curve between two tool positions W_p and W_{p+1} extracted from the machined surface $S(u,v)$, where $t \in [s_p, s_{p+1}]$ is a parametric coordinate along the curve. The total KE is the Hausdorff distance (dist_H) between the desired trajectories $W^D_{p,p+1}(t) \equiv W^D(s_p, s_{p+1}, t)$ and the actual trajectories $W_{p,p+1}(t) = W(s_p, s_{p+1}, t)$ defined by

$$\varepsilon = \sum_p \text{dist}_H(W^D_{p,p+1}(t), W_{p,p+1}(t)).$$

The average KE is then defined by $\varepsilon^H_{av} = \dfrac{\varepsilon}{N_{CC} - 1}$, where N_{CC} is the number of the CC points. In many cases, the machining quality is defined by the maximum error given by

$$\varepsilon^H = \max_p(\text{dist}_H(W^D_{p,p+1}(t), W_{p,p+1}(t))).$$

The trajectories $W_{p,p+1}(t)$ are obtained by the corresponding inverse kinematics transformations[1]. An interactive classification the prescribed the levels of the KE indicated by color (see TABLE I). The experienced user can modify the levels of the KE for a particular part surface.

TABLE I. ERROR CLASSIFICATION TABLE

Error level	Range	Description
Blue	0 to <0.1	The error is acceptable
Green	0.1 to <0.05	The error is very small, delete a considerable # of the CC points or move them out of that area.
Violet	0.05 to <0.3	The error is too large, insert a few CC points or move CC points.
Red	≥0.3	The error is huge, insert considerable # of the CC points

4.Numerical example

As an example consider a manual adaptation of the KE for an STL surface displayed in Fig. 2. Five operators (Bachelor students of SIIT) were assigned to reduce the KE using the proposed IS. The students did not have any knowledge about five-axis machining. The system was treated as a game with a goal to reduce the visible KE using the available operations. The error classification TABLE I was set up by the administrator. The students were instructed to use the proposed menu to reduce the kinematic errors ε^H , ε^H_{av} while keeping or even reducing the number of CC points. If the student was able to reduce the error by 30% without increasing the number of CC points more than by 20%, the win was registered and the gaming time was recorded in Table II.

TABLE II shows results of the experiment, the KE after the manual adaptation has been reduced up to 91.73% of ε^H and 43.70% of ε^H_{av}.

The error reduction of ε^H and ε^H_{av} were evaluated by $\varepsilon_{reduction} = \frac{\varepsilon_{old} - \varepsilon_{new}}{\varepsilon_{old}} 100\%$

TABLE II. THE COMPARISON TABLE AFTER USING ISS

Operators	Operating time	The reduction of ε^H, %	The reduction of ε^H_{av}, %	CC points moved	CC points inserted	CC points deleted	Total # of CC points
Operator 1	7 min.	36.70%	16.86%	120	0	0	400
Operator 2	8 min.	33.20%	12.54%	85	15	15	400
Operator 3	12 min.	74.75%	29.34%	110	29	10	419
Operator 4	13 min.	86.19%	35.73%	150	38	0	438
Operator 5	10 min.	91.73%	43.70%	120	76	0	476

As an example consider a manual adaptation of the CC points for a surface displayed in Fig. 4 was created by operator 1 and operator 5

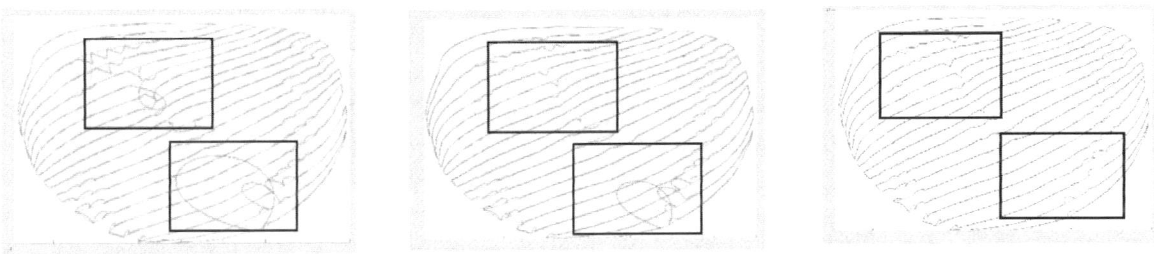

(a) The original tool trajectories (b) The tool trajectories by operator 1 (c) The tool trajectories by operator 5

Figure 4a-c. The tool trajectories

5. Conclusion

The preliminary experiments show that the proposed system allows an unexperienced user to substantially improve the quality of the tool path. For approximately ten minutes of an interactive adjustment, the users were able reduce ε^H by 33.20-91.73% and ε^H_{av} by 12.54-43.70%.

As for the future work, the real cutting experiment would provide further verification on the IS.

6. Acknowledgment

This research is supported by the Center of Excellence in Biomedical Engineering of Thammasat University.

7. References

[1] W. Anotaipaiboon and S. Makhanov, "Minimization of the kinematic error for five-axis machining", *Computer-Aided Design*, vol. 43, no. 12, pp. 1740-1757, 2011.

[2] J. Li, H. Zhouyang and Y. Lou, "Tool path optimization in postprocessor of five-axis machine tools", *The International Journal of Advanced Manufacturing Technology*, vol. 68, no. 9-12, pp. 2683-2691, 2013.

[3] Y. Lu, Q. Bi and L. Zhu, "Five-axis Flank Milling Tool Path Generation with Smooth Rotary Motions", *Procedia CIRP*, vol. 56, pp. 161-166, 2016.

[4] E. Bohez, B. Ariyajunya, C. Sinlapeecheewa, T. Shein, D. Lap and G. Belforte, "Systematic geometric rigid body error identification of 5-axis milling machines", *Computer-Aided Design*, vol. 39, no. 4, pp. 229-244, 2007.

[5] S. Makhanov, "Adaptable geometric patterns for five-axis machining: a survey", *The International Journal of Advanced Manufacturing Technology*, vol. 47, no. 9-12, pp. 1167-1208, 2009.

[6] S. Ivanenko, S. Makhanov and M. Munlin, "New numerical algorithms to optimize cutting operations of a five-axis milling machine", *Applied Numerical Mathematics*, vol. 49, no. 3-4, pp. 395-413, 2004.

[7] Information on https://www.mastercam.com/en-us/

[8] S. Samuel and A. Ericksen, *Basic to advanced computer aided design using NX11 modeling, drafting, assemblies, and sheet metal*. Design Visionaries, 2017.

[9] D. Planchard, *SolidWorks 2018 reference guide*. Sdc Pubns, 2017.

[10] Information on https://www.mathworks.com/products/matlab.html

[11] S. Yuwen, G. Dongming, J. zhenyuan and W. Haixia, "Iso-parametric tool path generation from triangular meshes for free-form surface machining", *The International Journal of Advanced Manufacturing Technology*, vol. 28, no. 7-8, pp. 721-726, 2005.

[12] C. Tournier and E. Duc, "Iso-scallop tool path generation in 5-axis milling", *The International Journal of Advanced Manufacturing Technology*, vol. 25, no. 9-10, pp. 867-875, 2004.

[13] M. Munlin, "Tool Path Optimization for Five-Axis Machining", *Journal of Information Science and Technology*, vol. 1,pp. 63-73, 2010.

[14] M. Munlin and S. Makhanov, "Iterative Tool Path Optimization for Five-Axis Machines with Optimal Point Insertion", *Computer-Aided Design and Applications*, vol. 9, no. 1, pp. 33-41, 2012.

IOP Conf. Series: Materials Science and Engineering **652** (2019) 012058 doi:10.1088/1757-899X/652/1/012058

Design and Manufacturing of a 3D Printer for Teaching and Research

Dr. Rafiq Noorani, Michael Lynch, Patrick Hodgkiss, Matt Fumo, Christopher Burke

Department of Mechanical Engineering, Loyola Marymount University, Los Angeles, CA 90045, USA
Email: rnoorani@lmu.edu

Abstract. The goal of this paper is to design and fabricate a 3D printer that will be the most accurate and cost-effective printer possible, while still being faster than the Rapid Prototyping (RP) Lab's Stratasys FDM-1650 at Loyola Marymount University. This project is done to greatly increase the capability of the RP lab in addition to exposing students to the design and fabrication of a 3D printer. To complete the construction of the printer, the design is originally generated in Solidworks to ensure the components would work together effectively and to aid with design selection. The frame is then constructed using V-slot aluminum and RP parts that are made in house. Following which, the motors and extruders are assembled according to design plans and the circuitry is wired through an Arduino based programming system. Further steps needed to be taken were reconstructing the Y-axis motion and calibrating all aspects of the machine to finalize the assembly of the printer. The printer has been completed recently and is being used for teaching and research on design and manufacturing applications.

1. Introduction

3D printing refers to the fabrication of a physical, three-dimensional part from a numerical description (such as a CAD model) by means of an automated process without any need for tooling [1-2]. Also known as rapid prototyping, 3D printing can aid in prototyping parts very quickly in a "layer by layer" manufacturing technique that builds up (or down) a model in discrete slices. Fused Deposition Modeling (FDM), a popular subset of 3D printing, achieves this by controlling the position and movement of a heated nozzle that can selectively deposit molten thermoplastic onto a model. While FDM lacks the resolution that other 3D printer systems can achieve (SLA, SLS, DLP, etc.), its main advantages are that it is far less expensive (about $2000 -$500) to operate, needs no hazardous chemicals, and is relatively fast at completing prints. These benefits make FDM a popular tool for designers who want to be able to prototype new designs quickly and inexpensively.

This paper sets out to explore the important parameters that influence the fabrication and function of a FDM printer. Design considerations such as frame rigidity, printing accuracy, and electrical component selection are detailed in such a way as to show just how these constraints impact the operation of 3D printers.

2. Necessary Parts

There are a multitude of necessary parts needed to make a 3D printer [3]. First there are the extruders. Extruders are the parts through which the printing material is deposited. Within a FDM printer, an extruder consists of a nozzle, a hot end, and some form of mechanical drive to push the material through the nozzle. The hot end is a piece of metal that heats up to melt the thermoplastic into a molten form before it is placed. The extruders need to be able to move in the *X*, *Y*, and *Z* planes to be able to print a part properly. There are multiple different ways to do this, but typically a FDM printer will consist of a combination of belt and screw driven motors. Motors actuate both the screws and the belts to provide motion in each direction. Screws are typically used for the *Z* motion since they provide very fine control over the layer thickness. In the *X* and *Y* planes belts are used to reduce complexity and size at the cost of some accuracy. All these components are held within some sort of ridged frame.

3. Functional Description and Design Analysis

To fabricate this 3D printer, many materials and processes were utilized. Most of the parts used for this project were ordered, however, many of the complex and or custom parts were 3D printed using the MakerBot 2 RP machine of the department.

The fabricated printer can print nearly every RP material including ABS and PLA, giving it a huge advantage over the other printers in the LMU RP lab. This is largely due to the hot end of the extruders. The hot end can print different materials due to the range of temperatures they can achieve. Each hot end in its current state can reach 300°C, and with few modifications as high as 400°C, allowing them to print plastics with extremely high melting points. Additionally, the RP machine is designed for dual extruders, allowing it to print two different materials at once. This allows the printer to extrude either two different colors or both build and support material at the same time.

Three common frame archetypes were considered for the 3D printer: square, delta, and A-frame [4-5]. Ultimately the A-frame was chosen due to its enhanced rigidity and larger build envelope as compared to the square and delta frames [1]. Rigidity improves accuracy of machines by preventing the machine from shaking while it is printing. In terms of durability, the machine can be picked up and moved around without needing to be recalibrated due to the strength of the triangular designs on the sides of the printer. Fig. 1 depicts the A-Frame Design.

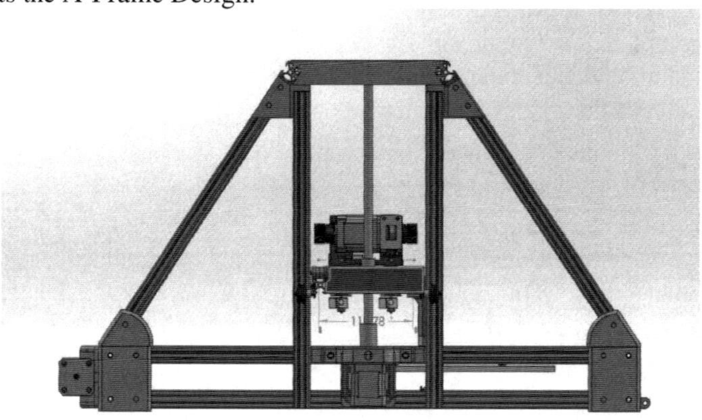

Fig. 1: A-Frame Design

Multiple extruder types were also considered for the design. A Bowden extruder takes the motor and extruder components and moves them from the *X-Y* carriage to the frame. There is better control of the filament through reducing weight of the *X-Y* carriage.

Direct drive extruders were also considered. Direct drive extruders keep the motor and other extruder components on the *X-Y* carriage, while driving the filament directly into the extruder. There is less fine control due to the extra weight on the carriage. This also typically requires a higher torque motor.

However, by adding gearing to a direct drive extruder, there is no need for a higher torque motor. Though direct drive extruders may have some drawbacks with respect to increased weight on the carriage, they are still very accurate in comparison to a Bowden extruder. Therefore, a geared direct driven extruder was picked for use within this design for accuracy. Using this type of extruder in combination with a .25mm nozzle, which is small for a 3-D printer, very accurate parts can be created at the cost of print speed. The gears used on the extruder also use herringbone teeth to prevent backlash from occurring, which would produce extra plastic when not necessary. A picture of a geared extruder can be seen in Fig. 2.

Fig. 2: Geared Extruders

Fig. 3: Heated Build Plate

Heated build plates are critical parts when it comes to printing certain materials such as ABS. A heated build plate prevents thermal gradients from forming within the printed material. Thermal gradients can cause deformation and warping, which is not ideal for the parts that need to be produced. A figure of a heated build plate can be seen in Fig. 3.

This printer uses an Arduino mega 2650 combine with a ramps shield to control the printing process. The ramps shield contains 5 pololu stepper drivers to run the X, Y, and Z axes in addition to two extruder motors. The use of an open source firmware called Marlin is uploaded to the controller and will give future students the ability to expand and tune the new machine. Extra code can be added to either augment or increase the capability of the machine, such as automatic bed leveling, additional lights, and fan inputs. All features of the rapid prototyping machine are controlled via the Arduino.

Fig. 4: CAD Model

4. Build Process

The 3D Printer was first designed in Solidworks, a Computer Aided Design program, in order to show compatibility of parts, size of the printer, logistics for construction, and to provide the builders with a reference. The Solidworks files operated as a blueprint for the construction of the printer and can be seen in Fig. 4. The first step to construction was to cut down the 1.5m V-slot stock on the horizontal band saw. There are 18 different V-slot extrusions in 6 different lengths needed for the whole assembly.

Using this model, the frame was constructed using V-slot aluminum, RP parts from the Makerbot Replicator 2, and varying fasteners. Extra fastening points were inserted into the frame for the various mounts for the *Y* and *Z*-axes. The base was constructed first to provide a sturdy platform for the rest of the construction of the printer. The top half of the frame was then assembled and mounted on top of the base. The bottom bracket holds the platform together and the top bracket holds the top half of the frame to the platform.

There are four rapid prototyped mounting brackets that mount the A-frame to the top supports. These mounting brackets hold the top of the assembly together. During construction, it was imperative to put T-slot nuts within the frame where more parts would be fastened, as T-slot nuts cannot be added or removed without disassembly. These T-slot nuts were used to fasten all the brackets to the V-Slot.

With the frame constructed, the Z-axis motor supports, and a Y-belt drive motor and Y-idler were mounted to the frame. The motor mounts and idler mounts are also designed and made with RP parts.

The next step was to build the frame for the *Z* carriage. This was assembled and mounted within the frame on the 2mm pitch Z-axis screws. With the vertical supports somewhat loose in the frame the z carriage can be used to move up and down to align the vertical supports. Once the vertical supports are aligned they can be tightened down to prevent them from moving. The *Z*-axis screws can be fastened to the Z-axis motors and mounted in the RP ends of the Z-axis carriage. The mounts for the V-slot wheels are slid into the supports and tightened down stabilize the *Z*-axis carriage. The *Y*-axis carriage is comprised of a ¼ inch acrylic with holes for V-slot wheels and acentric spacers to tension them. The acrylic also has holes to mount the spring suspended heated build plate. The final part of the axis assembly is the *X*-axis plate that has 4 V-slot wheels mounted in slots to tension them on the Z-axis carriage aluminum extrusions. Belts can be run for the *X*-axis and the *Y*-axis and zip tied to prevent them from slipping.

The extruders and extruder mounting plate were assembled next. The extruders work by gearing down the drive from the NEMA 17 stepper motor via herringbone gears to increase both torque and filament control. The gears drive hobbed bolts to create give smallest diameter filament drivers. 688 bearings that are spring loaded against the bolt act as idlers to force the filament against the hobbing on the bolts. The filament is forced through the hot ends using the stepper motors. Each hot end mounts into its corresponding extruder and has building instructions that can be found on the manufacture website. The two extruders are mounted to the plate and the entire dual extruder assembly is mounted to the x carriage on 4 springs to allow for hot end leveling. The mounted dual extruder tips can be seen in Fig. 5.

Fig. 5: Hot End Extruders

Fig. 6: Arduino Board

The next step is to mount the Arduino board, as seen in Fig. 6, and start the process of wiring. Each of the stepper motors has a corresponding location to plug. The heaters for the hot end and the relay input for the bed must both be properly secured to prevent short circuits and bad connections. The

IOP Conf. Series: Materials Science and Engineering **652** (2019) 012058 doi:10.1088/1757-899X/652/1/012058

thermistors on each hot end and bed must also be plugged into their corresponding locations. The final part of the wiring was the 12v rail of the ATX power supply to provide power to the whole ramps board. Endstops need to be mounted adjustably to each frame axis. Endstops allow the printer to find the origin in 3D space. Special care must be used when wiring the end stops to make sure they function properly.

5. Result and Future Improvements

The final fabrication of the newly built 3D printer is shown in Fig. 7. It is producing parts to meet the design and research needs of the students and faculty. This project has encouraged other students to participate in 3D printing technology. All industries of the world are using rapid prototyping and 3D printing to expedite their product development process and reduce costs of manufacturing and reduce waste.

Fig. 7: LMU 3D Printer

There are a multitude of different modifications that can be made to this design to further improve its ability to print. The printer is printing parts, but they are not as accurate as we would like them to be. The next improvement would be the addition of an enclosed build chamber. By enclosing the print area, the parts printed within the heated volume of the printer will not warp easily, as they will be of uniform temperature throughout the printing process. Another improvement that could be made to the machine would be increasing the size of the extruder nozzle. By increasing the size of the nozzle, faster dispensation of plastic can occur and create an overall faster build process for the 3-D parts. All technical specifications are listed in Table 1.

Table 1: Technical Specifications

Build volume (*X, Y, Z*)	220*220*230[mm]
Nozzle diameter	0.25[mm]
X/*Y* speed	150[mm/s]
Maximum nozzle temperature	300[°C]
Maximum bed temperature	110[°C]
File type	G-Code
Layer thickness	0.05-0.2[mm]
Power requirement	100-240[V] AC

Installing an automatic bed leveling system would also greatly reduce the maintenance time needed to ensure that the heated build plate is both level with the frame and offset from the *Z* origin. Finally, the wiring harness should be properly maintained in order to ensure that the electrical connections are secure and not likely to interfere with the printing process.

6. References

[1] R. Noorani, *Rapid Prototyping: Principles and Applications,* John Wiley and Sons, Englewood, NJ, 2006

[2] T. Wohlers, *Wholers Report 2016*; 3D printing and Additive Manufacturing State of the Industry.

[3] C. Cameron, *Idiot's Guide As Easy As It Gets 3D Printing*. Alpha by Penguin Group. 2015

[4] H. Kalani Kirk and R. Horne, *3D Printing for Dummies*. Hoboken: Joh Wiley & Sons, 2014.

[5] *Make: Ultimate Guide to 3D Printing* Winter 2014: 1-112.

[6] Jbrazio. "What Is Marlin?" *Marlin Firmware*, Jbrazio, 26 Mar. 2019, marlinfw.org/docs/basics/introduction.html.

[7] Schmitt, Betina Madeira, et al. "A Comparative Study of Cartesian and Delta 3D Printers on Producing PLA Parts." *Materials Research*, vol. 20, no. suppl 2, 5 Feb. 2018, pp. 883–886., doi:10.1590/1980-5373-mr-2016-1039.

[8] "3D Printer Calibration Guide – How to Calibrate Your 3D Printer." *All3DP*, 28 Sept. 2018, all3dp.com/2/how-to-calibrate-a-3d-printer-simply-explained/.

IOP Conf. Series: Materials Science and Engineering **652** (2019) 012059 doi:10.1088/1757-899X/652/1/012059

Milling of Precision Gear-shaped Parts

Yanqing Wu[1, a], Jinghua Chen[2,b]

[1,2] Research Institute of Physical and Chemical Engineering of Tianjin, Tianjin,China
[a,b] wuyanqing0821@126.com

Abstract: Precision gear-shaped part is an important part of a new type reducer. Due to its high overall hardness, precise design requirements, and semi-closed processing area, the precision machining technology based on UG is studied by using UG software in view of this processing characteristics of precision gear-shaped parts, including high resistance in X and Y direction, difficulty in chip removal, easy damage of cutter, and difficulty in workpiece quality. It solves the difficulties in cutting tool damage and precision control in precision gear parts machining, and provides reference value for similar parts machining.

1. Introduction

Precision gear-shaped part is an important part of a new type reducer. Due to its conditions of use requirements, the overall high hardness, design requirements, and semi-closed processing area, cutting tool in machining process of the X, Y, resistance and chip removal difficult, cutting tool is easy to damage, to guarantee the quality of workpiece precision tooth shape parts adopt gear shaper is the main processing method, grinding and EDM[1-3], but this three kinds of processing methods of workpiece structure has the certain requirement, can meet the processing requirements, all kinds of tooth structure with machining center machine tool overall performance and the improvement of process system, The method of using high speed milling to process tooth parts is gradually adopted in China, but the production cost of parts is higher by using this method. In this paper, UG software is used to study the milling process of precision tooth parts from the aspects of processing mode, processing type and cutting parameter setting, etc., which solves the processing problem of precision tooth parts and provides some reference value for similar parts processing.

2. CAD modeling of precision gear-shaped parts parts

UG software is a CAD/CAE/CAM integrated software, with strong 3d modeling and CNC programming capabilities, providing full digital support for product design, analysis, processing and assembly, and is widely used in general machinery, automotive transportation, aerospace and industrial equipment and other fields [2-6].

In the application of UG software for three-dimensional modeling of the gear-shaped parts, first of all, in the sketch drawing environment of UG software, according to its structure size, the section sketch of the gear-shaped parts is drew. Then, in the three-dimensional solid modeling environment, used the "stretch" command to stretch the sketch to generate the three-dimensional entity of gear shape. Finally, according to the part structure, the three-dimensional solid modeling of the tooth part is completed through commands such as bevel angle, hole feature and boolean operation, as shown in figure 1.

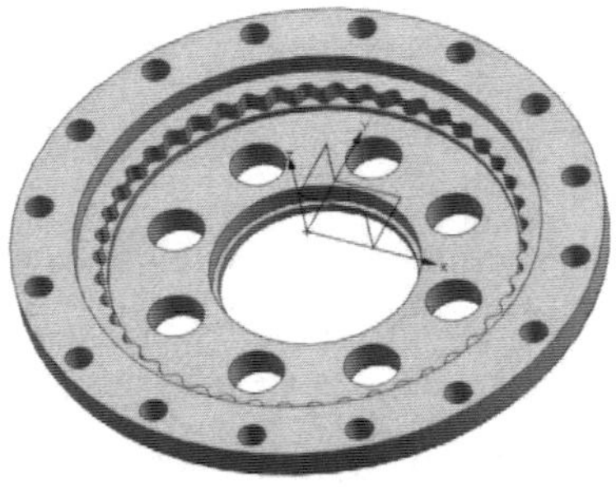

Figure 1. 3d model of the gear-shaped part

3. Process design of gear-shaped parts

3. 1 Processing technology analysis

It can be seen from figure 1 that this part has three processing elements: hole, gear shape and chamfering along gear shape edge. The hole is through hole, cutting processing is also good, the diameter of 8mm and 15mm, so can choose the right diameter of the end milling cutter direct processing. The interference between the cutter and the workpiece is mainly considered in the machining of chamfering along the gear edge to prevent over-cutting. The complex shape of the gear, high precision requirements, cutting technology is also poor, is the difficulty of the whole parts processing. According to the analysis and design of the CNC machining process of tooth parts as shown in table 1.

Table 1. machining process design of tooth parts

Processing steps	processing content	tool number	tool specification	process requirements
1	Gear rough machining	T01	Φ8 keyway milling cutter	To remove material, 0.5 mm allowance for semi-finishing, base amount
2	Gear semi-finishing	T02	Φ6 keyway milling cutter	To ensure the precision, 0.05 mm allowance for finishing, base amount
3	Gear finishing	T03	Φ6 keyway milling cutter	Ensure dimensional accuracy and roughness.
4	Chamfer along the gear	T04	Φ10 chamfer cutter	To ensure uniform chamfering
5	16-Φ8 hole machining	T05	Φ8 end milling cutter	
6	8-Φ15 hole machining	T06	Φ15 end milling cutter	

IOP Conf. Series: Materials Science and Engineering **652** (2019) 012059 doi:10.1088/1757-899X/652/1/012059

3. 2 Semi-finishing of gear shape

After rough machining, the rough machining allowance of the part shape is removed. Due to restrictions on the size of the parts, outline the minimum radius for R4mm, therefore, in order to ensure the tooth shape integrity and machining precision, semi-finishing choose 6 mm solid carbide keyway milling cutter, processing type also uses the UG type for deep processing is commonly used in finish machining contour (Zlevel_Profile), choose tooth part for machining area. Maximum cutting distance of 1.5 mm, choose suitable way of milling, cutting sequence set to depth-first, cutter path is set to the fairing, feed type is set to the arc, and in order to ensure consistency of machining dimension requirement, lay a foundation for finish machining quality, increase the cutting tool in cutting tool path compensation Settings, use the cutting parameters are: speed s of 5000 r/min, feed f for 200 mm/min, cutting depth a_p is 0.15 mm, tooth half finish machining tool path, as shown in figure 2.

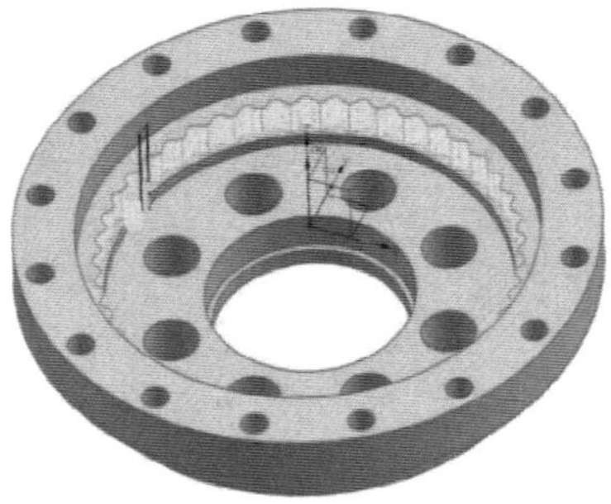

Figure 2. tool path for semi - precision machining of gear shape

3. 3 Finishing of gear shape

In order to keep the tooth shape integrity and machining precision, and machining efficiency of give attention to two or more things, finishing still chose and semi-finishing tool (Φ 6 mm solid carbide keyway milling cutter), also uses the UG processing type for deep processing is commonly used in finish machining contour (Zlevel_Profile), choose tooth part for machining area[7-8]. Maximum cutting distance of 9 mm, select inverse milling, cutting sequence set to depth-first, cutter path is set to the fairing, feed type is set to the arc, and to ensure consistency requirements of machining dimension and tolerance behavior such as vertical degree, coaxial degree requirements, increase the cutting tool compensation in cutting tool path Settings, and USES the whole tooth profile milling, cutting parameters for use: speed s of 4000 r/min, feed f 60 mm/min, cutting depth a_p is 0.05 mm, tooth profile machining tool path, as shown in figure 3.

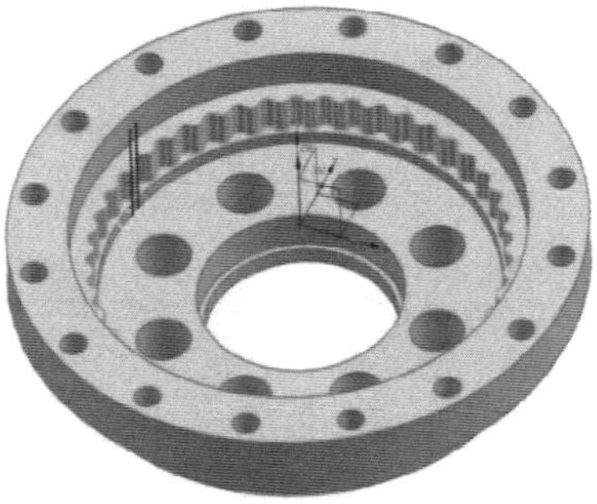

Figure 3. tool path of tooth finishing

3. 4 Chamfering along the gear shape

There is no special operation for chamfering in UG processing, and chamfering is often used for deburring, so the requirements are generally not high [9], but the chamfering size of this part is required to be high, because the processing quality of chamfering affects the efficiency of the machine after assembly, so it is necessary to achieve accurate chamfering, and chamfering along the tooth shape. The machining type is Face milling area, and the cutting mode is contour machining. The highest surface of the tooth shape is selected as the boundary of the part, and the highest surface is also selected as the bottom surface. The key to realize precise chamfering is to set the machining allowance, which is equal to - (D/4+C/2). The component allowance is the same as the bottom allowance. If want to see the processing effect, can change the angle of cutting tool angle to 45 °, chamfering processing cutting tool path as shown in figure 4.

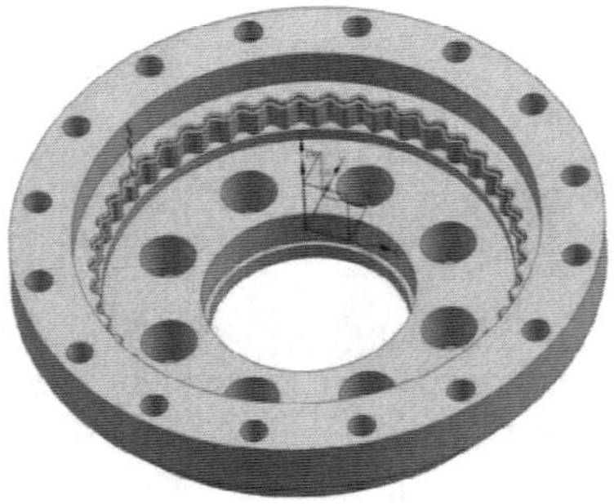

Figure 4. tool path of chamfering along tooth shape

3.5 Cutting simulation and machining

After completely generated the tool path, the processing simulation function of UG software is used to simulate the whole processing process of parts. And intuitively observe whether the tool line is reasonable, whether there is a collision in the processing process. After the simulation verification, select the tool path, use the post-processing function of UG software, import the special post-processing file of DMG 1035eco vertical machining center, generate the NC machining code, and then use the U disk to transfer the program to the machine tool for the actual processing. Small batch production shows that the process has high efficiency, stable quality and reasonable programming.

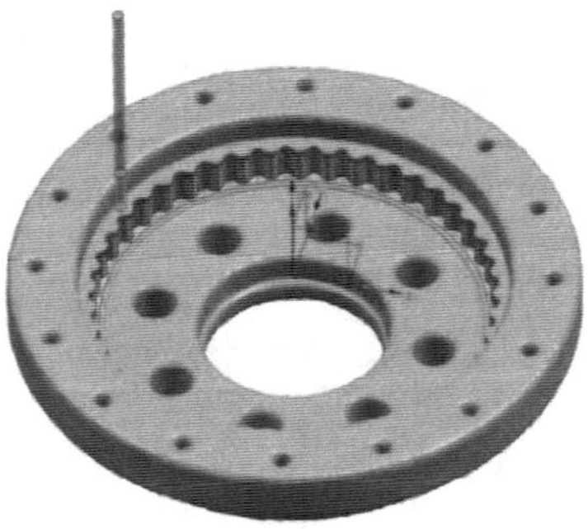

Figure 5. tool path simulation of precision tooth parts

4. Summary

IOP Conf. Series: Materials Science and Engineering **652** (2019) 012059 doi:10.1088/1757-899X/652/1/012059

This paper provides a general method and procedure for machining such precision gear-shaped parts. According to the structural characteristics of the gear-shaped part, the reasonable machining process was designed, and then the tool path and numerical control program were generated by using the CAM function of UG software. The rationality of the numerical control program was verified by the actual machining on the DMG 1035eco vertical machining center. It solves the difficulty of chip removal and quality control in the machining of precision gear-shaped parts, and provides a certain reference value for the numerical control machining of similar parts.

5. References

[1] Yuchuan Chu, Jianzhong Wei. Proficient in UG NX8.0 Chinese Version of CNC Machining [J]. Beijing: science press, 2013.

[2] Zhaofeng Zhou, Wensheng Xia, Chaoya Guo. Surface Parts are Processed by UG and NC Machining Center [J]. Coal mining machinery, 2014,35 (4) : 111-112.

[3] Guangming Chen. Research on Process Design Principles and Methods Based on Numerical Control Machining [J]. Manufacturing automation, 2005 (9) : 54-59.

[4] Jiping Lu, Shuiyuan Tang, Jianhua Zuo. Blade Modeling and Machining Based on UG [J]. Journal of Beijing institute of technology, 2007,24 (4) : 340-341.

[5] Xiaoying Chen, Cheng Xu. Application of UG Software in Numerical Control Machining [J]. Machine tools and hydraulics, 2006(1) : 64-66.

[6] Suyu Wang, Bin Jiang. Surface Design and Numerical Control Machining Simulation Based on UG [J]. Coal mining machinery, 2013,34 (2) : 141-143.

[7] Libo Zhou, Houjia Li. Processing Simulation and Post-processor Development Based on UG [J]. Manufacturing automation, 2009 (2) : 47-50.

[8] Xia Jin, Xiaojing Yang. Research on CNC Milling of Fixed Axis Surface Contour Based on UG/CAM Module [J]. Machine tools and hydraulics, 2007,35 (11) : 53-55.

[9] Guangqiang Dong. Research on Optimization of Relevant Parameters in CNC Milling Based on UG [J]. Machine tools and hydraulics, 2012,40 (3) : 80-83.

Utilization of fly ash: basic oxygen furnace slag as a raw material in geopolymerization

T Mashifana and N Sithole

University of Johannesburg, Department of Chemical Engineering Technology, P.O. Box 17011, Doornfontein 2088, South Africa

Email: tmashifana@uj.ac.za; nastassias@uj.ac.za

Abstract. The aim of this project was to develop a basic oxygen furnace slag (BOFS) and fly ash (FA) based geopolymer. The developed geopolymer were required to have strength to withstand hydraulic load. In respect to BOFS and FA, there are two main ways in which they can be handled following the best practices of conserving the environment. The two are disposal and utilisation, the disposal thereof is a problem because it increases the amount of FA annually since they are not consumed completely yearly. Results also showed that wastes beneficiate FA and BOFS can be utilised for other applications and this will reduce these wastes in the landfills. The development of the geopolymer to investigate the optimum conditions to enhance strength has resulted in a 20 % FA mass blend ratio, 9 M concentration of sodium hydroxide, 20 % liquids solid-liquid ratio and 80 °C curing temperature. The addition of hydrogen peroxide to the geopolymer increased the porosity of the geopolymer.

1. Introduction

In South Africa alone, 35 million tons of coal ash is generated annually, 29 million being fly ash and 6 million is bottom ash [1]. Kruger (2015) has stated that the main sources of fly ash (FA) are coal fired power stations and cement producers are the main consumers of FA. Kruger (2015) did a study of what FA is currently used for and has found that the cement making industry uses about 72% of the total FA volume (2.5 million tons) consumed yearly in South Africa. Out of the 29 million tons produces, only 2.5 million tons is consumed and the remaining annually piles up in our landfills. It is estimated that by 2020 42 million tons of coal ash will be generated yearly on average. Kruger (2015) has outlined the FA grades as unclassified, air-classified and twice-classified where they are can be characterised as sourced directly from power plant, sub 45μ and sub 8μ respectively. Generally the grades are based on the size of the FA particle. The finer FA is less likely to contain impurities, as opposed to bottom ash which is most likely to be mixed with other solids (e.g. any solid particle that could fall to the ash collection system). However, because it is too fine, it is the most problematic because there are limited ways in which one can handle it [2]. In South Africa, the process of steel making mainly takes place in a basic oxygen furnace. This process involves the mixing of iron ore, oxygen, limestone and coke as the feed materials [3]. While the goal is to produce steel, one of the waste by-product in this regard is basic oxygen furnace slag (BOFS). The BOFS results when the limestone is reduced to lime which reacts with the impurities like silica resulting in a slag which forms and floats on top of the molten iron [4]. BOFS is therefore a heterogeneous mixture of calcium, silica, manganese, sulphur and other impurities. Mahieux et. al (2009) has stated that the production of steel

results in 100 – 200kg of BOFS generation for every ton of crude steel produced [5]. This steel has a low market depand and mostly thrown into the landfills upon generation. According to [6], geopolymers are a new form of binder which means they can be considered as new materials for coatings and adhesives, new binders for composites and new cement for concrete. Yuanyuan (2015) has studied the removal of Cu2+ from aqueous solution using porous geopolymeric spheres [7]. His research has revealed that metakaolin-based geopolymeric spheres have been synthesised successfully. Other less frequent uses of FA include: water stabilization and solidification, mine reclamation, road sub base and geopolymerisation being the emerging use of FA [8]. Geopolymerisation involves chemically condensing the tetrahedral aluminosilicate units using alkali activators (metal ions), usually a soluble silicate, by virtue of balancing the charge associated with the tetrahedral aluminium molecule [9]. A pozzolanic reaction of the FA with moisture and an alkaline activator results in a FA slag, which could be relatively hard enough to be used as a membrane to treat mineral effluents was developed [10]. Novais, et al. (2016) have developed a FA based geopolymer for the adsorption of lead from waste water [11]. The experimental set up exhibited good results, with a removal capacity range of between 0.95 -6.34 mg lead per unit gram of geopolymer.
The synthesis of geopolymers follows one of two routes, acidic and as well as alkaline medium. The most commonly used route is the alkaline route [12] and this project will use sodium hydroxide. The process involves the dissolution of the aluminosilicate and then the condensation of the free aluminate and silicate units to form a structure of silico-aluminate which is three-dimensional [13]. This chemical reaction applies to both class F and class C FA with class C being the more spontaneous of the two [10]. Class C has some self-cementing properties, however, unlike class F, it contains reasonably higher alkali and sulphate contents and in the presence of water it hardens and gets stronger with time [10]. The application of geopolymers is categorized into three, namely, geopolymer resins and binder, geopolymer cements and concrete as well as arts and archaeology [6]. Owing to the pozzolanic nature of FA, the application explored in this project falls under the cement and concrete category because a slag will be casted using the procedure of brick making.
According to Davidovits (2015), geopolymers are a new form of binder which means they can be considered as new materials for coatings and adhesives, new binders for composites and new cement for concrete [6].
The production of FA and BOFS are also something that cannot be avoided, correct disposal measures or utilization needs to be studied in order to protect the environment. Rycroft (2017) has argued that the uses of FA are perceived as limited because of the lack of knowledge thereof. There are 150 known uses of FA (existing and emerging potential uses) [14]. These include, cement, concrete and brick making, geopolymers, zeolites, AME treatment and mine backfilling, rubber and plastic fillers, etc. [14]. Mahieux et.al. (2009) also argued the same for BOFS saying that it too has a low market demand [5]. The aim of this project is to develop a porous geopolymer which can withstand hydraulic load and also remove metals from acidic industrial effluents.

2. Experimental details

2.1. Material
There materials were used to synthesis the FA/BOFS based geopolymer. The FA was obtained from Camden Power Station and the BOF-slag from Arcelormittal South Africa. Deionised water was used and reacted with a strong alkaline (NaOH) and Sodium Silicate which (Na2SiO3) was supplied by Sigma Aldrich.

2.2. Sample Preparation
The two wastes fly ash and BOF slag were crushed then pulverised to ensure that the particle size is kept below 500μm. A solution of the NaOH and H_2O_2 was prepared using deionised water.

2.3. Geopolymer Synthesis

IOP Conf. Series: Materials Science and Engineering **652** (2019) 012060 doi:10.1088/1757-899X/652/1/012060

2.3.1. Effect of the blend ratio (FA: BOF-slag). This is the first and foremost step in the geopolymer synthesis. A total mass of 250g was weighed and the ratio varied from: 0 %, 20 %, 40 %, 60 %, 80 % and 100 % by mass of FA. 3 M solution of NaOH was prepared and mixed with the solids at a ratio of 25 % liquid to make a paste cast in the 50mm cube moulds. The paste was cast at room temperature for 24 hours and cured at the temperature of 80°C, over 6 days. The unconfined compressive strength (UCS) of the geopolymers was determined in order to find out the mix design that yielded the highest strength.

2.3.2. The effect of the sodium hydroxide (NaOH) concentration). After obtaining the optimum FA: BOF-slag ratio, this parameter was controlled throughout the experiment. The next step was to find the best sodium hydroxide concentration. The concentrations was varied from 3 M, 6 M and 9 M. These prepared solutions were mixed with the solids at a ratio of 25% to make a paste. The paste was also cast at room temperature for 24 hours and cured at 80°C for 6 days. The UCS of the geopolymers was determined to find out the NaOH concentration that yielded the highest strength.

2.3.3. The effect of solid liquid ratio. After finding obtaining the optimum NaOH concentration, this parameter was also controlled throughout the experiment. The FA and BOF-slag in their powder form was weighed and a solution of NaOH mixed at different solid liquid ratios. The ratios were selected based on workability of the paste. 15 %, 20 %, 25 % and 30 % by weight of NaOH solution was added to the solids to make the paste. This was conducted in order to vary the consistencies of the paste to test for the optimum solid-liquid ratio and this paste which will be poured inside the 50mm cube moulds.

2.3.4. The effect of curing temperature on strength. After determining the optimum solid-liquid ratio in respect to the obtained UCS, this parameter was controlled throughout the experiment. The curing temperature was investigated and the temperature varied to investigate the effect of temperature. The curing temperature was varied between 25 °C, 40 °C, 80 °C and 100 °C for a period of 6 days, thereafter the UCS the geopolymers was determined.

2.3.5. The effect of porosity on strength. The solid-liquid ratio that yielded the highest strength was cast again and varying volume of a blowing agent (hydrogen peroxide) was added to increase the porosity of the geopolymers. The volume of hydrogen peroxide was varied between 1 ml, 1.5 ml and 3 ml. The effect of open porosity was then analysed. The analysis was conducted by casting and curing the brick for a duration of seven days, weighing the brick and recording the mass as Mdg, soaking the geopolymer in deionised water for 24 hours, removing any access water on the surface of the geopolymer and weighing the geopolymer again to find Mwg. Then the open porosity was calculated using the following equation:

$$Open\ Porosity = \frac{M_{dg-}\ M_{wg}}{\rho_{water}\ x\ Volume\ of\ geopolymer} \quad (1)$$

The UCS tests were conducted again in order to determine the best strength when the geopolymer is porous.

2.3.6. Characterisation of fly ash and basic oxygen furnace slag. A uniaxial compressive strength (UCS) testing apparatus (Matest, spa treviolo) was used for testing the strength of the geopolymer. X-Ray fluoroscence (XRF: Rigaku ultima IV) was used to study the elemental composition of the geopolymer . The morphology of the geopolymer was studied by using a Scanning Electron Microscopy (SEM: Tescan Veg 3 XMU).

3. Results and discussion

3.1. Micrograph of FA and BOF slag (SEM analysis)

The samples were characterised individually, the SEM in figure 1 has shown that FA (A) exhibited a spherical shape while BOFS (B) had shapes ranging from subangular to sub-rounded. This agrees with literature as stated by [15-16].

3.2. Elemental composition of FA and BOF slag (XRF analysis)

Table 1 shows the elemental compositions of the two raw materials. According to Charles & Carraher (2016), the XRF results shows that the FA used is a Class C FA because it has a calcium oxide content greater than 10 wt % [10]. Li et.al. (2013) added that Class C FA results when a stronger coal is combusted [17]. This shows that the CaO content is important in enhancing the strength of the geopolymer. BOFS on the other hand has a greater CaO content, which plays a significant role in strength development of the material; resulting into a stronger geopolymer.

Table 1. Elemental composition of BOF slag and fly ash.

Component	Al_2O_3	SiO_2	CaO	Fe_2O_2	CuO	MgO	Other
Amount in FA (mass %)	20.34	45.00	12.55	14.60	0.01	1.60	5.90
Amount in BOFS (mass %)	4.21	10.15	47.50	27.73	0.00	3.63	6.78

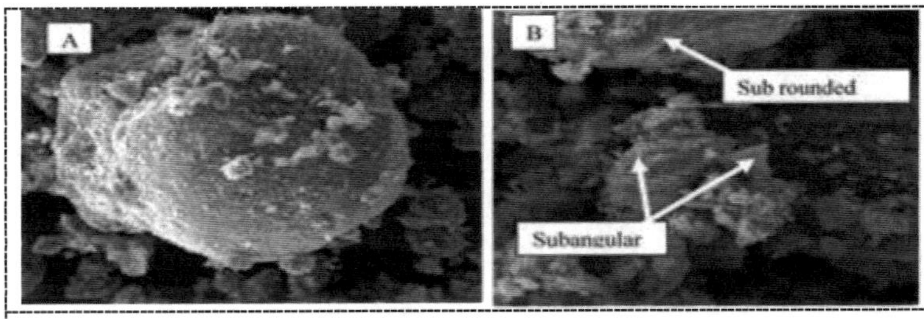

Figure 1. SEM image of (A) FA and (B) BOFS

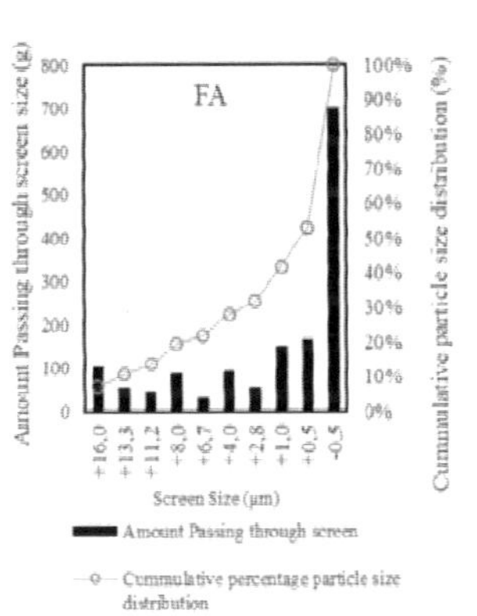

Figure 2(a). SEM image of FA..

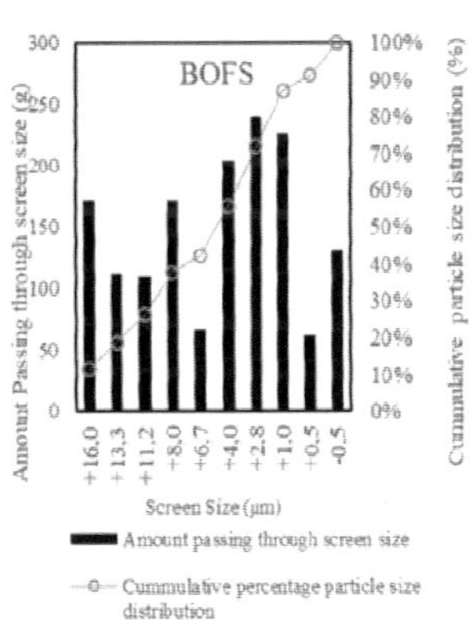

Figure 2(b). SEM image of BOFS.

3.3. Particle size distribution

The particle size distribution of the FA and the BOFS were analysed using a 1500g sample of each material. Figure 2(a) and figure 2 (b) shows the results of both the raw materials. FA was finer than the than the BOFS, with nearly half of the sample being less than 0.5μm. BOFS had approximately 100g of the 1500g sample being less than 0.5μm. This shows that there was a need to homogenise the samples in order to eliminate any error which could arise from the size distribution factor. The size reduction was achieved using a rod mill. The rod mill was loaded with 12 rods and to achieve size homogenising below 0.5μm the crushing was set for a duration of 30 minutes per 1000g sample. After the size reduction, all the sample managed to pass through a 0.5μm screen (i.e. 100 % recovery of desired product). In addition, crushing the BOFS results in more CaO being exposed and thus contributes to overall strength development of the geopolymer [16].

3.4. Effect of the BOFS: FA ratio, sodium hydroxide (NaOH) concentration and solid-liquid ratio on the UCS of the geopolymer

The results for the unconfined compressive strength (UCS) at different FA:BOFS, the effect of sodium hydroxide concentration and effect of solid-liquid ratio is shown in Figure 3, Figure 4 and Figure 5, respectively. The geopolymer developed with a 0 % FA (100 % BOFS) had a higher strength as opposed to the one with 100 % FA. The XRF results of these geopolymers have shown that 100 % FA has a high CaO content which is believed to enhance the strength of geopolymers [10]. The different ratios have resulted in 20 % FA yielding the highest strength and has a reasonably high amount of CaO as well. The increase in the concentration of sodium hydroxide has resulted in the increase of the UCS.

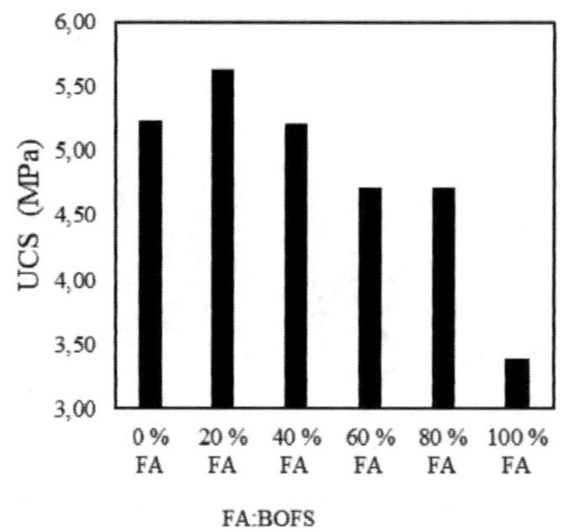

Figure 3. UCS at varying FA:BOFS.

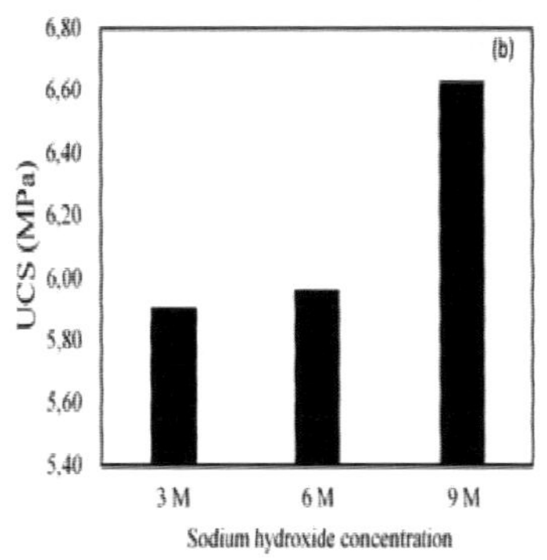

Figure 4. UCS at varying NaOH concentration.

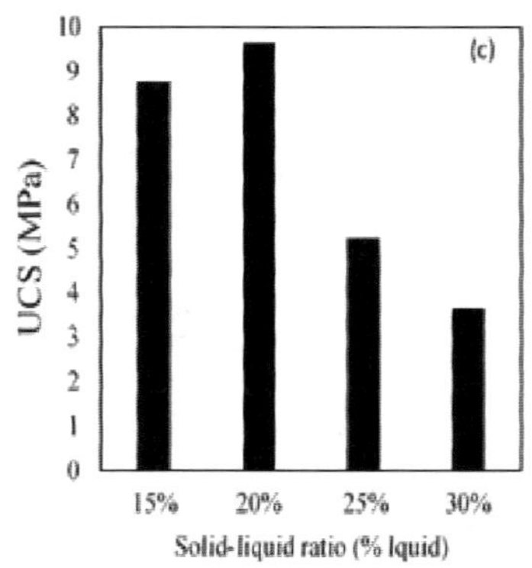

Figure 5. UCS at varying solid-liquid ratio.

This supports the conclusion reached by [10] who stated that the pozzolanic reaction of the FA with moisture and an alkaline activator results in a FA slag which could be relatively hard enough to be used as a backfill paste. The use of the alkaline medium (NaOH) promoted polymerisation of the aluminosilicates resulting in stronger bonds. In this experiment, the concentration was limited at 9M and greater concentrations were not investigated in order to make the project feasible when commercialised.

The solid-liquid of 20% resulted in the highest UCS. The SEM of the geopolymer resulting from the different solid-liquid ratios is presented in Figure 6.

The study of the optimum solid-liquid ratio is highly influenced by the workability of the paste created. When making the paste in the nominal solid-liquid ratios, 15 % liquid was too dry, and the liquid medium was not enough to allow interactions with the FA and BOFS and the alkaline solution. 20 % and 30 % solid-liquid ratios were reasonably workable and had a texture of plastering cement (construction industry).

The 30 % solid-liquid ratio was too wet and not workable, hence there was no need to investigate it any further. The 20 % solid-ratio was the optimum amongst the two workable solid-liquid ratios (20 % and 30 %), having the highest strength. The SEM results show increasing 'white particles' moving from A to D. These 'white particles' can be considered as the moisture and thus an increase in moisture is inversely proportional to the strength as supported by the workability and the UCS results in figure 5.

3.5. The effect of curing temperature on UCS and hydrogen peroxide on the open porosity and UCS

The UCS test was conducted on the geopolymers to find out which curing temperature gives the highest strength and to determine the effect of hydrogen peroxide on the open porosity and UCS. The obtained results are presented in Figure 7 (a) and Figure 7(b). The strength of the geopolymer tends to increase as the temperature increases. However, increasing the temperature beyond 80 °C decreases the strength again.

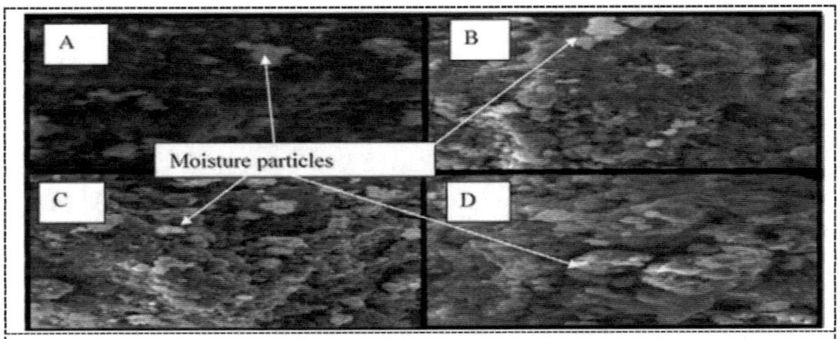

Figure 6. SEM image of BOFS.

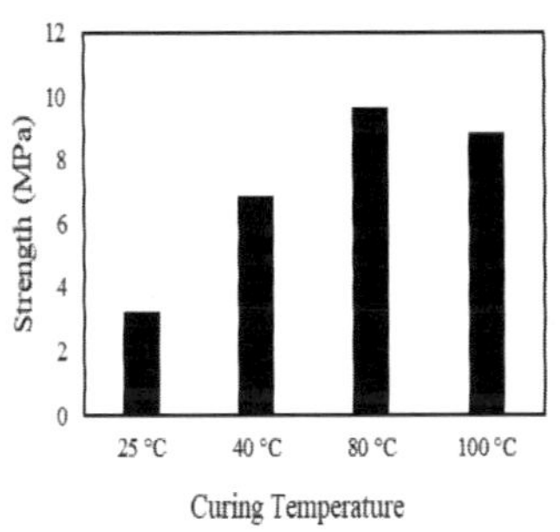

Figure 7(a). Effect of curing temperature on the UCS.

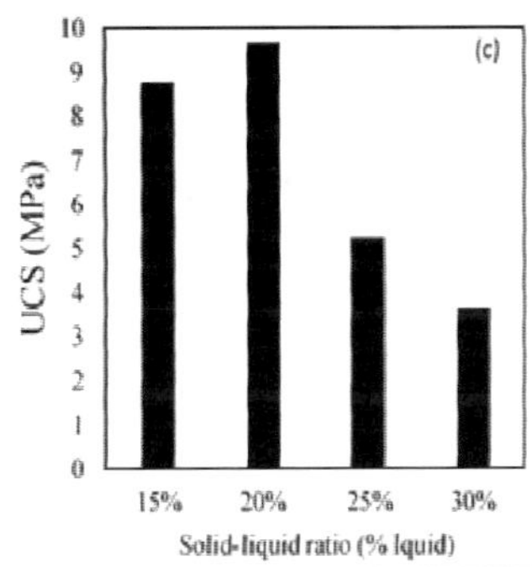

Figure 7(b). Effect of hydrogen peroxide on the open porosity and UCS.

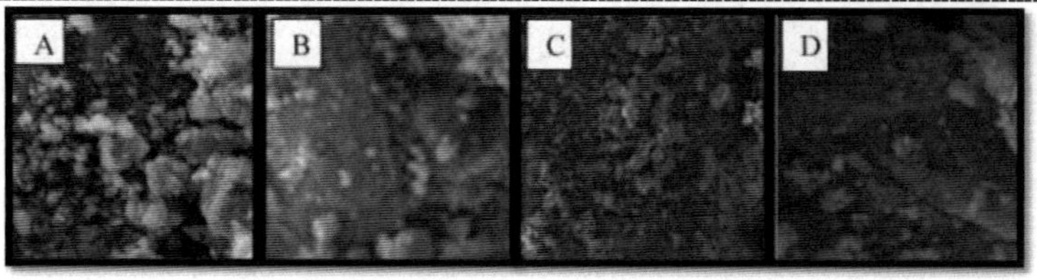

Figure 8. SEM of developed geopolymers.

This means that 80 °C is the optimum curing temperature for the geopolymerisation of FA: BOFS. The open porosity of the geopolymers generally increases with an increased hydrogen peroxide dosage. This is because hydrogen peroxide is considered a blowing agent in the process of geopolymerisation, therefore, increasing the dosage of hydrogen peroxide makes the geopolymer more porous.

The SEM of the developed geopolymers in Figure 8 shows the effect of temperature as it was increased. The moisture particles decrease from A to D due to the temperature. There were cracks observed on the geopolymer, the weaker geopolymers was seen to have cracks. No cracks were found on C which justifies why it had the highest strength.

4. Conclusion

The characterisation of the geopolymer has proven to align with literature. The SEM results has confirmed that FA is spherical and BOFS ranges from sub-rounded to sub-angular. The XRF results have shown that FA is an aluminosilicate material. BOFS had a high relative proportion of CaO, a constituent that contributes to material strength. The particle size distribution and the moisture content of the FA and BOFS have resulted in a need for further homogenising and to be cured at elevated temperatures respectively.

The geopolymer synthesis process has resulted in reliable optimum conditions and was supported by UCS, SEM and XRD analysis. The blend ratio proved to be influenced by the CaO content and the optimum was found to be 20 % FA, with a relatively high amount of CaO. Increasing the concentration of sodium hydroxide resulted into the increase in UCS, however, the maximum concentration was limited at 9 M for this project. The solid liquid ratio was influenced by the workability of the paste and out of the two most workable pastes (20 % and 25 %), 20 % had the highest strength. The optimum curing temperature that yielded the highest geopolymer UCS was at was 80 °C. It was observed that 25 °C, 40 °C and 100 °C had cracks on their surfaces which explains why the strength was lower.

The effect of porosity was negative on the development of the geopolymer as it decreased the strength when hydrogen peroxide was increased. The porosity increased with an increased dosage of hydrogen peroxide. South African waste fly ash and basic oxygen fly ash can be used to make a geopolymer. This results presented provide an opportunity for utilisation of wastes as surface capping material or as paste backfill. Curing over longer period at room temperatures should be investigated. This will allow for reduction in energy consumption.

5. Acknowledgement

The authors would like to thank the University of Johannesburg and National Research Foundation of South Africa for the financial support.

6. References

[1] Kruger RA 2015 *World of Coal Ash.*
[2] Nawaz I 2013 *Int J of Innov Res Sci Eng Techn,* **2** 10
[3] Simmons P, Ziemkiewicz P and Black DC 2002 *Mine Water Environ.* 21
[4] Westholm E 2014 *Environ. Sci. Pollut. Res. Int.* 21

[5] Mahieux PY, Aubert JE and Escadeillas G 2009 *Constr. Build. Mater.* **23** 2
[6] Davidovits J 2015 *Institut Géopolymère*
[7] Ge Y, Cui X, Kong Y, Li Z, He Y, Zhou Q 2015 *J. Hazard. Mater.*
[8] Behera RK 2010 *Min. Eng. Dept : Nat Inst Technol.*
[9] Feng D, Provis JL and Van Deventer JS 2012 *J. Am.Ceram Soc.* **95** 2
[10] Charles E, Carraher J 2016 *CRC Press.*
[11] Novais RM, Buruberry LH, Seabra MP and Labrincha JA 2016 *J. Hazard Mater.* 318
[12] Kim D, Lai H, Chilingar G and Yen T 2006 *Environ. Geol,* **51** 1
[13] Al-Bakri AM, Kamarudin H, Bnhussain M, Nizar IK, Rafiza AR, Izzat AM 2011 *Aust. J. Basic. Appl. Sci.*, **5** 7
[14] Rycroft M 2017
[15] Speight JG 2017 *USA: Butterworth-Heinemnn.* (2017)
[16] Yildirim H, Sümer M, Akyüncü V and Gürbüz E 2011 *Const. Build. Mater*, **25** 6
[17] Li X, Ma X, Zhang S and Zheng E 2013 *Mater* **6** 4

IOP Conf. Series: Materials Science and Engineering **652** (2019) 012061 doi:10.1088/1757-899X/652/1/012061

Alkaline solidification of gold mine tailings for production of lightweight masonry blocks

.N.T Sithole[*] and TP Mashifana

Chemical Engineering Technology, University of Johannesburg, South Africa
[*]Email: nastassias@uj.ac.za

Abstract. South Africa is the world's largest gold resource. This is due to the substantial amount of gold that exists in the Witwatersrand Basin. The processing of gold leads to the production of gold mine tailings. Gold mine tailings are generated from mineral processing of gold ore through which gold is separated. Mine tailings are generally waste materials and are normally disposed of in slurry form in storages constructed on huge areas of land, these storage facilities are called tailings dams. This study was conducted to strengthen and stabilize gold mine tailings by alkaline activation to produce masonry blocks that can be used for building and construction as per minimum requirements for ASTM C34-13, C129-14a and South African standard (SANS227: 2007). The effect of variation of Solid to liquid (S/L) ratio, NaOH concentration and temperature on the UCS was investigated. This was achieved by treating the gold mine tailings with NaOH solution. The feasibility of using gold mine tailings to produce lightweight masonry blocks was studied by conducting unconfined Unconfined Compressive Strength (UCS) tests, SEM imaging, and XRF and XRF analysis. The alkaline activation of Gold Mine Tailings (AU MT) at a S/L ratio of 14.31% and 9 M concentration had higher UCS than those produced at 3 M and 6 M. Curing temperature is an important factor affecting alkaline activation and the UCS of masonry block. The UCS increases with the curing temperature up to a certain level and then decreases with the curing temperature. For the AU MT herein, the optimum curing temperature was found to be 80 °C with bulk density of 1578 kg/m3. The AU MT based masonry block meet ASTM standards they can be used for building and construction purposes such as the building of structural clay load bearing wall tiles of all C34-03 ASTM designation grades and building blocks of ASTM designation C6210, type NW (negligible weathering) with the minimum UCS of 10.3 MPa.

1. Introduction

The mining of gold in South Africa continuously results in massive volumes of waste material, primarily in the form of tailings material [1]. Gold mine tailings are generated from mineral processing of gold ore through which gold is separated. Mine tailings are generally waste materials and are normally disposed of in slurry form in storages constructed on huge areas of land, these storage facilities are called tailings dams. The continuous storage of gold mine tailings in dams has foremost disadvantages that include failure of the storage dam, release of heavy metals due to acid mine drainage (AMD), surface erosion, costly construction, high maintenance and dust generation [2]. Gold mine tailings contain process water, gangue minerals, heavy metals and any harmful substances used in the beneficiation of ores. They are generally discarded into tailings dams in slurry shape with everything they contain. A Mintek database states that there are in excess 446 gold tailings dumps that extend to over 18 000 hectares crosswise over Johannesburg alone [3]. There are over 32 gold mines in the country, each generating their own fair share of waste mine tailings storing them in dams. The disposal of mine tailings is a liability as the mining activities are expanding and becoming bigger [4]. Mine tailings can cause water contamination, emerging from the release of water polluted with solids, heavy metals, process reagents, sulphur compounds, and so forth. Furthermore, failure of steep-sided tailing dams can occur because of seismic tremors, blasting vibration and extreme climate [4]. Previous research on alkaline activation of waste material to produce a building and construction material mostly focused most on using metakaolin, fly ash, and slag as the raw material for production of building material. Limited research has been reported on alkaline activation of AU MT for potential use in building and construction. Due to the high content of silica and alumina of Au MT, this study focuses on utilizing AU MT as the potential raw material to produce a building and construction material after alkaline activation. The very few researchers that investigated possible applications of beneficiation of the aforementioned waste materials in building and construction are; Kiventerä et al. [5] investigated the utilization of sulphidic gold mine tailings as a raw material in geopolymerization using NaOH as the alkaline activator. The results show that the it is possible to alkaline activate sulphidic mine tailings using 5 M NaOH without any co-binders, to produce building material with compressive strength up to 3.5 MPa. However, the downside is that water permeability is high indicating that the matrix is weak. Due to the chemical characteristics and geotechnical properties, most of the gold mine tailings require modification before the material can be used for any application. Zhang et al. [6] studied the synthesis of Fly Ash modified gold mine tailing geopolymers using NaOH as a alkaline activator. The results revealed that the compressive strength of the MT-based geopolymer can be controlled by adjusting the FA content and NaOH concentration. The MT-based geopolymers synthesized by varying these factors show UCS ranging from 1.37 to 21.2 MPa. So, the MT-based geopolymers are a viable and promising construction material which can be tailored for different applications. Rachman et al. investigated the development of Portland cement modified traditional gold mine tailings based solidification. The result demonstrated that, the optimum ratio to mix Portland cement and tailings was 10:90, with compression test of 257ton/m2 and Toxicity Characteristic Leaching Procedure (TCLP) test was 0.0069 mg/L. The compression test results were in accordance to US EPA Standard quality of 35 ton/m2. Based on these studies the feasibility to alkaline activate Western basin mine tailings (South Africa) was explored. Alkaline activation was employed as a technology to stabilize and solidify gold mine tailings to use them as a dominant if not the only material for the production of masonry blocks.

2. Materials and Methods

Gold tailings samples were obtained from a Gold Mining company in the Western basin of South Africa. The The alkaline activator NaOH was supplied by Rochelle. X-ray Fluorescence (XRF-Rigaku ZSX Primus II) was used to determine the elemental composition of the material, X-ray Diffraction (XRD-Rigaku) to study the mineralogy of the material and Scanning Electron Microscope (SEM, Tescan Vega 3 XMU) for the morphology. The samples of gold tailings were dried in an oven at the temperature of 50 oC before conducting any experimental work. The material was then pulverized to fine particles size of -75µm. Alkaline activation of the gold mine tailings were carried out varying the NaOH concentrations, S/L ratio stirring and curing temperature. The developed specimens were cast and cured at different curing ages.

3. Results and Discussion

3.1. Effect of NaOH and S/L ratio on UCS

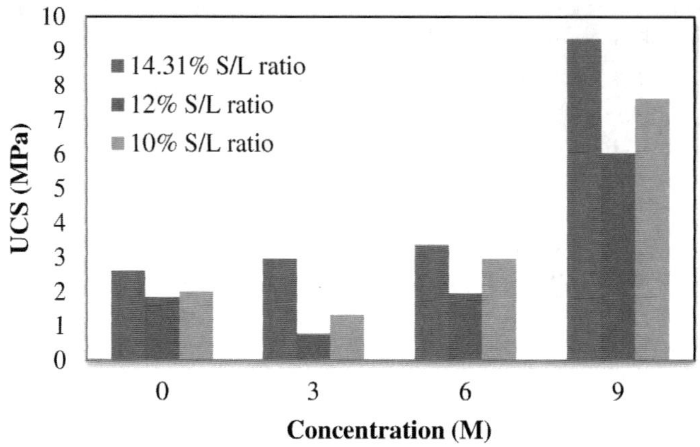

Figure 1. The effect of NaOH concentration and S/L on the UCS

Figure 1 shows the UCS of the masonry blocks after 7 days curing which shows that there was an increase in UCS with increase in NaOH concentration. The 9 M NaOH concentration gave the highest UCS; and was concluded to be the optimum concentration. This is due to more dissolution of aluminosilicates by using the 9 M NaOH solution. There was also an increase in UCS with an increase of water content. This was due to lower water content the paste was stiff reducing the time for dissolution of aluminosilicates and at higher water content the workability was high allowing more time for dissolution of aluminosilicate, which in turn enhances the UCS of the masonry block.

3.2. Effect of Temperature on UCS

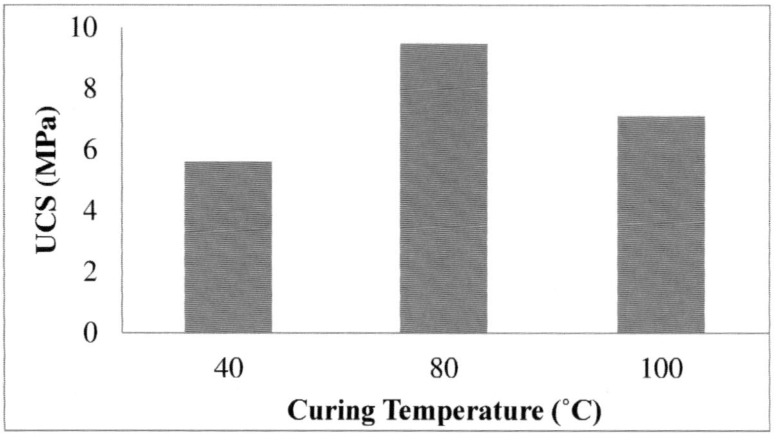

Figure 2. The effect of temperature on the UCS

As the 9 M NaOH concentration with 14. 86 % water content showed the highest UCS; these conditions were used to investigate the effect of temperature variation. Figure 2 shows that as the curing temperature increases so is the UCS up to 80 °C and then there was a decline in UCS when the temperature was increased to 100 °C. The change of UCS with curing temperature can be explained by the underlying mechanism in alkaline activation. This is due to the increase in curing temperature helps accelerate the dissolution of silica and alumina species and then polycondensation. As the temperature increases to 100°C, the polycondensation process becomes so fast and rapid formation of geopolymeric gel will hinder further dissolution of silica and alumina species and thus affect the UCS unfavorably [4]. Furthermore at elevated temperatures there rapid moisture loss that leads to incomplete alkaline activation. The masonry blocks cured at 40 ˚C still showed traces of moisture content which means the masonry block was not dry enough. Therefore the optimum drying temperature was 80 ˚C since it gave the highest UCS.

3.3. Effect of Curing Period on UCS

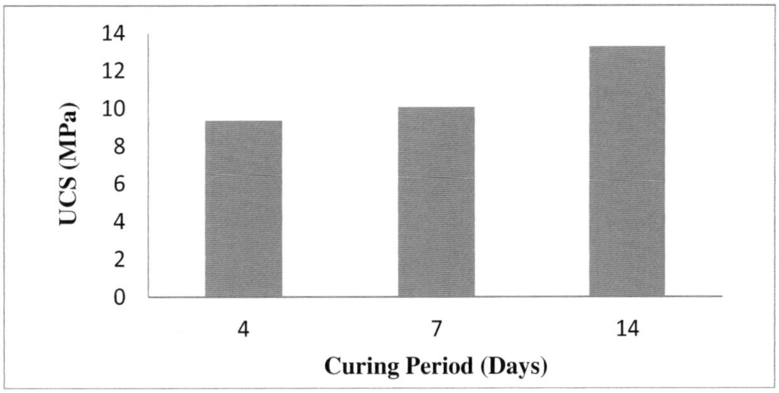

Figure 3. The effect of curing age on the UCS of the masonry blocks

Figure 3 shows that there was an increase in UCS with an increase in curing period. The results show that the masonry block reached their ultimate UCS in 14 days and a major portion of the UCS (about 76%) is gained within only 7 days and in 14 days the UCS reached 13.26 MPa. This means that enough time was allowed for dissolution of aluminosilicate in the geopolymeric gel hence the UCS was high compared to 4 and 14 days. The masonry block cured for 14 days at 80 °C satisfied the requirements of ASTM C34-13 with bulk density of 1578 kg/m^3.

3.4. XRF analysis before and after alkaline activation

Table 1. the effect of alkaline activation on chemical composition of Au MT

Component	Raw Au MT (Mass %)	Au MT after alkaline activation
Na_2O	0.3511	19.3434
MgO	3.7519	4.0103
AL_2O_3	10.9919	8.4790
SiO_2	45.5128	34.6241
P_2O_5	0.0836	0.0473
SO_3	4.0559	2.2666
Cl	0.0184	-
K_2O	4.0559	5.8760
CaO	8.7444	7.3728
TiO_2	0.9363	0.8128
Cr_2O_3	0.4254	0.4256
MnO	0.3998	0.4499
Fe_2O_3	14.8064	15.0894
CO_2O_3	0.0422	-
NpO	4.1577	-
CuO	0.0203	0.1230

ZnO	0.0265	0.0283
As_2O_3	0.7512	0.6644
Rb_2O	0.0361	0.0420
SrO	0.0272	0.0339
ZrO_2	0.0510	0.0446
Nb_2O_5	0.0068	-
BaO	0.1793	0.2479

Casting and curing at optimum conditions 9 M NaOH, 14 days curing period, 80 °C curing temperature and 14.37 % solid to liquid ratio shows that there was drastic change in the chemical composition of the masonry block. SiO2 changed from 45.51% to 34.62%. While Na2O changes from 0.3511% to 19.34%. The increase in Na2O was due to reasonable amount of a NaOH solution was introduced into the raw tailings during alkaline activation. These results reveal that there was enough dissolution of aluminosilicates, which resulted in more complete alkaline activation (Table 1).

3.5 FTIR analyses of gold mine tailings before and after alkaline activation

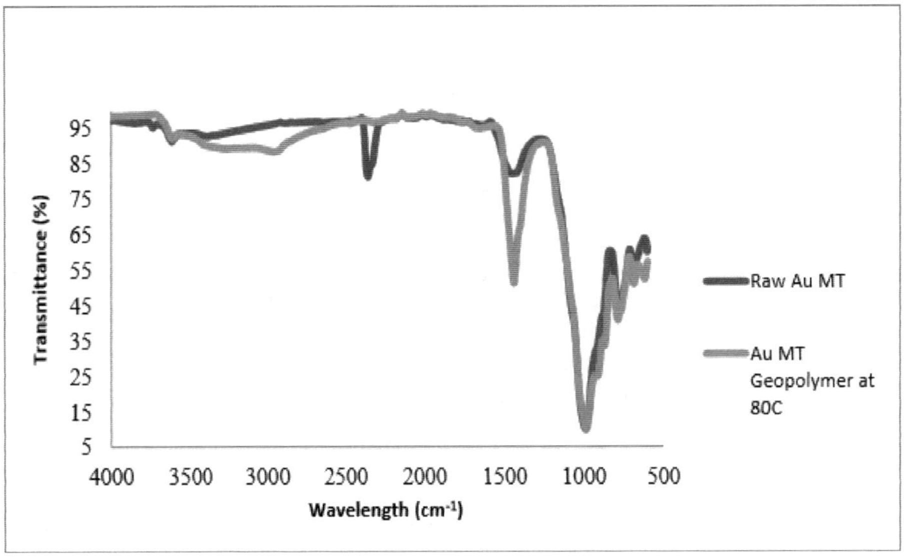

Figure 4. FTIR analyses of gold mine tailings before and after alkaline activation

Figure 4 shows the FTIR of gold mine tailings before and after alkaline activation. The raw gold mine tailings showed that the transmittance dropped until 81% at the wavelength of 2357 cm-1. The broad bands appearing at 1000 - 2535 cm−1 were due to stretching vibrations OH and HOH [5]. At 1442 cm^{-1} the transmittance for the alkaline activated sample was 52%.

3.6. The morphology of Gold mine tailings before and after alkaline activation

Figure 5 shows the SEM micrograph of Au tailings before alkaline activation. Figure 6 shows the effect of alkaline activation on the morphology of Au tailings. It can be seen that alkaline activation resulted in a dense construction block with some unreacted white particles revealing that alkaline activation was incomplete. The existence of pores was detected in this sample.

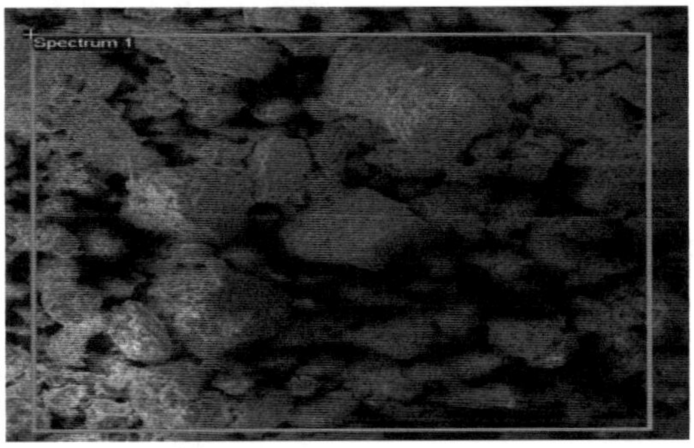

Figure 5. SEM for Raw Au Tailings

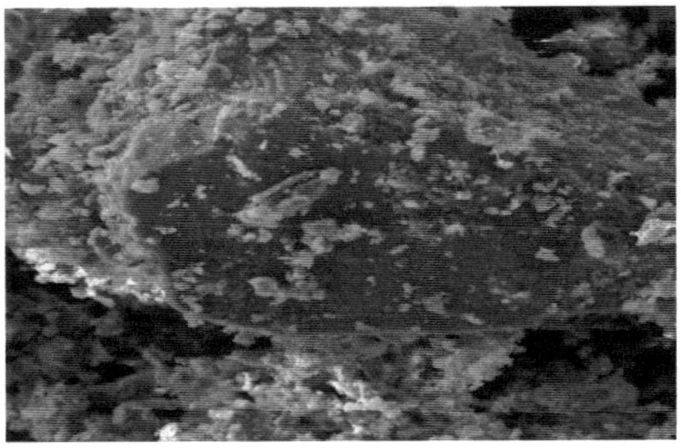

Figure 6. SEM after alkaline activation of Tailing at 80 °C

4. Conclusion

Based on the experimental results the optimum conditions to develop a stabilized and solidified mansonry block with high UCS was at 14.31% S/L and 9 M NaOH concentration and 80°C curing temperature. This is due to a decrease in S/L ratio results in low workability of the paste which in turn decrease the UCS. Higher NaOH concentration provides larger amount of NaOH at a certain initial water content required for alkaline activation of gold mine tailings. Curing temperature is an important factor affecting the alkaline activation process and the UCS of masonry block. The UCS increases with the curing temperature up to a 80°C na d decreases. With regards to curing period, masonry blocks reached their ultimate UCS within 14 days and a major portion of the ultimate UCS (about 90%) is gained within only 4 days. Therefore the produced masonry block meet ASTM standards and can be used for building and construction purposes such as the building of structural clay load bearing wall tiles of all C34-03 ASTM designation grades and building blocks of ASTM designation C6210, type NW (negligible weathering) with the minimum UCS of 10.3 MPa [6].

References

[1] T. Rosner., The environmental impact of seepage from gold mine tailings dams (1999).
[2] S. Ahmari, & L. Zhang, Constr. Build. Mater 40, 1002-1011 (2013).
[3] X.S. H Guo., Cem. Concr. Compos 32 (2), 142-7 (2010).
[4] M.S., Muñiz-Villarreal, A., Manzano-Ramírez, S., Sampieri-Bulbarela, J. R., Gasca-Tirado, J.L., Reyes-Araiza, J. C., Rubio-Ávalos, J.J., Pérez-Bueno, L.M., Apatiga, A., Zaldivar-Cadena, and V., Amigó-Borrás . Mater Letters 65(6), 995-998 (2011).
[5] J., Kiventerä, L., Golek, J., Yliniemi, V., Ferreira, J., Deja, and M., Illikainen,. Int. J. Miner. Process 149, 104-110 (2016).
[6] L., Zhang, S., Ahmari and J., Zhang,. Constr. Build. Mater 25(9), 3773-3781

Applications of UAS-obtained thermal images for vegetation coverage ratio monitoring of mudstone areas

Chien-Yuan Chen[1,*] **and Chai-Li Lin**[1]

[1] Dept. of Civil and Water Resources Engineering, National Chiayi University, Chiayi City 60004, Taiwan R.O.C.

Email: chienyuc@mail.ncyu.edu.tw; s1050428@alumni.ncyu.edu.tw

Abstract. Soil vegetation coverage is a crucial factor for those wishing to protect naked slopes. Thermal imaging enables monitoring of the changes in infrared temperature of a slope surface and calculating the vegetation coverage ratio (VCR). The VCR is an efficient method for determining infrared temperature changes in different slope lands, which include regions exposed to sunlight, vegetation areas, and shaded areas. The thermal images capture the infrared temperature distribution, boundary of vegetation of temperature, and barren vegetation areas. The VCR can be subsequently calculated. In this study, thermal images were obtained using an unmanned aircraft system (UAS) according to airline planning, frame size of an image, and overlapping of lateral and front images for taking pictures. Moreover, 3D images can compare the VCR by calculating the aggregation of the study area and vegetation area. A satisfactory agreement was obtained between the infrared temperature obtained through thermal imaging and the VCR. Thermal imaging can thus be used to monitor changes in the VCR of mudstone. Moreover, a 3D topographic map was created using the UAS-obtained images to overlap with the thermal images.

1. Introduction

Mudstone in Taiwan is a type of mudrock, which is a component of clayey materials of sedimentary rock. Mudstone has poor cementation and is prone to water-induced collapse, inducing mudflow. Taiwan is located in a track of typhoons, and the mudstone in southwestern Taiwan is prone to collapse when typhoons bring torrential rainfall. Weathered-mudstone soil contains highly soluble salts, causes soil layer dispersion, and is prone to erosion by rainfall. Vegetation is the primary measure used to prevent bald mudstone-induced disasters. The vegetation coverage ratio (VCR) is a crucial index indicating the degree of protection of mudstone slopes.

The southern aspect of a mudstone slope has the highest frequency of erosion among all aspects for mudstone in the southwestern area in Taiwan. The temperature changes in naked mudstone depend on the intensity of solar radiation. When the incident angle of sunshine is approximately equal to 90°, a mudstone slope receives the most solar radiation, which abruptly increases infrared temperature [1]. Figure 1 shows the maximum surficial temperature at different incident sunlight in mudstone area [2]. The normalized difference vegetation index (NDVI) is a simple graphical indicator used to analyze remote sensing images. The NDVI quantifies vegetation by measuring the difference between spectral reflectance in the near-infrared (NIR) region and that in the red light (RED) region. Vegetation strongly reflects NIR light but absorbs red light. The formula for calculating the NDVI is as follows [3],

$$\text{NDVI} = (\text{NIR} - \text{RED}) / (\text{NIR} + \text{RED}) \quad (1)$$

The index's value is between −1 and 1. The index is 0 for naked areas, indicating a reflection rate close to those of soil and rock. The NDVI value for vegetation areas is between 0 and 1 and increases with vegetation density. Nonvegetation areas such as areas of cloud, water, and ice have negative NDVI values.

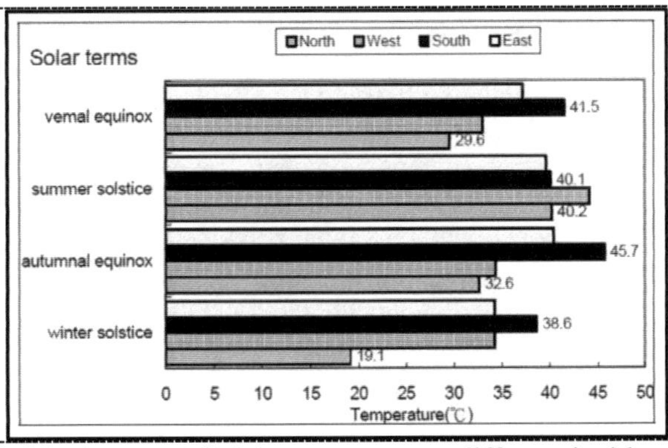

Figure 1. Maximum surficial temperature at different incident sunlight in mudstone area [2].

NDVI interpretation of satellite images has coarse spatial resolution (20 m in SPOT images), and these interpretations are easily affected by clouds and shadows [4]. However, evaluating how the NDVI changes over time is impossible at this stage. This study aimed to monitor the NDVI by using unmanned aircraft system (UAS)-obtained thermal images for high precision and rapid interpretation, thus enabling determination of the VCR.

A thermal image presents energy transfer through imaging technology by detecting infrared thermal radiation and projecting the energy onto a detector by using a lens. The detector then converts the infrared energy into a digital signal, and a thermal image is obtained on a screen [5]. Testing using a thermal imager is a noncontact and nondestructive rapid testing methodology for detecting surficial infrared temperature changes in objects. Objects exposed to sunshine in the morning exhibit a surficial increase in temperature until noon, and temperature decreases during the night until sunrise. The use of thermal image technologies for infrared temperature monitoring has several advantages, including remote, large-area monitoring, high visibility, quick response, light weight, and digital operation [6].

The applications of thermal imagers range from military industries to medicine, science, and engineering. The technology of the thermal imager has been used for landslide monitoring and is combined with remote sensing [7]. Moreover, it is used for geologic investigation and shore cliff slope stability evaluation [8].

UASs have been used for military purposes since 1917. The two primary categories of UASs are those for fixed-wing planes and those for rotor-type planes. UASs have been widely used for the investigation of postdisaster events and disaster prevention monitoring. The systems are lightweight, small, highly mobile and safe, easy to repair, and low-cost. A UAS can be used in high-risk areas.

UASs have been widely used in several fields, such as for search and rescue, forest soil and water conservation [9], and investigation of flood and debris-flow disasters [10–11]. UASs fitted with thermal imagers have been developed in recent years and could provide useful information for landslide monitoring and analysis.

2. Study area

The study area was located in the mudstone area in Tainan city, southern Taiwan (Figure 2), in the range between 22°55'33.3" N and 120°24'19.5" E. The area has been designed a special soil and water conservation area since 1998. The elevation of the study area is between 100 and 150 m.

Remote sensing images showed that the area had sparse vegetation between 2010 and 2013, resulting in erosion caused by torrential rains. The vegetation coverage increased in 2015; however, the increase was not substantial.

2.1. Geologic and hydrologic conditions

The geological strata of the watershed are primarily sandy yellow soil and mudstone and were created during the Gutingkeng Formation between the Late Miocene epoch and the Pleistocene (Figure 3). The primary contents of the formation are fine granules of grey mudstone and loose bedding sandstone and interbedded limestone. The mudstone is impermeable and the land became badlands after the mudstone underwent weathering.

Taiwan is located in a subtropical area and has frequent typhoon landings. It experiences monsoon-rain-induced disasters. The rainfall starts to increase in May and is highest in August. January to March is the dry season.

Figure 2. Location of the study area in Tainan City, southern Taiwan (Google Earth).

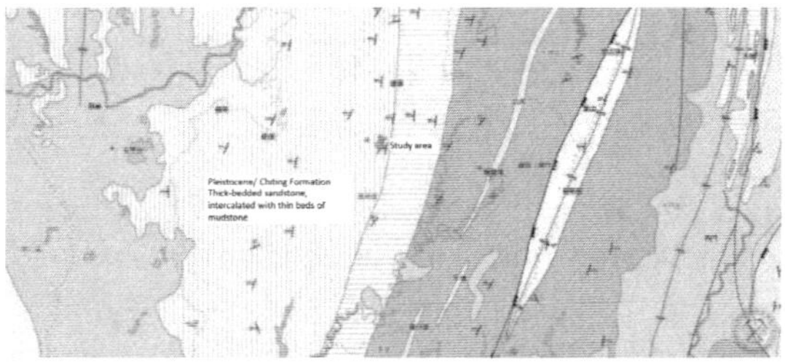

Figure 3. Geological map of the study area [12].

3. Method

NEC F30W was the thermal imager used in this study. The size of the imager was 100 × 65 × 45 mm^3, and its weight was 300 g. The UAS fitted with the NEC F30W thermal imager could measure infrared temperatures between −20 and 350°C. The two global positioning systems (GPSs) of the UAS had an autopilot function and predefined flight path planning (Figure 4).

Figure 4. Map of the study area and flight path of the UAS.

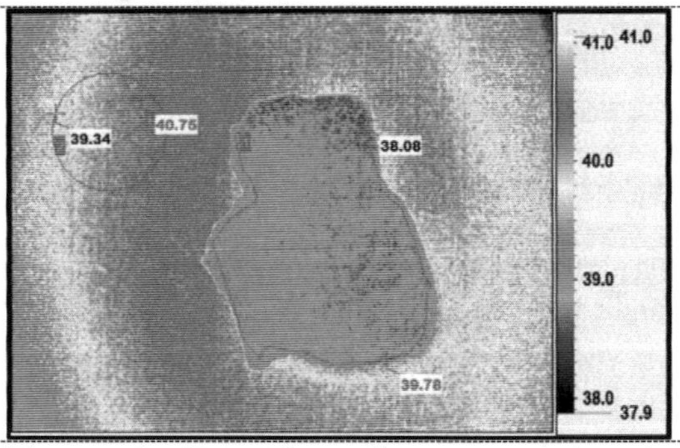

Figure 5. Examples of postprocessing in thermal image analysis.

A postprocessing package was used for thermal image analysis. The package provides thermal image emissivity correction, image correction, autocalculation of the emissivity of the entire area, partial enlargement, 90° rotation, multipoint temperature measurement, and graphic demonstration. The figures provided by the package include the regional thermal distribution curve, isotherm diagram, profile thermal curve, dynamic thermal curve, and 3D thermal trend diagram. The study used multiple regions and single-region thermal image analysis for determining the highest, average, and lowest infrared temperatures (Figure 5).

4. Results and discussion

4.1. Field monitoring by UAS taken of thermal images

Thermal images were obtained by the UAS on 16 August, 2017, at 13:00 at an elevation of 51 m by using a positioned GPS. Figure 6 shows overlapped UAS-obtained thermal images and satellite images (Google Earth) in the study area.

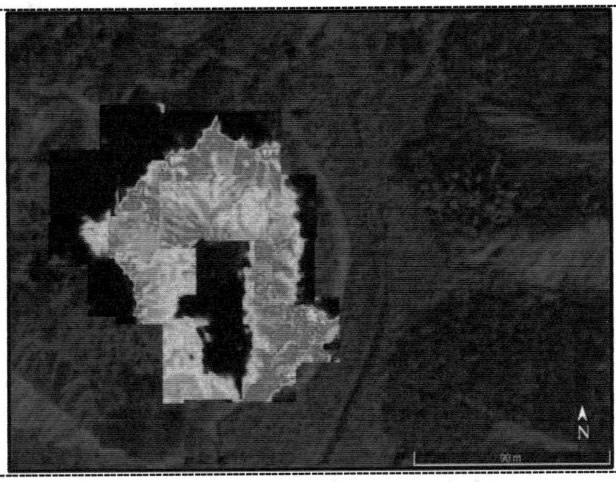

Figure 6. Overlap of UAS-obtained thermal images and satellite image (Google Earth) for the study area.

A threshold infrared temperature was employed to separate vegetation and naked areas. In general, the infrared temperature was between 30°C and 50°C in the study area. For example, image T00911 shows that the vegetation area had an infrared temperature less than 36°C and thus this temperature is suitable for separating the naked and vegetation areas (Figure 7). This temperature was thus used in the analysis. Equations should be centred and should be numbered with the number on the right-hand side.

Regional temperature analysis was performed on the UAS-obtained thermal images separated at 36°C for the following aerial images of the study area: T00874, T00885, T00900, T00907, T00909, and T00911 (Figure 8).

A histogram of the number of grid points in the thermal images was created after performing the single-region thermal image analysis. Each thermal image was separated by the threshold infrared temperature (36°C) into naked and vegetation areas. The VCR was calculated using the number of grid points below 36° divided by the total number of grid points in the thermal image.

The study separated images using the infrared temperature of 36°C and the distribution of temperature by using the number of points in the thermal image. The histogram was separated into two parts: those of temperature 35.99° or lower and those of temperature 36.00°C or higher. The obtained histogram is presented in Figure 9. The number of grid points in image T00874 at 35.99°C or lower and at 36.00°C or higher were 6957 and 12243, respectively. The corresponding VCR was thus 36%.

Similarly, the VCR for the thermal images T00885, T00900, T00907, T00909, and T00911 (Figure 8) were 3%, 47%, 15%, 57%, and 48%, respectively.

In this study, 295651 grid points were below 36°C and 180725 grid points were above 36°. The number of grid points below 36°C divided by the total number of grid points is the VCR, which was 62.1% in the study area.

4.2. Vegetation area calculation using a digital terrain model

A 3D digital terrain model (DTM) was constructed in this study by using Pix4D (professional drone mapping and photogrammetry software). The study area and vegetation cover area were directly calculated for VCR estimation. The UAS-obtained thermal images were overlapped with the remote image obtained from Google Earth (Figure 6). The 3D study area, naked area, and vegetation cover area were 24,443, 8,359, and 16,084 m2, respectively. The VCR in the study area was 65.8%.

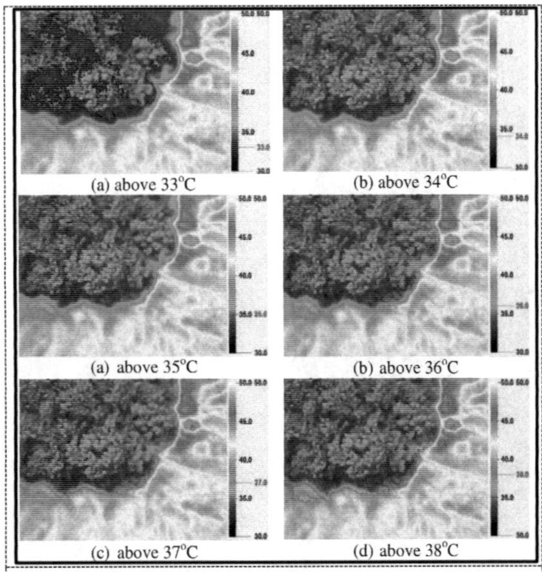

Figure 7. Example of thermal image T00796 with regional analysis, separated at different infrared temperatures between 33°C and 38°C.

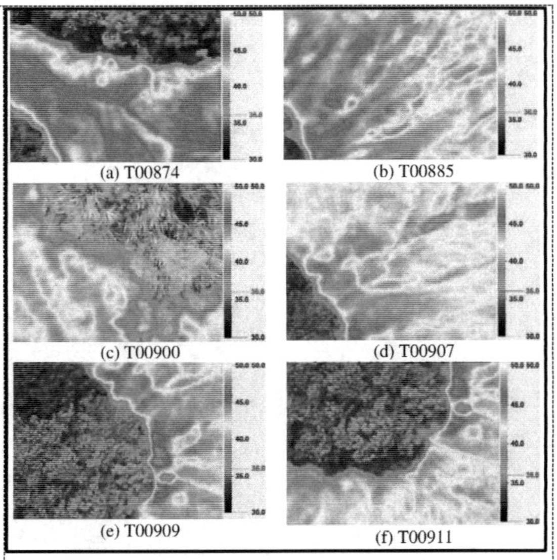

Figure 8. Regional analysis of UAS-obtained thermal images separated at 36°C.

5. Conclusion

This study used UAS-obtained thermal images for monitoring the vegetation ratio in the study mudstone area. The thermal images show distinct infrared temperature differences between naked and vegetation areas. Moreover, the eroded gully and ground crack areas show a lower infrared temperature than other areas. The thermal images clearly separate nonvegetation and vegetation areas, with these areas having different infrared temperatures, thus enabling calculation of the VCR.

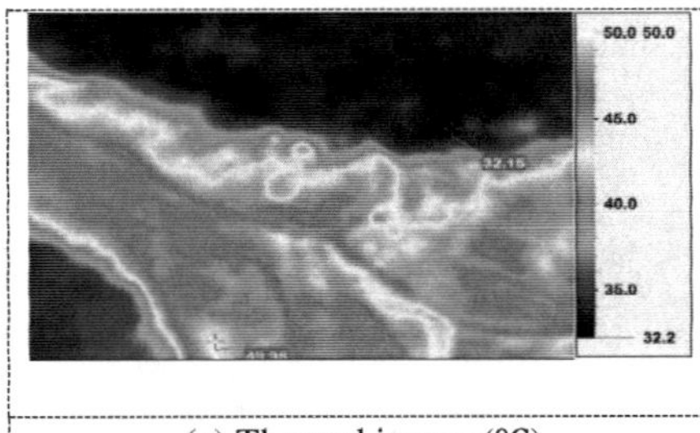

(a) Thermal image (°C)

Range	Points
31.00-31.99	0
32.00-32.99	913
33.00-33.99	4049
34.00-34.99	1491
35.00-35.99	504
36.00-36.99	349
37.00-37.99	256
38.00-38.99	239
39.00-39.99	309
40.00-40.99	489
41.00-41.99	709
42.00-42.99	998
43.00-43.99	1418
44.00-44.99	1990
45.00-45.99	2554
46.00-46.99	1806
47.00-47.99	795
48.00-48.99	252
49.00-49.99	79
50.00-50.99	0
	19200

(b) Grid points

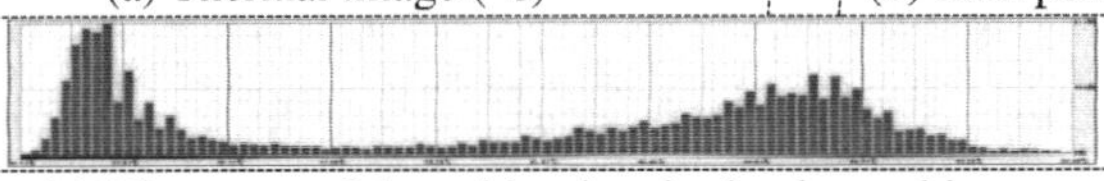

(c) Histogram of the grid points in the thermal image

Figure 9. Example of statistic histogram of grid points in UAV-obtained thermal image T00847.

The VCR in the study area, obtained using thermal image interpretation, was 62% and comparable to the 65.8% calculated using the UAS-obtained thermal images in the studied area. Thermal images can be effectively used to estimate the VCR in mudstone areas. This study provides another approach for monitoring changes in the VCR in real time. UAS-obtained thermal images provide the VCR with high precision, which is required in the regional area.

Acknowledgments
Financial supports under contract MOST 107-2625-M-415-001 of Ministry of Science and Technology in Taiwan are appreciated.

References

[1] Hsu S H, Huang K H and Liao S Y 2005 Application of thermo-tracer on studying surface temperature changes on mudstone slopes *Crop, Environment & Bioinformatics* **2** 148–156
[2] Hsu S H, Huang K H and Liao S Y 2005 Estimation of solar radiation on the mudstone slopes *Crop, Environment & Bioinformatics* **2** 211–220
[3] Elvidge C D, Chen Z 1995 Comparison of broad-band and narrow-band red and near-infrared vegetation indices *Remote Sensing of Environment* **54** 38–48
[4] Huang L C, Chuang C W, Ho S H and Lin C Y 2008 Selection of optimum vegetation index from spot satellite image Journal of Soil and Water Conservation 40 39–50
[5] Lee C F, Chang S Y and Lin Z P 2004 Infrared thermal image applied on debris flow monitoring Journal of Chinese Soil and Water Conservation 35 261–274
[6] Clark M R, McCann D M and Forde M C 2003 Application of infrared thermography to the non-destructive testing of concrete and masonry bridges NDT International 36 265–275
[7] Mondini A, Carlà R, Reichenbach P, Cardinali M and Guzzetti F 2009 Use of a remote sensing approach to detect landslide thermal behavior 11 EGU2009-7790
[8] Martino S, Mazzanti P 2014 Integrating geomechanical surveys and remote sensing for sea cliff slope stability analysis: the Mt. Pucci case study (Italy) *Nat. Hazards Earth Syst. Sci.* **14** 831–848
[9] Chang W, Hsiao Y S, Chang J C 2017 Feasibility analyses of river sediment estimation and mountain mapping by UAV techniques *Journal of Chinese Soil and Water Conservation* **48** 1–13
[10] Erdelj M, Natalizio E 2016 UAV-assisted disaster management: Applications and open issues International Conference on Computing Networking and Communications (ICNC 2016), Feb 2016, Kauai, United States, 1–5
[11] Liu C C, Shieh M C, Ke M S, Wang K H 2018 Flood prevention and emergency response system powered by google earth engine *Remote Sens.* **10** 1283
[12] CGS Central Geologic Survey in Taiwan http://gis.moeacgs.gov.tw/gwh/gsb97-1/sys8/index.cfm

IOP Conf. Series: Materials Science and Engineering **652** (2019) 012063 doi:10.1088/1757-899X/652/1/012063

Detecting device and technology of pavement texture depth based on high precision 3D laser scanning technology

Jia Liang[1], Xingyu Gu[1,2,*], Hanyu Deng[1], and Fujian Ni[1]

[1] School of Transportation, Southeast University, Nanjing 211189, P.R. China
[2] Institute of technology, Tibet University, Lhasa 850012, P.R. China

Email: guxingyu1976@163.com

Abstract. The pavement means texture depth (MTD) is part of the most significant indexes to the characteristics of the pavement macrostructure. Traditional measuring MTD methods are easily impacted by the human factors with inefficiency and low-accuracy. In order to realize the automatic detection of the MTD, a high precision asphalt pavement MTD detection system is built based on three-dimensional line laser scanning, and a MTD calculation method based on the line laser three-dimensional (3-D) data is proposed. Firstly, the 3-D elevation data are extracted from the obtained linear structure light images. Secondly, a denoising algorithm based on two-way standard deviation filter is presented for the 3-D elevation data. Finally, a MTD calculation method based on the line laser 3-D elevation data is designed. The MTD of five asphalt specimens with different roughness was measured by the calculation method designed in this study and sand patch measurements, respectively, and the results were comparable. The research demonstrates that the error between the calculation result of this algorithm and that of sand patch measurements is not more than 5%, which further proves the feasibility of this algorithm.

1. Introduction

Nowadays, the development of the pavement has been transformed from construction to detection and maintenance. How to maintain and manage such a large-scale road network and how to test the performance of the road has become an important issue in road construction in China. The complex road condition and the underdeveloped image acquisition software and hardware technology make the achieved pavement images with high noise; it brings difficult image processing and low detection. Many researchers devote themselves to the fast, correct and efficient evaluation of the safety performance of the pavement surface, among which the texture depth (TD) as the most important indicator has been widely studied.

Numerous studies show that the pavement texture depth is one of the important indexes for evaluating pavement performance[1-4]. Within the wavelength range of macro-texture, 0.5-50mm, the most significant influence of texture is in the fields of rolling resistance, tire-road friction and noise[5]. The evaluation of pavement structure can be used to identify areas where friction performance of pavement may not meet the friction requirements, thus supporting the reduction of accidents[6]. For asphalt pavement, areas with uneven surface texture are segregation indicators, which often lead to poor pavement performance[7]. The maturity and development of digital image technology provide a new way for texture detection of asphalt pavement. Although digital image processing technology improves the efficiency of pavement damage detection, it depends on the accuracy of imaging equipment, and cannot be used for high-precision identification. Based on the principle of laser triangulation[8,9], three-dimensional machine vision measurement using structured light and planar CCD cameras is gradually emerging and applied to damage detection and three-dimensional image reconstruction and other fields. It is noted that the measuring range of this method is 50μm[9]. Based on three-dimensional laser scanning pavement data, it can effectively detect typical damages under different road conditions and environments, and the detection accuracy is over 98%[10]. The traditional sand patch measurements method is used to measure the mean texture depth of asphalt pavement, but its accuracy is low and it is difficult to satisfy the needs of practical engineering. Therefore, more innovative methods are needed to characterize the three-dimensional texture of asphalt pavement[11].

Based on the problems of low efficiency and low precision of pavement detection, the paper combines digital image technology with three-dimensional laser scanning technology to measure and calculate the pavement texture depth, which provides a reference for fast and accurate calculation of pavement texture depth. An algorithm for automatically estimating road roughness from three-dimensional laser scanning data is proposed. The algorithm is helpful for a more comprehensive measurement of road structure depth. By comparing and analyzing several kinds of pavement data acquisition sensors, this paper puts forward the pavement crack identification technology based on CCD camera, and studies how to use digital image processing technology to solve the problem of pavement damage automatic identification.

2. Experimental

2.1 3D laser scanning measurement principle

Laser triangular method is one of the most common optical three-dimensional measurement techniques that calculate the displacement on the camera sensor due to the position of the pavement surface reflection points. Line laser is projected into a bright line on the object surface under test. The camera acquires an image for each forward 0.1mm, which can obtain all the three-dimensional data on the object surface contour, and calculates the three-dimensional height information by the triangulation method. The laser emits a laser that is reflected on the object surface, then the laser beam goes through the lens, the light energy is collected through the photosensitive element, and finally, the imaging system is imaged. When the position of the object surface is changed, the corresponding displacement occurs in the imaging system. According to the relationship between the image displacement and the real displacement, the actual object displacement is calculated by detecting the image displacement. The principle of the laser triangulation method is illustrated in Figure 1(a). The angle between the axis of the line laser and the axis of the camera is theta θ, the focal length of the camera is f, the distance between the transmitting lens and the receiving lens is *L*, the distance from the receiving lens to the reference point is a, and the variation displacement of the real depth of the object surface is *x*. The change in the displacement of the depth change on the target surface of the camera is represented by *x'*,

In triangle AOB, according to sine law:

$$\frac{a}{\sin(\theta\text{-}\alpha)}=\frac{x}{\alpha} \quad (1)$$

where *a* is the distance from the receiving lens to the reference point, *x* is the displacement of the real depth of the object surface, *θ* is the angle between the line laser axis and the camera axis, *α* indicates the angle due to the change of object position.

After reorganizing the formula, one gets the Equation (2):

$$x=\frac{a}{\sin\theta\cot\alpha-\cos\theta} \quad (2)$$

In right triangle COD, substituting parameters $\cot\alpha=f/x'$、 $\sin\theta=L/a$、 $\cos\theta=\sqrt{a^2-L^2}/a$ into Equation (2), then the variation displacement *x* of the real depth of the object surface is obtained,

$$x=\frac{a^2x'}{Lf\text{-}x'\sqrt{a^2\text{-}L^2}} \quad (3)$$

Where *f* is the focal length of the camera, *L* is the distance between the transmitting lens and the receiving lens, *x'* is the displacement on the camera sensor due to the position of the pavement surface reflection points, and $Lf>>x'\sqrt{a^2-L^2}$, therefore,

$$x = \frac{a^2 x'}{Lf} \quad (4)$$

2.2 Pavement surface images acquisitions

A quasi-static and high-precision pavement damage detection device is set up in the laboratory. The detection device includes a movable detection vehicle, an electric linear screw, a surface scan camera, a linear laser illumination device, a photoelectric rotary encoder, an uninterrupted power supply (UPS) device and an Industrial Personal Computer (IPC), as shown in Figure 1(b). The scan camera is used to obtain the optical image of asphalt pavement and the three-dimensional elevation data of the pavement surface is extracted according to laser three-dimensional scanning technology. The line laser illuminator in this device is a near-infrared light laser which can effectively prevent the influence of other light sources on image uniformity, and it provides auxiliary lighting for the camera.

The rotary encoder outputs 100 pulse signals per turn. When the camera moves forward 0.1mm, the rotary encoder sends out a pulse signal that triggers the camera to work, and then the camera collects a light image of the pavement surface. The UPS provides power for the detection device. The IPC as the core part of the detection device undertakes the main task of image data acquisition, image processing, and image data statistics. In order to avoid the inconvenience caused by the disassembly of the camera and line laser, they are fixed as a whole. The electric linear screw module is the power system of the detection device. The stepper motor of electric linear screw module drives the screw, and the reciprocating motion of the slider on the screw is controlled by the control switch. The moving slider on the screw keeps the camera and the line laser maintains relative movement to the object. At the same time, the axis of the camera and the axis of the line laser are in an angle of 30°, and their axes are coplanar and the plane is perpendicular to the horizontal plane.

The pavement data collection steps of the pavement detection device are as follows:

(1) Five kinds of asphalt specimens were prepared, which were the maximum particle size of 9.5mm micro surface (MS), 9.5mm ultra-thin wearing course (UTWC), and 13.2mm stone matrix asphalt (SMA), 16mm rubber asphalt concrete (RAC) and 16mm dense graded asphalt concrete (AC), respectively. Their sizes are all 300mm×300mm×50mm. The experimental location is in the civil engineering and transportation engineering laboratory of Nanjing, Jiangsu Southeast University Jiulonghu Campus (31.89°N, 118.82°E);

(2) The reset switches are installed on both ends of the module screw because the reset switches control the effective displacement of the slider. The stepper motor, the device driver, the controller, and the reset switch is connected correctly. The power supply equipment was started to make the electric linear screw module in a standby state after making sure it is correct;

(3) The reset switches of the electric linear screw are controlled by manual control to make the slider reciprocate, and the effective displacement of the tested pavement could be roughly determined;

(4) The encoder is connected to the camera through the data line, same as the camera is connected to the Industrial Personal Computer. The encoder, the camera, the Industrial Personal Computer, and the linear laser are all connected to the power supply device;

(5) The inspection vehicle was placed on the damaged road surface and the small sealing handle equipped with the camera and the line laser was rigidly attached to the slider;

(6) Open the industrial computer, open the image information collection software in the industrial computer, and set the parameters;

(7) The reset switch of the electric linear screw is controlled by a manual control method, so that the slider drives the small sealed box equipped with the camera and the line laser to reciprocate and collect road information.

(8) The road surface information collected by the camera is converted into a hard disk of the industrial computer through the conversion of the Gigabit Ethernet card;

(9) The three-dimensional height data were calculated from the structured light image stored in the hard disk according to the three-dimensional laser scanning technique.

IOP Conf. Series: Materials Science and Engineering **652** (2019) 012063 doi:10.1088/1757-899X/652/1/012063

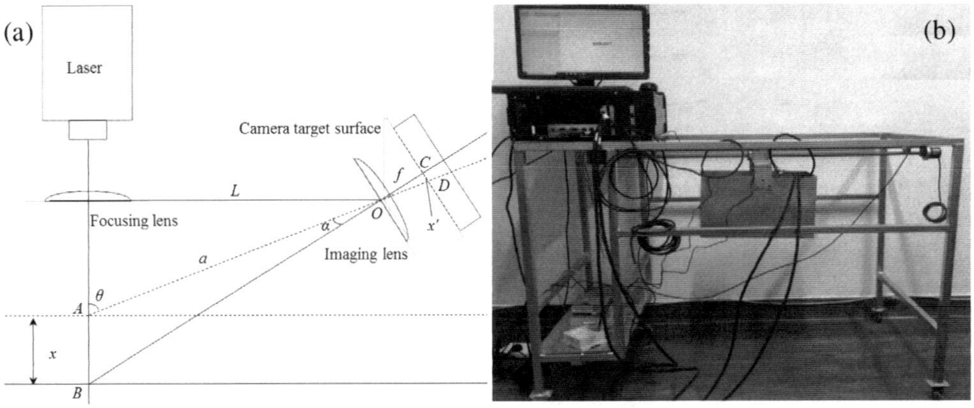

Figure 1. Detection Systems.
(a) Principle of laser triangulation. (b) Schematic diagram of pavement detection system device.

2.3 A digital image processing methods
Due to factors such as diffuse reflection of the exposed pavement aggregates and jitter of the detection system equipment, the collected pavement data are inevitably affected by noise, which causes the collected 3D data to deviate from the real depth value at different levels. Noise distribution contains the following characteristics. One is that the standard deviation of noise is much greater than the standard deviation of useful data, and the other is that the noise has strong randomness, and it exists in the whole data space, not only in the linear range of single-dimensional and on the two-dimensional plane. In view of this, this paper uses the "biphasic standard deviation filtering method" to perform row and column filtering on 3D data. The main idea of bidirectional standard deviation filtering is to scan the data matrix from two directions respectively and to use certain filtering conditions to judge and filter the original 3D data containing noise, and to retain the useful data to filter out the noise data so as to obtain the filtered data matrix. Specific steps are as follows,
Firstly, the data of the three-dimensional image data matrix $A_{m\times n}$ are read row by row, and the data in the i row are recorded as $R_i=(Z_{i1},Z_{i2},...,Z_{in})$ and $i=(1,2,...,n)$, and the arithmetic mean and standard deviation of R_i are calculated from equations (5) and (6), respectively:

$$\overline{R_i}=\frac{\sum_{i=1}^{n}(Z_{i1}+Z_{i2}+\ldots+Z_{in})}{n} \tag{5}$$

$$S_i=\sqrt{\frac{(Z_{i1}-\overline{R_i})^2+(Z_{i2}-\overline{R_i})^2+\ldots+(Z_{in}-\overline{R_i})^2}{(n-1)}} \tag{6}$$

Secondly, each data z_{ij} in the row is taken in turn. If $|z_{ij}-\overline{R}i|/s_i>k$, the arithmetic mean value Ri is used instead of the point data value z_{ij}, k represents the filtering factor, and k is 1.6.
Finally, noise points are eliminated again by column processing in the same way. After all, rows are processed; the 3D data matrix E after bidirectional standard deviation filtering is obtained.

2.4 A principles of texture depth calculation
The pavement means texture depth is the most important indicator of the anti-sliding performance. The basic idea of the MTD calculation method based on 3D line laser scanning is presented. First, a datum is determined, then the volume of the irregular graph formed by pavement and datum is calculated according to the idea of traditional volume method, and then the MTD value can be obtained by splitting the volume by the area of the region. By building structural elements to fit the datum, the selection of the base level determines the feasibility of the algorithm. For the fitting of pavement datum, the structural elements selected by morphological filtering are first determined and used to generate the datum. In order to consider the rationality of the selection of the structural elements, two kinds of structural elements are selected for comparison, that is, the rectangular structural element (R) and the

spherical structural element (S). The operation mode is the closing operation, which is the first dilation and then erosion:

$$m \odot n = (m \oplus n) \ominus n \tag{7}$$

Firstly, the size of the structural elements is established according to the size of the matrix. For the rectangular structural element, the parameters of the length l and the width b should be determined. For the spherical structural element, the radius r of the circle projected on the x-y plane and the maximum offset height h should be determined. Secondly, according to the generated structural element, the gray scale image is expanded first and then corroded by the morphological closing operation, and the dilation and erosion are operated by the same structural element to form the pavement fitting surface. Then, the volume between the simulated datum and the irregular three-dimensional surface after filtering is calculated, and the MTD value can be obtained by dividing the volume by the area of the region. The spherical structure element is utilized to generate a spacing surface fitted with the original pavement, but this has a high requirement for the parameter selection of the spherical structure element.
By filtering and plane fitting, a new spatial model function is obtained,

$$Z=F(x,y) \tag{8}$$

Where x and y are the horizontal and vertical coordinates of the spatial point. A circular region D is selected optionally mapped by a spacing surface in the X-Y plane, and the volume size of the spatial surface built by the spatial model in D of the circular region and the spatial region mapped to the datum Z_0 is calculated, namely,

$$V=\iint_D [Z_0-F(x,y)]dxdy \tag{9}$$

Where V is represented as the volume of the space surface and the datum Z_0, and the ratio of volume V to area D is the MTD value of the area to be measured, that is,

$$MTD=\frac{V}{D} \tag{10}$$

3. Results

3.1 Pointed cloud data filtering

The pointed cloud data denoising process is realized by the MATLAB secondary development program based on the bidirectional standard deviation filtering. Figure 2(a) is a three-dimensional model diagram of the surface of an ultra-thin wearing course (UTWC) with a maximum aggregate size of 9.5 mm. The macroscopic performance is that the surface is rough and the "void" is more. (b) is a three-dimensional model diagram of the surface after filtered., the surface depth is more uniform after the filtering process, which is more satisfied with the actual situation of the pavement. The surface depth at different positions is within 2 mm. However, the study concluded that the filtering factor directly affects the pavement mean texture depth value. The larger the filtering factor, the larger the mean texture depth value. This is because the points satisfying the equation $|zij- \bar{R}i|/si>k$ are less, and the filtering has less influence on the original data.
After many calculations, UTWC selects the filtering factor k=1.6. When the filtering factor k=1.6, the error between the calculated texture depth value and the value calculated by the ellipsoid element fitting datum plane is within 5%.

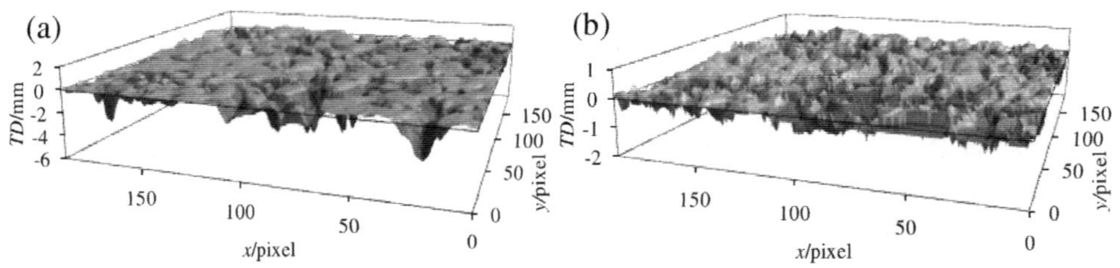

Figure 2. Comparison of 3D models before and after filtering.

The program design is completed as required. Each group of data transferred to the program is filtered in turn by the program, and the three-dimensional model of the pavement surface is built according to the filtered data, and the datum is built at the same time. The fitting datum of different structural elements is shown Figure 3.

(a) Plane(Max) (b) Rectangle (c) Ball

Figure 3. Fitting planes of different structural elements.

3.2 Texture depth calculation

Figure 4 shows the pavement surface means texture depth values of different types of asphalt mixture specimens. These values are calculated on the basis of digital image technology. Figure 4(a) shows that the measured value is compared with the calculated value, and the results show the calculated mean texture depth value of the calculation method is consistent with the measured value, so the MTD of asphalt pavement can be calculated by using the method of fitting the datum surface. Figure 4(b) represents the MTD value based on the datum of the morphological filter rectangular structure element and spherical structure element. At the same time, different filter factors are selected for different asphalt mixture specimens. It is found that the MTD values calculated by the two methods are similar, which shows that the datum of spherical structure element can also be used to calculate the MTD of the pavement surface, and achieves good results. However, the pavement datum formed by the fitting of the rectangular structure element is terraced, which is not consistent with the actual situation. Therefore, the method of using the rectangular element to fit pavement datum is debatable. (b) shows the pavement surface MTD fitting curve calculated by two different methods. The study found that the two calculation results had a high degree of linear correlation. And the correlation coefficient R2=0.998. At the same time, it is found that the ultra-thin wearing course (UTWC) with a maximum particle size of 9.5mm has the largest MTD value of 1.7986mm. The MTD value of the maximum particle size of 13.2mm stone matrix asphalt (SMA) is minimal, only 0.8516mm. In the dense asphalt concrete mixture, whether there is a certain relationship between the MTD value of the pavement surface and the size of the grading, more experimental verification is needed.

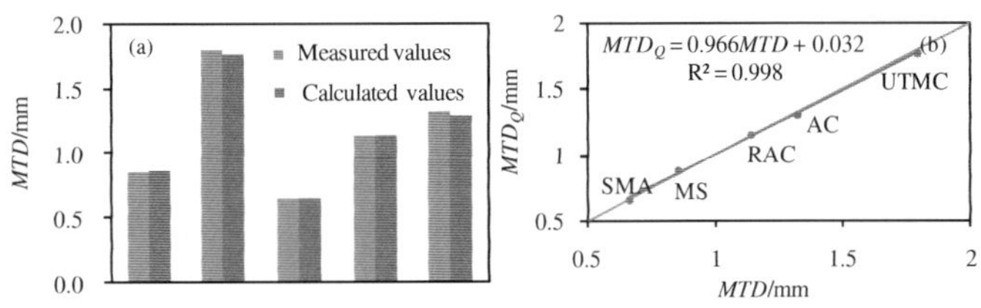

Figure 4. Comparison of texture depth values calculated by two different methods.

4. Conclusions

This research is mainly divided into three distinct parts. First of all, according to the research is required to determine the equipment parameters and build indoor detection device. Secondly, three-dimensional data are obtained. Then, the image filtering processing is carried out by MATLAB and the pavement surface 3D Reconstruction for the region to be measured is built. Finally, the mean texture depth of the pavement structure is calculated according to the fitting datum of the rectangular element and spherical element, which is compared with the MTD value measured by the volume method. It is well known that the pavement surface texture can affect surface properties, tire-to-vehicle interactions, and road safety. According to the experimental and calculation model, the following conclusions are drawn,

(1) The indoor quasi-static high-precision pavement detection device was built, and the experiment was carried out by five different types of asphalt mixture specimens. The research shows that the experimental device can meet the indoor testing requirements and can be used to detect the pavement distress;
(2) According to the standard deviation comparison before and after filtering, it can be found that the bidirectional standard deviation filtering can filter out noise better, but the filtering factors selected by different road types are different, and the pavement mean texture depth calculated by different filtering factors is different;
(3) The calculation method of asphalt pavement texture depth based on three-dimensional information is designed. The experimental results demonstrate that the algorithm has good stability and high reliability.

Acknowledgments
This work is supported by National Key R&D Program of China (Grant No.2017YFF0205600) and the Fundamental Research Funds for the Central Universities (Grant No.3221008101). The first author thanks the members of the research group for their help during my experimental.

References

[1] Praticò F G and Vaiana R 2015 "A study on the relationship between mean texture depth and mean profile depth of asphalt pavements" Constr. Build Mater. 101 72-9
[2] Meegoda J N and Gao S 2015 "Evaluation of pavement skid resistance using high speed texture measurement" J.Traffic Transp.Eng.(Engl.Ed.) 2 (6) 382-390
[3] Cui X, Zhou X, Lou J, Zhang J and Ran M 2017 "Measurement method of asphalt pavement mean texture depth based on multi-line laser and binocular vision" Int. J. Pavement ENG. 18 (5) 459-471
[4] Gendy A E, Shalaby A, Saleh M and Flintsch G W 2011 "Stereo-vision applications to reconstruct the 3D texture of pavement surface" Int. J. Pavement ENG. 12 (3) 263-273
[5] ISO 13473-1 1997 "Characterization of pavement texture by use of surface profiles–Part 1: Determination of mean profile depth"
[6] Britton S, Ledbetter W and Gallaway B 1974 "Estimation of Skid Numbers from Surface Texture Parameters in the Rational Design of Standard Reference Pavements for Test Equipment Calibration"
[7] Flintsch G W, Mcghee K K and De Leon E 2005"Field Validation of Macrotexture-Based Hot Mix Asphalt Segregation Detection Methods" J.A.A.P.T. 74
[8] Chen W, Ni Z, Hu X and Lu X F 2016 "Research on pavement roughness based on the laser triangulation" Photonic Sens. 6 (2) 177-180
[9] Cigada A, Mancosu F, Manzoni S and Zappa E 2010 "Laser-triangulation device for in-line measurement of road texture at medium and high speed" Mech. Syst. Signal Pr. 24 (7) 2225-34.
[10] Zhang D, Zou Q, Lin H, Xu X, He L, Gui R and Li Q 2018 "Automatic pavement defect detection using 3D laser profiling technology" Automat. Constr. 96 350-365
[11] Annual Book of ASTM Standards 2005 " ASTM. Road and paving materials; vehicle pavement systems" E1845 E965-96 E2157

IOP Conf. Series: Materials Science and Engineering **652** (2019) 012064 doi:10.1088/1757-899X/652/1/012064

The dilemma and situation of detention pond in terms of extreme climate - a case study of Baoyeli detention pond

Min-chia Hsueh[1] **and Yao-shun Hsueh**[2,*]

[1] Department of Hydraulic and Ocean Engineering, National Cheng Kung University, Tainan
[2] Department of Physical Education, Health & Recreation, National Chiayi University, Taiwan

Email: a0933596895@yahoo.com.tw

Abstract. Constructing detention ponds to prevent flood has become a trend in recent years, but the utility of detention pond is yet to be confirmed. Therefore, in this study, we investigate detention pond's function in terms of flood solving by analyzing the rainfall data, and comprehending the dilemma of detention pond in technology, also, we interview the user of detention pond to know the ordeal of detention pond in politic. We discover that detention pond is very effective in making up for the weaknesses in the drainage system, however, detention pond isn't very effective in improving the overall flood control capacity of the drainage system. Furthermore, government needs to outreach policies to solve misunderstanding of public on infrastructures.

1. Introduction

In the heavy rain of August 28, 2018, the flood control system of Kaohsiung City faced a strong challenge. The extreme climate is normal nowadays. If the coping ability to flood control of the city cannot be improved, it will easily lead to the loss of people's lives and property. The government must adjust the policy according to the current development situation, to prevent the disasters. Therefore, this paper takes Baoyeli detention pond as the object, which was built in 2010. And the research region is Baoye Village, Sanmin District, Kaohsiung City, which Baoyeli detention pond located in, and Wenshan area, which is the watershed of the detention pond. Wenshan area includes Wende Village, Wenheng Village and Wenshan Village, Fengshan District, Kaohsiung City. In this study, we use hydrological data to analyze whether urban multi-purpose detention pond reduces flood occurrence under extreme rainfall conditions, and the function of detention pond in drainage system. Also, we investigate the opinion of the detention pond's users, to understand the ordeal of detention pond in terms of extreme climate.

2. What is Baoyeli detention pond

2.1. Definition and function of detention pond

Everettwa City Government [1] indicates that detention pond treats rain to protect our creeks and streams, and during severe storms, helping prevent flooding. We define a facility that uses an artificial structure to store floods and discharge stored floods according to management needs.

With the progress of hydraulic engineering, the detention pond has gradually increased its functions. Not only can detention pond absorb the carbon dioxide by its vegetation, forming the cold island effect in the city, but also can it provide habitat for the wild animals.[2][3] In general, the multi-purpose detention pool has gradually replaced the traditional detention ponds.

2.2. Background information of Baoyeli detention pond

The northern part of Fengshan District where is nearby Niaosong District had been being still in the waters before 1898. Due to the urbanization, the surface of the land is gradually covered by concrete. The rainwater accumulated in the area slops westward along the terrain, emitting into Yihua road sewer. Kaohsiung City Government [4] pointed out the sewer can't release the run-off caused by heavy rain, causing the flood in Chenqing-Yihua intersection. To solve the flood, Kaohsiung County Government was suggested to construct Chishan detention pond, but they canceled the plan because of the protest of residents. Kaohsiung City Government considered when the heavy rain happened the run-off would flow into Takao River, causing overflow, so Kaohsiung City Government still decided to build Baoyeli detention pond instead of NSTM detention pond. [5] The detention pond completed in 2012.

2.3. Design of Baoyeli detention pond

Baoyeli detention pond locates in Kaohsiung City 41 Urban Land Readjustment Implementation by the Private Sector. The detention pond should at least tolerate of 72mm/hr, 5-year return period, for an hour. [4] Kaohsiung City Government Water Resources Bureau [6] assessed that the detention pond could exceed a 50-years return period rainfall, 87.5mm/hr, for 2 hours.

According to the paper [6], the area of the detention pond is 4.46 ha, the detention pond comprise north pond and south pond. It can contain 10^5 t ton of water. The detention pond launches when the flows speed of Yihua road sewer come over 12cms. The entrance of detention pond connects to Chishan A main, and the exit connects to Yihua road sewer. The watershed is 100 ha.

3. Flood control capacity analyzing of Baoyeli detention pond

3.1. Selection of rainfall data

3.1.1. Analysis of rainfall frequency. According to the data [7], the rainfall frequency in Fengshan and Chenqing Lake is analyzed by generalized extreme value distribution as shown in Figure 1.

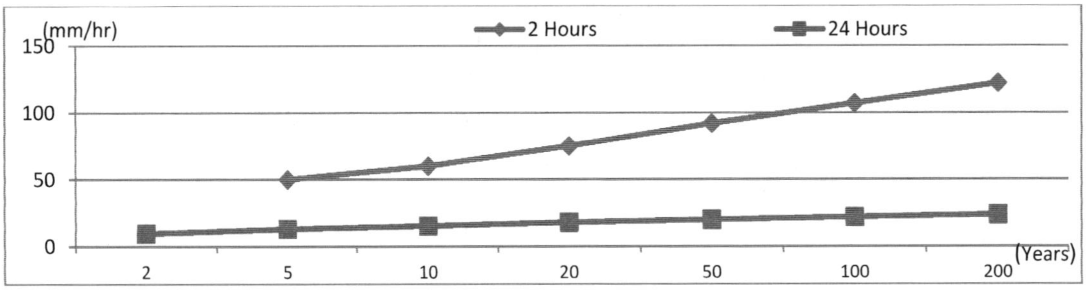

Figure 1. IDF polyline of Fengshan and Chenqing Lake

3.1.2. Selection of rainfall station and rainy event.

Selecting rainfall data of Fengshan Rainfall Station (No. 4, Yuanyi Lane, Wenshan Road, Fengshan District, Kaohsiung City) , which is located in the watershed of the Baoyeli detention pond, and under jurisdiction of TCWB.

Collecting Taiwanese news and government reports, analyze the flooding of Chenqing road and the watershed of the Baoyeli detention pond, and selected the 2 hours and 24 hours during rainy event whose rainfall are largest. The data are shown in Table 1 and Table 2.

3.2. Before and after comparison of the construction of the detention pond

Table 1. Rainy event of 2 hours rainfall duration

Rainy event	Started time	Ended time
Haitang 05	2005/7/19 6:00	2005/7/19 8:00
Kong-rey	2013/8/29 8:00	2013/8/29 10:00
Haitang 17	2017/7/30 23:00	2017/7/31 1:00
828 flood	2018/8/28 5:00	2018/8/28 7:00

Table 2. Rainy event of 24 hours rainfall duration

Rainy event	Started time	Ended time
Haitang 05	2005/7/18 9:00	2005/7/19 9:00
Kong-rey	2013/8/28 13:00	2013/8/29 13:00
Haitang 17	2017/7/30 17:00	2017/7/31 17:00
828 flood	2018/8/27 20:00	2018/8/28 20:00

The rainfall data of 2 hours rainfall duration are shown in Table 3. And the data of 24 hours rainfall duration are shown in Table 4 and Figure 2. The larger average value of rainfall per hour showed the larger total rainfall. If the event has higher standard deviation and maximum of rainfall, it will have more concentrated rainfall time. Then we can know the rainfall type by analyzing these data.

In terms of short-rainfall-duration rainfall, according to Haitang 05, Kong-rey and Haitang 17, we can knew that facing of a rainfall which is not only short-term but also high-intensity, detention pond can effectively reduce the incidence of flooding However, in terms of long-rainfall-duration rainfall, it can be found in the 828 flood, detention pond isn't able to solve the long-term and high-intensity rainfall.

According to the design of the detention pond, the detention pond should at least tolerate of 72mm/hr, 5-year return period, for an hour. The rainfall of the 828 flood did exceed the capacity and caused flooding. However, Kaohsiung City Government [6] originally assessed that the detention pond could exceed a 50-years return period rainfall, 87.5mm/hr, for 2 hours. Because the happened of 828 flood, we can know the assessing is wrong.

It is also worth noting that the rainfall in Fengshan is different from that in Kaohsiung. In terms of rainfall in Fengshan, although the rainfall of 5-year return period is higher than that in Kaohsiung, Fengshan's rainfall will be higher in 50-year return period. It is still unclear whether Kaohsiung City Government use of the rainfall in Kaohsiung as the standard for the design of the detention pond is right. In summary, the detention pond still has indeed improved the flood control capacity in research region.

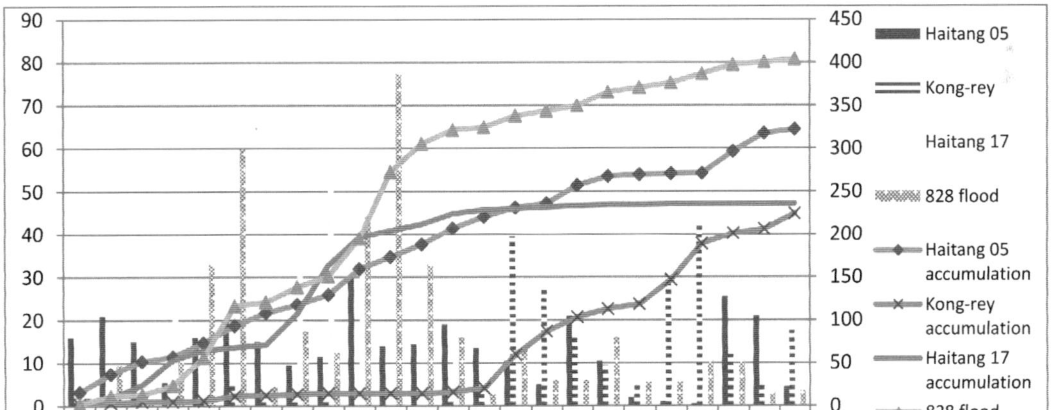

Figure 2. Analysis of 24 hours rainfall duration

Table 3. Analysis of 2 hours rainfall duration

	Haitang 05 (Before)	Kong-rey (After)	Haitang 17 (After)	828 flood (After)
1st hour	25.5 mm	28.5 mm	55.5 mm	44 mm
2nd hour	21 mm	42 mm	33 mm	77.5 mm

Table 4. Analysis of 24 hours rainfall duration

Rainy event	Haitang 05 (Before)	Kong-rey (After)	Haitang 17 (After)	828 flood (After)
Average value	13.4 mm	9.3 mm	9.8 mm	16.8mm
Standard Deviation	7.6 mm	12.5 mm	14.1 mm	19.0mm
Maximum	30 mm	42 mm	55.5 mm	77.5mm
Rainfall Type	Large, Dispersion	Medium, Concentrative	Medium, Concentrative	Very Large, Concentrative
Flooding	yes	no	no	yes

3.3. Analysis of 828 flood

When the watershed is small than 10 km^2, because rainfall duration will be longer than catchment basin, we can use rational formula:

$$Q = \frac{CIA}{360} \tag{1}$$

Q: Acre-inches/hour (m^3/s), C: Runoff coefficient (no unit), I: Rainfall Intensity (mm/hr), A: Area of watershed (ha)

Because research region are almost cover by concrete, and return period of 24 hours rainfall duration in 828 flood is over 10 years, according to the advice [9], we chose 0.83 for runoff coefficient. The rainfall and hydrograph [Q(t)] calculated by modified rational formula are shown in Figure 3. The catchment which is $V(t) = \int [Q(t)]dt$ is shown in Figure 4. The effect of detention pond on research region analyzed from runoff and catchment is shown in Table 5.

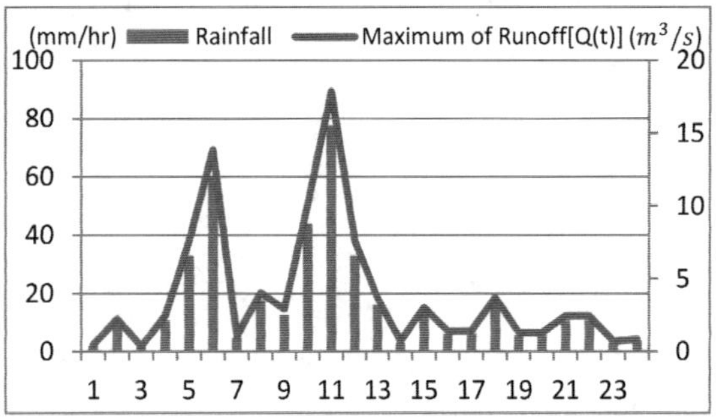

Figure 3. The rainfall and hydrograph of 828 flood

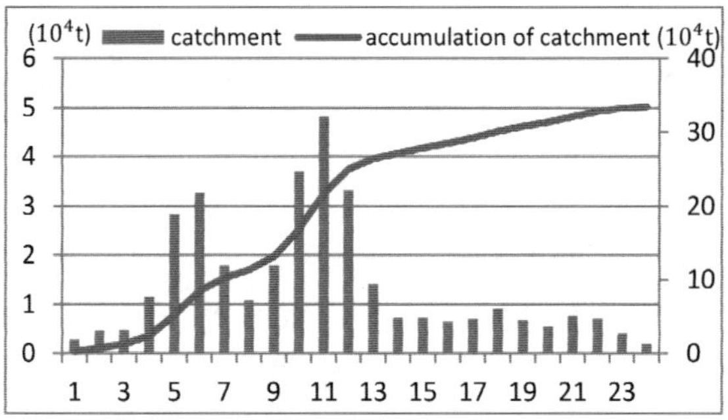

Figure 4. The catchment of 828 flood

Table 5. The effect of detention pond under the ideal condition on flood in research region

Flooding		Drain outlet	
		Existence	Not Existence
Detention pond	Existence	1. No Flooding	3. 2018/8/28 02:00
	Not Existence	2. No Flooding	4. 2018/8/27 20:00

3.3.1. The effect of detention pond in research region.

Table 5 show that, (1) If both Drain outlet and detention pond exist, since the runoff doesn't exceed the allowable flow of the drain outlet, it will not cause flooding, and the detention pond will be activated at 2018/8/28 06:00 to control the flow of Yihua road sewer, letting the flow of the sewer be less than two-thirds of the allowable flow. (2) If only drain outlet existed, although it wouldn't flood as stated in above, it will cause the flow of Yihua road sewer become 13.4 m^3/s at 2018/8/28 06:00, higher than two-thirds of the allowable flow. (3) If only detention pond existed, detention pond would be full at 2018/8/28 02:00, the 9th hours after the rain started, causing Flood. (4) If both Drain outlet and detention pond didn't exist, it would be flood when the rain started.

The above is in the condition of ideal. Actually, it flooded in Chenqing road at 2018/8/28 06:20 [8], He, Quan-feng indicated the flood swamped at 2018/8/28 09:00 [8]. It is different to the result we analyzed above, so it shows the hypothesis of the ideal condition may be wrong.

The ideal condition above was the condition which is that Yihua road sewer was empty, but rainfall is continuous in everywhere, it not only rained in research region, but also rained in the watershed of Yihua road sewer, causing the run-off. So we conjecture that when the torrential rain first happened, namely, 2018/8/27 24:00, it was almost full at Yihua road sewer. Making the flow of Chishan A main line can't release, launching the detention pond early. When the heaviest rain happened at 2018/8/28 06:00, detention pond was totally full, the run-off can't release by Yihua road sewer and became overland flow. Because the evaluation [8] also supposed that Yihua road sewer were empty, it will overestimate the flood control capacity of detention pond.

In conclusion, although it flooded at 2018/8/28 06:20, according the analysis above, if detention pond hadn't been exist, it would have flooded at 2018/8/27 24:00. It showed that detention pond shortened the time that flooding affected users, and reduced the damage in research region. And the return period of 24 hours rainfall duration in this rainy event is 15 years, it has exceed the design guidelines that is 5-years return period.

4. The opinion of the detention pond's users and the ordeal of detention pond

This chapter is divided into five main sections, each of which presents the results relating to flooding or detention pond. To understand the cognition of the detention pond's users on flood controlling, we interviewed the Chief of Baoye Village, Mr. Lin Huan-ran, in 2018/12/7. To obtain the reliability, we wrote interview transcript before analyzing, and confirming with the interviewee whether the analysis of the results is different from what he want expressed.

4.1. The experience of flooding

According to what interviewee said, the relief of Baoye Village, where detention pond located in, is higher than others. "The flooding came from Benyang Village, then Yihua road and, Juemin road and Chenqing road" he said. Different from the neighbourhood, it won't flood in village when the rain isn't heavy. He also said that some stores in village have floodgates because of the flooding in their basement.

In the case of 828 flood, there is no residents apply the subsidy for flood, it didn't cause the catastrophe, only caused 10cm depth puddle on road. However, interviewee also said, "Benyang Village and Wende Village flooded", the detention pond was originally designed to solve the flooding problem of Chenqing-Yihua intersection, and Chenqing-Yihua intersection is located in Benyang Village.

4.2. The cognition of flooding

When we talked about the definition of flooding, interviewee told author, "Flooding is that water comes up to sidewalk, and comes into houses, if it doesn't come into houses, I call it puddle". He considers that whether flood or not shouldn't be decided by flooding depth, it should be decided by the risk to people's life and property. He also pointed out, it's important to dredge the ditch usually to prevent flooding, and he thought the possibility of a flood like 828 flood happen in future is very high.

4.3. The cognition of detention pond

When the detention pond was planned to be built, interviewee campaigned for the Chief of Baoye Village. Interviewee said, "I agreed to the construction of detention pond at that time, but the ex-chief disagreed to this." He also said, "The ex-chief thought that flooding is more severe in the nearby village, but it didn't cause large damage in our village. If the detention pond is built in our village, it will make our village flood."

As explained earlier, same as the people who don't know the operation of detention pond, the ex-chief thought detention pond is a facility which helps others at ours own expense, and then protest the building. For this, interviewee also said that Kaohsiung City Government has hosted a seminar, and invited residents for joining the seminar. From the previous discussion, it can be seen that Kaohsiung City Government has received the objection of residents, and propagandized the functions and concept of detention pond to residents.

During the interview, interviewee emphasized constantly that the construction of detention pond will not make the flooding severe, also, it is impossible that water overflow from detention pond. He said, "All of residents in our village know that detention pond won't make flooding severe, unlike those media." He also knew detention pond's function is adjusting or detaining the flood, Detention pond can't completely put an end to flooding, but reduce the probability of flooding. As explained in this section, it is clear that the advocacy to residents which has done by Kaohsiung City Government is effective.

4.4. The expectation of detention pond

Interviewee said, "The residents in village hope detention pond is also a park in ordinary." Due to this, the detention pond was designed to construct some versatile facilities additionally, for example, open-air wooden stands and cement trails. The open-air wooden stands were used for holding open-air concerts originally, but interviewee thought the relief around the open-air wooden stands is uneven, so it is unsuitable to be used for holding open-air concerts. He also told the author with a little worry, building cement trails is a good intention originally, however, because the underground water level in the detention pond is a little high, it possibly appears moss due to humidity, it is a potential danger to walker when the moss doesn't be removed immediately. Furthermore, since the water in the detention pond is the sewage on the road, if we let residents immediately enter detention pond when clearing up, it may cause health concerns. Returning briefly to the subject of constructing versatile facilities, although interviewee thought these facilities caused some problems on management or using, he still agreed to take detention pond as park in ordinary.

During the interview, interviewee showed the confidence in flood control of detention pond repeatedly. He said, "Everyone said detention pond is good, because they (the residents) has compared, it didn't flooded during the heavy rain after detention pond has been constructed." He also said, "Actually, in the villages where flooded, the chiefs are stressful. So they also want the second detention pond can be constructed." Interviewee thought the budget which Kaohsiung City Government has spent on detention pond is still not enough.

4.5. The dilemma of detention pond

This section discusses the findings which emerged from the analysis presented in the previous sections and previous chapter. After detention pond has been built, the residents in watershed think flooding is decrease. Most of residents have correct concepts of detention pond. In addition, despite detention pond can help the drainage system become more efficient, detention pond still has a fatal flaw, it will flood when the detention pond has been filled with water.

Hence, we consider detention pond is an expedient measure base on the considering funds and the effectiveness, it can been revealed by a comparison of budget spent on detention pond and other flood controlling facilities. Taipei area flood control plan which started from 1960 has spent over sixteen hundred billion New Taiwan Dollars, although it has spent such much money, its return period still less than 200 years [10], it even still flooded in Taipei after the plan has been executed completely. And Kaohsiung City even larger than Taipei City, to achieve the flood control effectiveness now, Kaohsiung City Government only spent sixty point eight billion New Taiwan Dollars. Under the condition of considering inflation, budget spent by Taipeiese is more than thirty times as many as

IOP Conf. Series: Materials Science and Engineering **652** (2019) 012064 doi:10.1088/1757-899X/652/1/012064

budget spending on detention pond in Kaohsiung. Thus, we consider constructing detention pond is very economical.

5. Conclusions

5.1. It's important to consider the continuity of rainfall on space when we are doing hydrologic analysis.

When we are doing hydrologic analysis, not only the rainfalls inside watershed but also the rainfalls outside watershed cause runoff, and this runoff makes the allowable flow decreased. Therefore, if the continuity of rainfall on space is neglected, the capacity of flood discharge and detention will be overestimated.

5.2. Detention pond is efficient to make up for the weaknesses in the drainage system.

Detention pond can solve the flood in small region effectively, and helps the drainage system become more efficient. Also, because of its low cost, it is suitable for constructing in areas which don't have so many budgets. However, detention pond can't work well without other facility, like drain outlet. It still will flood when detention pond has been filled with water, thus, detention pond's improvement of the whole flood control capacity of the drainage system is limited. The case study in Kaohsiung City reveals that drain outlet is more important than detention pond, and the function of detention pond is affected easily by underground water level and relief. Hence, under the situation of unlimited budget, if we want to do long-term planning, drainage system should be upgraded comprehensively to keep sewer expedite. Only in the case which allowable flow of sewers are not allow to be improved due to budget or other factor, detention ponds are used for flood controlling.

5.3. Government needs to outreach policies to solve misunderstanding of public on infrastructures.

People may resist new infrastructures or systems, this phenomenon should be improved through the publicity of government. However, government shouldn't outreach policies unilaterally to public, but commutating with people to understand the reasons why they opposite, and trying solving the reasons. Also, if the infrastructures or systems have significant positive results after implementation, people will eager to accept these infrastructures and systems, even support it to keep implementing.

References

[1]. Everettwa City Government. "What is a Detention Pond?" (2016)
[2]. Tetsuya, and Nakada Tomoko, and Murakami Masahiro. "Estimation of reduction in CO2 emissions due to thermal environment mitigation effect of detention pond with diverse vegetation,"(2004). presented at 17th Japan Society of Hydrology and Water Resources Research presentation, 2004/8/20
[3]. Yu-jia, Qiu, and Shang-su, Shi. "Flood control, ecological, landscape, urban area, detention pond, three-in-one function," Construction News Record, 374, 27-40. (2014)
[4]. Sewage Systems Office, Public Works Bureau, Kaohsiung City Government. "The implementation of the plan for the detention pond in Baoyeli, Sanmin District, Kaohsiung City." (2009)
[5]. Che-Shi, Ding, and Zhen-kun, Zhong, and Han Meyer. "The Research of Kaohsiung Waterfront Redevelopment Under the Effect of Climate Change," National Pingtung University of Science and Technology. (2012)
[6]. Keng-Ta, Lin, and Jun-Jie, Liao. "A brief introduction of typical detention pond in ultra-urban areas: a case study of detention pond in the Bao-Ye, Kaohsiung," Water Resources Bureau, Kaohsiung City Government. (2012)
[7]. NCKU Research and Development Foundation. "The flood control review of the Kaohsiung Metropolitan Area MRT System." (2003)
[8]. Li, Yi. 2018/8/29. "The road turns into a river! The heavy rainfall is fierce," China Times.

IOP Conf. Series: Materials Science and Engineering **652** (2019) 012064 doi:10.1088/1757-899X/652/1/012064

[9]. Chow, V. T., and D.R Maidment, and L. W. Mays, Applied Hydrology, McGraw-Hill Book Co., New York. (1988)
[10]. Taipei Area Flood Control Project Implementation Center, Taiwan Province Government. "Taipei area flood control plan." (1983)

IOP Conf. Series: Materials Science and Engineering **652** (2019) 012065 doi:10.1088/1757-899X/652/1/012065

Research on Wire-Broken Monitoring of Bridge cable based on Acoustic Emission technique

Zhiqin Zhu[1] , Dawei Wang[1] ,Chen Liu[2], Bingjian Wang[3]

[1] Guangxi GuiHai Expressway Co., Ltd, Guangxi Communications Investment Group Co Ltd, Nanning 530000, Guangxi, China;
[2] School of Civil Engineering, Beijing jiaotong University, Beijijng , 100044, China
[3] Research Institute of Highway Ministry of Transport, Beijijng, 100088, China
Email: zz@yeah.net; dw@yeah.net; 17121056@bjtu.edu.cn; 45889572@qq.com

Abstract. Acoustic emission signals generated by broken prestressed steel strand of cable can transmit up to 29.5 meters without maintaining significant attenuation of the amplitude. Therefore, the installation of a few acoustic emission sensors can comprehensively monitor the damage of the entire prestressed steel strand. Thereby, the detection of unknown defects is realized, and the structural health status can be quantitatively and quantitatively evaluated. In this paper, the acoustic emission technology is used to monitor the damage and fracture of prestressed steel strand under over tension. The acoustic emission technology is used to test the broken wires of cable of a concrete-filled steel tubular arch bridge. The correlation of the broken wires test results of the bridge is qualitatively analysed, including the stress state of the cable, the location of the measuring points and whether there is a PVC sheath at the measuring points. Therefore, a safe, reliable and efficient method is sought to monitor the damage status of the in-service cable-supported bridge.

1. Introduction

Acoustic emission is a stress wave in the ultrasonic (generally 20~1000KHz) frequency range generated by the release of transient energy when the material reaches micro or macroscopic damage under stress. The amplitude, energy, and count of the wave are directly related to the material defect in the propagation[1,2], and the dynamic process of the defect itself can be monitored[3,4,5]. Casey's tests show that the acoustic emission signal generated by the broken prestressed steel strands can transmit up to 29.5 meters without maintaining a significant attenuation of the amplitude[6]. Therefore, the acoustic emission sensor with a few layouts can fully monitor the damage of the entire prestressed steel strand[7]. Thereby, the detection of unknown defects is realized, and the structural health status can be quantitatively and quantitatively evaluated[8,9]. Li Dong sheng et al. introduced the fractal dimension of time series fractal features into the acoustic emission monitoring of prestressed steel strand failure, and studied the acoustic emission identification technology of multi-age cable and corrosion damage[10-11].

Based on this purpose, the tensile test of prestressed steel strands is designed in this paper. The characteristic parameters of acoustic emission signals related to stress damage and fracture are obtained, which can provide relevant monitoring data for the tension of prestressed steel strands in the future.

2. Acoustic emission monitoring test scheme for stress damage of prestressed steel strands

2.1. Model Description

In this test, the sprayed prestressed steel strand is used as the test piece, and the prestressed steel strand is composed of 7 steel wires wound, the total length is 3000mm, which is passed through the connecting cylinder and the jack, and then the prestressed steel strand is clamped by the clip to perform tensioning. The test device is shown in Figure 1. The acoustic emission sensor is arranged on the anchors at both ends, wherein the sensors 1 and 2 are at the right chuck; the 3rd and 4th channels are at the left chuck, as shown in picture 2.

Figure 1. Photo of the test machine **Figure 2. Anchorage and sensor**

The line positioning technology is used to determine the position of the broken wire. The specific acoustic emission parameters are set as follows: the center frequency of the sensor is 150**kHz**. The gain of the preamplifier is 40 dB. The filtering frequency is 100~600 **kHz**, and the system threshold first tests the ambient noise level. The noise level was found to be below 45 dB, so the threshold was set to 45 dB and the main gain was 20**dB**. Impact signal timing parameter setting: peak definition time PDT is 300μs, impact definition time is 600**μs**, and impact lock time is 1000**μs**. This test uses acoustic emission parameter analysis methods to reveal the acoustic emission characteristics of steel strand damage.The test was carried out in ten stages by loading 40kN per step. When loading to 400**kN**, it is considered that theprestressed steel strand has no working ability and the acoustic emission monitoring is carried out in the whole process.

2.2. Attenuation and coupling test

Before the test begins, the test device and the acquisition device are first subjected to signal attenuation test. With the centre punch, the impact is three times on the left side of the prestressed steel strand near the 3rd and 4th channels, and the amplitude of the received signal is compared with the acoustic emission channel. After comparison, the signal attenuation is small. The sensor and test device have been arranged reasonably to meet the test conditions. The result is shown in Figure 3 below:

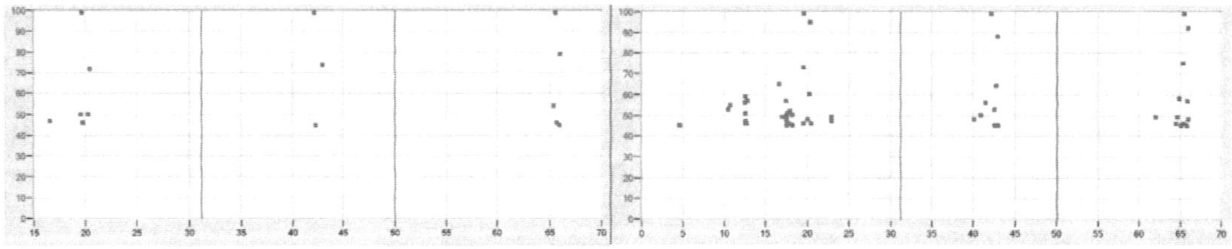

(a) (b)

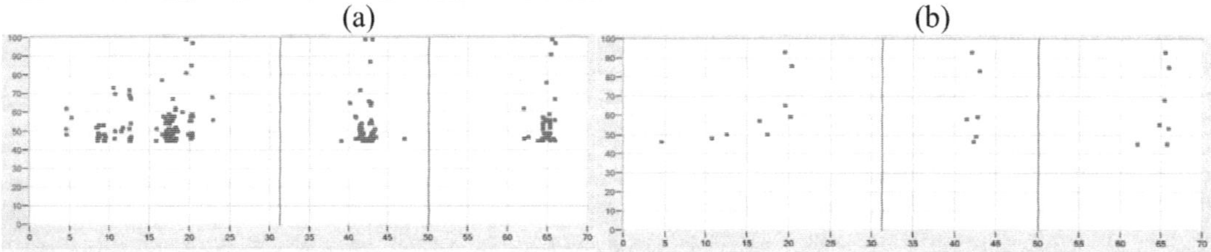

(c) (d)

Figure 3. attenuation, coupling test (amplitude vs time)

2.3. Prestressed stranded wire breaking test

Damage monitoring of prestressed steel strands is the most important purpose of applying acoustic emission technology. In order to quantify the damage process, the most appropriate parameter analysis method is related-point-graph analysis method.

In order to judge the degree of damage of the broken wire, the broken wire factor diagram is used to monitor the breakage of the prestressed steel strand during the loading process. When the test is carried out to 750s, the remaining six wires are broken except only the one in the center remained intact. The break position is at the right chuck. During the 700 seconds monitoring period, the noise levels included due to tension and friction between the wire and the clamp during stretching continue to present low signal levels. However, at the moment when the prestressed steel strand is broken, the breaking factor is increased by at least two orders of magnitude.

In order to judge the position of the broken wire, the acoustic emission linear positioning method is used to predict the fracture position of the prestressed steel strand. The position of the source determined by it is on the line or arc of the line connecting the two transducers. The positional coordinates of the acoustic emission source $P(x)$ can be determined by the following formula (1):

$$P(x) = \mathrm{sgn}u(\Delta t) \cdot \frac{\Delta t}{2} \cdot v \qquad (1)$$

Where $\mathrm{sgn}u(\Delta t)$ is the sign function; Δt is the time difference between the arrival of the two transducers; v is the speed of sound[]. The acoustic emission location map is shown in Figure 5:

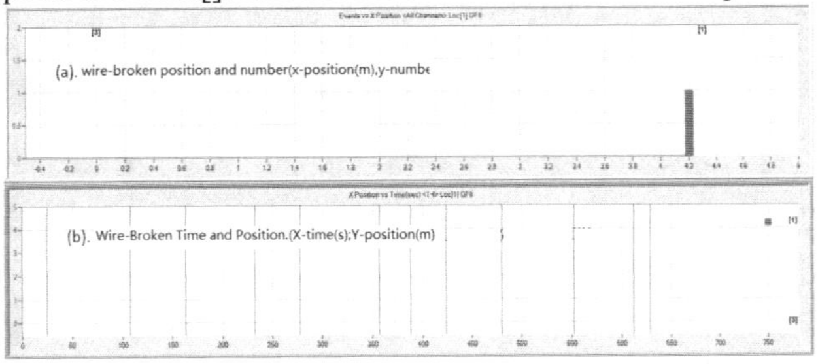

Figure 4. Location map(The red column position in the figure is the fracture position of the software.)

Figure 4 shows the positioning of the broken wire, wherein Figure 4(a) shows the number of positioning and the positioning position, that is, the positioning occurs at the sensors 1 and 2 near the right. Since the six broken wires are simultaneous, the number of times of positioning is one. Figure 4(b) shows the time and location of the positioning, that is, the broken wire occurs 750 seconds after the start of monitoring, and its position is close to 1st, 2nd sensors.

3. Engineering Application

In a concrete-filled steel tubular arch bridge as shown in Figure 5, the original cable of the bridge needs to be replaced in the bridge deck replacement and repair and reinforcement project. Monitoring the broken wire that will be removed on the bridge cable. In order to simulate the evolution process of the broken wire damage of the cable, oxygen cutting is adopted to break the wire.

IOP Conf. Series: Materials Science and Engineering **652** (2019) 012065 doi:10.1088/1757-899X/652/1/012065

Figure 5. Photo of the deck and bridge side

The bridge cable uses PES7-61 high strength parallel steel wire as shown in Figure 6. The process of removing the cable is as follows: firstly, the bridge deck near the cable is cut off, and then the pulling force of the cables on both sides is controlled by the hoist to reduce the tension of the cable until the drawbar is bent and does not bear the force, and then the oxygen is cut respectively. Cut off the cable and bridge arch at the junction with the bridge deck to complete the disassembly of the cable.

Figure 6. Disassemble the cable (the middle orange is the cable to be removed)

The sensor is placed at three positions of the cable, the upper anchor head (position 3), and the bridge deck (position 1, position 2). As shown in Figure 7~8:

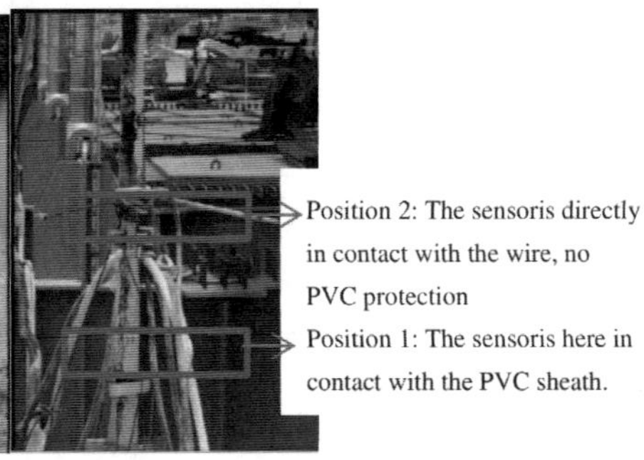

Figure 7. Upper anchor head (position 3) **Figure 8.** Close to the deck(0.7-1m higher than anchorage)

The probe at the upper anchor head and the position near the bridge deck is directly in contact with the steel wire, and is protected by the PVC sheath. The probe is in contact with the PVC sheath at position 2 near the deck.

3.1. Noise test

During the test, both ends of the bridge were closed and no vehicles were allowed to pass, and there were cutting bridges near the cable. From the data collected, the noise is relatively low and the monitoring conditions are met as shown in Figure 9.

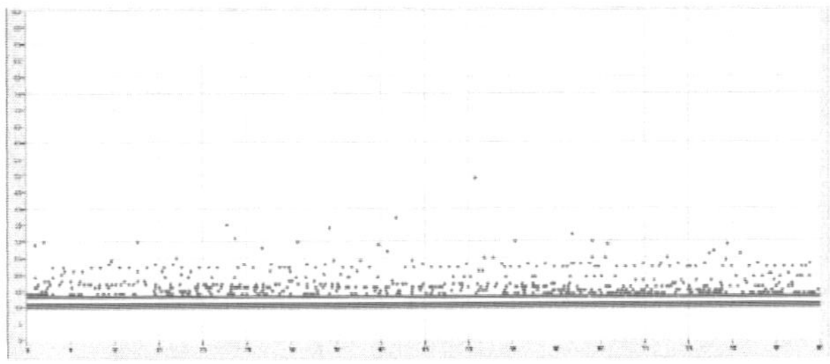

Figure 9. Noise test (magnitude vs time)

3.2. Attenuation Experiments

During the test, the center punch is used to perform the impact at position 2 (removing the PVC sheath), and each impact is equivalent to transmitting a signal. It can be seen from the test comparison that the presence of the PVC sheath has a large attenuation of the signal. Therefore, when installing the sensor, the sensor should be placed as far as possible in the metal contact with the tie rod wire or the tie rod fastening member, such as position 3. The collection threshold is 40dB. The two sets of data are as Table 1 and Figure 10. The position of the determined signal is accurate as figure 11.

Table 1. Sling Acoustic Emission Test Signal Attenuation Data Unit: dB

First group(sling natural extension state)	Position 2●	97	97	90	96
	Position 1	69	89	57	88
	Position 3	<40	<40	<40	<40
Second Group(the sling force is initially tightened)	Position 2●	93	96	99	
	Position 1	74	75	88	
	Position 3	53	53	51	

Note: ● is the impact position

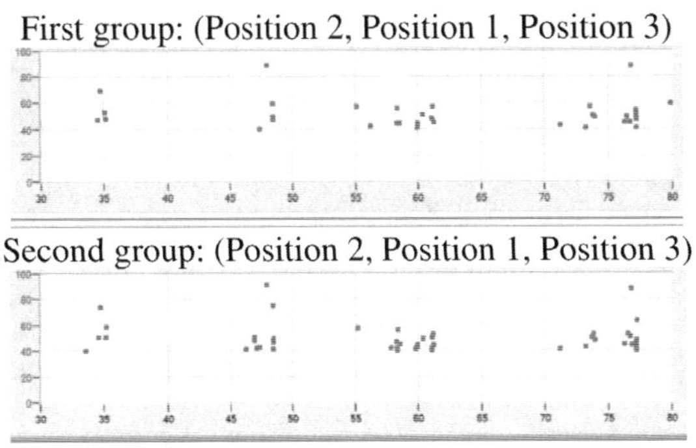

Figure 10. Attenuation test (magnitude vs time)

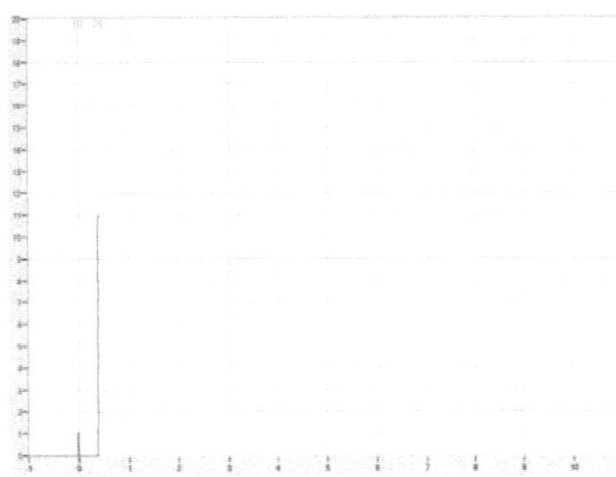

Figure 11. Location map

3.3. Wire-broken monitoring

When the wire-broken monitoring, the hoist is used to control the cable until the cable is in a straight state from bending to initial tension, and the position of the hoist is kept fixed. At this time, the wire is cut by oxygen welding, and a total of 7 wires are cut. The specific test results are shown in the following figure 12.

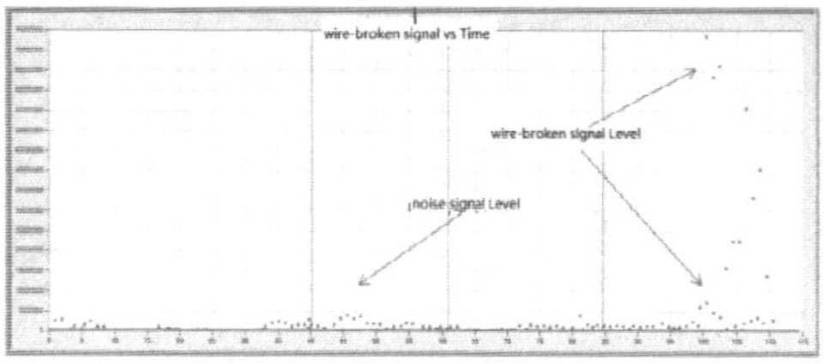

Figure 12. Broken wire factor monitoring

As can be seen from the above figure, the acoustic emission signal corresponding to the wire being fusing is very significant. Moreover, the wire-fused acoustic emission signals corresponding to different stress levels are significantly different. The normally used high-stress state sling wire breakage acoustic emission signal is more significant and easy to identify.

4. Conclusion

The main purpose of this paper is to judge the damage degree and position of the steel strand, and to conduct the wire-broken monitoring and evaluation by the broken wire factor and the position criterion. Finally, the test results are in good agreement with the monitoring results, and the error is about 5%. Acoustic emission technology is used to test the broken wire of a concrete-filled steel tubular arch bridge. The correlation qualitative analysis between the cable stress state, the position of the measuring point and the measuring point of the PVC sheath on the wire-broken monitoring result of the acoustic emission suspension system is studied. The experimental results show that in the low noise environment, the force of the cable is positively correlated with the apparent effect of the test. It fully demonstrates that the whole process of monitoring the cable damage by using acoustic emission technology has a good tracking effect, which can not only determine the moment of occurrence of the cabledamage, but also determine the location of the cable-broken.

Acknowledgement

We'd like to take the opportunity to thanks Guangxi Program of Science and Technology (AC16380126) for their great support to this research.

Reference

[1] Wang Pingguang. Acoustic emission monitoring and pattern recognition of bridge cable corrosion damage [D]. Dalian: Dalian University of Technology, 2015.
[2] Farnam Y, Geiker M R, Bentz D, et al. Acoustic emission waveform characterization of crackorigin and mode in fractured and ASR damaged concrete[J]. Cement and Concrete Composites,2015, 60: 135-145.
[3] Dan E J, Shen H w, Finlayson R D, et al. Acoustic emission evaluation of reinforced concretebridge beam with graphite composite laminate[J]. SPIE, 2001, 4330: 498-504.
[4] Duke J c, Sison M. Acoustic emission monitoring of steel bridge members final report[J]. TheVirginia Transportation Research Council, 1995: 9-11.
[5] Liu x, Hiroshi H Y, Inaba T, et al. Damage evaluation of railway structures by usingtrain-induced AE[J]. Construction and Building Materials, 2004, 18(3): 215-223.
[6] Degroot P J, Wijnen P, Janssen R. Real-time frequency determination of acoustic emission fordifferent fracture mechanisms in carbon epoxy composites[J]. Cmoposites Science andTechnology, 1995, 55(4): 405-412
[7] Zhao Guangyu. Study on acoustic emission characteristics of meso-damage process of ductile metal [D]. Daqing: Northeast Petroleum University, 2014. Zhang Ping. Research on integrated acoustic emission signal processing platform [D]. Beijing: Tsinghua University, 2002

[8] Sunil P, Jason w. Assessment of localized damage in concrete under compression using acousticemission[J]. Journal of Material in Civil Engineering, 2006, 18(3): 325-333.
[9] Chen H L, Cheng c T, Chen s E. Determination of fracture paramaters of mortar and concretebeams by using acoustic emission[J]. Materials Evaluation, 1992, 50(7): 888-894.
[10] Li Dongsheng. Monitoring and Evaluation of Acoustic Emission in Damage of Civil Engineering Structures—Theories, Methods and Applications [M]. Science Press, 2017.9
[11] Li Dongsheng, Ou Jinping. Acoustic Emission Monitoring of Bridge Cable Damage and Its Critical Damage Judgment[J].Prestressing Technology,2009(5):3-6.

IOP Conf. Series: Materials Science and Engineering **652** (2019) 012066 doi:10.1088/1757-899X/652/1/012066

Key factors affecting the rent of office buildings: case study on Taipei 101

Tse-Hsiung Lin* and Yeng-Horng Perng

Department of Architecture, National Taiwan University of Science and Technology, Taipei City 10607, Taiwan

*Email: barry.lin@tfc101.com.tw

Abstract. An exploration of the rent of office buildings in metropolises worldwide indicates that in response to the level of overall economic development, tenants often need to consider numerous factors when selecting the location of their offices, whereas landlords attempt to increase the rent of their buildings by using high-grade building materials to renovate the buildings and offering new facilities and property management services to meet the various needs of tenants. This study adopted the Delphic hierarchy process and analytic hierarchy process to explore office building rent in Taipei City by distributing a questionnaire to experts in the real estate field who possess practical experience in executing real estate projects. Specifically, the study investigated how office building rent is affected by factors such as geographical location and public facilities, thereby generating a reference for identifying related key factors. Analyzing the data collected from the questionnaire and the weighted indictor scores revealed that location in the main business district is the most crucial factor, followed by proximity to a Taipei Mass Rapid Transit station and the office building being a renowned landmark. By contrast, adequate property management is not a primary factor considered by the experts.

1. Introduction

After a country or region achieves notable economic development goals, the development focus tends to shift toward constructing unique buildings as landmarks. Therefore, office buildings have become a crucial indicator of the extent of regional economic development and a symbol representing the wealth and prosperity of a region. Fluctuations in economic development exert a direct effect on the management of office buildings. Of multiple real estate development types (e.g., office buildings, residential buildings, commercial buildings, hotels, and factories), office buildings are considered an investment item with a relatively high return because of their favorable lease performance over the long term and stable income generation. However, office building rent is affected by factors such as the economy, geographical location, transportation convenience, and market positioning, all of which cause fluctuations in rent [1]. The conventional approach for leasing office buildings entails landlords determining the rent according to their market experience. However, such an approach easily overlooks geographical factors, inhibiting the ability of a landlord to find adequate tenants.

In contrast to regular buildings, an office building can be viewed as a vertical city that contains various problems concerning safety, transportation, hygiene, and energy consumption that are difficult to resolve. Office buildings with a higher number of floors tend to exhibit more problems related to safety, durability, and comfort, and property management requirements for such buildings are also extensive in the areas of physical structure, building features, electromechanical components, air conditioning, and elevators.

The rental market for office buildings in Taipei City is deeply affected by the overall economy and political activities. In addition, the demand for office buildings in Taiwan has decreased in recent years, but developers and builders have continued to erect additional buildings. From 2008 to 2015, the total floor areas of unrented A-grade office buildings in Taipei City increased from 18 million ft2 to 24 million ft2 (an increase by 6.5 million ft2 in seven years). Amidst an excessive supply of office buildings lingering on the rental market for an extensive period, landlords must also contend with the negative effect of economic recession when attempting to lease their buildings.

2. Research methods

This study conducted in-depth interviews [2] with experts who have participated in and executed real estate projects. After the experts' opinions were collected, related literature and actual cases of office rentals in Taipei 101 were reviewed to establish indicators of factors that affect rent. Finally, an anonymous questionnaire survey was conducted. The Delphic hierarchy process (DHP) and analytical hierarchy process (AHP; Satty, 1980) were employed to construct a core indicator framework and rank the weighted indicators.

2.1. Expert interviews

The experts consisted of members of management teams in domestic and international enterprises as well as professional consultants who have participated in real estate projects. According to their professional backgrounds, the experts were divided into two groups, namely the management group and consultant group. A total of 26 experts were invited. Table 1 summarizes the composition of the experts.

Table 1. Backgrounds of experts in each group.

Group	Type	No. of experts	Ratio	Total
Management group	Taipei 101	4	15.4%	26.93%
	Local enterprise	1	3.84%	
	Foreign enterprise	2	7.69%	
Consultant group	Tenant coordinator	12	46.15%	73.07%
	Real estate appraiser	4	15.38%	
	Property management consultant	3	11.54%	
	Total	26	100%	100%

Hartman (1981) proposed an anonymous written discussion that could help experts of different professional backgrounds reach a consensus [3]. In addition, in-depth interviews can be conducted with experts to identify new core values associated with the rent of office buildings.

2.2. Literature review

Various theories have been proposed regarding the determination of rent for office buildings, including the bid rent function theory proposed by Alonso in 1960 [4]. Alonso's study concurs with that of Huang and Chang (2005) [5], who explored the office location shifts of large-scale enterprises in the Taipei Metropolitan Area and reported that accessibility to comprehensive facilities and distance from the central business area affected the amount of rent that tenants were willing to pay. Holloway and Wheeler [6] used the data of top 500 enterprises compiled by Fortune magazine to analyze office location shifts between 1980 and 1897. Lin [7] divided factors affecting the determination of office building rent into regular, regional, and individual factors; in addition, physical factors such as design quality, building materials, facilities, and property management also affect the rent. According to these prior works, factors affecting office building rent can be divided into economic, location, and physical factors. Economic factors include economic growth [8] and the consumer price index, location factors include transportation convenience and business district characteristics, and physical factors include building area and building age. [9] On the basis of these three major factor types, this study compiled 25 core indicators to establish an evaluation framework for office building rent (Table 2).

Table 2. Key factors affecting office building rent

Goal	Level 1	Level 2
Key factors affecting office building rent	Building facilities	1. Ceiling height
		2. Seismic damper
		3. High-grade building materials
		4. Independent air conditioning
		5. Backup power supply
	Geographical location	6. Main business district
		7. Shopping district
		8. Restaurant selection
		9. Zoning
		10. Recreational space
	Nearby facilities	11. Public facilities
		12. Business facilities
		13. Two power stations supply
		14. Healthcare resources
		15. Accommodation selection
	Transportation convenience	16. Proximity to Taipei Mass Rapid Transit (MRT) stations
		17. Proximity to highway interchanges
		18. Parking space
		19. Public transportation
		20. Road connection to major networks
	Image	21. Renowned landmark
		22. Green building and smart building[10]
		23. Friendly neighbors
		24. Social image
		25. Adequate property management

3. Questionnaire survey and DHP/AHP Analysis

Also called the expert judgement method, the Delphi method is a type of group decision-making approach commonly used in qualitative research. In this method, experts' experience and knowledge are adopted to address a specific topic through multiple rounds of feedback and questioning, thereby minimizing the discrepancy between the experts' opinions. The AHP is mainly employed when addressing a decision problem that involves unknown or uncertain situations with numerous evaluative criteria. This method is prevalently used in behavioral science, marketing strategies, and investment portfolios.

Khorramshahgol and Moustakis proposed the DHP in 1988 [11]. In this method, the establishment of the hierarchy and comparison matrix is accomplished through the Delphi method, and the remaining process is identical to that of the AHP. The AHP is a multicriteria decision theory developed by Saaty in 1990 [12] to systemize complex problems and facilitate the selection of the optimal decision. When a problem entails more than one evaluation factor, multicriteria decision-making can be adopted to simplify the problem. The AHP not only helps experts reach a consensus, but also uses standardized criteria to enhance the reliability of the results, thereby ensuring the rigidity of the evaluation. Through using the AHP, the evaluative value and weight of each factor can be identified to serve as a reference for decision-making. The DHP not only features the advantages of the AHP but also utilizes the group

decision-making model of the Delphi method to reduce the discrepancy between experts' subjective opinions that occur when the AHP is used to establish a hierarchy and perform pair-wise comparisons of different individuals or groups [13].
This study adopted the AHP to analyze data collected from the questionnaire. Microsoft Excel was used to calculate the weight of each indicator. A total of 40 expert questionnaires were distributed; 36 responses were returned, and 26 were valid (Table 3).

Table 3. Statistics of expert questionnaires.

Group	Target	No. of distributed copies	No. of returned copies	No. of valid copies
Management group	Finance department	5	4	4
	Engineering department	5	5	1
	Management department	5	5	2
Consultant group	Tenant coordinator	10	10	12
	Real estate appraiser	5	5	4
	Property management consultant	10	7	3
	Total	40	36	26

The questionnaire return rate was 90%. Analysis of the consistency of the valid questionnaire responses yielded a consistency ratio of ≤0.1 when all collected data were included in the analysis. As shown in Table 4, the weight ranking of the Level 1 constructs was in the following descending order: geographical location (29.99%), transportation convenience (24.25%), image (20.41%), building facilities (15.06%), and nearby facilities (10.29%). This indicates that geographical location is the most crucial factor affecting office building rent, followed by transportation convenience.

Table 4. Analysis of Level 1 constructs.

Rank	Item	Weight of the management group	Weight of the consultant group	Combined weight
1	Geographical location	29.92%	30.07%	29.99%
2	Transportation convenience	17.23%	31.28%	24.25%
3	Image	27.54%	13.28%	20.41%
4	Building facilities	19.52%	10.60%	15.06%
5	Nearby facilities	5.80%	14.77%	10.29%
	Total	100%	100%	100%

After the data were summed, the five most essential indicators were identified from among the 25 indicators established in this study. In descending order of importance, these indicators were main business district (11.57%), proximity to MRT stations (9.97%), renowned landmark (7.65%), restaurant selection (7.46%), and public transportation (6.29%). The ranking of the weighted indicators is summarized in Table 5.

Table 5. Ranking of the weighted indicators using the AHP.

Level 1	Level 2	Cumulative weight	Rank
Building facilities 15.06%	1. Ceiling height	1.17%	25
	2. Seismic damper	3.22%	10
	3. High-grade building materials	1.30%	14
	4. Independent air conditioning	2.37%	20
	5. Backup power supply	2.69%	16
Geographical location 29.99%	6. Main business district	11.57%	1
	7. Shopping district	3.12%	11
	8. Restaurant selection	7.46%	4
	9. Zoning	3.54%	9
	10. Recreational space	2.67%	17
Nearby facilities 10.29%	11. Public facilities	2.84%	14
	12. Business facilities	2.83%	15
	13. Dual power supply	2.48%	18
	14. Healthcare resources	1.49%	23
	15. Accommodation selection	1.78%	22
Transportation convenience 24.25%	16. Proximity to MRT stations	9.97%	2
	17. Proximity to highway interchanges	2.26%	21
	18. Parking space	4.03%	8
	19. Public transportation	6.29%	5
	20. Road connection to major networks	3.12%	12
Image 20.21%	21. Renowned landmark	7.65%	3
	22. Green building and smart building	2.92%	13
	23. Friendly neighbors	4.98%	7
	24. Social image	2.38%	19
	25. Adequate property management	5.88%	6

4. Conclusions

Conventional approaches for determining office building rent have overlooked the habits and intentions of tenants, resulting in a large gap between landlords and tenants regarding the lease rate, and tenants usually need to spend a considerable amount of time persuading landlords to reduce their rent. However, the occupancy rate of rented offices in Taipei 101, which was once the tallest building in the world and has received the Leadership in Energy and Environmental Design platinum certification, has continued to increase each year. In-depth analysis in this study indicated that the rental strategies adopted by Taipei 101 differ from those of other office buildings. In line with the egg theory proposed by Kostolany [14], the strategies of Taipei 101 involve regularly observing changes in market demand and supply and adjusting rent on an irregular basis. This supports its ability to meet the needs of tenants; hence, the occupancy rate of Taipei 101 remains high.

Analysis of the data collected from the expert questionnaire revealed that of the factors affecting office building rent, the main business district indicator in the geographical location construct is most crucial, followed by proximity to MRT stations, being a renowned landmark, and restaurant selection. By contrast, hardware-related indicators such as building materials and business facilities, which are

IOP Conf. Series: Materials Science and Engineering **652** (2019) 012066 doi:10.1088/1757-899X/652/1/012066

considered essential factors in conventional rental strategies, are actually not as crucial as other indicators. This analysis result accorded with the conclusion reached by the experts in this study.

This study discovered that because of the flourishing economic activity in the main business district, enterprises are more willing to accept high rent for office buildings in this area; hence, this consideration becomes the primary factor affecting office building rent. In addition, increasing awareness of employee benefits also increases the importance of factors such as transportation convenience, restaurant selection, and environmental safety.

In response to the changes in the overall economic trend and business district scale, how to prevent subjective factors from affecting rental strategies is a problem to be resolved by landlords. By using evaluative indicators that are reasonable and logical, landlords can devise rental strategies in line with market needs, thereby achieving their operational goals and generating maximal profits within the shortest period of time.

References

[1] Hekman JS 1985 *Real Estate Econ.* **13** 32-47

[2] Jin Z and Yanhong D 2013 *Journal of Puyang Vocational and Technical College* **26**

[3] Hartman A 1981 *Educ. Leadersh.* **38**(6) 495-497

[4] Alonso W 1960 *Big rent function theory*

[5] Huang MY and Chang CO 2005 *Manage. Rev.* **24**(4)

[6] Holloway S and Wheeler J 1991 *Econ. Geogr.* **67**(1) 55-72

[7] Lin YY 2006 *Real-estate appraisal Q&A* (Taipei: Wensheng Book Store)

[8] Glindro ET, Subhanij T, Szeto J and Zhu H 2011 *Int. J. Cent. Bank.* **7**(3) 163-204

[9] Cheng-Yao J 2018 A Study of the Interactions between Taipei Office Submarkets

[10] Eichholtz, P and Quigley JM 2012 *Europ. Econ. Rev.* **56**(5) 903-904

[11] Khorramshahgol R and Moustakis VS 1988 *Eur. J. Oper. Res.* **37** 347-354

[12] Saaty TL 1990 *Eur. J. Oper. Res.* **48**(1), 9-21

[13] Yang DJ, Chen Y and Yeh CL 2005 *J. Public Aff. Rev.* **6**(1) 76-110

[14] Kostolany A 1906-1999 *Egg theory*

Exploration of key commercial space design indicators

Mu-Fa Lin[1,*] and Yeng-Horng Perng[2]

[1]Department of Architecture, National Taiwan University of Science and Technology, Taiwan
[2]Department of Architecture, National Taiwan University of Science and Technology, Taiwan
*Corresponding author, Email: D10613013@mail.ntust.edu.tw

Abstract. The designs of commercial spaces change with time. This study explored competitive factors that affect commercial space development by using the Delphic hierarchy process with the aim of enhancing the overall performance of commercial spaces in a highly competitive retail market. The findings provide key factors that can be used by shopping center operators when selecting commercial space designers. Competitiveness indices were proposed by referring to the literature and conducting interviews with industry experts, scholars, and practitioners. Subsequently, hierarchy structures and indices were established, and relative weights were allocated to each dimension and index. Finally, the evaluation data were calculated and consistency tests were conducted. This study adopted 18 performance indicators to demonstrate the validity of the overall competitiveness indicators, which were verified as possessing high validity.

1. Introduction

Construction of commercial spaces in buildings is a worldwide trend, including in the United States and Europe; these spaces serve as representative and prominent landmarks of a city or country. Presentations of commercial and corporate cultures in commercial spaces vary depending on the shopping center operator, with the design of commercial spaces being a topic that warrants investigation. This study sought to determine the key indicators for commercial space design in buildings, with the results able to serve as a reference for shopping center operators.

2. Research Motivation and Objectives

Commercial space design is critical for attracting businesses to shopping centers and maintaining profitable operations. Shopping centers not only serve as commercial hubs but also as diverse cultural spaces. Therefore, each shopping center has unique characteristics and a sense of place that integrates local ethos. Consumers may visit shopping centers for both the products available and the sense of place conveyed. Maximizing a center's sense of place is becoming a crucial strategy for management because shopping centers increasingly emphasize their integration with local cultures and highlight their distinctive features to attract a diverse customer base [1].

The present study can serve as a useful reference for investors and operators of shopping centers to select appropriate commercial space designers capable of creating landmarks that are unique while combining local and corporate culture.

Alongside a literature review, this study conducted in- depth interviews with experts and practitioners in various fields related to interior design and referred to actual cases of interior design for commercial spaces to construct an indicator framework, and the weights of the indicators were subsequently analyzed. The results can serve as a reference for shopping center investors and operators to evaluate weight distributions and formulate management strategies. Investors and operators of well- known shopping centers in Taiwan, international brand extension project operators, and commercial space designers were invited to share their experiences and provide opinions.

The Delphic hierarchy process was adopted to establish an evaluation indicator framework for use by shopping centers in selecting commercial space designs. The percentage distribution of the evaluation indicators was determined for shopping centers to select designers for interior space design. The objectives of this study were as follows:

To explore the evaluation indicators for commercial space design.

To establish a design framework for commercial space and analyze the priorities and weights of key indicators.

3. Literature Review

Shopping centers adopt a variety of methods to create connections with the local sense of place [1]. Place attachment is formed through material and local characteristics [2]. Place meaning can be realized through the creation of an attractive space, and sense of place and social awareness can be achieved through social and promotional activities.

That these consumers live between two worlds span two cultures or have to continuously cross borders.[2].

A commercial space can be adjusted and modularized dynamically in response to interference and thresholds on a spatial scale [3].

As one of the important components of comprehensive space development, the mall. Played an increasingly important role in the development process in recent years, and its design. The program has been continuously improved and improved, and the huge social effects it produces and economic effects have a great impact on consumer behavior .[4]

Macao is adjusting its tourism appeal. It should also adjust Macau's shopping methods and habits to create a variety of shopping malls. This is not far from the goal of "suitable tourism". After all, shopping is indispensable for tourists. [5]

Considering the idle business in Macau, what is the current situation, whether it has potential, and whether it has its own personality. These are the factors that should be referenced in the future. [6]

Another strategy adopted by shopping centers is multiple combinations of stores. This strategy is primarily used to control which stores can open in a shopping center in order to attract a specific type of consumer. Guimarães [7] noted that the design of a shopping center must consider the following factors: the (1) innovation, (2) uniqueness, (3) ease of maintenance, (4) maintenance cost, and (5) modularizability of a space.

3.1 Delphi method

Captions should be typed in 9-point Times. They should be centred above the tables and flush left beneath the figures.

3.1.1 Characteristics of the Delphi method.

1. Anonymity.
2. Controlled feedback.
3. Statistical group response.

3.1.2 Steps of the Delphi method.

1. Experts who specialize in topics related to the questions to be investigated are invited as questionnaire respondents.
2. A questionnaire is designed for the first stage of the questionnaire survey.
3. Expert opinions collected from the first stage of the questionnaire survey are compiled and the median of these opinions is determined. The compiled data are provided for the experts as a reference, and they are then asked to complete the questionnaire again based on these reference data.
4. The experts' revised opinions are summarized and explained.

IOP Conf. Series: Materials Science and Engineering **652** (2019) 012067 doi:10.1088/1757-899X/652/1/012067

5. Whether the experts' opinions converge within an acceptable range is determined.
Steps 3–5 are repeated until consensus is reached or the participants are unable to further revise their opinions. Questionnaire results Table 1.

3.1.3 Shortcomings of the Delphi method.
1. Cognitive aspects.
During the questionnaire survey, the ambiguity in questions may cause experts to interpret the questions differently. In addition, the experts are not allowed to communicate with one another when completing the questionnaire. Therefore, their opinions may not reach consensus. When compiling expert opinions, the decision makers may have preconceptions that cause them to filter out correct opinions. That is, different opinions may be systematically reduced or even eliminated. When seeking consensus, the experts are asked to revise their opinions based on the median of the group's opinions. This may distort the experts' original thoughts.
2. Cost-related aspects.
When consensus cannot be reached among the expert opinions, the questionnaire survey must be conducted several times. However, this may increase the time and monetary costs and reduce the response rate, which can result in an excessively low response rate or the inability to retrieve responses.

3.2 Analytic hierarchy process
The analytic hierarchy process (AHP) systematizes complex decisions into a hierarchical system of simple elements through collection and evaluation of expert opinions. Operators must decompose complex decision problems and construct a hierarchical relationship to compare and evaluate the elements of said problem. After expert evaluation and mathematical calculation, solutions are prioritized based on the numerical values of factors to serve as a reference for operators to make final decisions [8]. However, when operators face highly diversified alternative solutions and must consider many factors, they may have difficulty making correct decisions through simple analysis or according to rule of thumb. Therefore, the objective of the AHP is to decompose a problem into a tree-structure hierarchy and to establish a hierarchical structure outlining the mutual influence of the constituent elements. Through quantitative judgment, the context can be identified and evaluated to provide sufficient information for decision makers to choose optimal solutions, thereby reducing the risk of errant decision-making [9].
Questionnaire results Table 2. AHP questionnaire implementation process **Figure 1**.

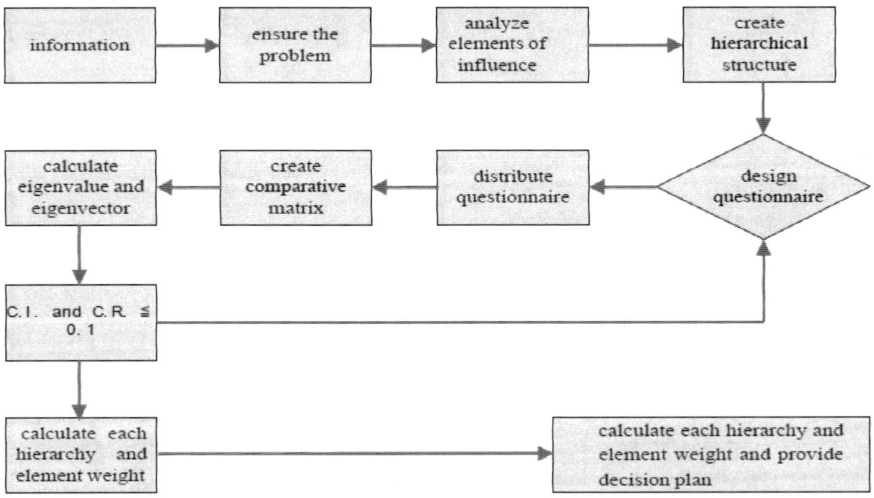

Figure 1. AHP questionnaire implementation process [8].

IOP Conf. Series: Materials Science and Engineering **652** (2019) 012067 doi:10.1088/1757-899X/652/1/012067

4. Data analysis and reviews

Table 1. AHP questionnaire data.

	Criteria	Expert number / subcriteria	Total T	Mean M	Standard deviation SD	Mode MO	Absolute value |MO-M|
1	Value creation	Corporate identity system	134	4.47	0.6288	5	0.533
2		Cross-industry cooperation	136	4.53	0.6814	5	0.467
3		Neighborhood connection	137	4.57	0.504	5	0.433
4	Economic advantages	Introducing resources	137	4.57	0.5683	5	0.433
5		Creative revenue streams	136	4.53	0.5074	5	0.467
6		Budget and schedule control	139	4.63	0.5561	5	0.367
7		Easy to maintain	134	4.47	0.7303	5	0.533
8		Low operation cost	134	4.47	0.5713	5	0.533
9		Space modularization and easy adjustment	132	4.40	0.724	5	0.600
10		Potential for renting for multiple purposes	128	4.27	0.6915	4	0.267
11	Tender characteristics	design company's experience	132	4.40	0.6215	5	0.600
12		Impressive presentation	131	4.37	0.7649	5	0.633
13		regulation compliance	134	4.47	0.5713	5	0.533
14		word of mouth	134	4.47	0.5074	4	0.467
15	Environmental consciousness	green material	129	4.30	0.7497	5	0.700
16		energy saving	134	4.47	0.5713	5	0.533
17		water saving	133	4.43	0.6261	5	0.567
18		Material recycling and reuse	135	4.50	0.5724	5	0.500
					0.622		**0.509**

Table 2. AHP questionnaire data.

Criteria	Level (1) Wi		Sub-Criteria	Level (2) Wi	Overall Wi	Ov erall	Criteria
Value creation	29.5297%	1	Corporate identity system	35.0960%	8.7740%	2	2.00
		2	Cross-industry cooperation	23.4047%	5.8512%	7	
		3	Neighborhood connection	18.6948%	4.6737%	12	
		4	Introducing resources	13.0362%	3.2591%	13	
		5	Creative revenue streams	9.7682%	2.4421%	16	
Economic advantages	35.3070%	6	Budget and schedule control	24.8005%	6.2001%	6	1.00
		7	Easy to maintain	32.5157%	8.1289%	5	
		8	Low operation cost	22.9786%	5.7447%	8	
		9	Space modularization and easy adjustment	12.8343%	3.2086%	14	
		10	Potential for renting for multiple purposes	6.8708%	1.7177%	18	
Tender characteristi cs	24.2879%	11	design company's experience	34.5932%	8.6483%	3	3.00
		12	Impressive presentation	32.9946%	8.2486%	4	
		13	regulation compliance	21.9549%	5.4887%	9	
		14	word of mouth	10.4573%	2.6143%	15	
environment al conscious	10.8755%	15	green material	21.4589%	5.3647%	10	4.00
		16	energy saving	46.8084%	11.7021%	1	
		17	water saving	21.4107%	5.3527%	11	
		18	Material recycling and reuse	10.3219%	2.5805%	15	

5. Conclusion

5.1 Strength of the top five most crucial factors

Overall Sequence
NO.01 energy conservation ---------------(11.7021%)
NO.02 Corporate identity system---------(8.7740%)
NO.03 design company's experience----(8.6483%)
NO.04 Impressive presentation-----------(8.2486%)
NO.05 Easy to maintain--------------------(8.1289%)

According to the results of the questionnaire analysis, commercial space designers and operators contended that the most crucial factor for the design and management of commercial space was energy conservation, followed by a strong corporate identity system and the experience of the design company. This indicates that both designers and shopping center operators recognize the importance of environmental consciousness and pay attention to the effectiveness of corporate identity systems. The experience of the design company was also a vital factor, which was expected to be supplemented by an impressive presentation to set out the content and benefits of the design proposal. The experts also noted the importance of maintenance because related costs influence future operations. Therefore, in addition to environmental consciousness and corporate identity, design companies should consider shopping center operating costs and cases in other industries to propose unique and novel planning suggestions that can create value for shopping centers. Only by implementing such measures are design companies likely to win design bids.

5.2 Sixth to tenth most crucial factors

The sixth to tenth most crucial factors were budget and schedule control (6.2001%), cross-industry cooperation (5.8512%), low operation cost (5.7447%), regulation compliance (5.4887%), and green building materials (5.3647%). Budget control directly influences investor costs and profits. Cooperation with other industries is a modern trend; building cross-industry alliances can yield considerable benefits. Operation cost in conjunction with ease of maintenance (the fifth priority) can substantially reduce regular maintenance costs. Codes and regulations in different countries regarding shopping centers vary; running shopping centers in a manner that conforms to local codes and regulations is a basic requirement for operators. Environmental protection serves to protect the natural environment at individual, organizational, and governmental levels for the well-being of nature and humans. Industrial development has caused serious environmental pollution and damaged ecoystems to an almost irreparable degree. This has resulted in many industrialized countries emphasizing environmental protection by imposing laws to regulate and control pollution and by raising social awareness of the profound influence of pollution on the environment. Since the 1960s, environmental movements have raised public awareness regarding numerous environmental problems.

Academic institutes provide courses regarding the history and methods of **environmental protection**, such as environmental studies, environmental management, and environmental engineering. Environmental protection requires the public to participate in various related activities. Waste products, air pollution, and the loss of biodiversity (due to invasive alien species and extinction) are all environmental protection–related topics. The three main factors of environmental protection, namely environmental legislation, morality, and education, all influence countries' environment-related decisions and individual people's environmental values and behaviors.

5.3 Comprehensive analysis

In this study, the key factors for commercial space design were determined using the Delphic hierarchy process. These methods are explorative and quantitative, with the aim of substantially increasing the use efficiency and popularity of commercial space. The questionnaire results revealed that the key factors for commercial space design considerably influence the success of shopping center development and operation. In addition to selling products, shopping centers typically provide various leisure facilities or even dining services, high-end accommodation, wedding and banquet venues, and conference rooms. They can also serve as major resorts and landmarks.

The rationality of circulation, the convenience of the various facilities on offer, and the refinedness of the spaces within all serve to increase the popularity of commercial spaces and provide customers with a sense of added value. Using customer satisfaction to enhance customer retention and application of information technology is highly conducive to the development of commercial spaces. This study determined the key factors of commercial space design that influence renovation of commercial spaces; based on the results, a sustainable operation model for commercial space can be constructed.

5.4 Research limitations and directions for future research

The importance of integrating water conservation with the concept of environmental protection should not be overlooked. Determining how to create a unique shopping center that incorporates local culture

IOP Conf. Series: Materials Science and Engineering **652** (2019) 012067 doi:10.1088/1757-899X/652/1/012067

and can be integrated into the surrounding environment while introducing external resources to increase revenue from other sources is a potential direction for research on space modularization. Regarding the increasing trend for shopping centers, future studies should use real-world cases for evaluation (e.g., Taipei 101) and investigate shopping centers throughout Asia. Taiwan's application of design aesthetics is frequently recognized by the international community, showing that the country is a design center capable of knowledge management in the Asia–Pacific region and a leader of design development among global Chinese communities. Shopping center design is manifested in Taiwan's strong design capacity and enables Taiwan to compete in the global market.

References

[1] Morris M 1998 *Things to do with shopping centre: Too soon too late: History in popular culture* (Bloomington: Indiana University Press).
[2] Silhouette-Dercourt V, Lassus C and Darpy D 2014 How second-generation consumers choose where to shop: A cross-cultural semiotic analysis P3-P4(2014).
[3] Najafi M, Shariff B M, IJHSS 6, 187-193 (2011).
[4] Wu C J, Macau Monthly 12, 30-31 (2015).
[5] Wang W, Macau Monthly 1, 13 (2016).
[6] Jiang yuan Liu, Macau Monthly 6, 10 (2017).
[7] P.P.C. Guimaraes, MGR 26, 160-172 (2018).
[8] Saaty T L, The analytic hierarchy process (McGraw-Hill, New York, 1980).
[9] Liu K S, Hsueh S L, Wu W C, Chen Y L, Energies 5, 4462-4480 (2012).

IOP Conf. Series: Materials Science and Engineering **652** (2019) 012068 doi:10.1088/1757-899X/652/1/012068

Voids distribution of pavement filters subjected to permeating fines: a coupled DEM-statistical inference

Sandun M. Dassanayake, Ahmad Mousa*and Daniel Kong

Monash University, School of Engineering, Malaysia
*E-mail: ahmad.mousa@monash.edu

Abstract. Granular soils in pavement systems, insidiously experience fines migration during their expected lifespan. Experimental monitoring of particles movement in a granular matrix is extremely challenging due to its concealed nature. Numerically modelling the permeation process demands a high computational power and requires additional conditions, which are frequently difficult to be obtained. This paper proposes an integrated numerical approach to estimate the conditional probability of observing a given voids ratio of a filter subjected to different fractions of permeating fines. The discrete element method (DEM) has been used for developing numerical samples of random voxels to estimate the void ratio distribution of the fines-free filter (i.e. virgin filter). Qualitative results obtained from a pilot scale experimental program have been presented to validate the hypotheses on which the novel methodology has been developed. The results further show that the distribution of the voids ratio in the filters subjected to permeating fines follow a gamma distribution, with a 0.9 probability of observing a voids ratio lower than the average value in the virgin filter.

1. Introduction

Seepage induced fine particle migration detrimentally affects the hydraulic performance and the life expectancy of the drainage systems in highway pavements. During the pavement construction, granular filter layers are designed using empirical criteria to minimise this fines migration [1-3]. The widely-used US Army Corps of Engineers filter design criteria (Eq. 1), for instance, consider ratios of determining sizes belonging to the particle size distributions (PSDs) of the filter and fine soil layers (i.e. fines). The left and the ride hand side of the criteria given in Eq. 1 are known as the clogging and permeability criterion respectively. However, even after designing` the filter following these criteria, still, spatiotemporal factors, such as inadequate compaction or segregation during compaction, excessive cyclic traffic loads and extreme weather conditions can result in undesired particle migration during the lifetime of a pavement filter layer [4-6].

$$\frac{D_{15,filter}}{D_{85,fines}} \leq 5 < \frac{D_{15,filter}}{D_{15,fines}} \quad (1)$$

where $D_{15,filter}$, $D_{85,fines}$, $D_{15,filter}$, and $D_{15,fines}$ are the diameters of corresponding to cumulative weight percentages of 15%, 85% and 15% of the filter and soil PSDs, respectively.
Migrated fines occupy the void spaces in the virgin filter (i.e. filter fabric free of fines) following a highly randomised pattern. This affects the mechanical and hydraulic performance of the filter soil, such as the compressional behaviour, strength, critical state line, static liquefaction, undrained fragility, and also the hydraulic conductivity [7,8]. It is extremely challenging to monitor (or correlate) the

IOP Conf. Series: Materials Science and Engineering **652** (2019) 012068 doi:10.1088/1757-899X/652/1/012068

resulting real-time variations of mechanical and hydraulic performances of these filters due to the concealed nature of the fines migration phenomenon. Very little experimental information, therefore, is available on the real-time variations of the void size distribution in the filter under the continuous infiltration of fines.

Due to these experimental challenges, the conventional numerical modelling techniques, which assume the continuum nature of materials, cannot comprehensively model this migration problem [4]. On the contrary, the discrete element method (DEM) [9], which simulates the particle movements using Newton's second law of motion, has shown potentials in modelling this phenomenon. However, DEM simulations are only possible for an assembly containing a limited number of particles [10]. The computational cost associated with DEM simulations limits them being applied to infinitely many numbers of particles, which ideally represent the large-scale applications. The scale issue can be one way addressed by coupling the DEM with continuum models; however, this significantly increases the demand for computational power. Additionally, improved algorithms are required to accurately couple the two techniques [11]. Therefore, a requirement exists for an alternative feasible robust inference technique that can be readily used in large-scale applications.

This study sets out to develop a probabilistic approach to infer the voids distribution in a filter-fines mixture after the migration of fines into the virgin filter. A pilot scale experimental program has been carried out to investigate the progression of fines migration in a well-designed (i.e. following Eq. 1) filter-soil assembly. A database of potential void ratios has been developed for the tested filter using the open source DEM software, YADE [12], to perform the statistical analysis. Monte-Carlo simulation technique has been employed to develop the conditional probability of the void ratio distribution of the filter-fines mixture upon plausible infiltrations of different fines contents.

2. Methodology

This study comprises of a pilot scale experimental program and a numerical simulation using DEM, which was followed by statistical analysis. In the experimental approach, a constant hydraulic gradient was used to induce base particle migration through a uniform granular filter. The numerical tests were used to develop the void ratio distribution of the tested filter soil, which was free of fines. The void ratio distribution was then employed to infer the conditional probability distribution of the void ratio of the filter-fines mixture.

2.1 Unidirectional flow tests

A custom-made transparent acrylic cell was used to perform the unidirectional flow tests under a constant hydraulic gradient of 5 (fig. 1a). A comparatively extreme hydraulic gradient was employed, following the literature [8,13], to reduce the experimental duration and enable a rapid migration of fines. Filter and the fine soil was constituted to have uniform distributions spanning from 1.18 mm to 4.75 mm and 63 μm to 600 μm respectively (fig. 1b). Their respective PSDs were designed according to the design criteria specified in Eq. 1. The stainless steel sieve plate that holds the sample in the cell has uniform 2mm size apertures to retain the filter particles while allowing only the water and fines to pass through the assembly during the flow tests.

First, the filter soil was placed on the sieve plate, and the layer was saturated at 8% and compacted up to 90-95% of the optimum proctor compaction. The base material was carefully placed on top of the filter using a funnel, such that the particles are evenly and loosely distributed. The filter and base height were taken as 70 mm and 35 mm (2:1) respectively. Marbles with 16mm particle size were used to provide and maintain overburden stress of approximately 1kPa.

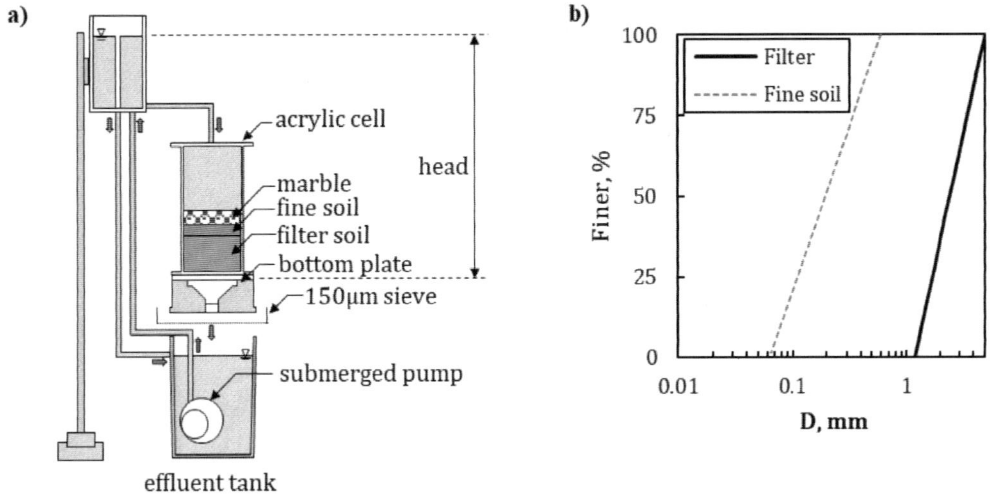

Figure 1. Experimental set-up for the unidirectional flow tests

The cell was filled from the bottom-up with filtered de-ionised water at an average temperature of 26.2°C to avoid dispersing the sample. The top air release valve was used to minimise the presence of trapped air in the cell. Then it was allowed to saturate for a duration of 60 minutes. After saturation, the flow direction was reversed subjecting the sample to a downward flow. The hydraulic gradient was kept constant (at 5), and the volumetric flow was measured during the test. When the hydraulic flow stabilized (approximately 90 minutes for this sample), the test was terminated. After the test, EPO-TEK 301 resin was injected to the filter-fines mixture. A representative core sample was taken from this mixture after the resin got fully crystalized to observe its micro-structure from the scanning electron microscope (SEM). The remaining filter soil was equally divided (as top and bottom sections) and their respective post-test PSDs were estimated using sieve analysis (ASTM D422).

2.2 Statistical estimation of the voids distribution: DEM approach

Open source DEM software, YADE, was employed to generate the tested filter sample numerically. In this method, a cloud of 10000 grains belonging to the filter PSD was generated [12] with the simulation parameters specified in table 1. The particle cloud is then allowed to free-fall under gravity (as given in fig 2a and b) into a bounding box with 1000x1000x1000 mm dimensions (D x D x D in Figure 2c). When the particle assembly gets fully settled in the bounding box, an overburden stress of 1000 kPa was applied on the sample to mimic the experimental conditions. The compaction consolidates the sample (fig 2d). After completing the consolidation, the bounding box was divided into identical square voxels with 10x10x10 mm dimensions (Figure 2c and d). The voxel samples were used to create the database for estimating the void ratio distribution of the filter before the fines migration.

Table 1. Simulation parameters used in DEM.

Property	Number of grains	Grain shape	Density (kg/m3)	Young's modulus (Pa)	Poisson's ratio	Friction angle (deg)
Value	10000	Spherical	2600	10^8	0.3	28

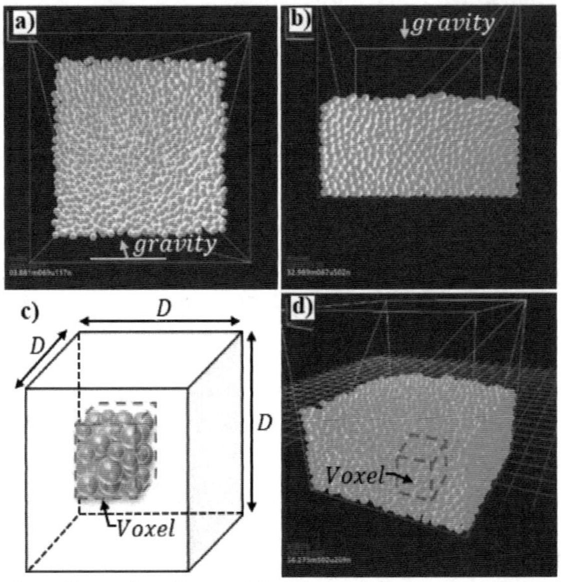

Figure 2. DEM simulations for estimating void ratio distribution: a) particle cloud generation; b) gravity deposition; c) dimension of the bounding box; d) random voxel porosity estimation.

Previous studies have proposed an index known as the skeleton void ratio (e_s) hypothesising that the fines content (f_c) in a soil mixture would only occupy the void space of the coarse grain material, which transfers the effective stress of the soil assembly [14]. This hypothesis can be also applied to the migration phenomenon discussed in this study since the virgin filter act as the load bearing skeleton of the resulting filter-fines mixture. Hence, the statistical void ratio distribution should be identical to the distribution of es pertinent to the filter-fines mixture. Using this analogy, the resulting void ratio of the mixture (e^F) can be therefore obtained using Eq. 3.

$$e_s = \frac{e^F + f_c}{1 - f_c} \tag{2}$$

$$e^F = e_s(1 - f_c) - f_c \tag{3}$$

The fc in Eq. 3 is a real number that can have any positive value from 0 to 1. Hence, the random number generating algorithm in MATLAB 2017 software is used to generate a random sample of 10000 fc values to perform a Monte-Carlo simulation using the Eq. 3. The cumulative probability distribution of possible different e_F values subjected to different fines contents $(P(e^F \mid f_c))$ (i.e. the conditional probability distribution [15]) has been generated using this approach.

3. Results

The post-grain size distribution of the filter reveals that the top and bottom section of the filter has received an approximately equal amount of fines (fig 3a). This also suggests that the duration of the experiments were sufficient for the fines to equally distribute throughout the filter. The fines fraction has increased approximately by 5% on average. The SEM image of the filter fabric shows that the fines have entered the coarse matrix and they have resided in the voids of the filter soil matrix (fig 3b). This reduces the void volume in the fabric and hence the void ratio of the mixture soil. Also this supports the validity of using eq. 3 in developing the mixture void distribution.

The void ratio distribution obtained using the voxel samples in the DEM simulations is shown in figure 4a. The void ratios follow a generalised extreme value distribution with a mean value of 0.798 and a standard deviation of 0.225. The location, scale and the shape parameters of this distribution are 0.692, 0.115 and 0.262 respectively. The estimated void ratio distribution of the post filter matrix was found to follow a gamma distribution with shape parameters α and β being equal to 1.56 and 0.26 (fig 4b). The probability of observing a void ratio lesser than the void ratio of the virgin filter was 0.9.

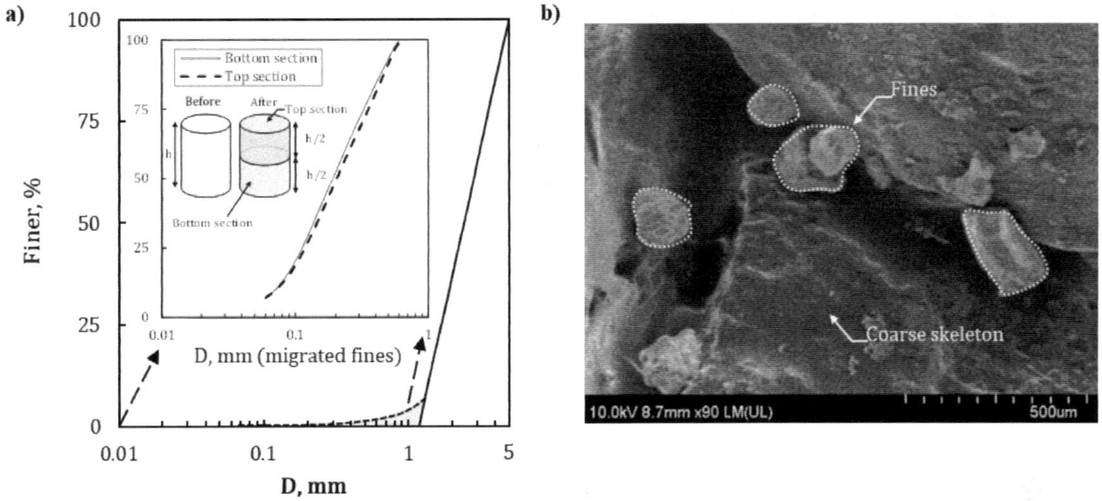

Figure 3. Experimental results: a) PSD of migrated fines fraction and the virgin filter; b) SEM image of the micro-structure of a sample taken from fines-filter mixture after the flow tests.

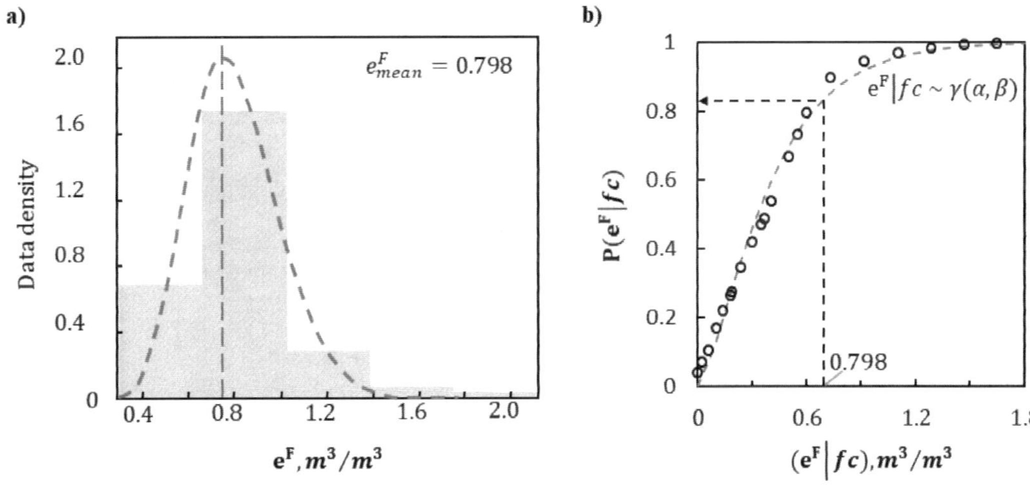

Figure 4. Statistical and probability distributions: a) void ratio distribution of the virgin filter; b) cumulative probability distribution of the void ratio of the fines-filter mixtures.

4. Future directions

Bayesian statistical inference (e.g. as given in [15]) can be employed to integrate the conditional probability distribution of the mixture void ratio, given in fig 4b, to probabilistically quantify the variations of hydraulic and mechanical responses of the pavement filters. This will enable a probabilistic lifetime assessment for the pavement filters. The proposed methodology simplifies the complex physics involved in the fines migration phenomenon in pavement filters. A comprehensive database of the void ratio distribution should be developed in the future using DEM with more emphasis on the typical relative compactions found in pavement applications. In addition, the particle shape effects and intrinsic material properties of the filter medium should be closely investigated in these DEM simulations to improve this technique to be a convenient tool for the practising engineers.

IOP Conf. Series: Materials Science and Engineering **652** (2019) 012068 doi:10.1088/1757-899X/652/1/012068

References

[1] Indraratna, B. and Raut, A.K., 2006. Enhanced criterion for base soil retention in embankment dam filters. Journal of geotechnical and geoenvironmental engineering, **132**(12), pp.1621-1627.

[2] Sherard, J.L., Dunnigan, L.P. and Talbot, J.R., 1984. Basic properties of sand and gravel filters. Journal of Geotechnical Engineering, **110**(6), pp.684-700.

[3] USACE 2000 *Engineering and Design: design and construction of levees.*

[4] Jeyisanker K. 2008 *Analysis of water seepage through earthen structures using particulate approach* (University of South Florida)

[5] Chapuis, R.P., Contant, A. and Baass, K.A., 1996. Migration of fines in 0 20 mm crushed base during placement, compaction, and seepage under laboratory conditions. Canadian Geotechnical Journal, **33**(1), pp.168-176.

[6] Locke, M., Indraratna, B. and Adikari, G., 2001. Time-dependent particle transport through granular filters. Journal of geotechnical and geoenvironmental engineering, **127**(6), pp.521-529.

[7] Chang, C.S., Wang, J.Y. and Ge, L., 2016. Maximum and minimum void ratios for sand-silt mixtures. Engineering geology, 211, pp.7-18.

[8] Rasheed A. K., Dassanayake, S. M. and Mousa A. 2018 in *Fifteenth International Conference on Structural and Geotechnical Engineering, Cairo*

[9] Cundall, P.A. and Strack, O.D., 1979. A discrete numerical model for granular assemblies. geotechnique, **29**(1), pp.47-65.

[10] Shire, T. and O'Sullivan, C., 2013. Micromechanical assessment of an internal stability criterion. Acta Geotechnica, **8**(1), pp.81-90.

[11] Kloss, C., Goniva, C., Hager, A., Amberger, S. and Pirker, S., 2012. Models, algorithms and validation for opensource DEM and CFD–DEM. Progress in Computational Fluid Dynamics, an International Journal, **12**(2-3), pp.140-152.

[12] Kozicki, J. and Donze, F.V., 2009. Yade-open dem: an open-source software using a discrete element method to simulate granular material. Engineering Computations, **26**(7), pp.786-805.

[13] Chang, D.S. and Zhang, L.M., 2013. Extended internal stability criteria for soils under seepage. Soils and Foundations, **53**(4), pp.569-583.

[14] Yang, J., Wei, L.M. and Dai, B.B., 2015. State variables for silty sands: Global void ratio or skeleton void ratio?. Soils and Foundations, **55**(1), pp.99-111.

[15] Dassanayake, S.M. and Mousa, A., 2018. Probabilistic stability evaluation for wildlife-damaged earth dams: a Bayesian approach. Georisk: Assessment and Management of Risk for Engineered Systems and Geohazards, pp.1-15.

IOP Conf. Series: Materials Science and Engineering **652** (2019) 012069 doi:10.1088/1757-899X/652/1/012069

Earthquake Hazard Safety Assessment of Buildings via Smartphone App: A Comparative Study

Ehsan Harirchian[1,*] and *Tom* Lahmer[1]

[1]The Institute of Structural Mechanics, Bauhaus University Weimar

Abstract. The failure of man-made structures is the main cause of more injuries during an earthquake and more economic losses. In addition, as urban areas continue to develop, the need for an effective disaster management system to develop resilient communities increased. To estimate the severity of a building's damage, several seismic assessment methods have been introduced. Rapid visual screening methods (RVS) are the fastest and simplest methods developed to evaluate the initial vulnerability profile of the existing building stock, constructed before or after the adoption and enforcement of seismic codes. In this study, FEMA P-154, Indian RVS (IITK-GSDMA), Turkish RVS (EMPI), and EHSAPP method have been utilized on some buildings and their results compared to each other with regard to the real occurring damage after an earthquake on the same buildings. It was observed that the EHSAPP method has a better and more realistic seismic risk assessment than others.

1 Introduction

The importance of studies and research into the prevention of the effects of earthquake has increased after destructive earthquakes and it is necessary to assess the seismic vulnerability of buildings in urban areas as an essential parameter of the earthquake disaster management policy. Therefore, there is an urgent need for reliable vulnerability assessments, appropriate seismic risk management strategies, and rapid response after the disaster to significantly reduce possible losses in future severe earthquakes. The term "seismic vulnerability" is defined as "the susceptibility of a population of buildings to undergo damage due to seismic ground motion" [1]. Rapid Visual Screening (RVS) is the simplest method among other vulnerability assessment methods, performing basic structural calculations for quick evaluation of a large building stock [2]. However, more detailed analysis would give a better assessment, but such an approach would entail unprecedented difficulties when an urban scale mitigation campaign is considered. In this manner, some efforts have been made to develop a tablet-based version of FEMA 154 [3], also to transmit GIS visualization and mobile data for database compiling [4]. Rapid Observation of Vulnerability and Estimation of Risk (ROVER v.1), which is a mobile-device and server-based software for both pre- and post-earthquake building safety screening [5] and ROVER v.2 [6] is the current version of it. Moreover, Wahyuni [7] implemented the FEMA 154 into a smartphone application with regional modification to conduct a quick check on the buildings in Surabaya-Indonesia. Research by Nassirpour [8] evaluates rapidly school buildings in the Philippines against multi-hazard conditions. The Urban Rapid

*e-mail: ehsan.harirchian@uni-weimar.de

Assessment Tool (RAT)[9] is another developed smartphone suite that includes seismic performance attributes based on FEMA 154 [10] and FEMA 310 [11]. The aim of this study is to compare our proposed EHSAPP [12] method with the most popular existing regional building vulnerability assessment methods such as FEMA P-154, Indian RVS (IITK-GSDMA), and Turkish RVS (EMPI). For the applicability and comparison of the proposed methodology, the city of Bingöl in Turkey, which is in a highly active seismic region and has experienced a severe earthquake recently, has been selected. The data collection of nine damaged reinforced concrete buildings have evaluating through FEMA P-154, IITK-GSDMA, EMPI and EHSAPP, and the estimation of each method has been compared with the actual damage of the same building.

2 Review on Existing Rapid Visual Screening methods

As was previously mentioned in this paper, a comparison between a new RVS method and other existing methods has been carried out, which is described in the following paragraphs. The initial Rapid Visual Screening (RVS) methodology was proposed by Federal Emergency Management Agency (FEMA), U.S.A in 1988 as Rapid Visual Screening of Buildings for Potential Seismic Hazards: A Handbook and modified in 2002 [13]. Since the publication of the second edition of FEMA 154, there have been many improvements in rapid visual screening of buildings, one of which efforts was the development of the FEMA P-154 [14]. In RVS methodology, a visual inspection involves collecting structural and nonstructural characteristics of a building such as building type, number of stories, soft and weak stories, short columns, openings etc. Based on the data collected during visual inspection, a structural score will be calculated without performing any structural calculations to define the expected damage of a building and whether the building needs the next step of assessment or not. This will lead to saving time and resources that can be respectively used for the buildings that need detailed assessment[14, 15]. In this paper, only the RC buildings have been considered and since the case study is on the high seismic area, the RVS data collection form for high seismic has been selected.

2.1 FEMA P-154

FEMA P-154 provided data forms for different levels of seismicity. Depending on the classification of seismicity, the relevant data form will be taken. The seismicity of the site is determined using site-specific values of seismic hazard for risk-targeted maximum considered earthquake ground motions (MCER) and Soil Type B as provided by the U.S. Geological Survey (USGS). The scoring procedure in FEMA P-154 begins with the selection of the appropriate basic score (B) for buildings that will be modified using modifier scores (M).

2.2 IITK-GSDMA

Indian Institute of Technology, Kanpur (IITK) developed RVS methodology based on FEMA 154 with the support of Gujarat State Disaster Mitigation Authority (GSDMA) in "Seismic Evaluation and Strengthening of Existing Buildings" (IITK-GSDMA-EQ06-V4.0) [16]. It was developed by implementing suitable modifications to FEMA 154 based on Indian conditions. Some building types, like un-reinforced masonry and non-ductile RC frame buildings, have been given special consideration for the assessment within this method.

IOP Conf. Series: Materials Science and Engineering **652** (2019) 012069 doi:10.1088/1757-899X/652/1/012069

2.3 EMPI

Turkish RVS methodology was developed by conducting the survey on buildings in Istanbul and is called Earthquake Master Plan of Istanbul (EMPI), which can also be used for other parts of the country [17]. EMPI provided different RVS methods for different types of construction.

2.4 EHSAPP

The Earthquake Hazard Safety Assessment of Buildings via Smartphone App (EHSAPP) is based on the implementation of Analytic Hierarchy Process (AHP) methods to achieve an optimized earthquake hazard safety assessment [12]. For this method, initially a sensitivity analysis has been performed on FEMA P-154 to prioritize and assign weight to criteria. Furthermore, a multi-criteria decision making tool AHP has been utilized to assess seismic vulnerability of buildings. The screen shot of some pages of EHSAPP have been shown on Figure 1 and 2.

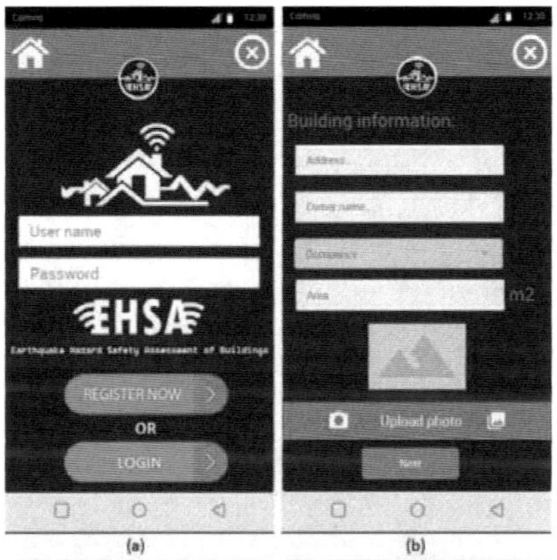

Figure 1. Screenshot of EHSAPP: (a) first page, (b) building information

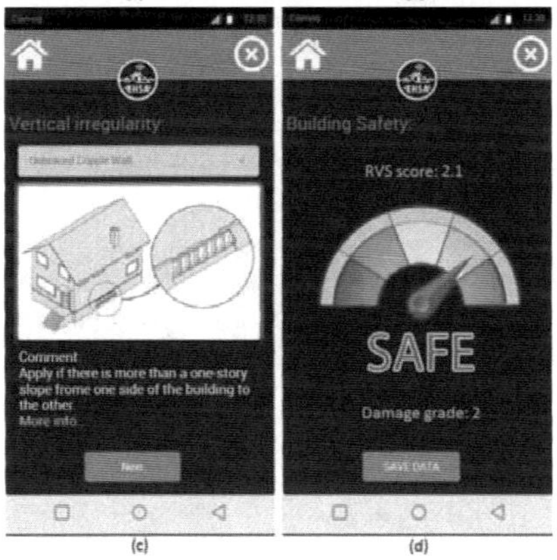

Figure 2. Screenshot of EHSAPP: (c) entering buildings irregularities, (d) building vulnerability assessment

3 Study area

After May 1, 2003, Bingöl earthquake in the province of the east Anatolian region of Turkey, with a magnitude of 6.4 (Mw), building damage data was collected by a team of researchers from Middle East Technical University [18]. From that report, nine damaged reinforced concrete buildings have been selected to study. The importance of this area is due to the Bingöl faults under high seismic zone with 10 percent probability in 50 years with PGA of 0.4g. The Vs 30 velocity of soil in Bingöl varies from 620 m/s to 760 m/s with an average Peak Ground Velocity (PGV) of 78.4 m/s.

4 Results and discussion

The main purpose of this paper is to compare the newly developed RVS method via smartphone app (EHSAPP) to the other existing RVS methods. Figure 3 compares the observed damage of the case study buildings (a-i) to FEMA P-154, EHSAPP, IITK-GSDMA, and EMPI based on EMS scale. There are some differences and similarities in the assessment and actual damage scale, and EHSAPP shows a more accurate valuation, as 6 out of 9 buildings had the same assessment as actual damage. In contrast, most of assessments by FEMA P-154 and IITK-GDSMA were overestimated, which are neither economical nor logical. The assessment, based on EMPI were also not satisfactory. In brief, it can be concluded that EHSAPP has a more accurate and realistic assessment compared to others.

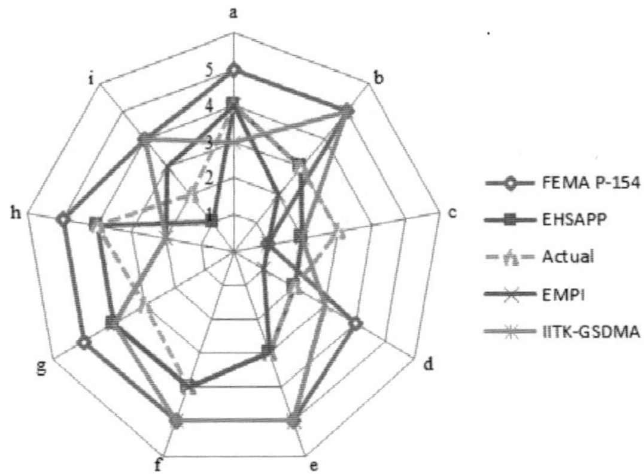

Figure 3. Damage grade comparison based on EMS-98

5 Conclusion

The results of this study suggest that the proposed EHSAPP method, which is based on the application of multi-criteria decisions and its implementation on smartphone apps, generally present a better assessment of the seismic vulnerability of buildings compared to other existing methods. Although the study examined only reinforced concrete buildings in one area, it could be developed for the study of another type and seismicity area.

References

[1] A. Shahriar, M. Modirzadeh, R. Sadiq, S. Tesfamariam, Earthquake and Structures **3**, 695 (2012)

[2] P. Kapetana, S. Dritsos, WIT Transactions on the Built Environment **93**, 409 (2007)

[3] Y. Yang, K.A. Goettel, *Enhanced Rapid Visual Screening (E-RVS) Method for Prioritization of Seismic Retrofits in Oregon* (Citeseer, 2007)

IOP Conf. Series: Materials Science and Engineering **652** (2019) 012069 doi:10.1088/1757-899X/652/1/012069

[4] A.G. Sextos, A. Kappos, K. Stylianidis, Computer-Aided Civil and Infrastructure Engineering **23**, 59 (2008)

[5] K. Porter, *Rapid observation of vulnerability and estimation of risk (rover): End-to-end seismic risk management software*, in *9th US national and 10th Canadian conference on earthquake engineering* (2010), pp. 3078–3087

[6] K. Porter, S. Hellman, A. Hortacsu, in *Improving the Seismic Performance of Existing Buildings and Other Structures 2015* (2015), pp. 787–796

[7] E. Wahyuni, D. Iranata, B. Suswanto, C.B. Nurcahyo, W. Sutrisno, International Journal on Advanced Science, Engineering and Information Technology **8**, 567 (2018)

[8] A. Nassirpour, C. Galasso, D.F. D'Ayala, 16th European Conference on Earthquake Engineering (16ECEE) (2018)

[9] S. Ploeger, M. Sawada, A. Elsabbagh, M. Saatcioglu, M. Nastev, E. Rosetti, Journal of Computing in Civil Engineering **30**, 04015006 (2015)

[10] F.E.M.A. (US), *Rapid visual screening of buildings for potential seismic hazards: A handbook* (Government Printing Office, 2012)

[11] A.S. of Civil Engineers (ASCE), *Handbook for the Seismic Evaluation of Buildings: A Prestandard* (Federal Emergency Management Agency, 1998)

[12] E. Harirchian, A. Harirchian, *Earthquake Hazard Safety Assessment of Buildings via Smartphone App: An Introduction to the Prototype Features- 30. Forum Bauinformatik: von jungen Forschenden für junge Forschende: September 2018, Informatik im Bauwesen* (Professur Informatik im Bauwesen, Bauhaus-Universität Weimar, 2018), `https://e-pub.uni-weimar.de/opus4/frontdoor/index/index/docId/3785`

[13] M. Yadollahi, A. Adnan, R. Zin, Archives of Civil Engineering **58**, 363 (2012)

[14] F.E.M.A. (US), Applied Technological Council (ATC) p. 388 (2015)

[15] R.P. Nanda, D.R. Majhi, KSCE Journal of Civil Engineering **18**, 2218 (2014)

[16] S.K. Jain, O. Jaiswal, National Information Centre of Earthquake Engineering, Kanpur (2007)

[17] M. Balamir, *Urban seismic risk management: the earthquake master plan of Istanbul (EMPI)*, in *Proceedings of the 13th World Conference on Earthquake Engineering, Vancouver, BC, Canada* (2004)

[18] C. Sim, C. Song, N. Skok, A. Irfanoglu, S. Pujol, M. Sozen, *Database of low-rise reinforced concrete buildings with earthquake damage* (2015), `https://datacenterhub.org/resources/123`

IOP Conf. Series: Materials Science and Engineering **652** (2019) 012070 doi:10.1088/1757-899X/652/1/012070

Coping with the Information Age: Development of a Data Flow Diagram-Based Knowledge Management System for Mitigating Delays for Construction

Roderick N. Macariola[1] and Dr. Dante L. Silva[2]

[1] School of Graduate Studies, Mapua University, 658 Muralla St., Intramuros, Manila, 1002, Metro Manila, Philippines
[2] Mathematics Department, Mapua University, 658 Muralla St., Intramuros, Manila, 1002, Metro Manila, Philippines
Email: macariolaroderick@gmail.com; dlsilva2000@yahoo.com

Abstract. The construction industry adopts itself to the paradigm shift of today's technology as evident in the numerous researches being conducted to create a more sustainable process and risk-resilient systems in management of projects. In this study, the documentation system of private construction companies in the Philippines served as the primary focus with an intention to scrutinize and improve the data management system of these construction projects. Data was gathered from three(3) different companies (A, AA, and AAA) involving six of their ongoing projects. After the conduct of two(2) Likert Scale surveys, contractor-based factors are the most common delay issues in projects namely financial difficulties and late deliveries with a mean of 4.32 and 4.26 respectively; while, communication and proper data management were deemed by the respondents as most effective with a mean of 4.66 and 4.51 respectively. The analysis of variance from the three companies does not show much difference as it was able to meet the 95% confidence level implying that the hypothesis is valid in all three sets and same as the analysis of variance among the three construction entities—contractor, client, and consultant. The DFD-Simulation Method showed a significant 2-3 days of schedule improvement valued from Php 10,000.00 – Php 125,000.00. Therefore, construction data management needs to apply DFD and other emerging technologies such as BIM to improve construction delay mitigation.

Index terms: *Knowledge Management System, Data Flow Diagrams (DFDs), Construction Data Classification, Process Sustainability, Construction Delay.*

1. Introduction

The engineering profession is technical in nature. With this, most engineers, especially construction managers, are highly expected to be well-versed in handling data in their respective projects. More often than not, these major construction personnel find documentation a challenging portion of their daily tasks. Another barrier includes the unwillingness of employees to acquire new methods and procedures in a project [1]. This is evident in the need to hire document clerks and technical assistants. Documentation management, utility and control are major factors in all processes of any project. It is with this that the system remains standardized. Focusing in the architecture, engineering, construction and owner-operated (AECO) industries, a structured data management system is vital in meeting all design specification, budgeting and scheduling requirements. In this primary reason that the researcher deem it necessary that the current data management system in the country be revisited to scrutinize its effectivity in probably becoming a helpful aid in addressing one of the major concerns in most construction projects—construction delays. Consequently, data management is intended to be revised in a manner that the documentation system will be fortified and restructured to transform projects' knowledge management methods to be more sustainable, flexible, and risk-oriented. Besides, managers and supervisors can take advantage of technologies like this as it will help them minimize supervision time for their respective projects and assure that these projects are on time and cost-effective. True that there is no single solution to construction problems like delays but it is always a case-to-case basis when it comes to solving project disputes. Most studies focusing in expediting or bettering the construction process utilized process alteration and improvement instead of looking at data as a powerful tool in mitigating delays. This then is where this paper rests its grounds. Besides, construction workers and even contractors will benefit in this system as it will provide them concrete delay-mitigation solution through document management. As a prediction, this paper aimed to contribute significantly in minimizing contingency expenses due to delays. This study expects a mental and workplace behavioral progress to document clerks, technical assistants, site engineers, and construction managers. Naturally, these construction key players in documentation shall be resistant at first. But, in the long run, this innovation in knowledge management system for construction projects shall be considered as an efficient part in maintaining and assuring delay-free execution of projects. Furthermore, this study opens doors to finally acknowledging the technological engineering advancements like the utilization of Building Information Modelling (BIM) in developing the current Philippine construction set-up.

2. Construction Data and Categorization

Knowledge utilization is defined very technically in different fields and sectors of the society. This is simply because we live in the Information Age where these are very vital in businesses and profit-making. Knowledge management as a scientific discipline was believed to have started just recently, estimated to be the early 1990s, with constantly growing research interest especially in the past two decades [2]. This paper highlights the importance and impact of adopting an innovatively structured knowledge management system in light with the related studies. Some researchers found that growth and success of companies are primarily based on the effectiveness and efficiency and of knowledge management applied in the whole construction process [3]. Today, numerous management techniques have been studied and implemented to better construction processes. It is imperative to do so since the industry is one of the highest contributors to national economic growth. In a study done in Malaysia by Hasmori, et al. [4], the paper consolidated construction delay factors as categorized by the contributing parties which is used in this study as major reference. Besides the delay factors, mitigation processes are also identified in order to assert the effective ways of addressing these factors.

After conducting the project immersion, the proponent was able to consolidate all the important in construction projects. These were identified in order to revisit and maximize their use and potential to be delay-mitigating factors. These specific data enumerated are Bidding Documents**,** Permits and Contracts**,** Project Specifications, Cost Specifications, Contractor Scope, Design Plan, Stock Requisitions, Equipment Specifications and Inventory, Materials Take-Off/Materials Schedule Construction Project Time Line (Schedule of Activities), Bill of Quantities/Liquidation, Testing Results, Accomplishment Reports, Delay Reports, Incident Reports, Change Orders, Minutes of the Meetings, and Revisions and Consultant Approval Reports.

In this paper, data shall be categorized as per use, type and flow order. For the first column of category, documents shall be identified as one(1) if they are being used by a single entity only (i.e. contractor, client, or consultant), two(2) if two entities use them and three(3) if all entities are expected to use, manage, or store these data. The second column categorizes data by their type as referred in the phase or stage of the construction where they are being used. Flow order, which is indicated in the third column, is based on whether they shall be used at the beginning, end or at the storage or processing

IOP Conf. Series: Materials Science and Engineering **652** (2019) 012070 doi:10.1088/1757-899X/652/1/012070

part of the DFD. The summarized categorization of data is being shown in Table 1 with corresponding notes for the legend of the used symbols.

Table 1. Data Categorization

Construction Documents	Use	Type	Flow Order
Bidding Documents	**2**	**I**[a]	**A**[f] **S**[g]
Permits and Contracts	**2**	**I**	**A P**[h] **S**
Project Specifications	**3**	**I / Ct**[b]	**A P S**
Cost Specifications	**2**	**I / Ct**	**A P**
Contractor Scope	**2**	**Ct**	**A S**
Design Plan	**3**	**I / Ct / Cn**[c]	**A P S**
Stock requisition	**2**	**Cn / D**[d]	**P S**
Equipment Specifications and Inventory	**2**	**Ct / D**	**A P S**
Materials Take-Off/Materials Schedule	**2**	**Ct / Cn**	**A P S**
Construction Project Time Line	**3**	**Ct**	**A P S**
Bill of Quantities/Liquidations	**2**	**Cn / D**	**P S Z**[i]
Testing Results	**3**	**Cn**	**P S Z**
Accomplishment Reports	**2**	**Cn**	**P S Z**
Delay Reports	**2**	**Cn / E**[e] **or Cn / D**	**P S Z**
Incident Reports	**3**	**Cn / E**	**P S Z**
Change Orders from Client	**3**	**E**	**P S Z**
Minutes of the Meetings	**3**	**Cn**	**P S Z**
Revisions and Consultant Approval	**3**	**E**	**P S Z**

[a] **I** for Initiating (Pre-construction Phase)
[b] **Ct** for Conceptual (Planning Stage)
[c] **Cn** for Conceptual (Construction Phase)
[d] **D** for Delivery (Materials and Labor)
[e] **E** for Emergent (AdHoc Documents)
[f] **A** for beginning of DFD
[g] **S** for storage
[h] **P** for processing
[i] **Z** for ending of DFD

3. Survey Results and Analysis

Three(3) companies were identified as the sources for the data of this project. These comprised of A, AA, and AAA companies who currently are working on low-rise residential projects, mid-rise buildings and high-rise structures. The two sets of survey (construction delays and mitigation processes) were designed in a form of a Likert Scale accomplished by technical personnel of a construction. As a result, contractor-based factors are the most common delay issues namely financial difficulties and late deliveries with a mean of 4.32 and 4.26 respectively. There were 34 respondents which comprised of 11, 14, and 9 respondents from the three (3) firms. On the other hand, communication and proper data management were deemed by the engineers and technical personnel as the most effective mitigation process with a mean of 4.66 and 4.51 respectively. To emphasize, site productivity is dependent on effective communication among the personnel [5].

As a point of comparison, one-way analysis of variance was applied to check whether project size, construction management, and contractor accreditation ranks have bearing in the more commonly occurring delay factors and mitigation processes deemed effective by its workforce. Considering a

IOP Conf. Series: Materials Science and Engineering **652** (2019) 012070 doi:10.1088/1757-899X/652/1/012070

degree of freedom of (2,31), the author was able to come up with a concrete comparison of variance from the three sets of data gathered from three companies.
The initial parameters of ANOVA and F-test include:

- Null Hypothesis, $H_O : \mu_1 = \mu_2 = \mu_3$
- Alternative Hypothesis, H_a : with at least 1 difference among means
- α= 0.005 (Level of Confidence = 95%)
- Both survey sets also took into account the following computed variables:
- Degree of Freedom Between Sets, ∂F_{BET}=2
- Degree of Freedom Within Sets, ∂F_{WITH}=31
- Critical value in the F-Test considering degrees of freedom 2, 31 for α = 0.005,
- F_{crit}= 4.161 (Interpolated value between 30 and 40)

After applying the specified statistical treatment, the F_{crit} values for company-based analysis of variance in the delay factor and mitigation process were 0.195 and 1.317 respectively. The same analysis was conducted comparing the data from the different construction entities and resulted to the following F_{crit} values, 0.798 and 1.336 respectively for the two(2) survey sets. Since all values are below 4.161, all tests failed to reject the null hypothesis. This only means that there is no significant difference in the data gathered for this study with respect to the company of respondents and entity they belong to.

4. Data Flow Diagrams

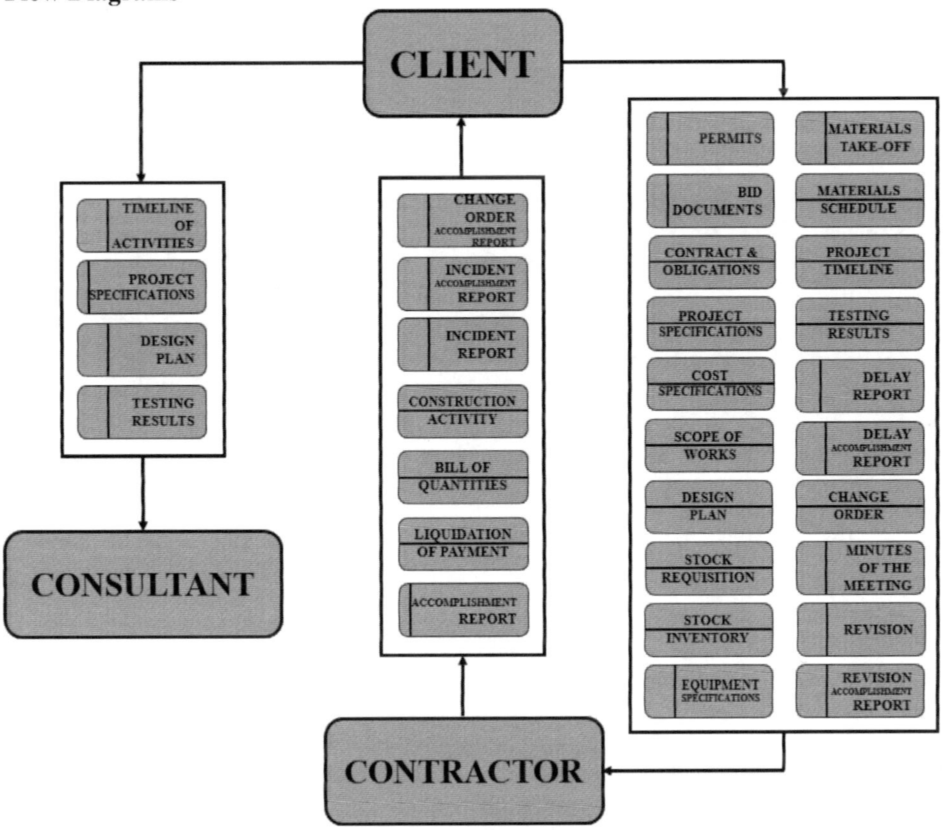

Figure 1. General Data Flow Diagram for Construction Documents

Each construction document was designed with their corresponding data flow diagrams summarized in Figure 1. This shows the flow of data with respect to the three construction entities emphasizing the responsible entity for document storage, processing and utilization.

Out of all the data flow diagrams patterned per construction data, the author has identified that the DFDs with higher levels have higher criticality value. Having shown these diagrams to the companies, they have asserted that Bill of Quantities/Liquidations, Delay Reports, Change Order, Minutes of the Meeting, and Revisions and Consultant Approval are the most crucial data that needs attention. It is highly important that managers assure that project completion is met on time and guided with the appropriated budget as to be able to cope in the current competitive environment of the industry [6].

5. Conclusion

Knowledge management system (KMS) has a powerful potential to enable continuous capture of knowledge from different stakeholders, especially knowledge that are tacit in nature and are being gathered on-site on a daily basis [7]. As for this paper, it was able to suggest a new method of categorizing construction data which is a vital aspect of a KMS. Besides, it was able to identify the documents that can help a lot in mitigating construction delays through data flow diagrams. To evaluate its effectivity, a system simulation was conducted in the projects to concretize the effect of the system as summarized by Table 2.

Table 2. Summary of Improved Delays after DFD-KMS Implementation

COMPANY	PROJECTS	*Delay (Days)*	*Improved Delay*	*Value (Php)*
A	High-Rise 1	15	13	125,542.77
A	Mid-Rise 1	7	6	36,669.94
B	High-Rise 2	24	21	65,179.63
B	Residential 1	3	1	10,274.81
C	Mid-Rise 2	14	12	43,329.02
C	Residential 2	6	4	12,087.56

This underlines the need to expedite the ability of the engineer in-charge of site engineer to report issues like this. Delaying reports will definitely be a burden and will pile up in the form of cost overruns. This signifies that communication, through updated and on-time reports is necessary to mitigate construction delays. Occurrences of late materials deliveries from the contractors were improved by earlier submission of requests and reports. Only that, financial difficulties cannot be solved by an improved management system.

Congruent to the standards set by ISO and Building Information Modeling (BIM), this paper suggests a way of focusing in critical data that will be helpful in improving process sustainability. As found in this paper, data flow diagrams are suggestive of a method that will better knowledge management system not only for the construction industry but also in other fields.

References

[1] Ahmad, H. et al (2017). Investigation of Document Management Systems in Small Size Construction Companies in Jordan. Procedia Engineering, 182, pp.3-9.

[2] Breznik, K. (2018). Knowledge Management – from its Inception to the Innovation Linkage. Procedia - Social And Behavioral Sciences, 238, 141-148. doi: 10.1016/j.sbspro.2018.03.017

[3] Canals, J. (2001). How to think about corporate growth? European Management Journal 19(6), pp. 587-601.

[4] Hasmori, M., Said, I., Deraman, R., Abas, N., Nagapan, S., & Ismail, M. et al. (2018). Significant Factors of Construction Delays Among Contractors in Klang Valley and its Mitigation. International Journal of Integrated Engineering, 10(2). doi: 10.30880/ijie.2018.10.02.007

[5] Olanrewaju, A., Tan, S., & Kwan, L. (2017). Roles of Communication on Performance of the Construction Sector. Procedia Engineering, 196, 763-770. doi: 10.1016/j.proeng.2017.08.005
[6] Senouci, A., Ismail, A. and Eldin, N. (2016). Time Delay and Cost Overrun in Qatari Public Construction Projects. Procedia Engineering, 164, pp.368-375.
[7] Grover, R. and Froese, T. (2016). Knowledge Management in Construction Using a SocioBIM Platform: A Case Study of AYO Smart Home Project. Procedia Engineering, 145, pp.1283-1290.